THE INFINITY PROJECT

Multimedia and Information Engineering

Preliminary Version

Pearson Education, Inc
Upper Saddle River, New Jersey 07458

THE INFINITY PROJECT™

Multimedia and Information Engineering

Preliminary Version

Geoffrey C. Orsak
Southern Methodist University

Ravindra Athale
George Mason University

Scott C. Douglas
Southern Methodist University

David C. Munson, Jr.
University of Illinois

John R. Treichler
Applied Signal Technology

Sally L. Wood
Santa Clara University

Mark A. Yoder
Rose-Hulman Institute of Technology

Pearson Education, Inc
Upper Saddle River, New Jersey 07458

Preface

The Infinity Project is an innovative nationwide program sponsored by the Southern Methodist University School of Engineering and Texas Instruments and designed by leading college engineering professors, in cooperation with education experts. The Project introduces state-of-the-art engineering and advanced technology covering a wide variety of topics relevant to Internet technologies, entertainment, medicine, and communications. In a world in which engineering and technology have become an integral part of our lives, *The Infinity Project* reaches out to students and affords them the opportunity to understand the relevance of technology in the 21st century and realize the career opportunities ahead of them.

The Infinity Project offers a complete turn-key solution for effectively and easily implementing engineering and technology in standard curricula today: innovative curricular content, comprehensive teacher training, high-tech lab equipment, an outstanding supplements package and on-line Web support.

CURRICULUM

The curriculum is covered in a year-long class. The team of curriculum designers and advisors includes leaders in engineering research and education drawn from academia, industry, and government. This group contains five past Presidents of major technical societies, two Presidents and past Presidents of universities, one of the Department of Defense's leaders in technology, science, and engineering, and two engineering Department Chairs.

STUDENT PREREQUISITES

The Infinity project engineering curriculum has been designed for students who have completed mathematics through a second course in algebra (Algebra II), and who have had a least one laboratory science course. These prerequisites will allow students to see firsthand the applications of math and science to engineering and technology early enough in their studies to allow them to pursue more advanced courses, and to begin to consider future careers in technology. The class focuses on the math and science fundamentals of the information revolution and teaches students how engineers create and design the technology around them. Applications are drawn from a wide array of modern devices and systems seen today.

PRELIMINARY COPY

This preliminary copy of *Multimedia And Information Engineering* covers the required curriculum for the program. The final edition is now in development and will be available soon. The material in this preliminary version has been used in schools countrywide.

PEDAGOGICAL FEATURES

- Notes and facts in the margins emphasize important points and interesting facts
- VAB computer-based designs and experiments are boxed throughout the text with icons in the margins for easy location
- Hundreds of 4-color illustrations
- Interesting applications boxed
- End of chapter exercises
- Challenging assignments

We have attempted to utilize many hands-on experiments to demonstrate the basic concepts described in the textbook. Each experiment is based on the VAB™ graphical software interface and utilizes the DSP hardware in the Infinity Technology Kit. Nearly all of the experiments and designs include video, audio, or graphics that are fun for both students and teachers.

SUPPLEMENTS

The Infinity Project supplements offer comprehensive additional resources and classroom support. The supplements are available through the *Infinity Project*:

- Instructor's manual including sample syllabi
- PowerPoint lecture notes for each chapter
- Laboratory Manual
- Lab Exercise Handouts
- Test Item File
- Infinity Technology Kit

Please contact *The Infinity Project* at **ipmail@infinity-project.org** for more details, or visit *The Infinity Project* Web site.

WEB SITE

The Infinity Project Web site **http://www.infinity-project.org** provides ongoing classroom support and resources:

- Curriculum Updates
- Links to Interesting Web Sites
- VAB Experiments and Updates
- Student and Faculty VAB Worksheet Submissions
- Discussion Groups
- Training Materials
- Installation and Support of the **Infinity Technology Kit**

INFINITY TECHNOLOGY KIT

The **Infinity Technology Kit** is a multimedia hardware and software system for converting standard PCs into easy-to-use modern engineering design environments with a wide array of sophisticated capabilities. These capabilities bring to life the engineering concepts taught in *The Infinity Project's* engineering curriculum. The pre-designed lab experiments that come with the software allow students to see firsthand the full range of engineering experiences of envisioning, designing, and testing modern technology. The technology used in the Infinity Technology Kit is based upon Texas Instruments' advanced Digital Signal Processor (DSP) chips and a new and innovative graphical programming environment, called Visual Application Builder, designed and developed by one of *The Infinity Project's* partners, Hyperception.

Components:

- DSP Board: Texas Instruments DSP hardware board with TMS320C31 Digital Signal Processor (DSK board)
- DSP Software: Visual Application Builder graphical component-oriented DSP (VAB software)

- Accessories include: Web camera, PC Powered Speakers, PC Microphone with preamp, Jack Converter, Audio Cable, AC Power Supply, Glasses (Red/Blue), 9 Volt battery

Technical support for *The Infinity Technology Kit* is provided by the staff of *The Infinity Project* and Hyperception, Inc. To order *The Infinity Technology Kit*, please visit *The Infinity Project* Web site at **http://www.infinity-project.org**

ACKNOWLEDGMENTS

The authors would like to acknowledge the generous support of the many individuals and organizations that made *The Infinity Project* and this writing effort possible.

First, we would like to thank the Southern Methodist University School of Engineering, Dean Steve Szygenda, and Electrical Engineering Department Chair Jerry Gibson for their financial support and commitment to *The Infinity Project*.

We would also like to thank Texas Instruments for their generous underwriting and corporate vision. In particular, we owe a deep debt of gratitude to Torrence Robinson for his ever-present involvement, support, and insights. The involvement of Philip J. Ritter, Jeffrey S. McCreary, Leon D. Adams, Renee Hartshorn, Kim Quirk, Steve Leven, Gene Frantz, and Thomas J. Engibous were also instrumental in our success.

Additional financial support for *The Infinity Project* has been provided by the National Science Foundation.

Hands-on engineering design is a fundamental component of this book. and this curriculum. Many thanks go to Jim Zachman and the entire team at Hyperception, Inc., for developing a superb software environment and the packaging of *The Infinity Technology Kit*.

We are very much indebted to the fantastic group of teachers who piloted this new engineering curriculum during the 2000–2001 academic year: Andrew Brown, *W.T. White High School, Dallas, TX*, Sue Kile, *Sunset High School, Dallas, TX*, Kathleen Weaver, *Hillcrest High School, Dallas, TX*, Tina Branch, *D.W. Carter High School, Dallas, TX*, Richard Taylor, *The Hockaday School, Dallas, TX*, Doug Rummel, *St. Mark's School of Texas, Dallas, TX*, Susan Cinque and Bill Miller, *Clements High School, Sugarland, TX*, Sylvia San Pedro and John Spikerman, *B.T. Washington High School, Houston, TX*, Don Ruggles, *Hightower High School, Sugarland, TX*, and Kurt Oehler, *St. John's School, Houston, TX*. Their feedback was invaluable in improving the content and flow of the course. We would like to express our appreciation to the teachers who participated in the roundtable discussions for their significant contributions in sharing with us their insight, experiences, and enthusiasm: Karen Donathan, *George Washington High School*, Charleston, WV, Mark Connor, *Homewood High School, Homewood, AL*, Aurelia Weil, *Cor Jesu Academy, Fenton, MO*, and Debbie Thompson, *Oldham County High School, Crestwood, KY*.

Also, many of the figures and labs assignments were completed by an outstanding collection of graduate students at SMU who worked tireless hours on short deadlines. These students include Sumant Paranjpe, Amitabh Dixit, Mark Westerman, and Joji Phillip.

The authors also owe thanks to the distinguished Infinity Project Board for their perspective and direction: Dr. Delores M. Etter, *U.S. Naval Academy*, formerly Deputy Under Secretary of Defense for Science and Technology, Dr. Don H. Johnson, Chair, *Department of Electrical and Computer Engineering, Rice University*; Dr. Jerry D. Gibson, Chair, *Department of Electrical Engineering, Southern Methodist University*; Dr. William F. Tate, Scholar in Residence, *Dallas Independent School District*; Dr. Leah H. Jamieson, Professor, *Purdue University*; Dr. Andrè G. Vacroux, President, *National Technical University (NTU)*; Dr. Wendell H. Nedderman, Past President, *University of Texas at Arlington*; Bonnie McNemar, Educational Consultant.

And, last but most certainly not least, all of us associated with *The Infinity Project* are deeply indebted to Felicia Hopson for keeping us organized and energized.

About the Authors

Geoffrey Orsak, *Southern Methodist University*

Geoffrey C. Orsak received the B.S.E.E., M.E.E., and Ph.D. degrees in electrical and computer engineering from Rice University, Houston, TX in 1985, 1986, and 1990, respectively.

He is currently Associate Dean of the School of Engineering at Southern Methodist University, where he is also Associate Professor of Electrical Engineering. In addition, he is Director of The Infinity Project, a partnership between SMU, Texas Instruments, leading national universities, and high schools aimed at bringing advanced technology and engineering education to the high school classroom. Prior to coming to SMU, he was Associate Professor of Electrical and Computer Engineering at George Mason University, Fairfax, VA, where he also served as Presidential Fellow. His research interests are in the area of wireless communications, information theory, and statistical signal processing. In addition to this work, he has also been active in the use of high technology for novel forms of pedagogy. In 1995, together with Professor Delores M. Etter of the University of Colorado at Boulder, he cofounded SPEC—The Signal Processing Education Consortium, a geographically distributed consortium of faculty whose aim is to advance DSP education at the undergraduate level. He has been a past recipient of the NSF Research Initiation Award 1991–1994 and is a member of Eta Kappa Nu.

During 1998–1999, Dr. Orsak served as a member of the Defense Science Study Group, a program "that introduces outstanding young scientists and engineers to challenges facing national security," which is sponsored by DARPA and the Institute for Defense Analyses.

Ravi Athale, *George Mason University*

Ravi Athale received the B.Sc. degree in 1972 from University of Bombay and the M.Sc degree in 1974 from Indian Institute of Technology, Kanpur, both in physics. He received the Ph.D. degree in 1980 in electrical engineering from University of California, San Diego.

From 1981 to 1985 he worked as a Research Physicist at US Naval Research Laboratory, Washington, DC. His areas of research were optical signal and image processing systems. From

1985 to 1990 he was a Senior Principal Staff Member at BDM Corporation, McLean, VA, where he headed a group in Optical Computing. His research there concerned optical interconnects and multistage switching networks and optical neural network implementations. Since 1990 he has been an Associate Professor with the Electrical and Computer Engineering Department, George Mason University, Fairfax, VA. His research at GMU has been in the areas of fiber optic signal processing and the analysis of fundamental limitations in optical interconnection networks. At George Mason University, he was a joint developer of a freshman introductory course for electrical engineering students, which focused on information technology aspects of EE. He is currently developing a new course on the principles of information technology, which is aimed at non-science/engineering major students and is a part of the information technology minor at George Mason University.

Dr. Athale has been awarded several patents in optical processing and computing. He is a cofounder of HoloSpex™, Inc. and a co-inventor of HoloSpex™ glasses, the first consumer product that is based on far-field holograms. He was elected Fellow of the Optical Society of America in 1989.

Scott Douglas, *Southern Methodist University*

Scott C. Douglas received the B.S. degree (with distinction), and the M.S., and Ph.D. degrees, all in electrical engineering from Stanford University, Stanford, CA, in 1988, 1989, and 1992, respectively.

From 1992 to 1998 he was an Assistant Professor with the Department of Electrical Engineering, University of Utah, Salt Lake City. Since August 1998, he has been an Associate Professor with the Department of Electrical Engineering, School of Engineering and Applied Science, Southern Methodist University, Dallas, TX. His research activities include adaptive filtering, active noise control, blind deconvolution and source separation, and VLSI/hardware implementations of digital signal processing systems.

Dr. Douglas received the Hughes Masters Fellowship Award in 1988 and the NSF Graduate Fellowship Award in 1989. He was a recipient of the NSF CAREER Award in 1995. He is the author or coauthor of four book chapters and more than 80 articles in journals and conference proceedings. He also served as a section editor for *The Digital Signal Processing Handbook* (Boca Raton, FL: CRC Press, 1998).

David C. Munson, Jr., *University of Illinois at Urbana, Champaign*

David C. Munson, Jr. was born in Red Oak, IA, in 1952. He received the B.S. degree in electrical engineering (with distinction) from the University of Delaware, Newark, DE, in 1975, and the M.S., M.A., and Ph.D. degrees in electrical engineering from Princeton University, Princeton, NJ, in 1977, 1977, and 1979, respectively.

Since 1979, he has been with the University of Illinois at Urbana-Champaign, where he is currently a Professor with the Department of Electrical and Computer Engineering, a Research Professor with the Coordinated Science Laboratory, and a Research Professor with the Beckman Institute for Advanced Science and Technology. His research interests are in the general area of signal and image processing with current work focused on radar imaging, tomography, interferometry, interpolation, time-frequency analysis, and digital filtering.

Dr. Munson is a Fellow of the IEEE and a member of Eta Kappa Nu and Tau Beta Pi. In 1990, he received the Outstanding Professor Award from the Alpha Chapter of Eta Kappa Nu. In 1995, he received the Meritorious Service Award from the IEEE Signal Processing Society and an Outstanding Alumnus Award from the College of Engineering, University of Delaware. In 1998, he received the Outstanding Teaching Award from the Department of Electrical and Computer Engineering, University of Illinois. He was named an IEEE Signal Processing Society Distinguished Lecturer and he received an IEEE Third Millennium Medal in 2000.

John R. Treichler, *Applied Signal Technology, Inc.*

John R. Treichler was born in Velasco, TX, on September 22, 1947. He received the B.A. and M.S. degrees in electrical engineering from Rice University, Houston, TX, in 1970 and the Ph.D. degree from Stanford University in 1977.

From 1970 to 1974 he served as a line officer aboard destroyers in the U.S. Navy. From 1977 to 1983 he was with ARGOSystems, Inc. (now a subsidiary of Boeing). He served as a lecturer at Stanford between 1975 and 1983, teaching digital and adaptive signal processing, and spent the 1983–1984 academic year as an Associate Professor with the School of Electrical Engineering, Cornell University, Ithaca, NY. In 1984 he cofounded Applied Signal Technology, Inc., Sunnyvale, CA, with three collegues. He is currently the company's Chief Technology Officer and also serves on the company's board of directors. The company designs and builds advanced signal processing equipment that is used by the United States government and its allies. His research interests are in the area of digital and adaptive signal processing, particularly as applied to solving problems in communications systems.

Dr. Treichler was named a Fellow of the IEEE in 1991. In 1999 he received an IEEE Third Millenium Medal and was recently presented with a Technical Achievement Award from the IEEE Signal Processing Society for the year 2000.

Sally L. Wood, *Santa Clara University*

Sally L. Wood received the B.S.E.E. degree from Columbia University in 1969 and the M.S.E.E. Ph.D. degrees from Stanford University in 1975 and 1978, respectively. While engaged in her Ph.D. research she also completed a minor in physiological psychology.

She joined the faculty of Santa Clara University in 1985 and is currently a Professor and the Chair of the Electrical Engineering Department. At Santa Clara University she has developed and taught courses in signal and image processing at both the undergraduate and graduate level. In addition, she has developed and taught a freshman-level laboratory-based introductory electrical engineering course and a sophomore-level signal processing architecture course. Over the past 10 years she has developed interactive tutorials with dynamic visual presentation of basic concepts to supplement undergraduate engineering courses. This work has been supported by both industry donations and federal funding agencies. Prior to joining Santa Clara University, she had 12 years of experience in industry working on design and development of medical imaging and visualization systems, optical character recognition systems, and assistive devices for the disabled.

Prof. Wood received the Special Recognition Award from Santa Clara University in 1994 and the Research Award from the School of Engineering in 1995. Her current research interests include multiple source image analysis and nonlinear signal processing.

Mark A. Yoder, *Rose-Hulman Institute of Technology*

Mark A. Yoder was born in Ames, Iowa on December 24th, 1956. He received the B.S. degree in 1980 and the Ph.D. degree in 1984, both in electrical engineering, from Purdue University.

He is currently Associate Professor of Electrical and Computer Engineering at Rose- Hulman Institute of Technology in Terre Haute, Indiana. Since 1988 he has been teaching engineering at Rose-Hulman. His research interests include investigating ways to use technology to teach engineering more effectively. He pioneered, at Rose, the use of Computer Algebra Systems (such as Maple and Mathematica) in teaching electrical engineering. He also helped introduce the teaching of digital signal processing (DSP) early in the curriculum. He is the co-author of the book *DSP First: A Multimedia Approach* with Jim McClellan and Ron Schafer, published by Prentice-Hall in 1998. Dr. Yoder has also co-authored the book *Electrical Engineering Applications with the TI-89*, with David R. Voltmer, published by Texas Instruments in 1999.

Dr. Yoder is serving as General Co-Chair of the 2000 IEEE Digital Signal Processing in Education Workshop. He is a member of the IEEE Education Society Administrative Committee and a member and vice-chair of the IEEE Signal Processing Society Technical Committee on Education. He has served as an Executive Board Director for the ERM division of ASEE and a program co-chair for the 1996 Frontiers in Engineering conference. He is a two time winner of the Helen Plants award for the best non-traditional workshop at FIE.

Contents

The Infinity Project

Chapter 1: The World of Modern Engineering

Approximate Length: 1 Week

Section 1.1: Who created all this great stuff?

Where did all this great stuff come from? Your TV, your car, your comp uter and your cell phone, just to name a few.

Nearly everything you touch during the day, had to be thought of, designed, and built. Who did all of this? And how did they do it?

The answer is simple: *Engineers armed with their ingenuity.*

It might come as a surprise to learn that engineers as a group are some of the most creative and inventive people working today. Society calls upon engineers to not only envision what the world will look like tomorrow but to also make it happen.

Can you imagine what your life would be like if engineers hadn't thought up and built TVs, radios, recording studios, stereos, automobiles, etc? What would your day be like if you couldn't call or e-mail your friends at night? What would your day be like if there was no radio or CD's to entertain you? What if there was no X-ray or CAT scan to help doctors diagnose injuries and illnesses? And what if the only way for you to get to school was to walk or ride a horse?

Today, we all take these great creations for granted. So, what new forms of digital entertainment, transportation, or even manufacturing will we all take for granted in the coming years? Computers that talk to us and even "think"? Cars that drive themselves?…Thankfully engineers are working on these devices today!

Exercises 1.1

1. Identify 10 items designed by engineers. What do these items do that was new at the time of their creation? What items did these new creations replace? How is the world a better place because of these designs?

Section 1.2: Scientists and Engineers

1.2.1 Making Dreams a Reality

Lets go back in time, before the human race ever went into space. There were sci-fi movies that suggested what life could be like in space, what our space ships might look like, what the surfaces of planets might look like, and even what aliens might look like. Yet, until the 1960's these images were all largely figments of Hollywood's imagination.

However, on May 25, 1961, president John F. Kennedy announced that the United States would put a person on the moon "in this decade." Who did he think was going to actually make this dream a reality? Politicians, bankers, lawyers? No, he knew it was going to be engineers and scientists.

At the time of Kennedy's speech, did engineers and scientists know how they were going achieve this remarkable goal? No, but they had confidence that by working together, breaking the problem of space travel into manageable pieces, solving these individual problems, putting all the components together, and then testing the final solution that they would have a very good shot at placing a human on the moon before 1970.

Well, the engineers and scientists were right: through their hard work and the with the help of many others, on July 20, 1969 the Apollo 11 mission placed Neil Armstrong and Buzz Aldrin on the surface of the moon – culminating one of the greatest achievements in all of human history, and a triumph for engineering and science.

> The word "Engineer" stems from the French word "*Ingénieur*" which means ingenuity. Contrary to the popular belief, its root is not the English word "engine".

Figure 1.1: Photo of Buzz Aldrin taken by Neil Armstrong: July 20th, 1969.

1.2.2 Engineers and Scientists

What makes engineers so unique? And why are engineers different from scientists and mathematicians?

You have had years of experience taking math and science courses. These classes have helped you understand the your world (science) and have helped you learn an entirely new language to describe our world (math).

The primary purpose of science and math is to help humans *understand* the world: How do cells divide? What makes objects fall to the ground? What are the basic building blocks of life?

To answer these questions, scientists over history have followed the **scientific method**. This **"algorithm"** is the basic road map for discovery and understanding. The scientists who sought to answer the fundamental questions about our world have used the scientific method as their guidepost.

The Scientific Method

1. Observe some aspect of the universe.
2. Invent a tentative description (*hypothesis*) consistent with what you have observed.
3. Use the hypothesis to make predictions.
4. Test those predictions by experiments or further observations and modify the hypothesis in the light of your results.
5. Repeat steps 3 and 4 until there are no discrepancies between theory and experiment and/or observation.

What makes engineers different from scientists?
Engineers, unlike scientists and mathematicians seek to create new objects and new realities that are important to humans. Yet like scientists, engineers use mathematics as their basic language. But engineers are not valued by the insights they make about the world; rather they are valued for what they create for the world.

While scientists rely on the scientific method for discovery, engineers rely upon the **engineering design algorithm** to create nearly every object around you.

Constraints are limits that are placed on the design problem. For example: the design should not cost more that $x, or weight more than v pounds.

The Engineering Design Algorithm

1. Identify the problem or design objective
2. Define the goals and identify the constraints
3. Research and gather information
4. Create potential design solutions
5. Analyze the viability of solutions
6. Choose most appropriate solution
7. Build or implement the design
8. Test and evaluate the design
9. Repeat steps ALL steps as necessary

At the root of the algorithm (step 1), the engineer must answer the most remarkable question: *What do I want to create today?* Very few other professions place such a high premium on the creative spirit of the individual.

As a way of understanding the engineering design algorithm, let's apply it to a piece of existing technology as if we were the engineers about to begin the design process of the cell phone.

Step 1: *Identify the design objective.* We want to build "something" that will allow humans to communicate with one another at any location on the globe and at any time.

Step 2: *Design goals and constraints.* The devices created to communicate should not use wires (so we can move around), be too large (so that we can carry them in our hands), be too small (must be able to reach from our ear to our mouth), use too much energy (we don't want to have to recharge it after each call), and should be cost effective.

Step 3: *Research and gather information.* Has anyone ever done something like this before? Well, there was the CB radio craze and you can still buy walkie-talkies today. However, neither system reaches around the globe and both make it nearly impossible to contact a wide variety of individuals. So, a system as described in step 1 didn't exist prior to its creation in the

Steps 4 – 8: *Create, analyze, choose, build, and test.* Various companies such as Nokia, Ericsson, and Motorola have designed, built, and tested a wide variety of cell phones.

Your Analysis: How well do you think current cell phones meet the objectives specified in Step 1 and satisfy the constraints in Step 2? Are you pleased with current cell phone technology? Would you change anything about the goals or constraints?

Exercises 1.2

1. Apply the engineering design algorithm to
    a) Making the family dinner
    b) Creating new laws
    c) Treating illness
   Be specific about each step in the design algorithm.
2. Determine the likely constraints applied by the engineer in designing these items:
    a) Cash register
    b) Bicycle
    c) Office lamp
    d) Tennis shoe
3. Select a device designed by an engineer. Step through each stage of the engineering design approach and describe the likely path taken by engineers in creating the design.
4. Evaluate the effectiveness of these various engineering designs. Make sure to describe the strengths and weaknesses of the designs. Try to guess what capabilities these technologies will have in the future.

a) Conventional telephone
b) Medical CAT scan
c) Desktop computer
d) Cell phone
e) MP3 player

Section 1.3: Birth of the Digital Age

1.3.1 Before Digital there was Analog

To understand where engineers will be taking our world in the future, it is important that we briefly look back on where we came from.

Up until the middle of the twentieth century, the human-made world was primarily **analog**. The use of the word "analog" implies that the devices and tools society relies upon use physical forces for their basic operation rather than some abstract quantity such as numbers. For example: analog audio records use the physical bumps and indentations in the groves on albums to store music. The tone arm of a turntable rides in these grooves and vibrates nearly identically to the original sound waves of the music, thus creating an electrical version of the music that is subsequently amplified and played through speakers. The entire process of recreating music from bumps and indentations is purely physical and thus "analog."

Analog	Digital
LPs	CDs
Film Cameras	Digital Cameras
Dial watches	Digital watches
Standard TV	HDTV
VHS Camcorders	Digital Camcorders
Etc.	

Analog systems like albums and turntables are quite functional, they are generally large, consume lots of energy, prone to breakdowns, and highly susceptible to physical problems (e.g., scratches on the album or the bumping of a turntable can ruin a good listening experience).

Many analog electronic devices were built using an important component called a **vacuum tube**. Just like light bulbs, these vacuum tubes got very hot, burned out regularly, and couldn't be made very small. (Your parents will remember how often TV's used to break down due to vacuum tubes burning out.)

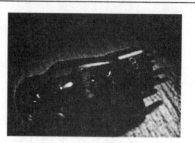

Vacuum tubes are still found today in very high-end audio systems and guitar amplifiers.

Figure 1.2: Vacuum Tube (shown on its side). Vacuum tubes were typically the size of a coke can and got as hot as a light bulb.

During the middle of the twentieth century, mathematicians and engineers discovered that most physical quantities found the world (sound waves, light intensity, forces, voltage, current, charge, etc.) could be converted to *numbers*. (This should not be surprising since scientists had been using mathematics to describe the physical world for centuries.) This remarkable yet simple realization that the physical world could often be equated to numbers gave birth to the **digital age**.

There are many advantages to "digitizing" analog quantities: "numbers" are much less sensitive to physical problems, and are easier to store and move. Let's revisit the recreation of music: today the sound of most music is converted to numbers and stored on the CD. The CD player simply reads the numbers and converts these numbers back to the original music (we will learn about the details of this process in a later chapter.) If you have ever compared the quality of music between an average turntable and an average CD player, there should be little doubt that digital technology is significantly better than the earlier analog technology. Also, can you imagine trying to take a jog with a turntable strapped to your waiste?

Unfortunately, there was a major problem in converting from the analog world into the new digital world: engineers just didn't have the right "digital" parts to built new digital systems. Not to be deterred, engineers working during the first half of the twentieth century tried the smart and reasonable thing: they attempted to use readily available vacuum tubes as basic digital building blocks. And in 1945, engineers successfully produced the first digital computer called the ENIAC that was built out of over 17,000 vacuum tubes, weighed 30 tons, and filled a 30-foot by 50-foot room. Just think of the heat produced by 17,000 light bulbs all burning in the same room!

While primitive by today's standards, the ENIAC was a major advance in engineering and technology. Never before in human history could we do math so fast and so accurately. The ENIAC opened up new digital horizons for society. Unfortunately, this first computer was so large and so expensive that only governments and the biggest companies could ever hope to own or even use one.

ENIAC or the Electronic Numerical Integrator and Computer was capable of multiplying 5000 numbers every second. Compare this to computers today which can execute billions of mathematical operations every second.

Interesting fact:
The typical human hair is 50 to 100 micro meters (microns) in diameter. While this may seem small, it is quite large when compared to the size of modern transistors. Today, transistors on computer chips are much smaller than a millionth of meter on a side.

What do transistors actually do?
Transistors behave just like electronic doors. They are either open or closed to either allow current to pass through or not. So from a mathematical perspective, we can assign the number 1 to the state that the transistor is open (current doesn't flow) and the number 0 to state that the transistor is closed (current does flow).

Figure 1.3: The ENIAC Computer

The Transistor

What the digital age needed was a truly digital component that could replace the vacuum tube, and run fast, use less power, and most importantly, be made small and inexpensive. Fortunately, we got that component, the **transistor**, in 1947 and its creation changed the world forever. Bill Shockley, Walter Brattain, and John Bardeen won the Nobel Prize in 1956 for their joint discovery and development of the transistor.

Now, engineers could unleash their imaginations to create smaller, portable devices that could run on small amounts of enery (batteries) and were rugged to normal use. For this reason, many people believe that the transistor is the most important invention of the 20th century. (Just look around you today to see the nearly infinite array of small gadgets and pieces of technology built from transistors.)

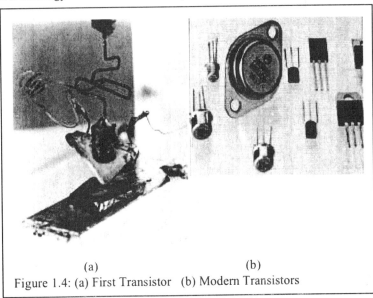

(a)   (b)
Figure 1.4: (a) First Transistor   (b) Modern Transistors

The Integrated Circuit (IC)

The next critical step forward into the digital age was the development of the technology necessary to put many transistors onto a single small part that could be used for a variety of complex tasks. Jack Kilby accomplished this remarkable feat in 1958, and for this discovery, Kilby was awarded the 2000 Nobel Prize in physics.

With the invention of the IC (**Integrated Circuit**), engineers were now able to undertake ever more complicated designs because they now had "digital parts" that could do significantly more complicated math on the newly digitized version of the real (analog) world. Interestingly, the integrated circuit has become so widely used in devices from computers to anti-lock breaks, that it is difficult to find individual transistors in modern technology.

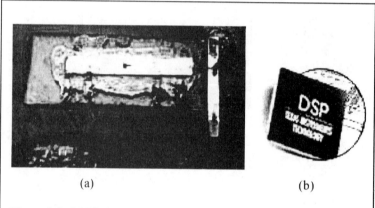

(a)   (b)

Figure 1.5: (a) First Integrated Circuit produced by Jack Kilby
(b) Modern Integrated Circuit produced by TI.

1.3.2 Why are bits so important?

Engineers and computer scientists seem to always be making reference to "bits". Why are bits so important?

The answer is very simple: as we discussed before, technology and engineering are steadily moving from an analog world to a digital world. And by doing this we get devices which are smaller, faster, more reliable, and more powerful in their capabilities.

The basis behind this shift from the analog to the digital is the conversion of the physical world into numbers (recall how music is stored by numbers on a CD). Nearly all numbers that are used today are from the base 10 system of numbers. That is, the traditional number system we have adopted is based on the powers of 10 and uses the digits 0, 1, 2, ... 9 to express quantities.

For example, the number "361" really means

$$3 - 100\text{'s or } 10^2$$
$$6 - 10\text{'s or } 10^1$$
$$1 - \text{digit or } 10^0$$

If we add these powers of 10 up appropriately scaled by 3 and 6 and 1, we get $3*100 + 6*10 + 1*1 = 361$.

This choice of base 10 was completely arbitrary and probably driven by fact that humans have 10 fingers. If humans had evolved with only two fingers, mathematicians would have likely chosen base 2 for our everyday number system. In this case, we would have only used the numbers "0" and "1" when expressing quantities, rather than the familiar "0, 1, 2, 3, ... , 9."

Well, as you read in the sidebar on the previous page, transistors, which form the building blocks of the digital age, have only two states (you can think of them as two fingers). So, it made perfect sense to the earliest digital engineers to choose the base 2 number system for digital technology, rather than the base 10 system. And, viola, the concept of the "bit" was born.

Bits are the digits of the digital age.

Now, this doesn't mean that you and I need to switch from the familiar base 10 system to the base 2 system. Computer and other digital systems today automatically convert digits to bits and back again. Think about your calculator – you type in digits because you are familiar with them. These digits are converted to bits (base 2), and then the appropriate binary mathematics is applied to these bits in accordance with your wishes (e.g., square root, sin) . The binary answer is then converted back to digits and displayed for your convenience.

So, what does the number 361 look like in binary form (base 2)?

$$361 \text{ (base 10)} = 101101001 \text{ (base 2)}$$

or $1\text{-}2^8$ or 256
$0\text{-}2^7$ or 128
$1\text{-}2^6$ or 64
$1\text{-}2^5$ or 32
$0\text{-}2^4$ or 16
$1\text{-}2^3$ or 8
$0\text{-}2^2$ or 4
$0\text{-}2^1$ or 2
$1\text{-}2^0$ or 1

By adding all these powers of two scaled by the appropriate bit value, we get
$$1*256+0*128+1*64+1*32+0*16+1*8+0*4+0*2+1*1=361.\text{ Simple.}$$

Bit: Short for binary digit. A bit is either a 0 or a 1, which in turn reflects whether a transistor is "open" or "closed." A byte is 8 consecutive bits.

Exercises 1.3

1. Determine whether the following designs are digital or analog?
   a) Car speedometer
   b) Car radio
   c) TV
   d) VCR
2. Why is modern technology based on binary mathematics?
3. Identify 5 modern systems that are completely analog. Do think these systems will stay analog or be converted to digital systems?
4. Write the following numbers in binary form (base 2)
   a) 271
   b) 42
   c) 18
   d) 167
5. Write the following binary numbers in base 10
   a) 111
   b) 01001
   c) 10010001
   d) 100100100

Section 1.4:
Moore's Law

1.4.1 Technology Projections

Have you ever wondered how engineers are able to predict what is going to happen in the future? A recent headline reads as follows: "Engineers predict that in ten years computers will be able to talk to us". How do they know this? What is the basis for their predictions?

Well, we are fortunate that history has shown that technological evolution often follows a fairly predictable path. For example, digital technologies such as computers are getting faster and faster at a regular rate and consequently engineers and computer scientists able to do more and more with this technology. Video games are better today than 10 years ago; special effects in movies today are better than they were 10 years ago.

1.4.2 Moore's Law

Is there a systematic way to predict the future of technology? Well, in 1965, Gordon Moore (founder of Intel) made a startling observation: Moore looked back in time and noticed that every two years his company's computer chips had twice as many transistors on them - which in turn meant that his company's ICs were twice as powerful or twice as fast. This observation has since been known as **Moore's Law**.

Moore's Law: The number of transistors on a computer chip (IC) will double every two years.

Equivalently stated, the computing power of ICs doubles every two years.

What was particularly bold about this prediction was that Moore said that the doubling of speed, power, or number of transistors on digital IC's would continue indefinitely. To fully understand the remarkable implication of Moores' Law, we need to first understand the power of doubling.

1.4.3 The Power of Doubling

Have you ever heard someone say "double or nothing" when gambling? If you have watched someone bet like this, then you no doubt noticed that you could either win a lot of money or loose a lot of money very fast. Doubling gets you to large numbers quite quickly. We can quantify this by developing some simple mathematical relationships.

Assume that you begin with X dollars. If you double it, you will have 2X dollars. If you double it again, you will have 2*2 X dollars, or 2^2 X dollars. Once more gives you 2*2*2 X dollars or 2^3 X dollars. From this it is very easy to show that if you double X dollars exactly N times, you will have

$$2^N \text{ X dollars.}$$

To simplify matters, lets assume that we start with one dollar (X=1). If N is 5 (double 5 times), you would have 32 dollars. If N is 10 (double 10 times), you would have 1024 dollars. And if N is 50, you would have 1.125×10^{15} dollars: that is more than a thousand trillion dollars. Wow!

Doubling just 50 times turns one dollar into a staggering amount of money. So, "doubling" clearly has tremendous power in taking small numbers (corresponding to dollars or even numbers of transistors) and turning them into absolutely gigantic numbers - fast.

We can plot the value of one dollar doubled N times for various choices of N to see first hand how fast money grows through the power of doubling. Notice from this figure that the monetary value grows at an exp onential rate (the function is linear when plotted on a semi-log plot.)

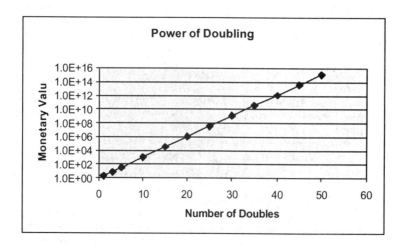

What does this tell us about Moore's Law? No matter how limited computer chips were when they were first introduced more than 40 years ago, by doubling in "computing power" every two years, we arrive at staggering amounts of digital power that will enable engineers to design systems that can do almost anything. If it can be dreamed of, Moore's' Law tells us that we can build it at some time in the future.

1.4.3 The Mathematics of Moore's Law

Now that we have a good understanding of the power of doubling, we can translate Moore's Law in simple mathematics.

Lets say that in year Y_1 there are N_1 transistors on a company's chip. Then following Moore's Law, in year $Y_1 + 2$, there should be 2N transistors on the next version of the chip. We can generalize this to calculate the number of transistors on an IC during some year Y_2 given that we know the number of transistors in year Y_1.

First, lets calculate the number of "doubles" which will occur between years Y_1 and Y_2 :

$$\text{Number of doubles} = \frac{Y_1 - Y_2}{2}$$

For example, if $Y_1 = 2000$ and $Y_2 = 2006$, then there would be (2006-2000)/2 =3 "doubles" in the number of transistors on computer chips between the years 2000 and 2006. Further, if there happened to be X transistors on a computer chip in the year 2000 then there would be

$$2.2.2.X = 2^3 X = 8X$$

transistors in the year 2006.

From this we are now able to identify the general relationship that predicts the number of transistors N_2 (or computing power) in any given year if we happen to know the number of transistors N_1 (or computing power) in any other year (such as today).

$$N_2 = 2^{\text{\# of doubles}} N_1$$

We can substitute in the equation for number of doubles in terms of years from above to obtain the number of transistors in year Y_2 given the number in year Y_1

$$N_2 = 2^{\frac{Y_2 - Y_1}{2}} . N_1$$

Gordon Moore made these same calculations roughly forty years ago. To verify his result, Moore plotted the actual number of transistors N_2 on his computer chips for various years and compared these values to those obtained from the formula given above. What he found was absolutely remarkable (and still true today), the predictions made by Moore's Law were nearly identical to the actual values.

To see how accurate Moore's Law has turned out to be, look at the figure below where we compare his predictions with reality.

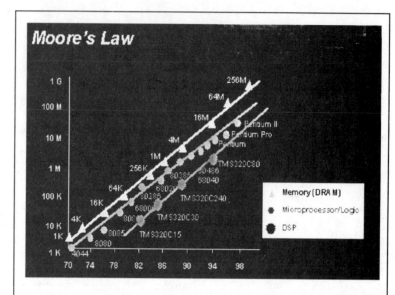

Figure 1.6: Moor's Law for various digital components. The lines represent the predictions, the marks represent the actual numbers of transistors.

This startling observation meant that computers and other digital technology were going to increase in speed and power at an exponential rate as time proceeds – and that the power of doubling would result in computer chips with such enormous power that nearly any design will be possible in the future.

Example: The Intel Pentium IV microprocessor (IC) has 47 million transistors in it. How many transistors will be in Intel's chip
    a) 10 years in the future
    b) 30 years in the future

Answer: The result is easy to find by applying Moore's Law.

Y_1= 2001 (at the time of writing the book)
N_1= 47,000,000

For part (a), Y_2=2011, and for part (b) Y_2=2031. From the formula above,

$$N_2 = e^{\left(\frac{2011-2001}{2}\right)\ln(2)} \cdot 47,000,000 \approx 1,503,000,000$$

and for part (b)

Useful fact:

$$2^x = e^{x\,\ln(2)}$$

where ln(2) is the natural log of 2.

$$N_2 = e^{\left(\frac{2031-2001}{2}\right)\ln(2)} \cdot 47{,}000{,}000 \approx 1{,}540{,}000{,}000{,}000$$

Exercises 1.4

1. What is more valuable, 1 penny doubled 30 times, or 1 million dollars?
2. Determine the equation that tells you the correct monetary value of a penny doubled N times.
3. How much money would you have at the end of a year if you had $1 on January 1 and
   a) You doubled it once a month   4096
   b) You doubled it once a week   4.5 v 10^15
   c) You doubled it every day   7.5 x 10^109
4. Using the Pentium IV as your reference point, determine the number of transistors on the version of the Pentium to be released in 2015.
5. If the Moore's law said that the number of transistors would double every 3 years, how many transistors would be there on the Pentium to be released in 2015?
6. Derive equations which predict the number of transistors on computer chips if:
   a) Moore's law said that the number of transistors tripled every two years.
   b) Moore's law said that the number of transistors doubled every three years.
7. How big are transistors today?
8. How many transistors would fit across a human hair today?   388 - 555
9. Moore's Law says that the number of transistors doubles every two years. Assume that these transistors are shrinking in size so that the resulting size of a computer IC stays the same over time. Find the mathematical equation that tells you how fast the edge of a square transistor is shrinking.
10. Assume that the side of a transistor is .18 microns and that the diameter of a human hair is 100 microns. Using Moore's Law, determine the number of transistors that would fit across a human hair in the year 2010?
11. Assume that you want to build a computer system that can carry on natural conversation with you. Engineers predict that it will take a very fast computer to achieve this goal. One prediction says that the computer will have to be able to execute 10^{12} instructions/sec. Predict the year that this will happen assuming that the instructions per second also double every two years and computers today run at 500 million instructions per second.
12. How big are the following atoms?
   d) Hydrogen 10^{-10} M
   e) Silicon   2.34 x 10^{-10} M
   f) Iron   2.54 x 10^{-10} M
   g) Uranium   2.76 x 10^{-10} M
13. Assuming Moore's law continues, predict the year that transistors will be the size of the atoms in problem 12.
14. What do you think are some barriers to Moore's law continuing?

15. Do you think engineers will be able to build systems which create
    h) Prize winning literature
    i) Academy award winning movies.
16. How has engineering technology changed popular music?
17. How has engineering technology changed popular films?
18. How has engineering technology changed the way people purchase products such as clothing?
19. Imagine three new technologies that you would like to see in the cars of the future? Can you predict when these might occur?
20. Is it reasonable to assume that you can get a peta of silicon atoms into a cubic millimeter of space?

Section 1.5: Engineering Design and the Infinity Technology Kit[SM]

Up to now, we have only *talked* about the creative aspect of engineering. It is now time for you to begin to express your own creativity through the engineering design process.

In this book, you will have the opportunity to create many new and interesting designs. Just to name a few, you will

- Design your own digital musical instruments and create unique sound effects
- Create special effects in images and movies
- Build your own cell phone
- Use state of the art encryption techniques to produce secret encoded messages
- And learn about the "guts" of the Internet.

To help bring your creations and designs to life, the authors of this book, working with engineers from Texas Instruments and Hyperception, have produced the **Infinity Technology Kit**[SM] that is part of your classroom materials.

This state of the art technology connects to standard personal computers and allows you to create, build, and test a wide variety of new devices. The Infinity Technology Kit[SM] uses a Texas Instrument's Digital Signal Processing IC to execute mathematical operations at blazing speeds. A block diagram based software environment named Visual Application Builder[TM], or VAB for short, created by Hyperception, Inc, controls the entire system.

The use of the kit in the design process is very simple:

1. After you have decided what you want to create (you will likely be assigned a specific task by your teacher), you will create a **block diagram** description of your design.
2. You will then input this block diagram design description into the VAB[TM] software environment through a set of very easy to use rules.
3. You will then run and test your design on the TI DSP hardware.

This may seem like a complicated task, however you will find that in little or no time, you will not only master this process, but actually have fun with it! So, lets get started and learn about block diagrams.

1.5.1 Block Diagrams

Suppose an engineer wanted to make video game box such as the Sony PlayStation 2[1]. Do you think that Sony engineers gathered some parts (transistors, IC's, resistors, capacitors) and sat down at a workbench, and couple of days later had a working PS2? No! The PS2 is a very complex machine. So complex that one person couldn't design the whole thing.

Complex designs are created by breaking them down into collections of simpler elements that are then organized through a **block diagram**. For example, here is a block diagram for the PS2.

PlayStation 2 Block Diagram from
http://arstechnica.com/reviews/1q00/playstation2/ee-1.html

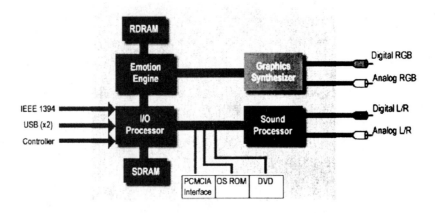

This design is typical of block diagrams in the way it is layed out and structured: The **inputs** come in from the left and flow to the right through the design, the the basic blocks of the design are organized in an easy to read manner and are color coded, and finally the **outputs** of the design exit on the righthand side.

You might recognize some of the inputs on the left. They are

1. IEEE 1394, better known as "FireWire." This is an interface that lets you download things fast, like live digital video.

---

[1] http://www.playstation.com/

2. USB, or universal serial bus. You may have this on your own computer at home or in the lab. It lets you attach a wide variety of things from keyboards and mice, to CCD cameras and printers.

3. Controller. For the PS2, the controller is what the player holds in their hands to play the game.

The four outputs are

1. *Digital RGB.* RGB stands for Red, Green, Blue, the three primary colors engineers use for creating color images or videos. (You'll learn more about why we use RGB in a later chapter.) This output goes to a computer display so that the player can see the game.

2. *Analog RGB.* This goes to a TV display in the case someone wants to watch the game on a television or more often use the DVD player that comes with the game to watch movies.

3. *Digital L/R.* This is an audio output (left/right) for playing the stereo sound effects of the game. (In a later chapter, you will have an opportunity to create your own digital audio sounds.)

4. *Analog L/R.* This goes to your speakers so that you can actually hear the sounds.

The blocks inside the PlayStation2 are

1. RDRAM, SDRAM and OS ROM: are all forms of **memory** to store the software and the data.

2. The *Emotion Engine* is the digital computer that determines where to place objects such as people, cars, buildings, etc. and determines how they are to move.

3. The *Graphics Synthesizer* takes the information from the Emotion Engine and creates the corresponding digital images for your video display.

4. The *Sound Processor* creates the sounds of the game.

5. The *I/O Processor* and the *PCMCIA Interface* control the Inputs and the Outputs.

6. Finally, the *DVD* talks to the DVD reader/player that contains all of the information about the specific game being played. (If you haven't heard of DVD or Digital Versatile Disk before, think of it as the next generation CD that can play both movies and sound. You will learn more about DVD's in a later chapter.)

1.5.2 VAB™

Just like the engineers from Sony, you will create your own devices by first constructing block diagrams of the design. The software environment that comes with the Infinity Technology Kit^{SM} and is used in this course takes your block diagram and automatically "builds" it for you on the programmable DSP hardware.

So, in essence, the Infinity Technology Kit^{SM} literally builds your block diagram without you having to go find or buy any parts or having to write thousands of lines of tedious software code.

With this kit, *you are able to go directly from Block Diagram to working design.*

One of the designs you will undertake during this course is to build a system which tracks moving objects. This is a very useful device because it can be used to calculate the speed of a moving car, or help sportscasts keep the camera locked onto flying objects such as golf balls or baseballs as they move through the sky.

This tracking system might appear at first to be a complicated design task, but like all designs, it can be made much simpler and more organized by first creating a block diagram. In our case, we create the block diagram in our software environment called the Visual Application Builder^{TM}. Block diagrams, first created on paper, can be directly translated into the VAB^{TM} software environment through a set of simple keystrokes and computer commands that you will master as part of this course.

A prototype design of the "object tracker" is shown in the VAB screen capture below:

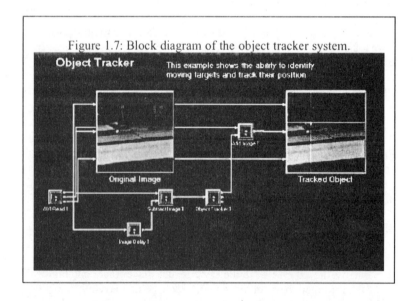

Figure 1.7: Block diagram of the object tracker system.

If you examine this figure closely, you will see the same features as those you saw in the block diagram of the PS2: the input (which in this case is the video from a camera or movie file) is on the left and it flows through the block diagram of the object tracker from left to right. The output, which in this case is a video display with a set of green crosshairs locked onto the moving object is displayed on the right hand side of the design.

Keep in mind that the software environment now only allows you to easily draw this block diagram, but it also "builds" it for you on the computer and

TI DSP hardware. So, as promised above, the kit converts your designs in block diagram form into a working prototype for you to test and modify.

As mentioned before, throughout this course, you will be asked to test an existing design, or create your own design to achieve a wide variety of objectives. You will recognize these design assignments in the book by the following standard graphic shown below. The title of the lab or design assignment appears in the top gray bar, with a very brief description in the lighter gray text. The complete details of the design assignment or lab are contained in web pages that are accessed through the VAB™ software or through the Infinity Project web space at www.infinity-project.org.

As a way of demonstrating to you the wide range of capabilities of the Infinity Technology Kit™, your first assignment is to run a small variety of demonstration designs – one of which is the object tracker we have been discussing. As you work with these designs, notice that each is built from a block diagram description that is then implemented on the TI DSP chip using the VAB™ software tools.

VAB Experiment 1.1: Demos
• Exposes students to the range of capabilities of the Infinity Technology Kit.
o   Object Tracker
o   Filtered Music
o   Touch Tone phone

Exercises 1.5

1. Construct a block diagram description of the following activities:
   a. getting dressed in the morning
   b. cooking dinner at night
   c. preparing for an exam.
2. Lets analyze a simple system: a portable CD player.
   a) Draw a block diagram that describes the functionality of this system.
   b) Determine which blocks are analog and which blocks are digital.
   c) Imagine that it is your job to improve this system. Draw a block diagram of your new design.

Section 1.6:
Conclusions

By now, it must be clear to you that engineers are creating our world of tomorrow with their know-how and ingenuity. During the remainder of this course, you will have the unique opportunity to participate in this remarkable endeavor.

You will be taught the skills and knowledge necessary to take your dreams and turn them into tomorrow's reality. And finally, as you move forward in this course, please remember the following statement made by Theodore Van Karman

"Scientists explore what is, engineers create what never has been."

Appendix

Show me the money

Generally speaking, employees are compensated in proportion to their value to society. Doctors extend the lives of humans, and as such are paid well. Lawyers protect our rights and also paid well. Entertainers bring joy to millions of people and in turn earn lots of money.

Engineers contribute mightily to the well being of society. As described in this chapter, engineers create all the technologies used by doctors, design safer cars and highways, and even produce the technology for all of the special effects that make actors look so glamorous.

So, it is not surprising that engineers are amongst the best-paid people today. And, what is particularly remarkable is that one only needs a four-year college degree to reach this level of earnings.

Occupation	Yearly Compensation ($)
President of the United States	390,000
Electrical Engineers	68,437
Computer Scientists	59,428
Scientists	54,558
Doctors	134,012
Teachers	55,330
Librarians, Archivists and curators	43,697
Social Scientists an Urban Planners	50,911
Social, Recreation and religious	29,612
Lawyers and Judges	80,731
Police and Detectives	42,265
Firefighting	41,607
Secretaries	27,763
Plumbers	41,205
Taxicab Drivers	18,476
Waiters and Waitresses	9,684
Janitors and Cleaners	20,017

The Units of Modern Technology

Just when you have become comfortable with the term "mega", here come "giga", and right behind giga is "tera." When will it ever stop?

The numbers in technology and modern engineering get so small *and* so large that it is much more convenient to refer to these numbers using words rather than digits.

For example, a typical DVD disks holds 4.7 giga bytes of data (a **byte** equal eight bits.) This is 37,600,000,000 bits.

Isn't it easier just to say "giga" than to say 1,000,000,000?

Multiplication Factor	Prefix	Symbol	Term
10^{18}	exa	E	One quintillion
10^{15}	peta	P	One quadrillion
10^{12}	tera	T	One trillion
10^{9}	giga	G	One billion
10^{6}	mega	M	One million
10^{3}	kilo	K	One thousand
10^{2}	hecto	h	One hundred
10^{1}	deca	da	Ten
10^{-1}	deci	d	One tenth
10^{-2}	centi	c	One hundredth
10^{-3}	milli	m	One thousandth
10^{-6}	micro	μ	One millionth
10^{-9}	nano	n	One billionth
10^{-12}	pico	p	One trillionth
10^{-15}	femto	f	One quadrillionth
10^{-18}	atto	a	One quintillionth

Even though you may know that peta means 10^{15}, it is very hard to put numbers of this scale into any human context. How big is 10^{10}? How small is 10^{-10}.

To help you get a grasp on big numbers and small numbers, we have prepared a chart below that relates items from your daily lives with some of these numbers.

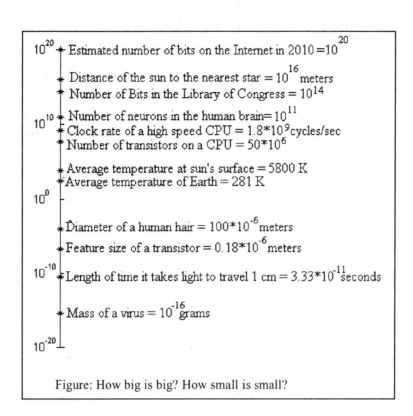

10^{20} — Estimated number of bits on the Internet in 2010 $= 10^{20}$

Distance of the sun to the nearest star $= 10^{16}$ meters
Number of Bits in the Library of Congress $= 10^{14}$

10^{10} — Number of neurons in the human brain $= 10^{11}$
Clock rate of a high speed CPU $= 1.8*10^{9}$ cycles/sec
Number of transistors on a CPU $= 50*10^{6}$

Average temperature at sun's surface $= 5800$ K
Average temperature of Earth $= 281$ K

10^{0}

Diameter of a human hair $= 100*10^{-6}$ meters

Feature size of a transistor $= 0.18*10^{-6}$ meters

10^{-10} — Length of time it takes light to travel 1 cm $= 3.33*10^{-11}$ seconds

Mass of a virus $= 10^{-16}$ grams

10^{-20}

Figure: How big is big? How small is small?

Section 1.7:
Glossary

algorithm: (1.2.2)
A step-by-step process to achieve some goal. In the case of this chapter, our goal is to design new products or devices.

analog: (1.3.1)
The real physical world made of forces and mass.

binary numbers: (1.3.2)
Mathematical representation of numbers using the base 2 system of numbers rather than the familiar base 10. For example: 6 in base 10 equals 110 in base 2.

bit: (1.3.2)
Short for **bi**nary digit. A bit only takes the values of 0 or 1.

block diagram: (1.6.1)
A graphical description of the overall operation of some system or design. Typically, the inputs to the system are on the left hand side and the outputs are on the right hand side of the block diagram.

byte: (1.3.2)
A byte is equivalent to eight consecutive bits.

constraints: (1.2.2)
Limits that are placed on the design problem.

digital age: (1.3.1)
The era born with the creation of the transistor. The digital age is generally thought to have begun at the time that computers became widespread during the middle of the 1980's. It also implies that most of the new devices being produced today and in the future will rely heavily on digital technology.

engineering design algorithm: (1.2.2)
The nine-step algorithm that is followed by nearly every engineer in creating nearly every object around you.

integrated circuit (IC): (1.3.1)
A single computer chip that is built from many different components. Typically, nearly all of the individual components on an IC are transistors.

input (1.5.1):
Inputs are instructions or data, which is used by a system to complete its task. For example, the CD and the controls on the CD player are both inputs to the CD system.

Moore's Law: (1.4.2)
The number of transistors on a computer chip doubles every two years.

output: (1.5.1)
Outputs are produced by devices. For example, the electrical version of music is the output of a CD player.

scientific method: (1.2.2)
The five step process by which scientists explain the universe.

transistor: (1.3.1)
The basic building block of all digital technology. When used in digital applications, transistors are devices which operate in only two states corresponding to the familiar "0" and "1" bit representation. Today, transistors found in computer chips are typically much smaller than 1 millionth of a meter on a side. The transistor is thought by many to be the most important invention of the 20th century.

vacuum tube: (1.3.1)
An early technology that was used in nearly every piece of electronics. It is rare to find vacuum tubes used today in any application other than very high-end audio systems and guitar amps.

The INFINITY Project

Chapter 2: Creating Digital Music

. . . .

Making and Playing Great Sounds

Approximate Length: 3 Weeks

Section 2.1:
Introduction

Approximate Length: 2 Lectures

2.1.1 Digital Music in a Rock Band

Technology is everywhere in today's world—we use it when we study, work, and even when we play. One place where technology has changed nearly all aspects of a field is in the arts, and specifically the musical arts. Almost everyone listens to music. Technology makes it easier for us to listen to, share, perform, and even compose the music that we and everybody else want to hear.

Because of the way sound works, music and mathematics are very closely tied together. The ties are so close between music and mathematics that a song can be created and stored entirely inside of a computing device using numbers and equations before it is turned into sound for everyone else to hear. This "digital composition" is used for background music for movie soundtracks, and the techniques even help make the pop stars of today sound as good as they do. Who knows—we are not too far away from creating a "digital singer" whose entire band shares her space inside of a computer. Of course, we humans still provide the artistic choices or control over the calculations, so that the sounds that are produced sound interesting to us.

More often, technology is used to enhance the abilities of a composer or performer to make music with other people. To motivate this idea, we're going to consider the following situation: Suppose a group of your friends have gotten a band together. Everyone is bringing a different musical instrument to play. You want to join in on the band, but you're not sure which instrument you want to play. In fact, different songs are going to require different instruments, so you want to be able to play a number of different ones. The only problem is, you don't have any of these instruments, and you probably don't know how to play all of them. How can digital technology still make it possible for you to be in the band's spotlight? We're going to make several digital instruments out of bits, using just the INFINITY Technology Kit hooked up to a computer. Later, we'll see how to add all the great effects that give singers and other performers the ability to shine in the spotlight of the stage. Along the way, we'll learn the important mathematical concepts behind the processing of *signals*. What are signals, you ask? A little digression is in order...

2.1.2 Signals are Everywhere

What do the voltage of a firing neuron, shown in Figure 2.1, and the temperature at DFW Airport, shown in Figure 2.2, have in common? The answer is that both are signals! Some signals, such as the chemical concentration within a cell or the sound of a person's voice, are created inside of our bodies. Some signals, such as the electromagnetic waves produced by a radio transmitter or cellular telephone, the sound of thunder, the vibration of an earthquake, or the light from the sun, are created in our surrounding environment.

As the nerve cell fires the voltage reaches a "spike" and then rapidly falls as the electrical signal is discharged to a target cell. The nerve then readies itself for another firing as shown by the curve leading up to the spike. The resting state of this nerve is at –60 millivolts.

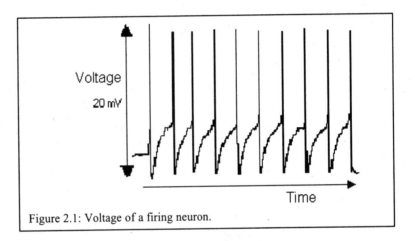

Figure 2.1: Voltage of a firing neuron.

What is a signal? From a higher-level point of view, a **signal** is a pattern of variation that contains information. A signal can be as simple as a sequence of numbers, such as your height as measured over every day of your lifetime or the temperature example above. Often, a signal represents some physical quantity that changes with time or space. The physical quantity can be in one of many forms—electrical (voltage), thermal (temperature), pressure (sound), or altitude (height). When the quantity depends on a single value or changes with time, we can plot or graph the signal as we have done in Figures 2.1 and 2.2. We call such plots **waveforms**. If the quantity depends on two different values or is a function of area, we call such signals **images**.

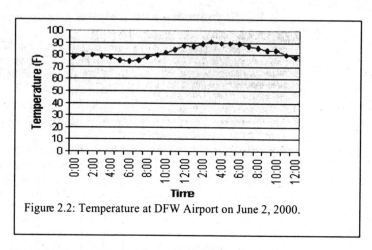

Figure 2.2: Temperature at DFW Airport on June 2, 2000.

Many signals are electrical in nature. For example, the heart produces electrical signals that can be measured by sensors attached to the chest. The set of waveforms or plots collected is shown in Figure 2.3 and is called an electrocardiogram (EKG). The EKG is a collection of waveforms. By finding differences in these signals from those of a healthy person, a cardiologist or heart doctor can diagnose problems and prescribe treatment if any is needed.

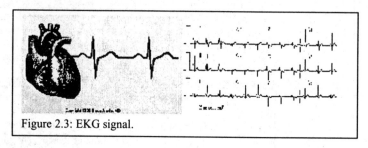

Figure 2.3: EKG signal.

Radio, television, and cellular phone stations broadcast electrical signals from an antenna as depicted in Figure 2.4. These signals, which represent variations in voltage versus time, can be captured by a receiving antenna and then converted to sound or images for human listening or viewing. These same signals also can be viewed directly by technicians or engineers on an instrument called an oscilloscope.

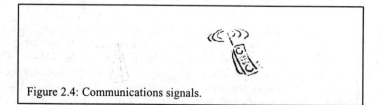

Figure 2.4: Communications signals.

Computers speak in the language of digital signals. This is shown stylistically in Figure 2.5 where the wide blocks represent one of two states (up or down) as a function of time. Analog signals can vary continuously like the sound of your voice or that of an instrument. A major objective of this course will be to show you exactly what a digital signal is and how digital signals are manipulated and communicated.

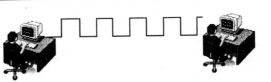

Figure 2.5: Digital signals – the language of computers.

Even though signals may have different forms and origins, they can all be converted to the digital language of computers and the computing devices inside a CD player, cellular telephone, TV, car stereo, and almost any device that you can think up. This is the main reason why signals are so important—they are the "messages" that make these devices work.

2.1.3 Analog Versus Digital

Most signals in the natural world are represented as waveforms that are continuous in time, or in the case of images, continuous in space. By this, we mean that at every point in time or space, the waveform or image has a specified value. Such signals are called **analog** or **continuous-time** signals. Figure 2.6 shows an analog signal. For each value of time on the horizontal axis, this waveform takes on a specified value that can be read from the plot as the corresponding height of the curve on the vertical axis.

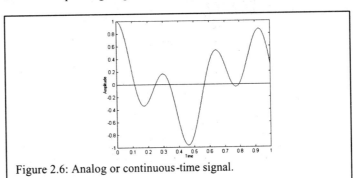

Figure 2.6: Analog or continuous-time signal.

Sampling:
the act of collecting values of a continuous-time signal at discrete points in time, in order to represent the signal as a set of numbers.

Computers cannot store analog signals directly. Computers only can store lists of numbers. Thus, a computer stores the above waveform by a list of numbers which are the heights of the waveform at a set of closely spaced points, as shown in Figure 2.7. The operation of creating these values is called **sampling**. The numbers that the computer stores to represent these samples are the heights of the dots in Figure 2.7 and are listed in Table 2.1. Signals that are stored in sampled form as a list of numbers or digits are called **digital**.

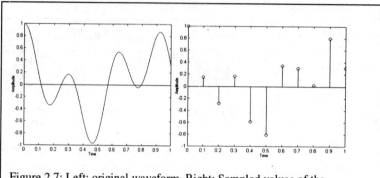

Figure 2.7: Left: original waveform. Right: Sampled values of the continuous curve.

1.0000,	0.1516,	-0.2873,	0.1656,	-0.5897,	-0.8090,	0.3445
0.2993,	0.0194,	0.8026,	0.3090			

Table 2.1 Digital Representation of Analog Waveform in Figure 2.7.

There are some signals that are simply lists of numbers to begin with, so that no sampling is required. For example, the closing Dow Jones Industrial Average on the New York Stock Exchange is a single number each day. This signal is inherently digital. Figure 2.8 shows a plot of this signal for one year from August 1999 to July 2000. The height of each dot represents the closing average on the day of the month indicated on the horizontal axis. Although the closing Dow Jones Industrial Average is a digital signal that arises naturally, most digital signals are obtained by sampling of continuous-time signals.

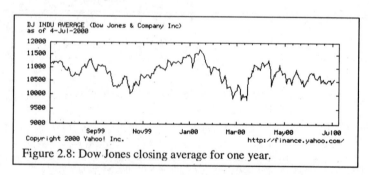

Figure 2.8: Dow Jones closing average for one year.

An advantage of digital signals is that they can be stored and manipulated by computers. Much of this course will deal with the storage, manipulation and communication of digital signals. On your computer, the main memory (RAM), the floppy disk, the hard disk, and the CD-ROM all store numbers; that is, they store digital signals.

2.1.4 Signals in Music

Music is made up of combinations of sounds. All sounds can be represented by waveforms as a function of time. In this case, the value of the sound at any point in time corresponds to the amount of pressure in the sound wave at a point in time and a point in

6

space. Sounds always have a value at every time instant (although this value can be zero at certain times), making them **analog** in nature. Musical sounds are usually associated with a particular type of instrument—a trumpet, saxophone, piano, or guitar, for example—although the quality of a musical sound also depends on the capabilities of the instrument player. Figure 2.9 shows an example of a sound—in this case, the sound of a guitar. The left-hand side shows the sound over its entire duration, although the line width of the plot doesn't show the intricate detail of the sound. The right-hand side shows a zoomed-in portion of the sound. Of course, if we use a digital device to make the sound, then this device converts the digital numbers into a continuous waveform, akin to a "connect-the-dots" drawing but smoother.

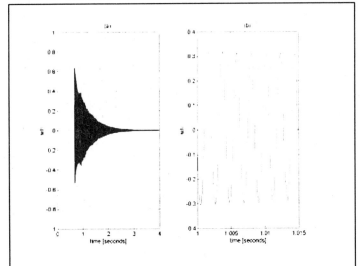

Figure 2.9. Plot of a guitar sound. (a) The guitar sound over 4 seconds. (b) The guitar sound over 0.015 seconds showing fine detail.

When computers and other digital devices process sound, they do so digitally. That is, they first convert the sound to digital form and then manipulate the numbers from the digital signal. To hear the sounds, these devices must then put them back into analog form again, because humans can only hear signals in analog form.

A **melody** is a sequence of notes that make up a particular song or part of a song. Notes are the musical "letters" that make up the "language" of melodies describing a piece of music. In the case of Western music, melodies are represented graphically using a five- or ten-line staff with specialized markings to indicate the pitch and duration of each particular note. This signal is actually a two-dimensional image, because it represents both the length of a sound (duration) as well as its pitch (frequency). Like the Dow Jones Industrial Average, Western melodies are clearly **digital** in nature because they represent specific pitch values and time lengths. Melodies do not depend on the instrument that plays them—they simply "carry the tune."

To create a song, a composer develops melodies for each instrument, and a group of musicians then play these melodies at the proper pitches and with the correct note lengths. Of course, the result is a lot better than any description we can give! This description, however, points out how we should think about creating music with technology: First, we need to figure out how to use technology to create interesting sounds—in effect, by building a digital musical instrument. Then, we need to figure out how to instruct this musical instrument to play interesting melodies. We'll first consider how to create sounds before creating melodies, so that we can start hearing the results of our experiments right away.

Exercise 2.1

1. Find examples of five different signals that are made up of waveforms. You should either create (by hand) or otherwise produce a graph that describes each of your signals. Be sure to label both axes of your graph. Which of your signals is an analog signal? Which of your signals is a digital signal?

2. To the next class period, bring pictures of both analog and digital systems. These are devices or systems that store, process (i.e. modify or change), or communicate digital and analog signals. Some systems are partly analog and partly digital. Label each of your pictures as analog, digital, or hybrid analog/digital. You may cut out pictures from magazines, newspapers, or catalogs, or download pictures from the Web.

3. While washing the dishes, you realize that a particular dish produces a "ping" when it is struck lightly. Is this an example of a sound or a melody?

4. Your friend has just heard a great song on the radio and is whistling the song. Is he recreating the sound of or the melody of the song?

5. A pianist sits down at a piano and plays the song *Rhapsody in Blue* by George Gershwin. She then plays the song *Jailhouse Rock* made popular by Elvis Presley. What has changed—the sound of the piano or the melody played by the piano?

6. Waveforms are **one-dimensional** in that they depend only on one quantity (usually time). Images are **two-dimensional** in that they depend on two quantities (usually x-y coordinates on a surface). Can you give two examples of a **three-dimensional** signal? How about a **four-dimensional** signal?

7. A melody is really a sequence of instructions for a musician to perform. Taking any task and breaking it down into a set of simple instructions is a useful part of many engineering tasks. Find a task or activity that you engage in at least once a week, and break this task or activity down into a sequence of instructions. Be as detailed as possible (so that a person following the instructions could do the task exactly as you would).

Section 2.2: Sound Synthesis

Approximate Length: 2 Lectures

2.2.1 What is Synthesis?

Our first task in the design of a digital musical instrument is creating the sound of the instrument. **Synthesis** is the creation of useful and complicated items from more basic ones. **Sound synthesis**, then, is the creation of useful and complicated sounds from more basic sounds. There are many ways to synthesize sound for music. In this chaper, we shall focus on two of the most popular techniques:

1) *Wavetable synthesis* (also known as **sampling**) attempts to build the sound waveform of the actual instrument being recreated. This synthesis method is the most straightforward, although can use a lot of digital memory unless clever techniques are employed.

2) *Additive synthesis* (also known as **Fourier synthesis**) uses sinusoidal functions as building blocks to create sounds. This synthesis method has some strong links to other signal-based problems in medicine and communications. This synthesis method was used in most early music synthesizers, although it is less popular in modern-day synthesizers.

In the next chapter, we'll talk about a third synthesis method, *physical modeling synthesis* (also known simply as **modeling**). This method uses a mathematical model of the physics of an instrument to compute the instrument's sound. This synthes is method is the most sophisticated and the newest; hence, few synthesizers use it. When it is properly implemented, however, this method can provide extremely realistic sounds.

Although we are focusing on making music in this chapter, you should realize that all of the above ideas can be connected to important problems in other fields. We will point out some important connections as we go along.

All sound synthesis methods use computing devices to create interesting-sounding waveforms. Since we'll be exploring many of these methods, we first need to learn about analog signals and how to plot them.

2.2.2 Plotting of Analog Signals

Our mathematical descriptions of signals will use the following notation.

The value of an analog signal (at time "t") = s(t). Analog signals that have been processed or altered after being recorded: *The value of a processed signal (at time "t") = y(t). The value of a digital signal (at sample number "n") = s[n]* Digital signals that have been processed or altered from the original: *The value of a processed digital signal (at sample number "n") = y[n]* Subscripts in s_1, s_2, s_3, etc. refer to different signals numbered 1 through 3.

You'll need a paper or a graphing calculator here.

As stated before, sounds can be represented as waveforms that, at first glance, seem similar to functions that you may have plotted in your math classes. While sounds might look more complicated in form, the same fundamental concepts apply. A **waveform** s(t) assigns a numerical value s to each value of t. We represent the values of s(t) as the height of a curve above the t-axis, which in our case is time. If the waveforms are simple enough, we can write down compact mathematical expressions that describe them. For example, a straight line with slope a and vertical intercept b is described by the equation

$$s(t) = at + b$$

and is plotted in Figure 2.10.

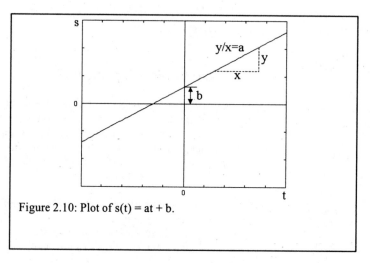

Figure 2.10: Plot of s(t) = at + b.

Since waveforms play a large role in making sounds, we need to gain some facility with plotting of waveforms and with simple manipulations of waveforms. As an aid in this, try plotting the following waveforms either by hand or on your calculator:

$$s_1(t) = 2t + 3$$
$$s_2(t) = 2s_1(t)$$
$$s_3(t) = s_1(t - 4)$$
$$s_4(t) = s_1(t + 5)$$

Here,

(a) $s_1(t) = 2t+3$ has a slope of 2 and crosses the y-axis at the value 3,

(b) $s_2(t) = 2s_1(t)$ scales up s_1 in amplitude by 2,

(c) $s_3(t) = s_1(t-4)$ is a delayed version of s_1 (shifted to the right) by 4,

(d) $s_4(t)$ is an advanced version of s_1 (shifted to the left) by 5.

Your plots should resemble those below in Figure 2.11.

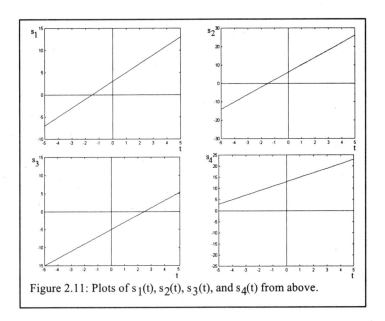

Figure 2.11: Plots of $s_1(t)$, $s_2(t)$, $s_3(t)$, and $s_4(t)$ from above.

Exercises 2.2.2

1. Plot $s_1(t)$ for the following and in each case also plot the corresponding waveforms $s_2(t)$, $s_3(t)$, and $s_4(t)$ defined as above.

$$s_1(t) = t^2$$

$$s_1(t) = -t - 2$$

$$s_1(t) = |t|$$

$$s_1(t) = 3 - 4t + t^2$$

2.2.3 Sound Signals: Envelope and Period

Most sound signals that we like to listen to are generally not as simple to describe as the signals above. Figure 2.12 shows the signals produced by three different musical instruments. The signals on the left and right are from the same sound. The only difference is the **scale** of the bottom axis of the plot, which is time. On the left, the signals are displayed over 4 seconds, whereas on the right, we've "zoomed in" to a short portion of each sound over 0.015 seconds or 15 milliseconds (ms).

Each plot shows a different structure of the sound. The plot on the left shows the overall **envelope** of the sound. Just like the paper envelopes that surround cards or letters, the envelope of a sound is a description of its general size or **amplitude** over time. The size of a sound determines its **loudness** or volume. Think of loudness as a physical quantity (it is), whereas amplitude corresponds to a mathematical quantity. The envelope of a sound is the way a signal's amplitude or a sound's loudness changes with time. So, when looking at the piano sound, we see that it has an envelope that starts at a large value at the beginning of the sound and gradually tapers off to a sma ll value over time. In comparison, we can see that the trumpet sound is large mainly during the time that the note is being played and is nearly zero otherwise. The guitar sound is similar to the piano sound, which makes sense because both sounds use the same physical object to make them—a vibrating string.

On the right of Figure 2.12 is a "zoomed in" version of both sounds starting at the time instant of t=1 second. Now we see something completely different; in this case, each sound has an up-and-down vibrating quality to it. In addition, each waveform is almost repetitive; the same waveform repeats at regular intervals or times. Signals that exactly repeat are called **periodic** signals. This repeating property can be described mathematically for a periodic signal p(t) as

$$p(t) = p(t+T)$$

where T is called the **period** of the signal . Periodic signals are uniquely described by the value of the waveform over any interval of time length T as well as the period T of the signal, because then we can simply repeat the signal every T seconds to get the entire signal. This fact makes it possible to nearly recreate sounds of instruments using a simple set of procedures— just by knowing the envelope and the periodic portion of a sound.

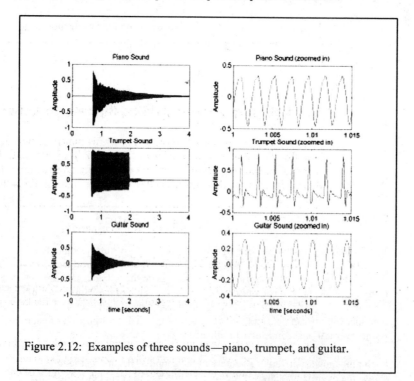

Figure 2.12: Examples of three sounds—piano, trumpet, and guitar.

Figure 2.13 below shows the waveforms illustrating the envelope and periodic structure of a speech signal, which in this case is a vowel sound (like "aaah", "eeeh", and so on). The plot on the right only shows one period, so you cannot see it repeating there, but the plot on the left clearly shows how the voice is repeating. As you can see, speech is like the musical instruments above, showing a similar structure. You can see the same structure in your own speech, as the VAB experiment shows.

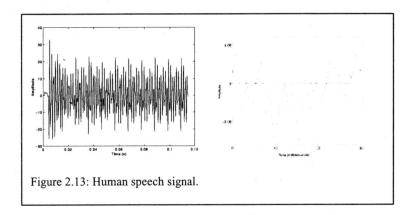

Figure 2.13: Human speech signal.

VAB Experiment 2.1: Plots of Speech

Students see their own voice.

2.2.4 Superposition: Adding Signals Together

More often than not, music consists of many instrument sounds playing together. What do such musical waveforms look like? Figure 2.14 shows example of such waveforms from several different types of songs (classical music, jazz music, R&B music, and (East) Indian music). From these plots, we see that

- Most music is not periodic over long periods.

- The waveforms look fairly complicated.

However, we know by listening to music that songs contain the sounds of different instruments playing together. How would we recreate such sounds on a computer?

One way to create such sounds is to combine the sounds of different instruments playing different notes. This combination amounts to simply adding the sounds together. Suppose $s_1(t)$ is the waveform of one instrument and $s_2(t)$ is the waveform of another instrument. Then, the sound of these two instruments playing together can be represented as

$$s(t) = s_1(t) + s_2(t)$$

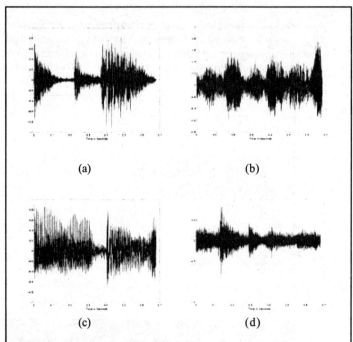

(a)   (b)

(c)   (d)

Figure 2.14: Sections of four different kinds of music:(a) Jazz
(b)Classical music (c) R&B music and (d) East Indian music

Mathematically, the sum of two signals can be found by considering the value of each signal at each individual time instant $t=t_0$ and adding the values of the signals together. For example, suppose a saxophone sound and trumpet sound are added together. The value of the saxophone sound waveform at time $t=1.5$ sec is found to be

$s_1(1.5) = 0.7$

and the value of the trumpet sound waveform at time $t=1.5$ sec is found to be

$s_2(1.5) = -0.4$.

Then, the value of the signal $s(t)$ at time $t=1.5$ sec is

$s(1.5) = s_1(1.5) + s_2(1.5) = 0.7 + (-0.4) = 0.3$.

To compute the waveform $s(t)$ for other time instants, we would have to do a similar calculation. Fortunately, computers can be used to add up the complex waveforms for us, but it is the mathematical operation and what is going on that is important.

The process by which we have created combinations of sounds is known in physics as the **principle of superposition**. As we'll describe later in this chapter, sound is made of variations in air pressure that travel through the air in waves. When two sound waves meet in air, they add together to produce a combined sound wave that is the sum of the individual sound waves. It is the principle of superposition that allows us to hear a saxophone-trumpet duet and recognize that it is two instruments playing

14

together. If sound waves didn't behave this way, then music wouldn't make sense. Imagine if adding a trumpet and a saxophone sound resulted in a violin sound or the engine sound of a truck! The principle of superposition is a physical property of the audio world that makes our job of synthesizing music possible.

Exercises 2.2.5

1. Suppose you are standing on a lonely highway, and a single truck appears on the horizon traveling toward you. The sound of its large diesel engine gets louder and louder as it comes toward you, and after 30 seconds, it races by you. Thirty seconds later, it disappears over the horizon again. Sketch the envelope of the truck sound that you heard, labeling the time axis carefully. Take an "educated guess" as to the shape of the envelope over the first 30 seconds (you need to know more about the physics of sound and the way the truck is traveling to get more specific), but sketch carefully the envelope of the latter 30 seconds.

2. Consider two periodic signals $p_1(t)$ and $p_2(t)$. The first signal has a period of $T_1=15$ milliseconds, and the second signal has a period of $T_2=20$ milliseconds. Is the sum of the two signals $s(t)=p_1(t)+p_2(t)$ periodic? Sketch an example to figure this problem out. If the signal is periodic, what is the new period?

3. Repeat Problem 2 for the periods $T_1=15$ milliseconds and $T_2=30$ milliseconds

4. Repeat Problem 2 for the periods $T_1=\sqrt{2}$ (square root of 2) milliseconds and $T_2 = 2$ milliseconds.

5. Can you think of a situation where adding two periodic signals together produces a signal whose period is the period of one of the signals? Do the periods of the two signals need to be the same in this case? Explain.

6. Can the envelope of a sound be periodic as well? Give an example of a signal with a periodic envelope.

7. Figure 2.15 shows the waveforms of two signals $s_1(t)$ and $s_2(t)$. Sketch the waveform resulting from the sum of these two signals $s(t) = s_1(t)+ s_2(t)$.

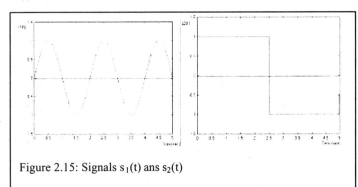

Figure 2.15: Signals $s_1(t)$ ans $s_2(t)$

Section 2.3: Waveform Synthesis

Approximate Length: 3 Lectures

2.3.1 Simple Waveform Synthesis

As we saw in the last section, many sounds have a coarse amplitude shape described by an envelope and a fine structure that is nearly periodic. We now describe how to construct a model of a sound from its envelope and periodic shape. We'll then used this method in a VAB experiment to synthesize sounds in real time.

Suppose we measure the sound of an instrument using a microphone connected to a computer, and that we have saved the sound so that we can measure its envelope as well as its finer repetitive structure, as shown in Figure 2.12. How do we make the computer play a similar sound? One way to do so is simply play back the recording. Of course, this "memorization" doesn't take any particular intelligence, nor does it give us the freedom to develop our own sounds. Instead, we use the following idea: Signals of the sort found in Figure 2.12 can be well approximated as

$$s(t) = e(t) \times p(t),$$

where $e(t)$ is the mathematical description of the envelope of the signal and $p(t)$ contains the underlying periodic structure of the signal. In this case, the product of the two signals $e(t)$ and $p(t)$ is taken at each time instant, so that if at time $t=1.5$ sec the values of $e(t)$ and $p(t)$ are $e(1.5) = 0.4$ and $p(1.5) = -0.3$, the value of $s(t)$ at $t=1.5$ sec is

$$s(1.5) = e(1.5) \times p(1.5)$$
$$= 0.4 \times (-0.3)$$
$$= -0.12 \quad .$$

To see how this works, we consider the following example.

Example: Approximating a plucked string.

A plucked string instrument, such as a guitar, mandolin, banjo, or ukelele, produces a sound that approximately has the structure of the $s(t) = e(t) \times p(t)$ model above. The form of $e(t)$ can be well-modeled by the **exponential function**

$$e(t) = \exp(-\alpha\, t) = (2.71828\ldots)^{-\alpha t}$$

where the positive constant α describes the **rate of decay** of the function over time. Figure 2.16(a) shows the shape of this function for the value α=0.92 over three seconds from t=0 to t=3 sec. Starting from a value of $\exp(-\alpha\, 0) = 1$, this function gradually decreases towards zero, becoming smaller and smaller over time. This function is quite similar to the positive portion of the guitar sound in Figure 2.9. As for the periodic portion p(t), it can be modeled to first order by the **sinusoidal function**

$$p(t) = \sin(2\,\pi t/T)$$

where $\sin(\theta)$ is the well-known sine function on your handheld calculator and π is the fundamental constant 3.1415926535…We'll go into more details about sinusoidal functions in the next section. For now, look at the plot in Figure 2.16(b), and compare it to the fine detail of the guitar sound in Figure 2.9. We have chosen a value of T=0.0024 sec, which makes p(t) look quite similar to the detailed guitar sound. Multiplying e(t) by p(t), we get the function in Figure 2.16(c), which compares favorably to that in Figure 2.9. It should be apparent that if two waveforms look the same, they should sound the same—otherwise the waveform wouldn't describe everything about the sound!

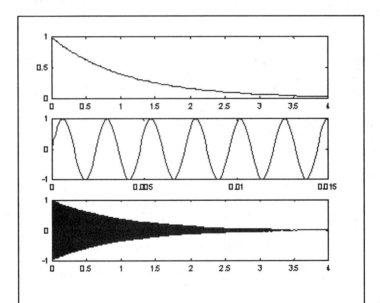

Figure 2.16: (a) exponential function (b)Sinusiod and (c) product of the exponential and sinusoid

We can use this basic idea to synthesize sounds using VAB. The worksheet that does this synthesis is called *SketchWave*. Like the name suggests, you sketch both one period of the waveform that makes up p(t) and the envelope e(t), and the system computes s(t) using the equation s(t) = e(t) × p(t).

The following notes can help you get the most out of the SketchWave VAB:

1. Our perception of the sound of an instrument is affected by its initial **attack**, or the way the note first begins. Instruments that are excited using blown air, such as flutes, clarinets, saxophones, trumpets, and the like, generally have more gradual attacks than pianos or guitars. For this reason, you should draw increase e(t) initially gradually to create realistic sounds.

2. The periodic part of a clarinet sound is well modeled by a **square wave**, which is a function that is (1) for about half its length and (-1) for about half its length. Draw this wave in the p(t) window and see whether you can create a realistic clarinet sound.

2.3.2. Frequency and Pitch

When a musician plays notes on an instrument, different notes sound different to us. Some notes sound "high", whereas others sound "low". This perceptual difference is called **pitch.** Figure 2.17 shows the envelope and nearly periodic fine structure of two notes on a piano keyboard corresponding to middle-C (or "doe") and the note of F above middle-C (or "fa"). The second note sounds higher as compared to the first one. Looking at their envelopes, however, the two sounds don't look different. Looking at their nearly-periodic portions, however, we see a difference: the second sound undulates or "wiggles" faster than the first.

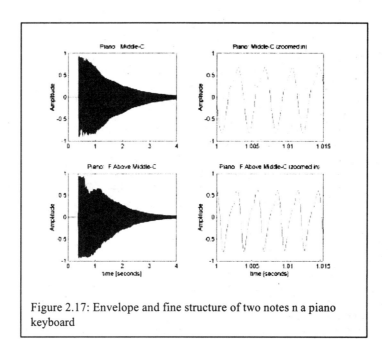

Figure 2.17: Envelope and fine structure of two notes n a piano keyboard

The pitch of a note has a mathematical counterpart that is related to the period of the sound waveform when we zoom in on its finer structure. This mathematical quantity, called the **fundamental frequency**, can be computed from the period T of the sound as

$$f = 1/T$$

All notes of a piano have a different fundamental frequency. Figure 2.18 shows the notes of a piano and the fundamental frequencies of each one.

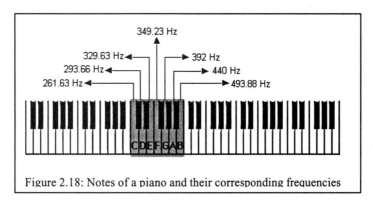

Figure 2.18: Notes of a piano and their corresponding frequencies

What are the units of frequency? Since the period T is in units of time (seconds), the units of frequency are in (seconds)$^{-1}$ or 1/(seconds) (read "one-over-seconds"). Engineers have given a special name to the unit of frequency: Hertz or Hz for short. This unit value is named after *Heinrich Hertz*, a famous physicist. You may have seen the units of Hz before in the context of high-quality audio equipment, e.g. a frequency response of 20Hz to 20kHz. In this case, 20Hz corresponds to a sound whose period is 1/20=0.05 seconds, whereas 20kHz=20000Hz corresponds to a sound whose period is 1/20000=0.00005 seconds.

The way pitch and fundamental frequency are related has an immediate implication for making sounds using a digital musical instrument. To make music, we have to figure out a way to produce signals with different periods. The simplest way to do this is to change the length of the period of the note in some way. There are several options open to us:

- We could simply remove a portion of the waveform at the beginning or end of the period of the sound if we're making a higher note. This operation is called **truncation.**
- We could add zeros to the waveform at the beginning or end of the period of the sound if were making a lower note. This operation is called **zero padding**.
- We could "stretch" or "compress" the period of the sound while keeping its overall shape intact. This operation is called **time warping**.

In SketchWave, the period of the sound is adjusted by a slider from 10 to 50, where 10 corresponds to a "high" pitch and low 50 corresponds to a "low" pitch. We'll relate these numbers to actual frequencies in Hz in a moment. For now, try adjusting this slider in your VAB exp eriment, then click on the p(t) and e(t) windows. What is going on? How is SketchWave altering the pitch of the sound it uses? Is it using truncation/zero padding or time warping?

Time-warping of sounds can be described mathematically as follows. Suppose a sound has a periodic structure $p_0(t)$ whose period in seconds is T. Then, a time-warped period signal with new period T_{NEW} can be created from the original periodic signal as

$$p_{NEW}(t) = p_0(t\ T/T_{NEW})$$

where T_{NEW} is the period of the new signal. Similar time-warping can be performed on other signals, although it usually doesn't make sense to time-warp the envelope signal e(t) or the final signal s(t) (Why not?).

In practice, the way the sound of an instrument changes with pitch (frequency) is much more complicated than the above two methods. In fact, to make an accurate sounding instrument, we would need to change the shape of the waveform for every note that the instrument can play. Of course, with proper control and an appropriate interface, we could employ such techniques to make sounds. These techniques are used by electronic musical instrument makers in their quest to create better-sounding instruments.

Exercise 2.3.2

1. When you call a friend, an answering machine picks up. You start recording your message, but after 30 seconds, the recorder stops. Is this an example of truncation or time-warping? How would you use time-warping of your voice in a telephone answering machine?

2.3.3 Sampling

The musical sounds that all digital devices produce—from the SketchWave worksheet in VAB to the music coming from a CD or MP3 player—cannot be stored as continuous waveforms inside of these devices. Computers by their very nature compute and store discrete numbers. How can such devices copy a musical sound? They do so by a process known as **analog-to-digital conversion** or **digitization**. Digitization has two parts:

- The analog signal is measured at regular intervals over time. This process is called **sampling.**
- The individual samples are rounded to some set of significant digits or fixed precision. This process is called **quantization.**

In a later chapter, you will learn more about the digitization process, along with **digital-to-analog conversion** whereby a digital signal is turned back into an analog one. For now, we shall focus on sampling and how sampling affects our ability to create musical instrument sounds.

Introduction to Sampling

The term **sampling** refers to the recording of values of points along a waveform, usually at uniformly spaced time instants. Figure 2.19 shows an analog waveform. Figure 2.20 shows samples of this same waveform taken every 0.1 seconds. The sample values are listed in Table 2.2 to the right of the figures. So, for example, the value of s(t) at t = 1.3 sec is 27748. We can plot the values of the samples as shown in Figure 2.21. Sometimes, this same type of plot is made without the vertical bars, just showing the dots.

Sample Values	Time (secs)	Sample Values	Time (secs)
2.0000	0.0	-0.7078	1.6
3.1674	0.1	-1.5621	1.7
3.8694	0.2	-1.6835	1.8
3.8289	0.3	-1.0707	1.9
3.0465	0.4	0.0000	2.0
1.8006	0.5	1.0807	2.1
0.5412	0.6	1.7233	2.2
-0.2814	0.7	1.6508	2.3
-0.3885	0.8	0.8637	2.4
0.2217	0.9	-0.3601	2.5
1.2732	1.0	-1.5718	2.6
2.3188	1.1	-2.3223	2.7
2.9112	1.2	-2.3346	2.8
2.7748	1.3	-1.6092	2.9
1.9113	1.4	-0.4244	3.0
0.6002	1.5		

Table 2.2: The sample values for the signal in Figure 2.20.

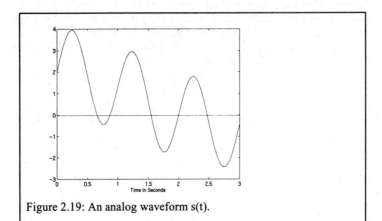

Figure 2.19: An analog waveform s(t).

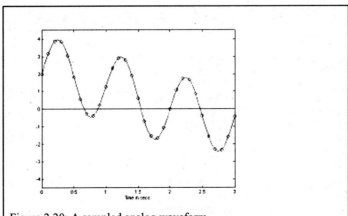

Figure 2.20: A sampled analog waveform.

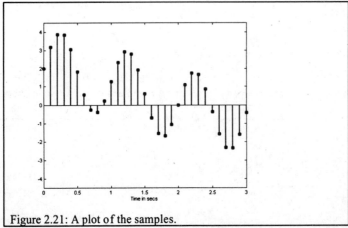

Figure 2.21: A plot of the samples.

For the analog signal above, a sample spacing of 0.1 seconds produces a sequence of sample values that accurately describe the original signal. In general, however, the sample spacing will be something other than 0.1 seconds. An analog signal s(t) that varies quickly must be sampled more frequently than a signal that varies slowly. This general situation is shown in Figure 2.22 where the sample spacing is T_s seconds.

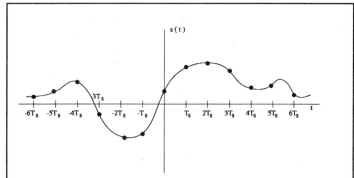

Figure 2.22: Sampled values of an analog waveform, with spacing T_s.

Here, the sampling operation may begin before the arbitrary time instant t = 0. The quantity T_s is referred to as the **sampling period**. The samples of the waveform are spaced every T_s seconds apart. The **sampling rate** or **sampling frequency** is $f_s = 1/T_s$. The samples of the waveform are a sequence of numbers. We denote these numbers by s[n] where

$$s[n] = s(nT_s).$$

The use of square brackets in s[ ] indicates that it is a sequence of numbers rather than a continuous waveform. We have

$$s[0] = s(0), \quad s[1] = s(T_s), \quad s[2] = s(2T_s), \quad s[3] = s(3T_s),$$

and so forth. Likewise,

$$s[-1] = s(-T_s), \quad s[-2] = s(-2T_s),$$

and so on.

Figure 2.23 shows an engineer's view of the sampling operation, where the input is the continuous waveform s(t), the output is the series of numbers s[n], and the sampling period is T_s.

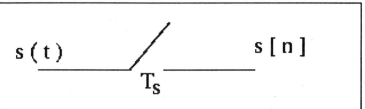

Figure 2.23: Symbolic depiction of the sampling operation.

We show the sampling operation as a switch, because we think of a switch closing for a brief instant every T_s seconds to measure the signal over that instant. In most digital devices, this switch is used to charge up an electrical device known as a capacitor, and the amount of charge is then measured. For the waveform in Figure 2.23, we can plot its samples as the **sequence** in Figure 2.24.

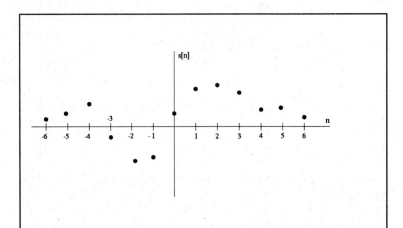

Figure 2.24: Samples of the waveform in Figure 2.23, plotted as a sequence.

Ordinarily, there is no simple mathematical function describing the signal s(t) and its samples s[n]. In the following examples, however, we consider signals that have some simple shapes for purposes of illustration..

Example

Suppose $s(t) = 2t - 4$ and $T_s = 0.5$. Find the values s[n] of the sampled waveform.

Solution:

$s[n] = s(nT_s) = s(0.5n) = 2(0.5n) - 4 = n - 4.$

Thus, s[0] = -4, s[1] = -3, s[-1] = -5, s[2] = -2, s[-2] = -6, etc. Try plotting this sequence to see that it outlines the straight-line shape of the analog waveform s(t) = 2t - 4.

Example

Suppose $s(t) = 7 - t^2$ and $T_s = 0.3$. Find the values s[n] of the sampled waveform.

Solution:

$s[n] = s(nT_s) = s(0.3n) = 7 - (0.3n)^2 = 7 - 0.09n^2.$

Thus, s[0] = 7, s[1] = 6.91, s[-1] = 6.91, s[2] = 6.64, s[-2] = 6.64, s[3] = 6.19, s[-3] = -6.19, etc. Try plotting this sequence to see that it outlines the upside-down parabolic shape of the analog waveform $s(t) = 7 - t^2$.

Sampling of Musical Signals

When sampling musical signals, the sampling period is generally much smaller than T_s =0.1 second as shown in the above examples. For example, the INFINITY Technology Kit uses a sampling period of T_s = 0.000125 second, or 1/8000 of a second. That means that this system computes and outputs 8000 numbers every second when making sounds, or f_s = 8000 Hz. This sampling rate is by no means large. If you consider a CD player, it employs a sampling rate of f_s = 44100 Hz. What is so special about these sampling rates?

As you would expect, a higher sampling rate generally results in a better-sounding device, because it is using more measurements of the sound waveform to produce its sound. The reasons for the higher fidelity, however, are not so simple as this in practical situations. There are some fundamental factors that determine a digital device's ability when the sampling rate is specified. These factors will be discussed more at length in Chapter 6. For now, we can consider f_s = 8000 samples/second as the minimum rate that we might want to use for recording and making musical sounds, because most musical sounds are "smooth enough" to be reasonably well-captured by 8000 samples a second.

If a higher sampling rate gives a better-sounding device, why not sample at a huge rate such as f_s = 100000Hz (100kHz) or even f_s = 1000000 Hz (1MHz)? There are two reasons to economize the number of samples:

1. More samples generally means more work in terms of processing and more memory in terms of storage. Think of carrying 100 pennies instead of four quarters in your pocket—both sets of change are worth the same, but the quarters are a lot lighter and easier to count!

2. There are many situations where the accuracy of a higher sampling rate is not needed. In the case of audio signals, the fundamental limitations of the human hearing system make sampling above f_s = 44.1kHz not very useful.

Converting Digital Periods to Analog Ones

One important issue in synthesizing musical sounds is producing signals with the right frequencies, because musical notes have specific frequencies. We know that a specific fundamental frequency corresponds to a specific period, through the formula f=1/T or, equivalently,

$$T = 1/f$$

For example, to make a sound with fundamental frequency f =400 Hz, we need a period of T=1/400 seconds or 0.0025 seconds. How do we make analog signals that have the right frequencies when dealing with digital signals created with a specific sampling rate? We can do this by using the right number of samples per period when defining the period of the sound.

For example, if we wanted to create a signal with a 400Hz fundamental frequency when using an 8kHz audio sampling rate, we need to use

$$T_{sound}/T_s = f_s/f_{sound} = 8000/400 = 20 \text{ samples per period}$$

when generating the sound. In other words, a signal whose period is 20 samples long at an 8kHz sampling rate has a period of 20/8000 = 0.0025 seconds and a fundamental frequency of 8000/20 = 400 Hz.

You can use this calculation to make and then listen to sounds of different frequencies in SketchWave. Use the SketchWave VAB worksheet to produce sounds with the following fundamental frequencies: 200Hz, 320 Hz, 400 Hz, 500 Hz, and 800 Hz. Try to sketch the same waveform for each signal (that is, try to draw the same shape for p(t) in the sketchpad window each time you change the period of the sound). How do the various sounds compare?

Exercises 2.3.3

1. Suppose $s(t) = 7 - 3t$ and $T_s = 0.2$. Find and plot the values of $s[n] = s(nT_s)$, $n = 0, 1, 2, 3, \ldots, 10$.

2. Suppose $s(t) = t^2 + 1$ and $T_s = 0.5$. Find and plot the values of the samples $s[n] = s(nT_s)$, $n = -4, -3, -2, -1, 0, 1, 2, 3, 4$.

3. Given the waveform, $s(t) = t^2 - 4t + 3$, plot its samples $s[n]$, for $n = -5, -4, -3, -2, -1, 0, 1, 2, 3, 4, 5$, assuming $T_s = 0.4$.

4. Given the waveform,

$$s(t) = \sqrt{t},$$

for $t > 0$, plot the samples $s[n]$ for $n = 0, 1, 2, 3, \ldots, 12$, assuming $T_s = 1.2$.

5. Suppose you recorded 4.5 seconds of sound using a sampling frequency of 8kHz. How many samples would you have measured? How about if you used a 44.1kHz sampling rate?

6. Suppose a signal that you've measured has 500 samples per period. The sampling rate is 20kHz. Find the frequency of the sound.

7. Determine the number of samples per period that you would need to create a 400 Hz sound when processing signals at a 10kHz sampling rate.

Section 2.4: Additive Synthesis

Approximate Length: 6 Lectures

2.4.1 Sinusoids Are Everywhere!

Most waveforms, including sound waveforms, have complicated shapes. Even so, we often have a way of artificially making them. Music signals are usually sums of the sounds of individual instruments playing different notes, so we can recreate music by recreating the sounds of the individual instruments and adding the signals together. In this section, we take this idea one step further: We recreate the sounds of individual instruments by summing together simple functions in specific ways. Each simple waveform serves as a "sound building block" in the construction of each instrument sound. The way these signals are combined together (such as what kinds and what amounts) is similar to a recipe in cooking, where tasty dishes are created from simple ingredients. It is usually difficult for us to break down the resulting sound into its individual "ingredients" with our ears — we simply recognize the sound as having its own character.

In the case of signals, there are many possible choices for simple waveforms that will serve as the building blocks for constructing complicated waveforms. We will choose our set of building blocks to be **sinusoids**. Figure 2.25 shows a sinusoid. Sinusoids all have the same simple shape shown in Figure 2.25, although they may wiggle more quickly or slowly and may be shifted to the left or right.

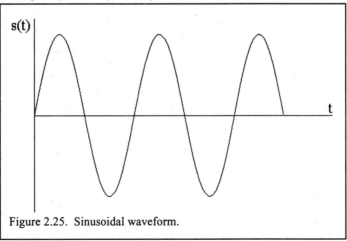

Figure 2.25. Sinusoidal waveform.

Sinusoids are very common waveforms. As just one example, the tuning fork in Figure 2.26 produces a change in acoustic pressure that varies nearly sinusoidally with time.

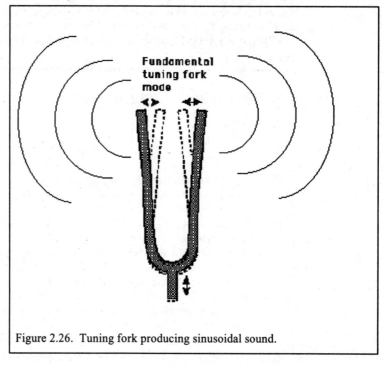

Figure 2.26. Tuning fork producing sinusoidal sound.

Amazingly, it is possible to construct nearly *any* waveform as a sum of different sinusoids! The technique for doing this in musical parlance is called **additive synthesis**. In more general engineering contexts, this task is called **Fourier synthesis,** after the mechanical engineer *Jean-Baptiste Fourier*. We'll illustrate how additive synthesis works after we've studied a bit more about sinusoids themselves.

2.4.2 Introduction to Sinusoids

Cosines and Sines as Functions

Sinusoids are the most important signals you will encounter in this course.

Perhaps you have already encountered sinusoids as cosines and sines in a trigonometry course. If not, don't worry. We will cover everything you need to know about cosines and sines. In trigonometry, cosines and sines are intimately related to angles. In this course, we will take a broader view where cosines and sines will be thought of as waveforms as in Figure 2.25. Let us begin, though, with the trigonometry viewpoint, which is what you may be familiar with. Consider Figure 2.27 where a point with coordinates (x,y) lies at a distance 1 from the origin and at an angle θ from the horizontal axis.

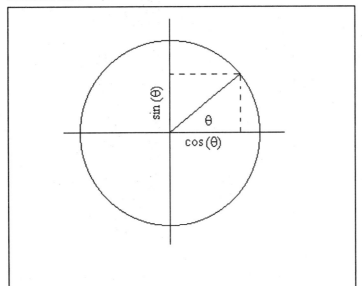

Figure 2.27: cos(θ) = x-coordinate of point on circle of radius 1. sin(θ) = y-coordinate.

Since the distance from the origin is fixed, the point lies on a circle and the location of the point can be described by its angle θ. The Cartesian coordinates x and y in turn depend on this same angle θ. The "x" value is given by cos(θ) and the "y" value is given by sin(θ).

We can get a good idea of what cos(θ) and sin(θ) look like as we vary θ by simply studying the diagram in Figure 2.27. We see that when θ = 0, we have x = 1 and y = 0, i.e. cos(0) = 1, and sin(0) = 0. Similarly, we see that if θ = 90° (90 degrees) then x = 0 and y = 1, or cos(90°) = 0 and sin(90°) = 1. Continuing in angle around the circle, we conclude that cos(180°) = -1, sin(180°) = 0, cos (270°) = 0, sin(270°) = -1. Considering some intermediate angles and getting out a ruler to precisely measure the corresponding values of x = cos(θ) and y = sin(θ) produces the data in the table below.

θ(°)	0	45	90	135	180	225	270	315	360
cos(θ)	1	0.707	0	-0.707	-1	-0.707	0	0 .707	1
sin(θ)	0	0.707	1	0.707	0	-0.707	-1	-0.707	0

Table 2.3. Values of x = cos(θ) and y = sin(θ) for Different Angles in Figure 2.28.

If we measure $\cos(\theta)$ and $\sin(\theta)$ for all values of θ between 0 and 360°, we can plot $\cos(\theta)$ and $\sin(\theta)$ as shown below. You can verify that for the values of θ in Table 2.3, the plots agree with the specified values of $\cos(\theta)$ and $\sin(\theta)$. From this figure, we clearly see that $\sin(\theta)$ is a shifted version, of $\cos(\theta)$ by 90°, or

$$\sin(\theta) = \cos(\theta - 90^{\circ})$$

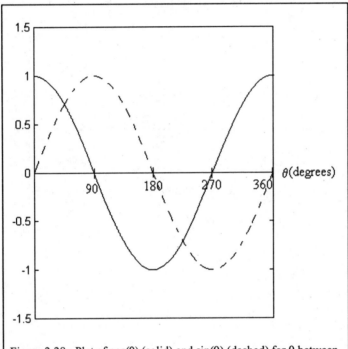

Figure 2.28: Plot of $\cos(\theta)$ (solid) and $\sin(\theta)$ (dashed) for θ between 0 and 360°.

Now, if we go back to Figure 2.27, clearly we could consider θ negative as well as positive. Likewise, we could consider θ larger than 360°, in which case the angle would wrap back around the circle. For this broader range of θ the plots of $\cos(\theta)$ and $\sin(\theta)$ are simply the repetitions of the plots in Figure 4.4, where the period is 360°. In other words,

$$\sin(\theta) = \sin(\theta + 360^{\circ})$$
$$\cos(\theta) = \cos(\theta + 360^{\circ})$$

Plots of $\cos(\theta)$ and $\sin(\theta)$ for an extended range of θ are shown individually in Figures 2.29 and 2.30.

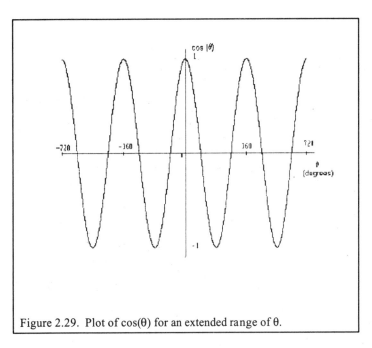

Figure 2.29. Plot of cos(θ) for an extended range of θ.

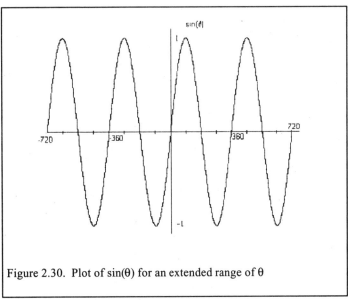

Figure 2.30. Plot of sin(θ) for an extended range of θ

Cosines and Sines as Waveforms —Sinusoids

To make music with sinusoids, we need to turn the sinusoidal function in Figure 2.25 into a sound waveform that is a function of time. The simplest

31

way to do this is to make θ a function of time t. We also need to give the function an amplitude parameter so that we can control its volume. The waveform that we obtain is given by

$$s(t) = A\sin(2\pi f t) = A\sin(\frac{2\pi}{T}t).$$

The parameter A is called the **amplitude** of the sinusoid, and the parameter f is the **frequency** of the sinusoid. Again, we have used the relationship f = 1/T to show the period of the sinusoid. In this case, we are using the units of **radians** to evaluate the sine function rather than degrees because it turns out this angle representation is more convenient for our application. The product 2 π is a scaling factor that makes the sine function evaluate its argument properly when using radians in the angular measurement.

Figure 2.31 shows a plot of this waveform. The period T of the sinusoid is shown on the plot as well as the amplitude A of the sinusoid. As we can see, both parameters make sense—A corresponds to the height of the sinusoid away from zero, and T is the repeating interval. We can verify through mathematics that T is the repeating interval of the sinusoid by noting that

s(t+T) = A sin(2 π (t + T)/T) = A sin(2 π t/T + 2 π) = A sin(2 π t/T),

because sin(θ + 2 π) = sin(θ) when radians are used for evaluating the sine function. The frequency of the sinusoid corresponds to the number of times the function repeats itself every second.

> **Fact:** The period T is equal to 1 divided by f. Likewise, the frequency f is 1 divided by the period T.

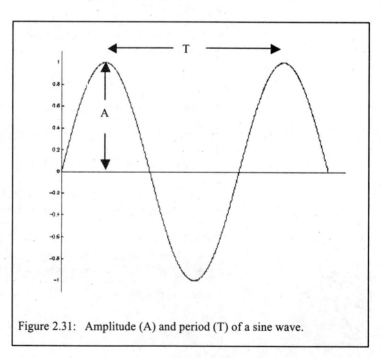

Figure 2.31: Amplitude (A) and period (T) of a sine wave.

Given the plot of a sinusoid, it is easy to determine its amplitude, period and frequency. This will be seen in the following experiment.

In the next VAB Experiment we explore how the frequency and amplitude
of a sinusoid affect its sound.

The following notes should help you interpret the sounds you hear and make in the VAB Experiment:

1) A doubling of frequency corresponds to one octave on the piano scale. Thus, 880 Hz should sound one octave above 440 Hz.

2) The frequency 440 Hz is sometimes called *concert A*, which is the A above middle C on the 88-key piano scale. Most modern Western music is referenced to this pitch.

3) The ratio between two adjacent notes on the musical scale is constant. Since an octave includes 12 different keys (including sharps or flats), the ratio of the fundamental frequencies of any two successive notes is always the 12th root of 2, which is approximately 1.05946. This rule lets you compute the fundamental frequency of any note on the scale, based on its distance (in number of piano keys) from concert A. For example, middle C is 9 notes below concert A. Thus, the fundamental frequency of middle C is $440/(1.05946)^9 = 261.6$ Hz. To go up in frequency, simply multiply by the 12th root of 2. For example, the B-flat above concert A, which is one half-step above concert A, has a fundamental frequency of $440(1.05946) = 466.16$ Hz.

4) The frequency 200 Hz is not a note on the musical scale (why is this)?. It lies in the octave below middle C, between G and G-sharp.

5) When you look at sinusoids using the time-domain tools in VAB, you might notice that the sinusoids do not start or end exactly like a sine or cosine function. The general situation is shown in Figure 2.32, where the start of a cosine function actually occurs at time $t=t_1$ rather than at time $t=0$. We can describe this situation using equations as

$$A\cos(2\pi ft + \phi) = A\cos(2\pi f(t + t_1))$$

where $\phi = 2\pi f t_1$ is called the **phase** of the sinusoid. Think of the phase of a sinusoid as the amount that the sinusoid is offset with respect to the origin, as scaled by the frequency parameter. Larger time shifts correspond to larger phase shifts, and vice versa.

The human hearing system has been shown to not be very sensitive to phase effects, so we shall largely ignore the phase of a sinusoid when constructing musical signals. Phase is an important property when constructing waveforms with arbitrary shapes, however.

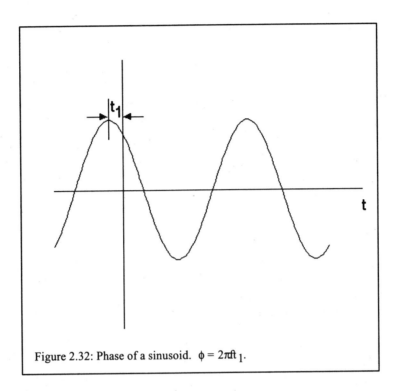

Figure 2.32: Phase of a sinusoid. $\phi = 2\pi f t_1$.

Exercises 2.4.3

1. In the exercises below, you are asked to specify the frequency, period, amplitude, and phase of each sinusoid from its plot. To assist you in this, let us use the information we have already covered to determine how you can specify these four parameters. To find the frequency, it is easiest to first measure the period, T. T is simply the length, in seconds, of one full period of the waveform. Then, we know that the frequency, measured in cycles per second or Hz, is

$$f = 1/T.$$

The amplitude, A, is the maximum height of the sinusoid, which can be read directly from the plot. The value of the advance t_1 easily can be measured from the plot. It is the amount that the first positive peak has been shifted to the left of the origin. Then, given t_1, the phase is calculated as

$$\phi = 2\pi f t_1$$

Now, estimate the frequencies, periods, amplitudes, and phases of the waveforms in Figure 4.10 below. In each case, assume you are to write the sinusoid as a cosine (rather than a sine).

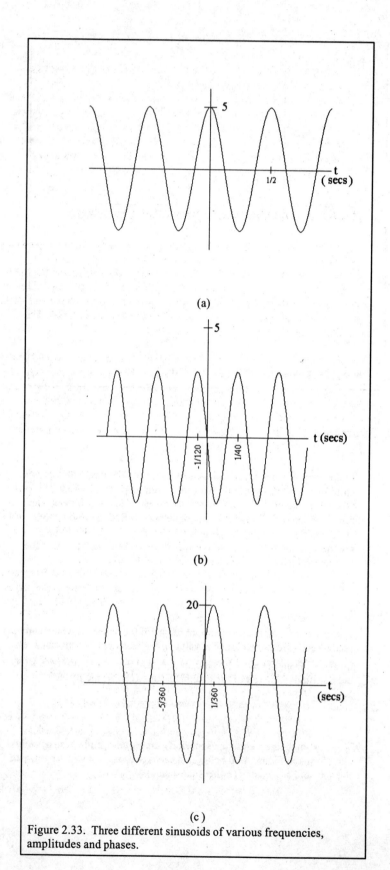

(a)

(b)

(c)

Figure 2.33. Three different sinusoids of various frequencies, amplitudes and phases.

2. Using the 12th root of 2, how many notes higher is 587.33Hz from 440Hz? How many notes lower is 220Hz?

3. In music, *chords* are groups of notes that are identifiable to musicians by their distinctive sounds. You can also figure out chords using the 12th root of 2. A *major chord* is a collection of notes consisting of a fundamental note, a note that is four half steps higher, and a note that is seven half steps higher. Find the frequencies of the major chord starting with concert-A (440 Hz) as the fundamental note.

2.4.4 Synthesizing Sounds With Sinusoids

As stated before, we can construct musical sounds using sums of sinusoids. Before we actually build sounds this way, it helps to do the opposite, namely, analyze individual sounds to see what sinusoidal signals of which they are made. The task of analyzing a nearly-periodic signal to determine the frequencies and amplitudes (as well as phases in the most general case) of its sum-of-sinusoids representation is called **Fourier analysis.** The resulting plot is called the signal's **spectrum.**

Figure 2.34 shows the zoomed-in portions of the piano, trumpet, and guitar sounds first shown in Figure 2.12. Shown on the right is the spectrum of each of the signals on the left. Each spectrum has been computed using a digital representation of each sound, and as such, the plots consist of discrete values as shown by the dots in each figure. We've also drawn the lines to connect the dots to give you an idea of how the spectrum tool in VAB might draw a similar figure.

To read the plots on the right, look along the x-axis, which in this case corresponds to frequency in Hz. The y-axis gives the amplitude of each individual sinusoid. Thus, taking both values, we have a frequency-amplitude pair $\{ f_i, A_i \}$ that identifies a particular sinusoid for every dot in each plot. Several features of each plot are striking and are worth discussing:

- Each spectrum differs from the other. In particular, the spectrum of the trumpet looks quite a bit different from those of the piano and guitar. This difference is clearly something we can hear and identify.
- In each spectrum, there are only a few non-zero amplitude points. In other words, each signal is made up of only a few sinusoids.
- The spectra are non-zero only at frequencies that are multiples of the lowest non-zero frequency term. Each of these non-zero sinusoids is called a **harmonic** of the sound.
- The lowest non-zero frequency term corresponds to the fundamental frequency of each waveform. To check this fact, note that the first non-zero point occurs at approximately f=466Hz, which corresponds to B-flat above concert-A in terms of piano pitch. Each of these waveforms were produced by instruments playing this note, and the corresponding period value $T = 1/466 = 0.00215$ seconds also matches the periods of the waveforms on the left.

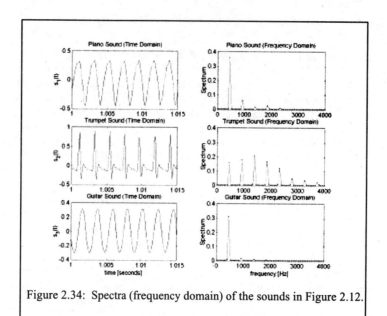

Figure 2.34: Spectra (frequency domain) of the sounds in Figure 2.12.

Putting all of these facts together, each individual instrument sound can be synthesized using the following equation:

$$s(t) = A_0 + A_1 \sin(2 \pi f_0 t) + A_2 \sin(2 \pi (2f_0 t)) + A_3 \sin(2 \pi (3f_0 t)) + A_4 \sin(2 \pi (4f_0 t)) + \ldots$$

where we continue adding terms until we have enough sinusoids to get an accurate representation of each particular sound. The values of each A_i can be read directly off of the plot. Table 2.4 lists the values of each frequency-amplitude pair for the three sounds in Figure 2.34, where $f_0 = 466$ Hz and $A_0 = 0$.

Instrument	A_1	A_2	A_3	A_4	A_5	A_6	A_7
Piano	0.1792	0.0324	0.0086	0.0125	0.0049	0.0010	0.0006
Trumpet	0.0798	0.0887	0.1039	0.0836	0.0598	0.0239	0.0172
Guitar	0.1557	0.0079	0.0043	0.0028	0.0016	0.0004	0.0013

Table 2.4: Amplitudes of the sinusoidal components of the sounds in Figure 2.34.

Before illustrating this synthesis method, we show that it can accurately recreate any periodic waveform—with enough sinusoids added together.

Example—Triangle Waveform

Consider the triangle waveform in Figure 2.35.

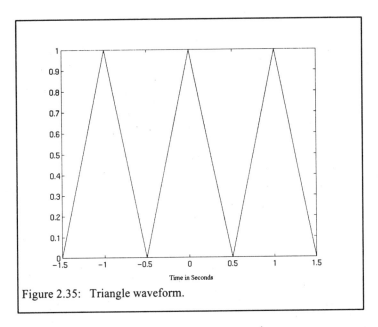

Figure 2.35: Triangle waveform.

Here the period is T=1.0 seconds, so the fundamental frequency is f_0 = 1/T Hz and the harmonics have frequencies 2, 3, 4 etc. If we do a Fourier analysis of one period of this waveform, either using a computer program or analytically, we discover that the amplitudes have a special structure:

$A_0 = 0.5$
$A_1 = 0.2026 = 2/\pi^2$
$A_2 = 0$
$A_3 = 0.0225 = 2/(\pi^2 3^2)$
$A_4 = 0$
$A_5 = 0.0081 = 2/(\pi^2 5^2)$

In other words, the values of A_i are zero for values of i that are both even and positive, and for odd values of i, they follow the formula

$$A_i = 2/(\pi^2 i^2)$$

If we try to build the triangle wave by adding together sinusoids of these frequencies and amplitudes (and just the right phases), we get the waveforms shown in Figures 2.36 (we show only a single period of the waveform in each figure). Notice that one sinusoid having the fundamental frequency comes fairly close to matching the triangle wave. Three sinusoids (the fundamental plus two harmonics) does a better job, and after adding together 100 sine waves, we have a nearly perfect match. As it turns out, to get a perfect match, we would have to add together an infinite number of sinusoids. But from an engineering perspective, we are more than close enough at 100, so that is as far as we will go.

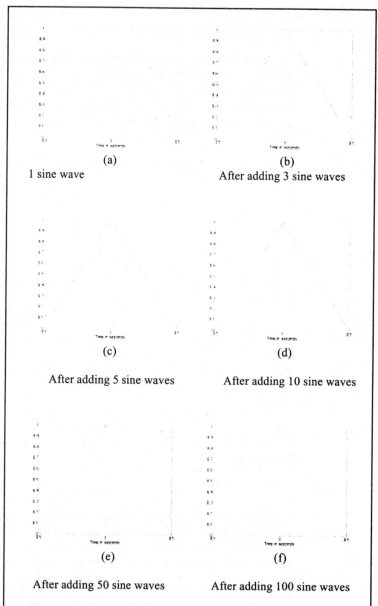

(a)
1 sine wave

(b)
After adding 3 sine waves

(c)
After adding 5 sine waves

(d)
After adding 10 sine waves

(e)
After adding 50 sine waves

(f)
After adding 100 sine waves

Figure 2.36: Triangle wave and its approximation using various numbers of sinusoids (various numbers of terms in the synthesis).

VAB Experiment 2.6: Additive Synthesis

Synthesize an approximation to a triangle wave using a sum of three sinusoids.

Example—Arbitrary Waveform

Let us next consider a waveform that is not so simple. Figure 2.37 shows a single period of a more complicated periodic waveform. Here the period is 2 seconds, so the fundamental frequency is $f_0 = 1/2$ Hz. The harmonics have frequencies 1, 3/2, 2 and so on. In this case, the amplitudes of the sinusoids don't have such a nice formula as the triangle wave example due to the funny shape of the waveform.

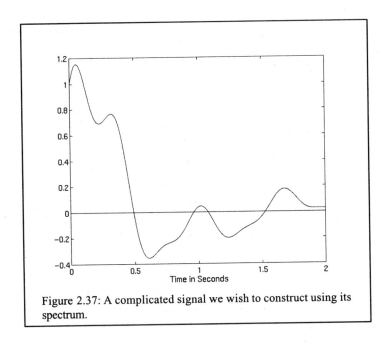

Figure 2.37: A complicated signal we wish to construct using its spectrum.

Figure 2.38 shows the approximation to the waveform in Figure 2.37 using various numbers of terms in the synthesis. Notice that for this signal, the approximation is quite good with just 20 sinusoids.

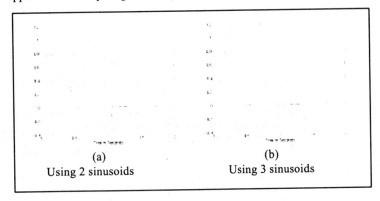

(a)
Using 2 sinusoids

(b)
Using 3 sinusoids

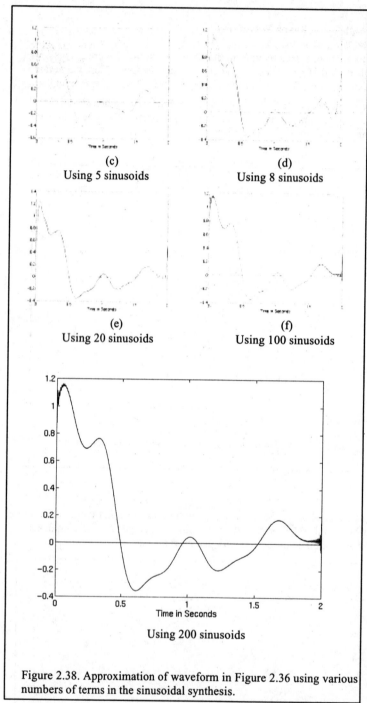

(c)
Using 5 sinusoids

(d)
Using 8 sinusoids

(e)
Using 20 sinusoids

(f)
Using 100 sinusoids

Using 200 sinusoids

Figure 2.38. Approximation of waveform in Figure 2.36 using various numbers of terms in the sinusoidal synthesis.

How would this work out in a real sound synthesis task? The following example illustrates some of the issues involved.

Example – Synthesizing a Saxophone Sound

Figure 2.39 shows the envelope of the sound of an alto saxophone playing B-flat above concert-A (466 Hz). The envelope of the sound can be seen,

42

but we need to zoom into portions of the sound to see its periodic structure. Four zoomed-in portions of the saxophone sound are shown on the left of Figure 2.40. Notice how each of the waveforms is slightly different at times t=1.0, t=2.0, t=3.0, and t=4.0 seconds, respectively. These slight differences are due to the way the saxophone is being played and are caused by slight differences in lip pressure and air pressure being exerted at the mouthpiece as well as by other performance factors. The spectra of the four sections of the sound are shown on the right-hand-side of this figure. Notice how the harmonics of the sound are represented at the frequencies of 466.67 Hz, 933.33 Hz, 1400 Hz, and so on in multiples of 466.67 Hz. In each plot, the amount of each sinusoid that makes up the entire sound is slightly different.

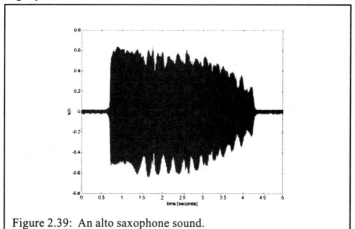

Figure 2.39: An alto saxophone sound.

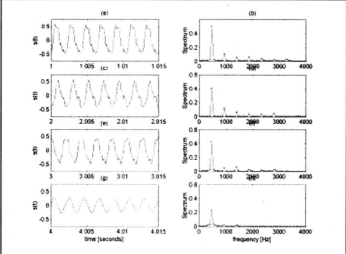

Figure 2.40: Zoomed-in portions of the alto saxophone sound and their spectra. (a) Sound around 1.0 sec. (b) Spectra of sound around sec.(c) Sound around 2.0 sec. (d) Spectra of sound around 2.0 sec. (e) Sound around 3.0 sec. (f) Spectra of sound around 3.0 sec. (g) Sound around 4.0 sec. (h) Spectra of sound around 4.0 sec.

Let's take the sound over the interval $2.0 \leq t \leq 2.015$ seconds, as shown in Figure 2.40(c). This sound has significant energy at frequencies f_0, $2f_0$, $3f_0$, $4f_0$, $5f_0$, and $6f_0$, where $f_0 = 466.67$ Hz. The amplitudes of the sinusoids can be read off of Figure 2.40(d) and are listed in the table below.

Instrument	A_1	A_2	A_3	A_4	A_5	A_6
Alto Sax	0.4082	0.1181	0.0680	0.0328	0.0338	0.0472

We then synthesize the alto saxophone sound using these amplitude values for each sinusoid. We also use the phase information about each sinusoid, although this information is not shown in the spectrum in Figure 2.40(d) (it can be computed using tools similar to that used to compute the amplitude spectrum). We construct the synthesized saxophone sound as

$$
\begin{aligned}
s_{synth}(t) = {} & 0.4082 \sin(2\pi\, 466.67\, t + \phi_1) + 0.1181 \sin(2\pi\, 933.33\, t + \phi_2) \\
& + 0.0680 \sin(2\pi\, 1400\, t + \phi_3) + 0.0328 \sin(2\pi\, 1866.67\, t + \phi_4) \\
& + 0.0338 \sin(2\pi\, 2333.33\, t + \phi_5) + 0.0472 \sin(2\pi\, 2800\, t + \phi_6)
\end{aligned}
$$

where the exact value of each ϕ_i has been left out for simplicity. Each of the sinusoids used to make up $s_{synth}(t)$ are shown in Figure 2.41.

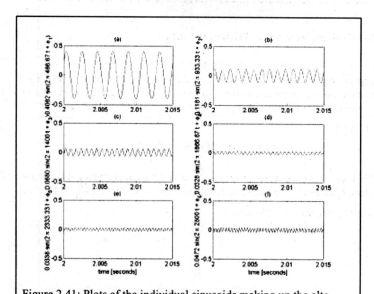

Figure 2.41: Plots of the individual sinusoids making up the alto saxophone sound. (a) 1st sinusoid (frequency f_0). (b) 2nd sinusoid (frequency $2f_0$). (c) 3rd sinusoid (frequency $3f_0$). (d) 4th sinusoid (frequency $4f_0$). (e) 5th sinusoid (frequency $5f_0$). (f) 6th sinusoid (frequency $6f_0$).

Summing these waveforms together, we get the results at the bottom of Figure 2.42. Looking at the two plots on the left, the waveforms look nearly identical except for a slight time shift. The spectra on the right also look remarkably similar.

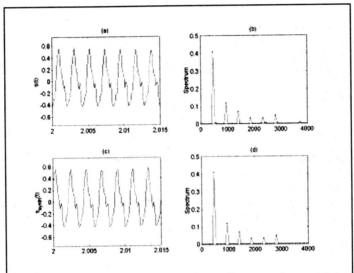

Figure 2.42: Original and synthesized saxophone sound around 2.0 sec. (a) Original sound. (b) Spectrum of original sound. (c) Synthesized sound. (d) Spectrum of synthesized sound.

Synthesizing Touchtone Telephone Sounds

So far, we primarily have considered waveforms that are periodic with a known fundamental frequency and hence a known set of harmonics. We need not be so restrictive in our choice of signals. Consider the sum of two cosines having frequencies f_1 and f_2 where the values of f_1 and f_2 are arbitrary:

$$s(t) = \cos(2\pi f_1 t) + \cos(2\pi f_2 t).$$

This waveform can take on very different appearances depending on how close f_1 and f_2 are to each other. Figure 2.43 shows s(t) for $(f_1, f_2) = (60, 5)$, $(60, 30)$, and $(60, 55)$ Hz..

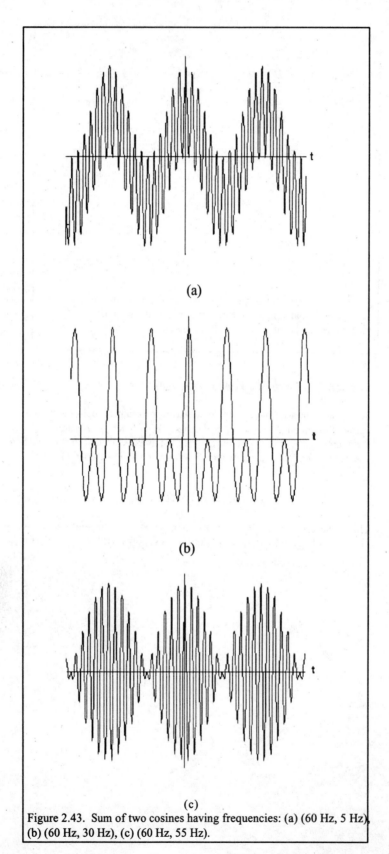

(a)

(b)

(c)

Figure 2.43. Sum of two cosines having frequencies: (a) (60 Hz, 5 Hz), (b) (60 Hz, 30 Hz), (c) (60 Hz, 55 Hz).

From the specified form of s(t), it is hard to guess the appearance of any of the waveforms in the above figures. The sum of two cosines can produce complicated waveforms!

Aside from the musical world, there are everyday electronic systems that employ sums of two tones. The keypad in your telephone is one such system. Each time you press a key, a sum of two tones is transmitted to the telephone switching office. The next time you make a phone call, listen as you push the buttons and see if you can recognize that each button push results in a sum of two tones! The signaling convention used is a subset of a convention called Dual-Tone Multi-Frequency (DTMF) for a 4 by 4 keypad, shown in Table 2.5. Normal telephones have has just the three left-most columns. Pressing a button causes the sum of the corresponding row and column frequencies to be sounded. Thus, for example, pressing the 4 button gives a 770 Hz tone plus a 1209 Hz tone. This signaling scheme replaced the old rotary dial telephone and was marketed as Touch Tone^{TM} by AT&T.

Table 2.5. DTMF signaling convention. Each button push results in a sum of the row and column frequencies.

	1209 Hz	1336 Hz	1477 Hz	1633 Hz
699 Hz	1	2	3	letter A
770 Hz	4	5	6	letter B
852 Hz	7	8	9	letter C
941 Hz	*	0	#	letter D

VAB Experiment 2.7: Touch Tone^{TM}

Generate the DTMF signals for telephone numbers by looking up the frequencies in Table 2.5 and adjusting the VAB cosine generators.

Section 2.5: MIDI and Spectrograms

Approximate Length: 2 Lectures

So far, we have focused our attention on synthesizing the sounds of particular instruments given a note that we want the instrument to play. We have ignored the issue of selecting the notes themselves. Creating pleasing melodies is a challenging task that is definitely outside the scope of this course. Instead, we shall focus on the task of taking a piece of music and translating the note information from the music to our sound synthesizing device. To do this task, we shall use a computer language for musical performance called *Musical Instrument Digital Interface*, or **MIDI** for short. MIDI has close ties to the spectrum of a sound, and we'll introduce the **spectrogram** as a way to generalize the spectrum for signals whose sinusoidal frequencies and/or amplitudes change with time.

2.5.1 A Little Musical Notation

To describe music as frequencies, we need to understand a little about musical scores and notation. Notes in music are specified using a system of lines and spaces known as **clefs**. Figure 2.44 shows a picture of the **treble clef** used for high-frequency musical notes, in which the names of the notes have been inscribed in the lines and spaces of the figure. The "a" that appears in the figure corresponds to a fundamental frequency of 440 Hz or concert-A.

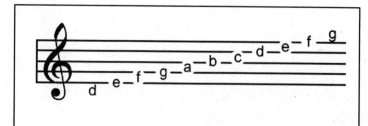

Figure 2.44: The treble clef in musical notation.

The **bass clef** (pronounced "base clef") shown in Figure 2.45 is used for lower-frequency musical notes, where the "c" that appears in the figure corresponds to one octave below middle-C (130.8 Hz).

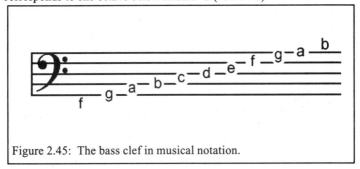

Figure 2.45: The bass clef in musical notation.

Figure 2.46 shows the note markings used to indicate the duration of notes, where half-, quarter-, eighth-, and sixteenth-notes are ½th , ¼th, 1/8th, and $1/16^{th}$ as long in duration as a whole note. The speed of a song, more commonly called its **tempo**, is specified at the beginning of the song and sets how long any whole note should last.

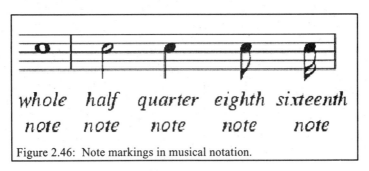

Figure 2.46: Note markings in musical notation.

To make music, we simply place the desired note markings indicating how long each note is in the lines or spaces of the bass or treble clef to indicate what notes should be played. The order of the notes proceeds from left to right on the page. Figure 2.47 shows a portion of the **musical score** to "Don Giovanni" by Wolfgang Amadeus Mozart. As you can see, the score can become pretty complicated—but all you need to remember is that each note is simply a *frequency* that needs to be played for a certain duration. Multiple notes appearing in the same vertical position means that multiple frequencies should be played; such combinations of notes are called **chords**.

Figure 2.47: A musical score.

2.5.2 MIDI—Musical Instrument Digital Interface

It should be fairly clear by the above description that written music simply describes the frequencies and durations of musical notes for one or more musicians to create. While written music serves a professional musician well, it is not the best interface for digital devices to use. Instead, a computer language has been developed to encode musical information into a set of commands for digital musical instruments. This computer language is called Musical Instrument Digital Interface, or **MIDI**. MIDI files on a computer, often denoted by the extension .mid, contain musical note information in digital form. MIDI can also be used to transmit musical information from one location to another, e.g. through a cable from a piano keyboard to a separate sound module or even by a connection over the Internet. With MIDI, several musicians can play music together without being in the same room, participating in a "virtual jam session".

What information can MIDI communicate? The most important information (at least for our discussion) are the frequencies, amplitudes, and durations of the music to be played. Figure 2.48 shows a listing of some of the instructions contained in a MIDI file. On the left is a column labeled "Timestamp" that describes when an instruction occurs (the information is encoded in a particular computer-based numerical format called *hexadecimal*, so it may not look very much like time to us). On the very right of the figure are three columns that describe the channel number, note value, and event associated with the time-stamped instruction. Each channel is assigned to a particular musical instrument (so the guitar sound of Channel 10 doesn't interfere with the trumpet sound of Channel 2). In this way, the MIDI instructions tell what instruments when to turn on and off which notes. Although not shown in this display, there are many other messages that MIDI information carries, such as note amplitudes, performance information such as velocity and pitch bend, and even instrument changes during the song.

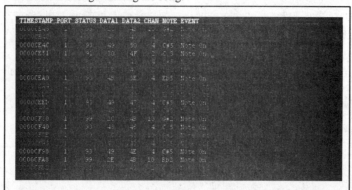

Figure 2.48: A listing of commands from a MIDI computer file.

Since MIDI is a conduit between musical information and a digital musical instrument, all we need to create music with our VAB-based digital musical instrument is to insert a block that reads MIDI information into the worksheet. The following VAB experiments give you a chance to both see and hear MIDI information as music in real-time by the Infinity Technology Kit.

Play standard MIDI .mid files using sinusoids as musical instruments. Study the resulting signals in the time and frequency domain to understand what is going on.

Play standard MIDI .mid files using the SketchWave musical instrument synthesizer.

Integer Periods and Musical Notes

As we have discussed previously, musical notes are defined by specific frequencies whose periods may not make up an integer number of sampling periods (see Comment 4 on page 32). Since MIDISketchWave employs an integer number of periods to create its sounds, it will sound out-of-tune when playing some MIDI files. A version of MIDISketchWave has been designed to play in tune—it is called, appropriately enough, MIDISketchWaveTune. Try out this version and compare it to MIDISketchWave. What do you notice?

Play standard MIDI .mid files using a tuned SketchWave musical instrument synthesizer.

2.5.3 The Spectrogram—A Sound's "Musical Score"

When dealing with sounds other than musical ones, we would still like a way to determine the sinusoidal "notes" that make up a sound. The spectrum of a signal tells us the amplitudes and frequencies of the sinusoids contained in a signal. Unfortunately, the spectrum doesn't tell us *when* these sinusoids occur. Such changes in spectra can be subtle—witness the differences in the spectra of the saxophone sound in Figure 2.39. We need a tool that tells us both what sinusoids are contained in a signal and when those sinusoids are "active."

Consider for a moment Figure 2.49, which shows a plot of just the melody of the opening bars of Goodnight My Someone. Time is represented by the horizontal axis (time is moving forward going to the right in the plot) and frequency is represented by the vertical dimension. The main differences between this plot and a musical score are (1) semi-circular notes have been replaced by straight lines of different lengths, and (2) the frequency axis is *linear* as opposed to the *logarithmic* scale used by the bass and treble clefs. No amplitude information is contained in this plot, just like the musical score.

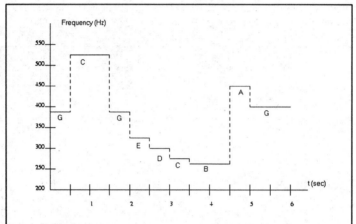

Figure 2.49. Melody of opening bars of Goodnight My Someone, from The Music Man.

We introduce the **spectrogram** as the analogous, but more general, signal processing tool to the musical score or plot above. It tells us graphically, exactly when each sinusoid or tone begins to appear and disappear in a arbitrary signal. Figure 2.50 shows the spectrogram of an 8-note Western musical scale being played.

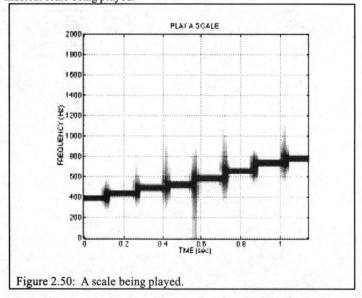

Figure 2.50: A scale being played.

The horizontal axis is time, the vertical axis is frequency, and the intensity of the plot at any point is the amplitude of a particular sinusoid that is present at time t. It is important to note that this picture was created by a mathematical procedure *without having the musical score!* In other words, we have "reverse engineered" a graph of the musical signal that was played.

Figure 2.51 shows the spectrogram of a short segment of Beethoven's Fifth Symphony. Notice how rich this spectrogram is in its frequency content, as compared to the spectrogram of the scale shown above. In this case, more than one sinusoid is present at several time instants throughout the selection. These sinusoids happen to correspond to different notes being played, but they could also correspond to different harmonics of the same note (so long as their frequencies are some multiple of a fundamental frequency).

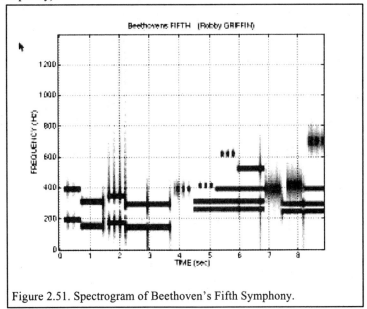

Figure 2.51. Spectrogram of Beethoven's Fifth Symphony.

In the real world, spectrograms can help with problems ranging from distinguishing one speaker from another, or one bird call from another, to identifying the type of signal a particular electronic device is transmitting.

Below are plotted two spectrograms of different kinds of sounds. In Figure 2.52 the amplitude of the sinusoids are encoded in the plot by color. Weak sinusoids are light blue, strong sinusoids are bright red, with yellow in between. In Figure 2.53 the intensity of the spectrogram is coded in gray scale.

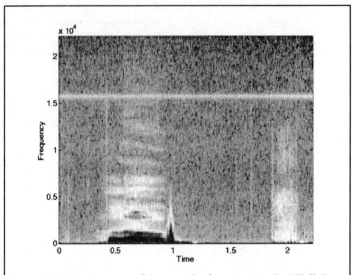

Figure 2.52: Spectrogram of the speech of a person saying "Hello".

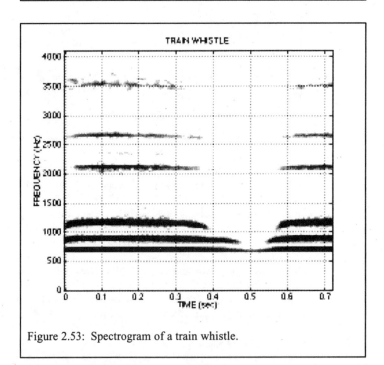

Figure 2.53: Spectrogram of a train whistle.

Did you expect what you see? You will have fun with spectrograms when you do the following VAB Experiment.

Section 2.6:
Glossary

additive synthesis
a synthesis technique that uses simpler functions as building blocks

amplitude
the size of a signal

analog signal
a signal that is defined over all time or space

analog-to-digital conversion
a process by which an analog signal is converted to a digital signal

attack
the way a musical note first begins

bass clef
a system of lines, spaces, and other markings that describe the lowest frequencies of a musical score

chord
a collection of simultaneous notes in a piece of music

clef
the system of lines, spaces, and other markings that denote a musical score

continuous-time signal
a signal that is defined over all time or space

digital signal
a signal that is defined only at discrete points in time or space

digital-to-analog conversion
a process by which a digital signal is converted to an analog signal

digitization
a process by which an analog signal is converted to a digital signal

envelope
a description of a signal's general size or amplitude over time

exponential function
a waveform that grows or decays in a particular increasing or decreasing pattern

Fourier analysis
the task of analyzing a nearly-periodic signal to determine the sinusoidal components that it contains

Fourier synthesis
a synthesis technique that uses sinusoidal functions as building blocks

frequency
the rate at which a periodic signal repeats, in cycles per second or Hertz (Hz)

fundamental frequency
a mathematical quantity describing the repetition rate of a periodic signal

harmonics
the individual sinusoids that make up a periodic signal

image
a two-dimensional signal that contains visual information

loudness
the perceived volume of an audio signal

MIDI
acronym for Musical Instrument Digital Interface; a specification for storing and transmitting musical information between digital devices

modeling
employing a mathematical description of a system to construct or determine
its behavior

musical score
a set of markings that are used to visually describe musical information

one-dimensional signal
a signal that is a function of one variable (usually time)

period
the repeating interval of a periodic signal

periodic signal
a signal that exactly repeats at regular intervals

phase
an angular offset in sinusoidal signals

pitch
the perceived frequency of a sound by our ears

principle of superposition
a process for creating a sound by adding several sounds together

quantization
a process by which numerical values are rounded to a particular precision

radians
a particular unit of angular measure, where 2 pi corresponds to one revolution

rate of decay
a number that describes the way an exponential function decreases over time

sampling
the process of creating a digital signal from an analog one

sampling frequency
the number of samples of a digital signal per unit time

sampling period
the spacing in seconds between samples of a digital signal

sampling rate
the number of samples of a digital signal per unit time

scale
1. (plots) the bottom axis of a plot. 2. (music) a collection of notes on a musical instrument

sequence
a series of numerical values

signal
a pattern or variation that contains information, usually denoted as s(t)

sinusoid or sinusoidal function
a simple oscillating waveform created from the sine or cosine function

sound synthesis
the creation of useful and complicated sounds from more basic sounds

spectrogram
a two-dimensional image describing the spectrum of a sound over time

spectrum
a plot of a periodic signal's sinusoidal components

square wave
a periodic function that is (1) for half its period and (-1) for the other half

synthesis
the creation of useful and complicated items from more basic ones

tempo
the speed of a song

time warping

the act of stretching or compressing the time extent of a periodic waveform

treble clef

a system of lines, spaces, and other markings that describe the lowest frequencies of a musical score

truncation

the act of removing a portion of a waveform

two-dimensional signal

a signal that is a function of two variables (usually x and y)

waveform

a plot or graph of a signal over time

zero padding

the act of lengthing a waveform by adding a zero-valued portion to it

The INFINITY Project

Chapter 3: Designing Digital Instruments

. . . .

Make Some Noise in the Real World

Approximate Length: 2 Weeks

Section 3.1: Introduction

Approximate Length: 3 Lectures

3.1.1 Your Career as a Solo Musical Artist

As you saw in the last chapter, one can make musical sounds digitally by drawing sampled waveforms using waveform synthesis and by adding sinusoids together using additive synthesis. These methods work well in copying sounds whose waveforms we already have. But what about situations where we *don't* already have sound examples of the instrument we want to play? In the Information Age, you might thing that getting such examples is pretty easy (e.g. from the World Wide Web). The problem is, how to we find *all* of the different sounds that an instrument can make? Think about the following:

- The sound of an instrument changes from one pitch to another. The changes are not just in its fundamental frequency. In other words, we can't just "shrink" or "stretch" a sound to get an accurate note of the same instrument. The entire waveform can change.

- The way an instrument is played can change the sounds that it produces. For example, we can pluck a guitar string aggressively, hammer on the string with our thumb, or even strum the string lightly with our fingertips. All of these playing styles change the sound of the guitar.

- The room in which the instrument is played affects its sound in a subtle way. The sound of an oboe played in a concert hall is certainly different from the sound of the same oboe played outside in a grassy field or inside a tiled bathroom.

Even if we had recordings of all of the different sounds that an instrument can make, how would we *play* the instrument? Finding the right sound for a particular note within a particular song would require endless trial-and-error looking through all of the sound recordings to get just the right one. Unlike the effects that a musician hears when changing the way an instrument is played, our catalog of sounds gives us no feedback as to what sounds right and what sounds wrong.

What we need is a way to make the instrument digitally, not just by copying its sound, but by figuring out *how* the instrument makes its sound. The procedure that we use to recreate the behavior of the instrument is called **physical modeling**, or modeling for short. Building a model is a way for us to get a handle on how something works—in this case, how a sound-making device creates sound. Modeling is a basic problem-solving method that can be used in many problems to figure out good solutions. In this chapter, we are going to use modeling to create new, realistic, and interesting sounds— and yet they only exist as numbers and equations inside of a digital device.

To motivate this idea, consider the following problem. You've gained enough experience from playing in your friend's band that you want to "strike out on your own" as a solo artist. You've been practicing your singing talents while in the shower and learned some tunes to play on guitar. The only problem is, you don't have the guitar, the sound enhancing equipment for your voice, or the PA (public address) speaker system. How can you use technology to make your solo rock star career take off? In this chapter, you will

- design your own digital model of a guitar that computes its sound entirely with equations and numbers rather than with vibrating strings and wood;

- add effects to your voice to recreate the echo and reverberation of a hall or stadium; and

- build your own loudspeaker using paper, wire, a magnet, a salad bowl, and some glue.

Along the way, we'll learn about **acoustics**, which is the study of sound and sound propagation. We'll also learn about the way many physical systems make and change sounds.

All systems in acoustics have the same structure shown in Figure 3.1. This block diagram contains three major blocks:

1. *Sound Source:* This is the system, device, or instrument that makes sound. These systems include a musical instrument, your vocal tract, or anything that makes sound.

2. *Propagation Path:* This is the system that carries the sound from its source to everywhere else. A room, a concert hall, and the inside of an automobile all have their unique propagation paths.

3. *Sound Sensor:* This is the system that listens to the sound. Your ears are sound sensors, as are microphones.

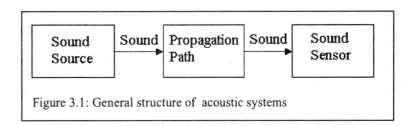

Figure 3.1: General structure of acoustic systems

1. Give three examples of acoustic systems that have the structure in Figure 3.1.

3.1.2 The Human Vocal Tract

Speech is an extremely important acoustic signal; without it we wouldn't communicate as effectively as we do. The physical cavity inside of our bodies that enables us to talk and sing is called the **vocal tract.** Figure 3.2 shows an X-ray of the vocal tract. The vocal tract begins at the opening of the vocal chords in the larynx and extends through the oral and nasal cavities. The nasal tract begins at the velum above the soft palate and ends at the nostrils. When the velum is lowered, the nasal tract is acoustically coupled to the vocal tract to produce the nasal sounds of speech. The speech signal is the acoustic waveform that is radiated from the mouth, nose, and cheek bones when air is expelled from the lungs and changed by the shape of the vocal tract.

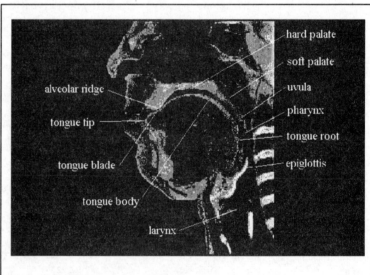

Figure 3.2: The human vocal tract .

> **Voiced Sound:**
> Sounds that are nearly periodic such as 'eee', 'oooh', 'eye', etc.

> **Unvoiced Sound:**
> Sounds formed by constricting the flow of air such as 'shh', 'fff','sss', etc.

Our vocal tracts can make two basic categories of sound. The first category is called **voiced sound**, which is created by forcing air through the vocal chords with the tension of the chords adjusted so that they vibrate at a steady rate or pitch. The result is a series of nearly periodic pulses of air that are then shaped by the vocal tract. Voiced sounds are almost periodic and include many vowel sounds like "eee", "oooh", "eye", and so on. These sounds have a "ring" to them.

The second category is **unvoiced sound**, sometimes called fricatives. Unvoiced sounds are produced by pinching the vocal tract at some point,

4

usually in the mouth with our tongue, teeth, and/or lips, and forcing the air through the pinched portion at a fast enough rate to produce "rough air" or **turbulence.** This activity creates a noisy sound that is recognizable. Examples of unvoiced sounds are the sounds of "fff", "sss", "shh", and so on.

While most natural speech contains a good mixture of voiced and unvoiced sounds, most singing styles tend to emphasize the voiced portion of human speech. The reason for this is because it is possible to make notes out of the vowel sounds. Try singing a note while making an "fff" or "sss" sound and you'll see what we mean! Another way of stating this fact is the following: voiced sounds are made up of nearly-periodic signals over short time periods (usually longer than 20 milliseconds), whereas unvoiced sounds are not periodic over short time periods. While unvoiced sounds don't "carry the tune," they are very important, because otherwise we wouldn't be able to understand the lyrics of a song. Unvoiced sounds are used extensively in some other applications, such as in computer speech recognition.

3.1.3 The Physics of Sound

We are all familiar with waves on the surface of a body of water. These waves are caused by wind or some other disturbance. Like waves in water, sound waves can only travel when there is some physical material in which to carry them (in other words, sound doesn't travel in the near-vacuum of outer space). Sound waves can travel in gasses, in liquids, and in solids. We'll only focus on sound waves in gasses, because we normally listen to sounds in air.

Sound travels in what is called a **longitudinal wave** pattern. We can illustrate how sounds travels in air using an analogy of springs. Figure 3.3 illustrates a longitudinal wave produced in a series of springs. Think of each connection between any two springs as a molecule of air and the spring as the effect that two adjacent molecules of air have on each other. If we suddenly push the left-most spring sideways to the right, it will push the second spring to the right, which will in turn push the third spring, and so on. Then, if we jerk the left-most spring back to its original position, it will pull on the second spring, which will pull on the third spring, and so forth. As this motion is transmitted from spring to spring, the disturbance will propagate along the row of springs. The resulting motion of the molecules is called a **traveling wave** .

How is the volume of the sound defined in this model? The amount that each molecule is moved from its "rest position" determines the amplitude of the wave. Bigger pushes-and-pulls on the end molecule make louder sounds than smaller pushes-and-pulls. Notice that although the wave pulse moves along the springs, the springs and molecules have very limited motion. They simply move back and forth. It is this key fact that makes sound wave motion different from the motion of physical objects. A sound wave leaves its air molecules behind.

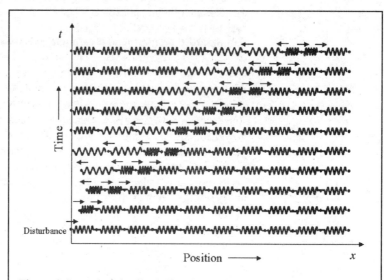

Figure 3.3: A longitudinal disturbance in a series of springs creates a wave that travels left to right. Arrows show molecule motion as time increases.

Sound waves don't look so much like a bunch of connected springs. If air were partially-opaque, a sound wave might look like that in Figure 3.4. Here, the dark regions show a concentration of air molecules, and the light regions indicate a lack of air molecules. The sound wave in this case would be traveling left-to-right, so that a short time later the entire drawing is shifted to the right by a small amount.

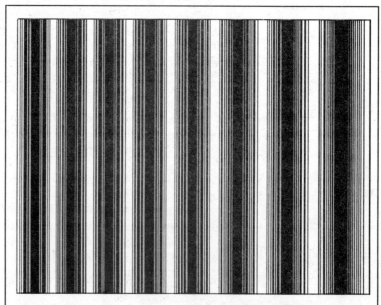

Figure 3.4: Compressions and rarefactions for a sinusoidal acoustic wave. Shade of gray indicates value of sound pressure.

The speed at which sound travels depends on the physical medium through which the wave is traveling. The velocity of sound in air at 0 degrees Centigrade and one atmosphere of pressure (sea level) is about 331 m/s. The speed of sound in several gasses and solids is shown in Table 3.1. Notice that the speed of sound in water is more than 4 times that in air, and the speed of sound in solids approaches 20 times that in air.

Table 3.1: Speed of Sound in Various Media

Medium	Meters/sec
Air	331
Hydrogen	1270
Carbon Dioxide	258
Water	1450
Iron	5100
Glass	5500
Granite	6000

For the case of a sinusoidal acoustic wave propagating through air, we picture a sinusoidal waveform flying through the air at 331 meters/second. At a fixed time, this waveform extends through space and therefore it has a period that can be measured in meters. This spatial period is called the **wavelength.** The wavelength, λ, is related to frequency and speed of propagation by the formula

$$\lambda = c/f$$

where c is the speed of propagation. This is illustrated in Figure 3.5. Notice that the wavelength is inversely proportional to frequency. Higher frequencies correspond to shorter wavelengths.

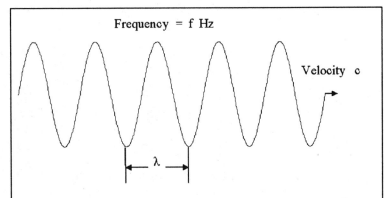

Figure 3.5: Relationship between wavelength, frequency, and velocity of propagation.

Exercises 3.1.3

1. How do sound waves in air differ from wind?
2. Find the wavelengths of a 200 Hz sine wave at 0 degrees Centigrade in air, water, and granite.
3. Find the frequency of sound in air whose wavelength corresponds to (a) one foot (0.305 meters) (b) one football field (91.44 meters) (c) 1 cm.
4. The crew of a submerged submarine hears an explosion from a source at the surface of the water 100 km away. The sub then takes two minutes to surface. Will the crew hear the explosion a second time? If so, how long after surfacing?
5. How fast is the speed of sound in km per hour? in miles per hour? Can one drive a standard car that fast? Can airplanes fly that fast? (Note: The speed of sound is also known as "Mach 1" in aviation.)
6. Suppose a sound has a 1 meter wavelength in air. What is the wavelength of the sound in water?

3.1.4 The Human Hearing System

The variations between molecule positions in sound waves in air is remarkably small; it is less than 1% of the average space between the air molecules even for loud sounds. In order for us to hear sound, our hearing system has to be extremely sensitive. Humans are endowed with remarkable listening devices—our ears—that convert acoustic energy (sound waves) to electrical pulses in the brain. The ear, shown in Figure 3.6, has three sections: the outer ear, middle ear, and inner ear. The outer ear directs acoustic energy to the eardrum where the middle ear transforms the pressure variations into motion of a series of three tiny bones. The last of these bones, called the **stapes**, transmits its motion to the snail-shaped **cochlea** within the inner ear. The cochlea is a fluid-filled tube that contains tiny hairs. The hairs bend from the vibrations and cause neurons to fire, sending electrical pulses through the auditory nerve to the brain.

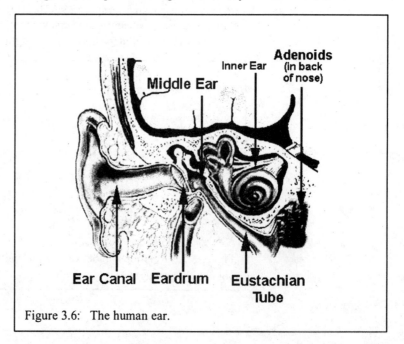

Figure 3.6: The human ear.

8

The human ear can respond to signals having a huge variation in amplitude. The largest signal you can hear is about one million times larger than the smallest! The ear has such a wide **dynamic range** because it does not respond to amplitudes in a linear way. The ear responds to ratios of signal amplitude, rather than differences in amplitude. Mathematically, our ears hear amplitudes on a **logarithmic scale** as shown in Figure 3.7. Here, we have plotted the *same* curve versus two different horizontal scales to best illustrate its shape. Because of the logarithmic profile involved in human hearing, signal levels in audio applications are usually expressed on the logarithmic scale, called **decibels** (dB). You'll learn more about decibels in the Digitization chapter that follows.

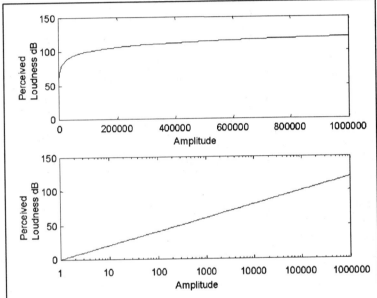

Figure 3.7. Apparent loudness versus signal amplitude for the human ear.(a) Plotted versus a linear scale, (b) Plotted versus a logarithmic scale.

Table 3.2 Some Common Sound Levels

Sound	Relative Level (on a linear scale; 1=threshold of hearing)	Listener's Perception
• Whisper • Human Breathing	10	Barely Audible
• Quiet Conversation	1,000	Faintly Heard
• Average Office	100,000	Moderate Level
• Summer Nocturnal Insects	1,000,000	Moderate Level

• Noisy Office • B-757 Cabin during Flight • Crackling of plastic food wrappers	10,000,000	Loud
• Average Street Traffic	320,000,000	Very Loud
• Crowd Noise at Football game	10,000,000,000	Very loud
• Accelerating motorcycle a few feet away • Nearby thunder	100,000,000,000	Deafening
• Hard Rock Band	1,000,000,000,000	Deafening
• Jet Aircraft Taking Off	10,000,000,000,000	Painful (and dangerous)

The frequency range of normal human hearing is approximately 20Hz to 20,000 Hz. That is, if a pure sinusoid were played at sufficient power through a high-quality speaker, most persons would be able to hear frequencies nearly as low as 20 Hz, which would seem almost like a vibration, and up to about 20,000 Hz. It is not unusual, however, for a person's hearing range to be somewhat worse than this, especially if he or she has suffered some form of hearing loss or is older.

What is the frequency range of your hearing? You can get a rough idea by using VAB to play sinusoids of varying frequencies through the computer's speakers. Before doing so, however, you should be aware that it is not possible to undertake a hearing test in a precise way without proper equipment. One reason for this is because every type of speaker has its own characteristic frequency response. That is, a speaker produces different frequencies more or less efficiently and therefore with different amplitudes. The frequency response of a typical small computer speaker might look like that shown in Figure 3.8.

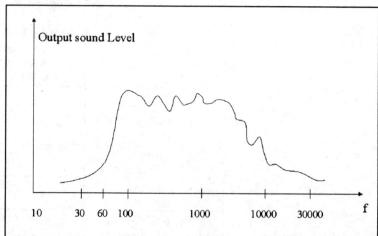

Figure 3.8. Output sound level as a function of frequency for a typical small computer speaker.

Here we have shown the horizontal axis on *a logarithmic scale*. Clearly there is "roll-off" in the response at low frequencies and some non-uniformity in response across all frequencies. For a valid hearing test we would need to present each frequency to the listener with the same amplitude.

VAB Experiment 3.1: Hearing Frequency Range

Given the limitations stated above, you can still get some idea of the frequency range of your hearing by slowly moving the frequency dial in VAB through a wide range of frequencies. Start with a low frequency. At 20 Hz can you hear anything at all? How about at 30 Hz or 40 Hz? Do you think this is a limitation of the speaker or your hearing? At what frequency do you first hear the signal? As you move up in frequency you will likely hear the signal very well in the 100 – 5,000 Hz range. At higher frequencies, the signal will be more difficult to discern. At what frequency does the signal seem to disappear?

Exercises 3.1.4

1. Where are the vocal chords located within the vocal tract?

2. What function is performed by the cochlea in the human ear?

3. The sound level of a hard rock band can reach 120 dB. The sound level of a quiet conversation is 30 dB. What is the ratio of the amplitudes of these two signals? Hint: The answer is *not* 120/30.

4. The sound level within a typical office is 50 dB. The sound level of a whisper is 10 dB. What is the ratio of the amplitudes of these signals?

Section 3.2: Physical Modeling

Approximate Length: 4 Lectures

3.2.1 Modeling A Guitar With Block Diagrams

As described in the introductory chapter of this book, block diagrams are a useful way to break down complicated systems into subsystems that are simple. In this section, we shall start the process of modeling the sound of a guitar by breaking down the way by which a guitar makes a sound.

Before making any block diagram, we must first think about the problem we are trying to solve. In the case of a guitar, here is the sequence of events it takes to makes a sound:

1. To start the guitar string vibrating, we first must bend the string to some shape, usually with our **finger** or guitar pick. This *displacement* is what gets the sound going.
2. The **guitar string** is let go, creating *vibrations* that travel up and down the string.
3. The string vibrations are transferred to the **soundboard**, which is the hourglass-shaped portion of the guitar.
4. The large area of the soundboard moves the air molecules near the soundboard, creating a *sound* that travels into the **room**.
5. The sound reaches our **ear**s, where we hear its distinctive "twang."

Our description suggests a block diagram with five modules. Figure 3.9 shows the block diagram of the guitar sound. The second and third modules can be lumped together into a single guitar block, and we've drawn a larger box around these blocks to show this fact. The arrows in between the blocks show what information flows from one block to another (displacement, vibrations, or sound).

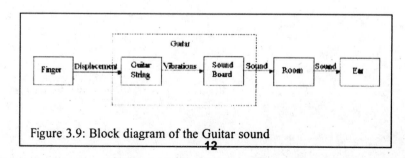

Figure 3.9: Block diagram of the Guitar sound

Since we don't need to model our fingers or our ears, we only need to deal with the two systems that make up the rest of the block diagram: the guitar and the room.

3.2.2 Physics of a Guitar String

Guitar strings are usually elastic pieces of plastic or metal that are stretched tight and attached at either end. A diagram of an acoustic guitar is shown in Figure 3.10. Two important parts of the guitar that so far haven't been discussed are the **saddle** and the **bridge**. The saddle is the grooved piece that holds the guitar string near the large end of the guitar, and the bridge sits inbetween the saddle and the soundboard.

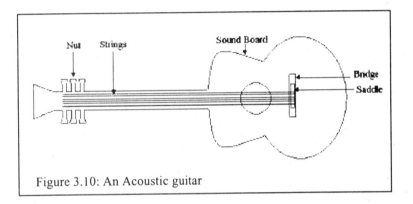

Figure 3.10: An Acoustic guitar

To create a sound with the guitar, one or more strings are plucked, which means that they are stretched from their rest position to some other position. Then, the string is let go and allowed to vibrate back and forth. This vibration creates **standing waves** that are the source of the guitar's distinctive sound. Shown in Figures 3.11 through 3.15 are the shapes of these standing waves in terms of the maximum displacement that the string undergoes (actually, the figures are a bit exaggerated to show the shapes better). Each of these figures describes a component of the vibrating string much like the sinusoidal components of periodic sounds in Chapter 2. The difference here, though, is that these are *vibrations* in the form of physical displacements and not sound waveforms.

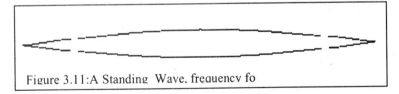

Figure 3.11:A Standing Wave. frequency fo

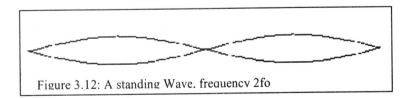

Figure 3.12: A standing Wave. frequency 2fo

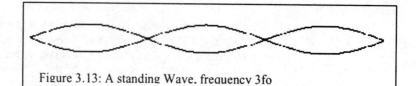

Figure 3.13: A standing Wave. frequency 3fo

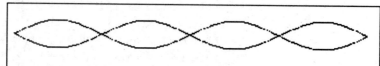

Figure 3.14: A standing Wave. frequency 4fo

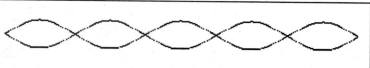

Figure 3.15: A standing Wave. frequency 5fo

The frequency associated with each standing wave component can be calculated from an equation that we've already seen:

$$f = c/\lambda$$

where c in this case is the speed of vibration in the string (in meters/second) and λ is the wavelength of the particular vibration component (in meters). Recall that the wavelength is basically the period or repeating interval of the physical wave. In the case of Figure 3.12, it is

$$\lambda_2 = L$$

where L is the length of the vibrating portion of the string. Remember, the period of the standing wave on the string can be found by following one curved line along its path until you get to a repeating point, which for Figure 3.12 is at either end of the string. In Figure 3.15, it is

$$\lambda_5 = 0.4\ L$$

because we have to move 40% along the length of the string to get to a repeating point in one of the curved displacement lines. In general, the wavelength for a given order of standing wave is

$$\lambda_n = 2L/n \quad for \quad n=\{1,2,3,4,5,\ldots\}$$

The speed of vibration in the string depends on two factors: the tension of the string and its mass per unit length. The formula defining c in this case turns out to be

$$c = (T/\mu)^{1/2}$$

14

where T is the tension and μ is the mass per unit length. Plugging in this formula and the one for wavelength into the overall formula for frequency, we find the frequencies of the various standing waves on the string are

$$f_n = m[\ T/(\mu\ 2L)\ ]^{1/2}.$$

There are several facts we can tell from this formula

- Each frequency f_n for $n>1$ is a multiple of the frequency f_1 of the first standing wave. Hence, the vibrations produced by the string are periodic with a fundamental frequency of $f_1=\sqrt{T}/(\mu\ 2L)$ or

$$f_1=c/(2L).$$

- Strings with higher tension have a higher fundamental frequency, but not by much. It takes four times the tension to get a doubling in fundamental frequency (why?). For this reason, we can't use tension too well in tuning the guitar because the strings will break and will also be harder to play.
- Strings with more mass per unit length produce lower frequencies. In bass guitars and the lower notes of a piano, the strings are wrapped with additional wire to make them heavier and therefore play lower notes.

The way the standing waves are made is easy to understand once you interpret the figures with a little know-how. Remember that all sounds and vibrations are waves that travel in a single direction (either left or right for purposes of our string). Standing waves are created when two waves of the same frequency and wavelength travel in opposite directions. Although standing waves work for any waveform, let's take the second harmonic frequency mode shown in Figure 3.12 for illustration. Call the displacement of the string $d_2(x,t)$, where x ranges from 0 on the left to L on the right and t is the time variable. We describe this function in terms of two variables because the function changes with both time t and distance x along the string. The form of $d_2(x,t)$ is then described as

$$d_2(x,t) = A \cos(2\pi(f_2 t - x/\lambda_2)) - A \cos(2\pi(f_2 t + x/\lambda_2))$$

where the first term is the sinusoidal wave traveling from left-to-right and the second term is the sinusoidal wave traveling from right-to-left. Using the trigonometric identities

$$\cos(\alpha - \beta) = \cos(\alpha)\cos(\beta) + \sin(\alpha)\sin(\beta)$$
$$\cos(\alpha + \beta) = \cos(\alpha)\cos(\beta) - \sin(\alpha)\sin(\beta)$$

we can simplify the expression to

$$d_2(x,t) = A \sin(2\pi f_2 t) \sin(2\pi\ x/\lambda_2)$$

The above equation says that the displacement is a product of two functions: one that depends on frequency, and one that depends on thedistance along the string. When combined together in this way, these two functions create a "wiggling waveform" whose overall shape along the string is the same at any particular point in time—only the amplitude of this shape varies.

To make different notes on the same string, we need to change one or more of its properties: length, tension, or mass. The mass of the string is usually fixed, so we can only change the tension and length. The **nut** at the end of each string on the guitar enables us to change tension by pulling or loosening the string. Usually, we use tension to **tune** each guitar string to the right frequencies corresponding to a standard set of notes when we pick up the guitar. Then, the guitar is played by placing fingers on the **frets**, which are the positions along the thin portion of the guitar. Pinching a string on a fret shortens the vibrating portion of the string, making a different fundamental frequency or note when we pluck the string.

In a real guitar string, the sound of the string is created by the way an initial displacement decays as it travels back and forth along the string. We need a more sophisticated model than that above to describe this behavior.

Exercises 3.2.2

1. A 1 meter guitar string has a fundamental frequency of 100 Hz. How fast is the speed of sound in the guitar string?
2. Suppose you are designing your own guitar with a 1 meter string length (from saddle to nut). You want to put a fret marking corresponding to a 200 Hz frequency on the neck of the guitar. What should be the distance from the marking to the saddle if the guitar string is tuned to 100 Hz? to 150 Hz?
3. How much tighter tension (in a percentage increase) is required to raise the frequency of a guitar string by 30% (so that a 200Hz string plays at 260 Hz)? How much shorter would the string have to be to raise the frequency of the string by the same amount?

3.2.3 Digital Model of a Guitar String

Digital technology enables us to build models of systems very accurately, using the power of bits to compute numbers on-the-fly. We can use the Infinity Technology Kit to build a model of a guitar string. Given what we know about the way a guitar string works, we can make a system that computes displacements like a guitar string and plays the vibrations as sounds through a loudspeaker.

First, we need a block diagram of the guitar string itself. Figure 3.16 shows a model that computes the displacement of a stretched string.

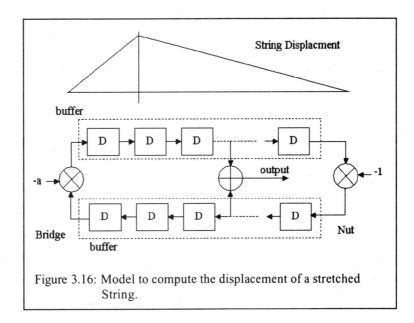

Figure 3.16: Model to compute the displacement of a stretched
          String.

This model is made up of several subsystems:

- The model is mainly made up of two long sets of memory, called
  buffers. Buffers hold information in a *first-in, first-out* fashion,
  so that information going into one end marches across the buffer
  and comes out the other end. The time it takes for a piece of
  information to travel across the buffer is called the **buffer delay**.
  Each buffer is actually made up of N discrete memory locations,
  called **delay elements**. The delay elements store the samples of a
  signal one-by-one, whereby the output of one delay element at any
  sample time is the input to the next delay element for the next
  sample time. In other words, if the input signal to a delay element
  is $s_{in}(n)$, its output signal is

$$s_{out}[n] = s_{in}[n-1]$$

  Therefore, if a digital signal $s_{in}(n)$ is sent to an N-element buffer,
  its output at the other end of the buffer is

$$s_{out}[n] = s_{in}[n-N]$$

  These delay elements model the way the displacement travels
  along the string. Since there are two functions that make up a
  standing wave (one going left-to-right and one going right-to-left),
  we need two buffers, one for each direction of travel.

- At the very right of the diagram is a **multiplier**, whose sole job is
  to multiply the signal coming into it by a number and send it back
  out again. The value of the multiplier in our example is (-1), so
  that the signal coming out of the multiplier is simply a flipped
  version of the signal going in, or

$$s_{out}[n] = -s_{in}[n]$$

  This multiplier models the effect of the string termination at the
  nut of the guitar, which is the place where one adjusts the tension

17

of the string. This nut causes the traveling wave to *reflect* and go back up the string.

- At the very right of the diagram is another multiplier. This multiplier models the effect that the bridge has on the string wave. The bridge is not rigid like the nut; instead, it transfers some of the energy to the guitar body, making its output a little smaller than its input. The value of the parameter (a) determines how much is kept in the traveling wave. In other words,

$$s_{out}[n] = -a \, s_{in}[n]$$

where (a) is a positive number a little less than one. For example, if (a=0.9), then 10% of the wave is lost every time the sound travels over the bridge, and 90% of the wave is inverted and passed back to the string.

- Near the center of the diagram is a device called an **adder.** This device takes two signals and simply adds them together . Two signals sent to the device produce the output
$$s_{out}[n] = s_{in,1}[n] + s_{in,2}[n]$$

The way the system works is as follows. We first set up a displacement on the string by preloading the buffers with a set of sample values. These sample values are simply the displacements of each point along the string. Then, we let the system run. The values in the buffers march left or right depending on which buffer they are in. At either end, the displacement values are multiplied by either (-1) or (-a) and sent back into the opposite buffer. Meanwhile, near the middle of both buffers, the displacement values are read out and subtracted from each other, and this signal is then sent to a loudspeaker for us to hear.

How well does this system simulate the sound of a guitar? Try it out yourself using this VAB experiment!

VAB Experiment 3.2: Guitar String Model

Simulate a guitar string using a digital physical model.

We can give some helpful insight into what is going on in the Guitar String VAB:

1. The number of delay elements in the buffer path determines the fundamental frequency that the Guitar String VAB produces when played. The value of the frequency can be figured out by determining the time it takes for one sample to travel completely through the system, i.e. to complete one cycle. Since there are N delay elements per buffer, the fundamental frequency of the string is
$$f = f_{sample}/(2N)$$

18

where f_{sample} is the sampling frequency of the VAB system (8 kHz). What is the fundamental frequency produced by the Guitar String VAB when N=50? How about when N=20?

2. The value of (a) controls how fast the sound decays. Making the value of (a) very, very close to one (a=0.9999) causes the sound to decay away slowly. Making (a) much less than one (a=0.9) makes the sound decay away quickly. Try both values and see what happens. Can you describe the sounds that you hear?

3. A typical initial displacement for a guitar string looks like the upper two sides of a triangle because we pull the string usually with one finger. However, you have complete freedom to pull the string any way you want. Try drawing a sine wave for the displacement and listen to the result. What do you hear? How about a notch? (A notch is a straight line except for a few samples where the value is close to one). What does the model sound like? If you do this right, you get the sound of a **piano**, because a piano hammer hitting a stretched string creates a notch displacement in the string!

4. We can control the position that we listen to the output of the string. Depending on where we place this position, we get different sounds from the string, because different amounts harmonics of the standing waves are being heard. Set the value of (a) very close to one (a=0.9999) and move the pickup position around. What do you hear? Does the sound of the guitar change significantly?

5. The harmonics produced by the string correspond to the different standing waves shown in Figures 3.11 through 3.15 and even higher (6th harmonic, 7th harmonic, and so on). We can use the spectrum analyzer in VAB to measure these harmonics. Try seeing what spectra are produced for different initial displacements. Are there significant differences?

As you can see, the guitar string model is very useful for capturing the sound of a real guitar string. It would be extremely difficult to make this sound any other way and still have precise and meaningful control over its characteristics. This shows the power of modeling—it is an important tool in a wide range of engineering and scientific problems.

Section 3.3:
Sound Effects

*Approximate Length: 3
Lectures*

3.3.1 Let Your Voice Be Heard

Now that you've seen how digital technology can make a great-sounding instrument, its time to see about making existing real-world sounds, like your voice, sound better. *Sound effects* is the generic term for techniques that change an existing sound in ways to make it more interesting, useful, and enjoyable. Sound effects have been in use in music and movie soundtracks for almost as long as they have existed. They are usually sonic "cheats" to fool the ear. Almost all recorded music today sounds like it is produced in a spacious place, but in reality the acoustical "spaciousness" is added after recording, usually after each performer has added her or his own sound to the recording in a detailed production process. In this section, we'll learn how three different audio effects   echo, reverberation, and flanging—can be created digitally. We'll then talk about changing the frequency content of signals using bass and treble controls which are known in engineering terms as **filters.**

3.3.2 Echo

Suppose you are standing on a flat plain out in the country, far away from any city or town. In front of you stands a majestic rock formation, jutting straight up from the ground and going up for hundreds of feet. You shout the familiar call "hello!" Waiting a little over a second, you hear your voice come back to you—"hello"—only slightly softer and maybe a little muffled. What is going on here? You've just experienced a physical phenomenon called **echo.**

Echo is caused by the time delay that sound experiences when traveling large distances. Unlike light which travels extremely fast, sound travels at a fast but still measurable rate: 331 meters per second. Moreover, when a sound wave hits a hard surface, like a wall or the rock formation, it bounces back to you much like a mirror reflects light rays. The end result is that you hear a copy of your voice delayed by the time it takes for the wave to make a round trip from you to the hard surface and back. We can calculate this delay. Suppose you are a distance L from a hard surface. The time delay sound undergoes in this situation is

$$T_{delay} = 2L/c$$

where c is in the units of meters/second and L is in the units of meters. The value of 2L is the round-trip distance that the sound takes in going from your mouth to the hard surface and back.

Example

Suppose you are 100m (a little over the length of one football field) away from the rock formation in the wilderness. How long does it take for your voice to come back to you? Since c=331 m/s, we calculate that

$$T_{delay} = 2 \ (100 \ m)/ \ (331 \ m/s) = 0.604 \ seconds$$

So, the sound comes back to you a little over half a second later.

Building a Single Echo Generator Digitally

Of course, trips to the wilderness are usually long and expensive, so we'd like a way to build a digital echo generator using the Infinity Technology Kit. How do we do this? One clue that we can gather from our experiment is the fact that the echo is simply a *delayed* copy of our voice. In other words, if s(t) were our original voice, the echo-only part would be given by

$$s_{echo}(t) = \alpha \ s(t\text{-}T_{delay})$$

where α is a number slightly less than one because the reflected wave loses some energy when it bounces off a hard surface. Since we also hear our voice when we create the sound, we need to add s(t) to $s_{echo}(t)$ to get the total sound

$$s_{tot}(t) = s(t) + s_{echo}(t) = s(t) + \alpha \ s(t\text{-} T_{delay})$$

We now need a way to simulate the effect of the delay, the loss of energy in the reflection, and the summing of the two signals.

Figure 3.17 shows the system we want to build. Our sampled voice s[n] enters on the left of this system and is split up into two copies. One copy passes through directly, whereas the other copy goes into a long buffer with N delay elements, each of which is denoted by a "boxed-D" in the figure. The output of this buffer is then multiplied by α before being added back together with the passed-through signal. Verify that the output of this system is given by

$$s_{out}[n] = s[n] + \alpha \ s[n\text{-}N]$$

Comparing this equation with the one for $s_{tot}(t)$ above, they are very similar. The only differences are the time index (n versus t) and the use of N in place of T_{delay}. To make the connection, remember that the time interval of each sample is T_s, so a buffer with N elements produces a total delay of

$$T_{delay} = NT_s.$$

So, we can figure out how long a buffer we need from the desired delay.

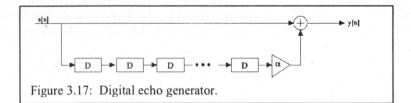

Figure 3.17: Digital echo generator.

Example

Suppose you wanted to simulate the effect of an echo caused by a surface 100 meters away and with a reflectivity of 90% using VAB. Then, we would use the block diagram in Figure 3.17 where

$$N = T_{delay}/T_s$$
$$= T_{delay} f_s$$
$$= (0.604 \text{ seconds}) (8000\text{Hz})$$
$$= 4832 \text{ buffer elements}$$

This amount of memory is pretty large. We cannot do it using the real-time hardware, but we can do it using the PC-based VAB modules. The following VAB shows you how it's done.

VAB Experiment 3.3: Digital Echo Generator

- Process your voice using digital echo. Experiment with different settings (delay time and amount of the reflection).
- Process music with digital echo. Who can make the coolest sounds?

Building a Digital Repetitive Echo Generator

Suppose now you're at the edge of a canyon looking downward. Shouting "hello" now produces not one but many copies of your voice, each copy getting softer and softer. An approximate mathematical model for this situation is

$$s_{tot}(t) = s(t) + \alpha\, s(t - T_{delay}) + \alpha^2\, s(t - 2T_{delay}) + \alpha^3\, s(t - 3T_{delay}) \ldots$$

where the (…) means terms that continue in the same pattern.
Why the α^2 constant on the third term? Because the echo of your voice careens off of the nearby canyon wall before going out to the far canyon wall and coming back again (actually, this would be three reflections and would require an α^3 value, but we've chosen this form for simplicity). Similarly, that echo echos yet again, creating a potentially infinite number of echoes that decay away.

How do we build a repetitive echo generator digitally? We could try to do it by copying the buffer structure in Figure 3.17 many many times, but we would need an infinite number of copies to make this happen! We also would use up memory (remember, each echo take a lot of memory because

22

it require so many sample instants to store). So, we want to figure out a more efficient way to do it. Figure 3.18 shows us how we can build this system easily. Instead of sending the input signal s[n] to the buffer, we send the output signal $s_{out}[n]$ to the buffer. That way, every part of the output signal gets copied again and again, an infinite number of times. Each time, the signal goes through a multiplication by α, creating the decay in the echoes as it progresses. Try out the VAB and see how it works!

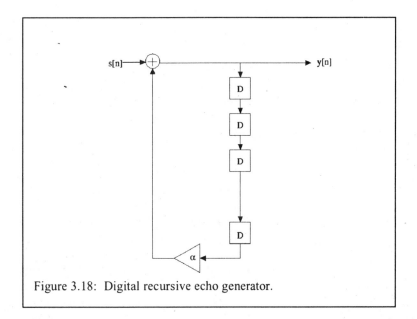

Figure 3.18: Digital recursive echo generator.

Exercises 3.3.2

1. You are at a baseball game, and your favorite baseball player is at bat. You see the player hit the baseball with the bat, and 0.45 seconds later, you hear the crack of the powerful hit. How far away are you from the player?

2. A car skids to a halt at an intersection, and you hear the skid 0.7 seconds later. If you can run 16km an hour, how long will it take you to run to the intersection?

3. A music company is building an echo generator to simulate the effect of the echo at a large canyon. The desired delay is large—3.5 seconds, and the sampling rate is 32kHz. How many sample delays will the echo generator need?

4. Suppose you wanted to simulate the echo of a 1.6km-wide canyon. How many delay elements would you need at an 8kHz sampling rate?

5. In some cases, we can approximate a recursive echo generator with a non-recursive one. Draw the block diagram of the non-recursive echo generator that implements the following equation:

$$s_{out}[n] = s[n] + \alpha\, s[n\text{-}N] + \alpha^2\, s[n\text{-}2N]$$

Try to minimize the number of delay elements in your implementation .

6. How would you fix the digital echo generator so that it has the right number of reflections per echo (that is, an attenuation of α for the first echo, α^3 for the second echo, α^5 for the third echo and so on)?

3.3.3 Reverberation

Reverberation, or reverb for short, is best described as echo but with a very short time delay or delays. Reverberation is usually caused by the reflective surfaces inside enclosed spaces, specifically the walls, ceiling, floor, and objects inside of a room. Since the distances between surfaces in a room are so small relative to the speed of sound, the reflected sounds come back to you quickly—so quickly, in fact, that you do not hear them as distinct sounds. Instead, the sounds simply sound a little "different." Usually, the sounds have a presence, warmth, or air that makes them more enjoyable to hear.

To get a sense about what reverb is, go into a large room and make a single loud clap with your hands (but don't do this in the library!) What you hear should sound like noise that decays slowly or quickly away. Figure 3.19 shows a plot of the signal that you might measure with a microphone in such an experiment. The very first signal that is nonzero is due to the clap, but everything else after it is due to the reverberation of the room. Note the time scale of the plot—sometimes the reverberation can last several seconds, especially for a large auditorium that has a lot of space inside of it. Concert halls are often designed to create certain reverberation patterns that make musical performers feel "at ease" on stage and add presence to their performances.

The model for reverberation is identical to that of a repetitive echo, except that we have to take more reflections into account. Typically this is done by designing several repetitive echo systems for several short (usually 20 milliseconds to 50 milliseconds) amounts of delay and then connecting them together one after another, as shown in Figure 3.18. Each of these systems then puts its own reflections onto the sound, producing a dense pattern like that shown in Figure 3.19. Reverberation is used everywhere in the entertainment world—to make speakers sound more important, to make singers sound more impressive, and to make music more enjoyable to hear.

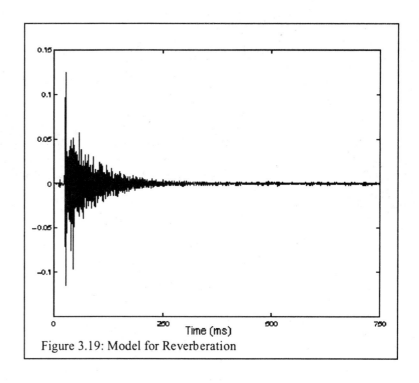

Figure 3.19: Model for Reverberation

Exercises 3.3.3

1. How long does it take for sound to travel from one end to another of a 5 meter-long room? Repeat this calculation for a 70m-wide concert hall and a 1.5m-wide bathroom stall.
2. How many delay elements would it require to simulate the round-trip time of a 5m-long reverberation path at a 20kHz sampling rate? Compare the number of elements to that needed for a 0.5 second echo at the same sampling rate. Which problem—reverb or echo—takes more memory?

3.3.4 Flanging

Flanging is another type of effect that musicians use to enhance the sound of instruments, usually electric guitars. Flanging amounts to a single echo generator with a small time-varying delay, or

$$s_{tot}(t) = s(t) + \alpha \, s(t - T_{delay}(t)).$$

Note that $T_{delay}(t)$ has now its own function of time. If we wanted to create a flanging effect acoustically, we would have to build a moving wall whose distance from us was

$$L(t) = c \, T_{delay}(t).$$

Imagine the mechanical and practical difficulties associated with a physical solution. Instead, we can use a digital implementation where $T_{delay}(t)$ is a number that we can vary every sample time if we want to.

The following VAB illustrates the flanging effect. Try blowing air into the microphone; what do you hear?

VAB Experiment 3.4: Digital Flanger
• Process your voice and music using the digital flanger. What happens to the sound of your voice? to the sound of music?

Exercises 3.3.4

1. Design a flanger that has two amounts of echo: one with 5 delays in it, and one with 10 delays in it, with a switch between them. How would you build it? Try to minimize the number of delays in your block diagram.
2. Can you build a recursive flanger? Why or why not?

Section 3.4: Speakers and Microphones

Approximate Length: 2 Lectures

3.4.1 How Speakers and Microphones Work

<table>
<tr><td>

Transducers:

are devices that convert energy from one form to another, e.g. speakers which convert electrical energy to acoustic energy, or microphones which do the reverse.

</td></tr>
</table>

Speakers and microphones are **transducers**, that is, they are devices that convert energy from one form to another. Speakers convert an electrical waveform to an acoustic one (sound pressure). Microphones do the opposite. Speakers and microphones are almost always needed in audio systems. Any audio signal in the real world can be converted to an electrical signal by a microphone. We then have the full array of electrical technologies at our disposal for changing the sound of our signal. Speakers and microphones are very similar in operation. In this section, we will focus primarily on speakers and only make some brief comments about microphones.

There are two basic types of speakers. The first and most common type is called **electromechanical** and consists of a flexible speaker cone whose back tip is wrapped with a coil of wire that moves back and forth around a permanent magnet. A second type of speaker works on an **electrostatic** or **capacitive** effect where the signal fed to the speaker charges two large parallel plates. As the voltage of the applied signal varies with time, the plates move toward or away from one another, which in turn moves air in front of the plates in a corresponding way to make sound.

We will focus on the electromechanical speaker that uses a permanent magnet and a coil of wire. A magnet has North and South poles with flux lines between them as shown in Figure 3.20.

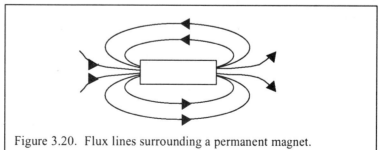

Figure 3.20. Flux lines surrounding a permanent magnet.

If we were to place the magnet below a thin piece of glass and pour iron filings onto the glass, the filings would line up in the directions of the flux lines. We call this aligning effect a *magnetic field*, B.

Now, suppose that an electric current flows in a wire that has been placed in a magnetic field, as shown in Figure 3.21.

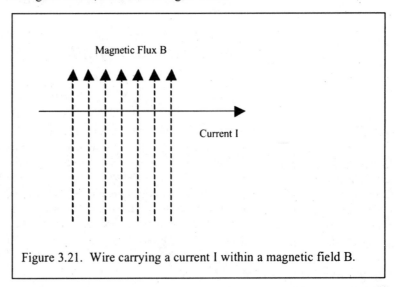

Figure 3.21. Wire carrying a current I within a magnetic field B.

Here, the current moves horizontally within the plane of the paper, and the magnetic flux lines are vertical within the plane of the paper. A physical principle known as **Oersted's Law** states that when such an electric current is located within a magnetic field, a force F is exerted on the wire with magnitude proportional to the product IB of the strengths of the current and the magnetic field. The direction of the force is perpendicular to both the direction of current flow and to the flux lines of the magnetic field. Thus, in Figure 3.22 the force on the wire is either into or out of the paper depending on the direction of current flow.

The force on the wire can be increased by looping the wire into a coil as shown in Figure 3.22.

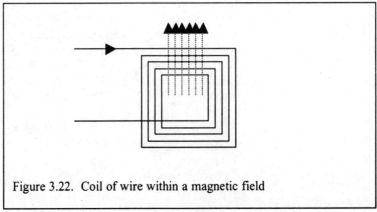

Figure 3.22. Coil of wire within a magnetic field

28

The force is now proportional to the current times the number of loops in the coil. Here we have shown just the top of the coil within the magnetic field. If, instead, the magnetic field cuts through both the top and the bottom of the coil then, since the current flows in opposite directions within the top and bottom of the loop, if the magnetic force pushes the top of the loop into the paper, the bottom will be pulled out of the paper. In other words, the coil will rotate. This is the basic principle behind a motor! Here our purpose is to study speakers, so we will not discuss motors any further.

Let us apply what we have learned to understanding the operation of a loudspeaker. A cut-away view of a typical loudspeaker is shown in Figure 3.23.

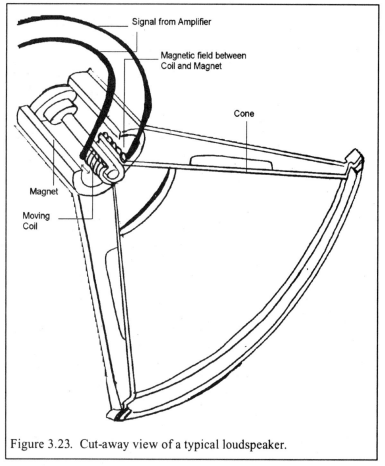

Figure 3.23. Cut-away view of a typical loudspeaker.

The electrical signal from an amplifier is passed through the coil of wire. One pole of a magnet has a cylindrical shape and passes up through the center of the coil. The other pole of the magnet is also cylindrical and fits around the outside of the coil. As the electrical signal changes with time, the coil moves up and down the axis of the magnet, moving the cone of the speaker, and in turn moving air in front of the speaker to create an acoustic signal.

The principle of operation is easier to understand if we look at a side view of the speaker as shown in Figure 3.24.

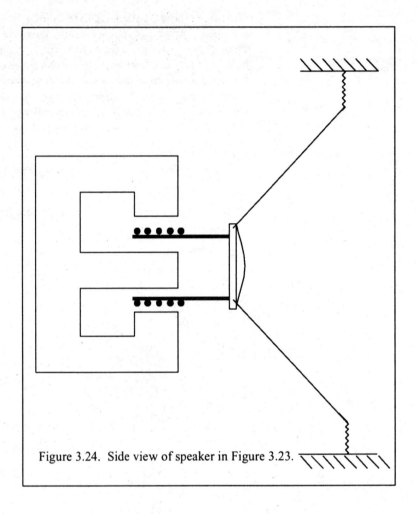

Figure 3.24. Side view of speaker in Figure 3.23.

The flux lines of the magnet are aligned vertically within the plane of the paper, pointing upward within the top half of the magnet and downward within the bottom half. The coil of wire is represented by the rows of dots. If current flows into the upper dots, then it flows outward from the bottom dots. Since *both* the magnetic field and the current change directions from the top half to the bottom half of the diagram, the force exerted on the coil by the magnet is the same in both halves of the diagram and is in a horizontal direction, perpendicular to both the magnetic flux and the current direction. The amount of force is proportional to the magnitude of the current. A changing current flow in the coil then moves the speaker cone in and out.

A microphone works on the same principle as a speaker, but in reverse. An acoustic pressure moves a small cone or membrane (called a **diaphragm** in a microphone) that is connected to a coil of wire that passes through the field of a permanent magnet. A current is induced in the wire by the motion of the coil, with larger motions causing larger currents. Microphones can be smaller than speakers, because the small electrical signal output by a small microphone can later be amplified electronically. Just as is the case with speakers, there is a second type of microphone that works on an electrostatic or capacitive principle. Such microphones are generally called **condenser** (another name for a capacitor) **microphones**.

Condenser Microphones:

are low-cost microphones that come with standard computers.

30

3.4.2 The Gosney Speaker

In this section we give instructions for you to construct a small working loudspeaker. The design of this speaker was provided courtesy of Prof. Milt Gosney, and hence we have named it the Gosney speaker. This same design also works very well as a microphone.

The Gosney speaker is diagrammed in Figure 3.25.

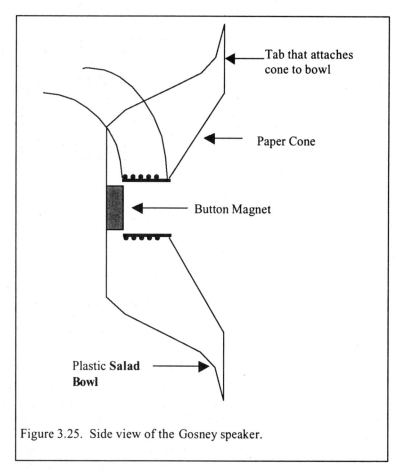

Figure 3.25. Side view of the Gosney speaker.

Here we use a button magnet, which is different from the more conventional speaker magnet shown in Figure 3.24. The button magnet produces the flux lines shown in Figure 3.26. The vertical parts of the flux lines to the right of the magnet are what interest us. In three dimensions, these flux lines will be perpendicular to the current flow in the coil. The resulting force on the coil will be perpendicular to both the magnetic flux lines and the direction of current flow. Thus, a horizontal force will be exerted on the speaker coil in Figure 3.25. That is, even though the arrangement of the magnet in Figure 3.25 is different from that in Figure 3.24, the force on the speaker coil will be in the same direction as that in Figure 3.23. The paper cone will move in and out of the salad bowl to produce sound.

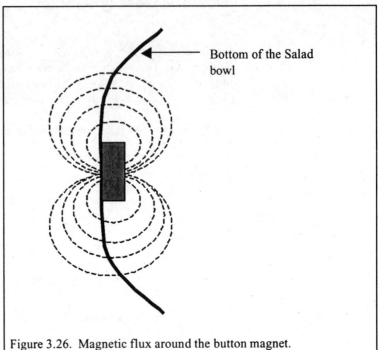

Figure 3.26. Magnetic flux around the button magnet.

The list of parts and supplies for the Gosney speaker is:

- Plastic salad bowl
- Button magnet
- Six feet of fine copper wire
- Paper
- Patterns for cone and coil cylinder (available at http://www.infinity-project.org)
- Scissors, straight pins, paper clips, tape, glue, AA battery as a form for coil

We will give instructions for a moving coil design (coil will be mounted on the cone and the magnet will be glued to the bottom of the bowl). A moving magnet design is also possible (magnet glued to the cone and the coil fastened to the bottom of the bowl). Most speakers are of the moving coil type since the coil is usually very light so that the cone will have a good response at high frequencies. However, the moving magnet design has the advantage of not requiring flexible connections to the cone. Figure 3.27 shows a photo of the Gosney speaker.

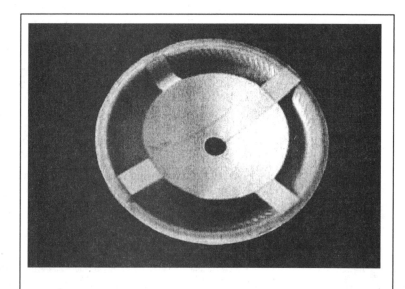

Figure 3.27. A photo of the Gosney speaker.

Construction Steps: (allow time for the glue to dry between steps)

1. Using scissors, cut out the cone pattern and the coil pattern.

2. Glue the cone edges together to form the cone assembly.

3. Glue the coil pattern into a cylindrical shape to serve as the coil cylinder. Be sure that your coil diameter is larger than your button magnet. Use a AA battery as a form. First cover the battery with clear plastic tape to keep the glue from sticking to the battery. Then apply glue to the coil pattern at the indicated area and wrap around the battery to form a cylinder. Let the glue dry.

4. While the coil cylinder is still mounted on the battery, glue it to the base of the cone and let it dry. Keep the coil cylinder on the battery.

5. Use your fingernails or a knife edge to strip about 1/2 inch of insulation from each end of the 6 foot length of wire. After the coil cylinder/cone assembly dries, smear glue around the outside of the coil cylinder and wrap the wire to form a multi-turn coil. Use all 6 feet of wire – the more turns the better, but be sure to leave about 6 inches of each of the two ends of the wire free for electrical connection. Smear glue over the outside of the coil to hold it in place. After the glue dries, remove the coil and cone assembly from the battery by slipping it off the end.

6. Using a straight pin, punch a series of small holes near the center of the bottom of the plastic salad bowl in the area where the magnet will be mounted. Then smear glue over the holes and mount the magnet, centered in the bottom of the bowl. (The holes give the glue something to grab onto since the glue assembly does not adhere well to the smooth surface of the plastic bowl.)

7. Punch two small holes through the salad bowl so that the coil connecting wires can be passed through the bowl.

8. Punch a series of small holes at the edge of the bowl in the areas where the cone tabs will attach. Smear glue over the attachment areas. Slip the connecting wires through the two small holes created in Step 7 and attach the speaker cone to the bowl. It is critical that the cone just clear the bottom of the bowl and that a small up and down motion of the coil be permitted. The coil cylinder should surround the magnet and extend directly above it. The cone motion should be free and the coil cylinder should not rub or hit the magnet. Use tape or paper clips to hold the cone in place while it dries.

9. Add a drop of glue to each wire hole to secure the wires, but allow some slack in the wires between the bowl and the coil to allow for movement of the coil.

VAB Experiment 3.5: Testing of the Gosney Speaker

It is time to test the speaker you have constructed! To do so, connect the speaker to the headphone jack of your computer speakers. This will cut off the computer speakers and the Gosney speaker will now be connected.

1) Use VAB to play a sinusoid through your speaker. How does it sound? Try both very low and very high frequencies. Roughl~ the frequency range of your speaker?

2) Record your speech through the microphone and play it back through your speaker. How does your voice sound?

3) Play music through your speaker. Does it sound as good as an inexpensive radio?

Section 3.5: Glossary

acoustics
the study of sound and sound propagation

adder
a digital device that performs addition on two numbers

buffer
a digital memory devices that holds information in a first-in, first-out
fashion

buffer delay
the total delay across a buffer

cochlea
a snail-shaped fluid-filled tube within the ear

delay element
the unit memory element within a buffer

diaphragm
a small cone or membrane inside of a loudspeaker or microphone

dynamic range
the range of smallest to largest waveforms for a particular physical
phenomenon (e.g. sound)

echo
a sound phenomenon caused by reflecting surfaces far away from the
listener

electromechanical device
a device, such as a speaker, that uses a combination of electrical and
magnetic
properties to function

electrostatic device
an acoustic transducer such as a loudspeaker or microphone that employs
charged parallel plates

filters
devices that change the frequency content of a signal

flanging
a sound effect that employs a time-varying echo

logarithmic scale
a function that follows a logarithmic spacing, such as loudness in human hearing

longitudinal wave
a wave pattern made up of expanding and contracting particles, such as a sound wave

multiplier
a digital device that performs a multiply operation on two numbers

Oersted's law
a physical principle that allows electromechanical speakers to function

physical modeling
in musical synthesis, employing a mathematical description of the physical behavior of an instrument in order to create its sound artificially

reverberation
a sound phenomenon caused by reflecting surfaces close to the listener

sound effects
a generic term for techniques that change an existing sound to some useful end

standing waves
physical waves that oscillate back and forth without moving in any particular direction

stapes
a bone within the middle ear that transmits motion

transducer
a devices that converts one form of physical energy into another

traveling wave
a wave pattern that moves

turbulence
"rough air" caused by forcing air through a pinched cavity

unvoiced sounds
sounds made by the vocal tract that are the result of turbulent air flow

vocal tract
the physical cavity inside our bodies that enables us to talk and sing

voiced sounds
sounds made by the vocal tract that have a rate or pitch

wavelength
the spatial period of a physical wave, in units of distance

The Infinity Project

Chapter 4: Making Digital Images

What is an image? What is a video?
Why is imaging going digital?

Approximate Length: 2 Weeks

The name of this course is **Multimedia and Information Engineering**. By **multimedia** we mean an integration of text, numbers, sound and still and moving pictures. The last part has been a relatively recent addition to the multimedia world and is the focus of next three chapters. We will study the general problem of acquiring still and moving images, processing them and displaying them. As with sound, introduction of digital technology to imaging systems has been nothing short of revolutionary. We will study the motivation for going digital, the consequence of going digital and the mathematics of going digital in this chapter.

Section 4.1: Problem: How to convince your skeptical friends that you *REALLY* were there?

Imagine this - you have gotten tickets to a hot rock concert. These are floor tickets and you are desperately hoping that you will get close enough to **XYZ** (fill in the name of your favorite rock star!) to make eye contact and maybe even get a wave from him / her. Then the day arrives, you go to the concert. The stage starts moving around, and wonders of all wonders, you *do* find yourself within 5 meters of the star and *do* get a wink in response to your hysteria. You come home walking on cloud nine and try telling your friends. Problem is – nobody believes you! They want some proof, like a **PICTURE**! So you confidently tell them to wait till you get the film developed. The pictures come back, and your heart sinks…. All the pictures of the close-up were too dark, blurred because you moved or just too fuzzy. Your friends remain skeptical as ever. What could you have done differently?

Another (less frivolous) problem involved the National Aeronautics and Space Agency (NASA) and its mission to explore earth and outer planets of our solar system. The problem was how to create high quality images of the objects from **VERY** large distances (ranging from hundreds of miles on up) and send them back to earth over even larger distances. Being in space, one doesn't get a second chance at doing something right, it has to be right the first time.

As it turns out, this is an age-old problem. To put it in a formal way –

"How does one share a visual experience with other human beings?"

Prehistoric humans faced the same question when they tried to share their latest exploits on a wooly mammoth hunt with their fellow cave dwellers. So they just picked up some sticks and other miscellaneous material and started drawing pictures on the cave walls. The results are there for all of us to see tens of thousands of years later.

3

The focus on visual experience is understandable since we derive so much of the information about the world through our visual senses. As a result, a large fraction of our brain (about 1/2) activity is dedicated to gathering, processing, and storing visual information. This, in turn, has led to common expressions like "Seeing is believing" or "A picture is worth a thousand words". It points to the need to develop technologies for image sensing, processing and storing in an efficient and flexible manner.

For thousands of years, the technology was basically the same as our prehistoric cave-dwelling ancestors. The paints, the surface on which the paintings were made and other implements got more varied and sophisticated over the years. But the principle of painting remained the primary means for storing and sharing our visual sensations right up to the middle of the nineteenth century. We know how George Washington looked only through paintings made by artists. Such paintings were hard to produce; once produced, hard to modify and almost never reproduced more than a few times. So they were hard or impossible to share and distribute.

The last 150 years since the invention of photography, we have seen an exponential growth in our ability to record, store and distribute images. Figure 4.1 shows the wide variety of different types of images that we can expect to encounter in our every day life. These images can be ordinary photographs of natural scenes or of people, black and white photographs produced by medical imaging systems such as X-rays, simple line drawings, or scientific images produced by various instruments.

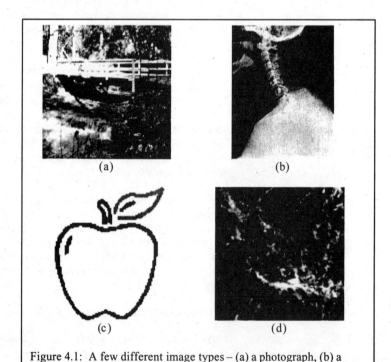

Figure 4.1: A few different image types – (a) a photograph, (b) a medical x-ray picture, (c) a simple line drawing and (d) an astronomical image.

This explosion in producing single images has been matched, and indeed exceeded, by the growth in movie production. A movie is simply a sequence of single images that are shown in time at a specific rate (images/second). The different images in the sequence are not random, but are closely related to each other with only small differences between subsequent images, commonly referred to as **frames**. Showing these closely related images quickly in time creates a sensation of movement in the image - hence the name "movies"! Single frame images and movies together form an important component of the present day "Multimedia Revolution".

The tremendous advances in photographic science and technology still do nothing to meet your need to impress your friends with your good fortune or to meet NASA's objectives of transmitting images of earth or other planets from its spacecrafts. It is clear that a different engineering solution needs to be found to attack these problems. As in any good engineering design, the first step is to identify the characteristics of an ideal solution. Let us make a list of these features of an imaging technology:

- It should be flexible and allow easy fixing of mistakes and defects

- It should be easy to make and distribute copies without losing quality

- It should offer a wide range of image quality

- It should be easy to use

- It should be cheap.

Before we delve into finding an engineering solution to our problem, let us engage in some exercises to see if we can get some clues.

Exercise 4.1

1. Bring several pictures in the class. Discuss how they were acquired (sensed or synthesized) and how they are used.

2. "Seeing is believing" - is it always true? If not, give examples where it is *NOT* true.

3. Take several pictures of varying complexity. Write a one-page description of what the picture looks like. Send it to your partner across the classroom. Then have him/her sketch the picture the best way they can. Compare the original with his/her drawing. Did your partner get all the significant details?

4. Take a black and white photograph to a copying machine. Make a copy, then copy of the copy and so on. How many generations of copies can you produce before the photo becomes unrecognizable?

5. Make a list of 32 random numbers between 0 and 9. Make a copy, then copy of the copy and so on. Make the same

number of generations as the previous exercise when the image becomes unrecognizable. Can you still recover the original list of 32 random numbers?

6. Take a drawing of a clock face showing time. Then write down the number indicating the time and send it to your partner across the room. Have the partner draw the picture of the clock. Compare the two pictures. Are the important details preserved?

4.1.1 Basic Approach

The last exercises gave us a clue as to what approach to take. We find that words are imprecise in describing something and that photographs when copied many times over lose their quality very quickly. We also find that numbers can be communicated very precisely. Furthermore, they can be duplicated over and over without losing the essential information contained in them. Now only if we could find a way to convert images into numbers, we would go a long way towards solving our problem. Since we represent numbers as "digits", we will call this approach **digital imaging**.

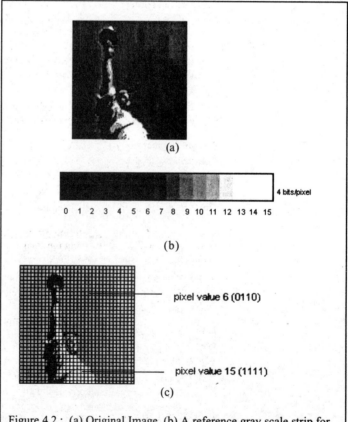

Figure 4.2 : (a) Original Image. (b) A reference gray scale strip for converting gray values to numbers between 0 and 15. (c) Image overlaid with a grid to show pixel values.

If we can somehow convert the picture information into a series of numbers and send the numbers along with the recipe for converting the numbers back into a picture, we would have a reliable way of sending and reproducing pictures. Figure 4.2 shows a way of accomplishing this operation.

Pixel:
Contraction of the phrase "Picture Element". A pixel corresponds to a smallest detail in the picture that one want to preserve.

The first step in converting a picture into a series of numbers is to create a grid with uniform cell size and superpose it on the picture. Each cell is now one element of a picture. The term **pixel** is created by contracting the compound phrase "Picture Element". The grid is a two dimensional array of pixels. Each pixel has an average "grayness" associated with it. Each gray value is associated with a number as shown in the reference strip with 16 discrete gray values and associated numbers. By performing a visual comparison (an admittedly error-prone operation), one can associate a number with each pixel. For example, the sky is somewhat dark and has a value of 6 associated with a pixel in the sky. On the other hand, the dress is white and therefore the number 15 is associated with a pixel on the dress.

In a subsequent chapter we will introduce the binary number system that uses a series of 1's and 0's to represent a number. We apply the same technique and convert the "6" into a series of four binary numbers (0110) and convert a "15" into another series of four binary numbers (1111). We will learn in a later chapter that if we have numbers that vary from 0 to 15, we need four binary bits to represent those numbers. Using the binary representation, each pixel gray value is now converted into a set of four binary numbers. The motivation for going with binary representation was discussed before, namely it is very easy to unambiguously represent and reproduce symbols that have only two options (0 and 1). So with binary numbers, the chances of making an error are far lower.

We can see that the grid consists of 32 rows and 32 columns. So the image of the Statue of Liberty has been converted into a table of numbers that take on values between 0 and 15. The image has 1024 pixels (32 * 32) and each pixel is represented by four binary digits (bits). So the image can be represented by a sequence of 4096 bits. These bits can now be stored, transmitted, reproduced and then converted back into the black and white image without any loss of quality. Furthermore, since the image now consists of numbers, processing the image means performing various arithmetic operations on these numbers (add, subtract, multiply, divide and numerous other operations). Since manipulating numbers is very easy with the help of a digital computer (which, after all, is designed to "compute" with numbers!), processing an image will be equally easy.

It seems that our approach of "digital imaging" has satisfied most of the original design goals. We can now take advantage of the digital computers, which have become cheaper, faster, more flexible, smaller and easier to use over the past 50 years to meet our objectives.

Exercise 4.2

1. Take a simple black and white picture. Create an 8x8 grid on top of it. Read the gray values as numbers between 0 and 15. Send the numbers

to your partner. Ask the partner to reproduce the picture according to numbers (sort of "color by numbers"!). Compare the pictures.

3. Take the 64 numbers provided by the teacher. Fill in the grid provided according to the numbers (black for 0, white for 15, intermediate gray for in between values). Collect the 16 grids filled by 16 students. Arrange them in an array of four rows and four columns to create a picture that has 32 rows and 32 columns.

4.1.2 Unintended Consequences

Any good invention or engineering design always has implications well beyond the original intent of the inventor or the engineer. Computers were originally invented with funding from the US Army to predict the trajectories of artillery shells. Now the same invention is powering the latest video games. The same statement is true for "digital imaging". Here we will explore some of these consequences.

Animation

One of the most spectacular results of digitization of imaging systems has been the advent of computer **animation**. Animation (creation of moving images) was demonstrated as far back as the early 1900's. In the early approaches, an artist carefully drew each frame in a movie. Walt Disney took this procedure to a masterful level in the 1930's with a series of full-length animated feature length films. This was of course a highly skilled profession and the whole operation was very time and money intensive. This can be called the analog age of animation. This changed after "Toy Story", which was the first full-length film that was produced completely using digital techniques. Figure 4.3 shows examples of a frame from the recent movies, "Toy Story 2" and "Bug's Life". Using a combination of computer generated animation and real live actors, movies like "Who Framed Roger Rabbit" and "Space Jam" were produced. All of this results from a representation of an image as an array of numbers.

Figure 4.3 : An animation picture – Toy Story2 and A Bug's Life screen shots.

Special Effects

A second consequence of a digital representation of an image is that by using computers to manipulate these numbers, a wide variety of special effects can be achieved. In the analog world, one had a limited set of tools based on novel optical devices to distort the image in a prescribed way. With images stored merely as an array of numbers, there is virtually no limit to the kind of manipulations that can be performed with the image. Figure 4.4 shows a couple of examples of digital special effects that would be impossible to achieve if the image were represented in conventional analog methods.

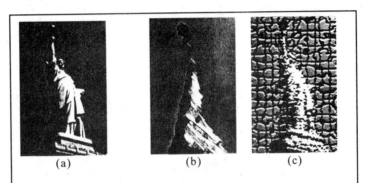

(a)   (b)   (c)

Figure 4.4 : Image processing for special effects – still image effects; (a) original image, (b,c) after special processing.

Image Enhancement

Once the images are represented by numbers, we can manipulate them to make them "better". For example, if we find the image brightness not to our liking, we can "lighten it" or make it darker.

Figure 4.5: Making the image lighter or darker by multiplying it by a single number

If we find it too grainy, we can make it smoother by performing special operations to be discussed in detail in the next chapter.

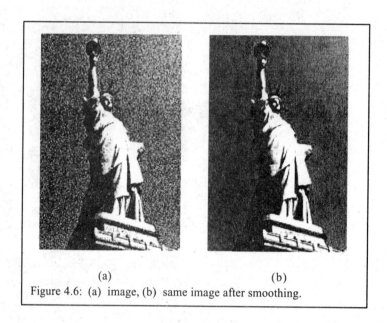

(a)   (b)

Figure 4.6: (a) image, (b) same image after smoothing.

If we find it too small, we can digitally magnify it as shown in Figure 4.7

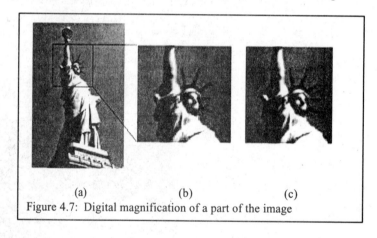

(a)   (b)   (c)

Figure 4.7: Digital magnification of a part of the image

Access Control

One example of a digital imaging system is one for access control as depicted in Figure 4.8. A secure facility will have an image sensor where a digital image of the person desiring entry is captured. The computer then compares that image with the list of people with access privileges (part (c) of the figure). If the person belongs to the database, the gate is lifted, allowing entry.

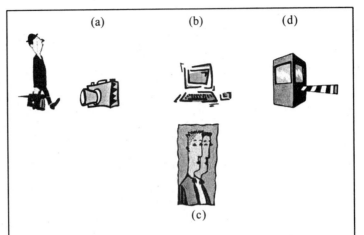

Figure 4.8 : An image recognition system, access control by facial verification. (a) image capture, (b) image comparison to (c) a template database, (d) decision, entry allowed yes/no.

Robot Vision

An autonomous robotic vehicle equipped with a digital imaging system will be able to develop the same awareness of the surroundings that a human operator of a vehicle is capable of. Based on this awareness and the goals that are programmed in the controlling computer of the robotic vehicle, it can move around in order to accomplish the goals. A spectacular example of this digital imaging system can be found in the Mars Pathfinder Rover shown in Figure 4.9.

Figure 4.9 : Robot vision – Mars pathfinder rover.

Medical

So far we discussed digital imaging systems where the image acquisition was more or less conventional. There is another way of acquiring image information where a significant amount of computation is needed before a recognizable image in available. The first and probably most important example of this kind of a digital image system is Computer Assisted Tomography (CAT). This system is capable of imaging a slice of a human body by taking measurements of several X-ray shadows from different angles. These shadows are combined together in a computer to generate a full three dimensional view of the human body. Figure 4.10 shows the photograph of a CAT scan system and an image of brain obtained by it.

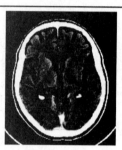

Figure 4.10 : A medical application of image processing – a CAT scan machine a CT image of brain.

Space

Telescopes have been used for several hundred years to make observations of stars and planets. Digital imaging systems allow one to further enhance the captured images in order to see the desired features. Figure 4.11 is an image of Saturn that was obtained by the Hubble telescope.

Figure 4.11 : Imaging in astronomy – a view of Saturn through the Hubble telescope.

Correction of Hubble images:
The main imaging mirror of the Hubble space telescope, as launched, had a microscopic flaw in its construction. That gave the telescope a blurry vision. A space shuttle mission was launched in order to add corrective optics to the telescope to produce sharp pictures (traditional analog solution). However, till the shuttle launch, the blurry pictures were corrected by using a digital computer to process the images.

Generic Digital Imaging System

All these digital imaging systems have varied designs, operations and objectives. However, they do have some common aspects to their design. Figure 4.12 shows a generic digital imaging system in its block diagram form. The three main stages in a digital imaging system are image acquisition from sensors, number manipulation via a digital computer and an output stage (hard copy via printers, live displays and storage). In the next chapter we will be studying some specific examples in greater details.

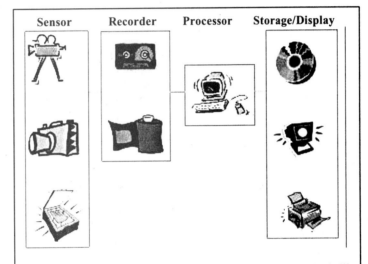

Figure 4.12 : A complete digital imaging system – (a) sensors (still camera/movie/document scanner), (b) recorders (film/tape), (c) processor (computer), (d) displays/storage (disk/monitor/printer).

4.1.3 How to Design a Digital Imaging System

We have identified the basic approach to solving the problem we outlined in the beginning of the chapter. Now we need to fill in the details. In order to perform a detailed design of a digital imaging system, we need to ask the following questions:

- What should be the size of the grid?

- What should be the size of the pixel?

- How many gray levels are needed to represent an image?

- How can we handle color?

- What should be the frame rate of a movie?

- What is the storage requirement for digital images?

In order to formalize these answers, we will extensively use the mathematical tools of matrices and simple rules of geometry. Another useful mathematical function we will need is the exponential and logarithm.

Exercise 4.3

1. Bring an unusual example of a digital imaging system other than the ones discussed in the book. Use the web, newspapers, and magazines. Write a short description of it and draw the block diagram

2. Think of a new digital imaging system. Write a short description of its purpose and a make block diagram. Don't assume any new physics principles – just new applications of future technologies.

3. In access control system, we would like the system to work even if the authorized person gets a hair cut, new glasses, or (for men) grows / shaves his beard. Which facial features do we use in recognizing a person?

4. Are there other ways of recognizing a person for achieving access control?

Section 4.2: Digitizing Images

In our simple example showing the digital representation of a picture, we constructed a grid that had 32 rows and 32 columns and the gray values were limited to one of sixteen values (0 to 15). We also considered representing the number corresponding to gray values in a binary representation giving us four bits per gray value. Putting all these numbers together, we came up with a 4096 binary bits representing that picture. Obviously there is nothing magical about the number "32" or "16" and we could have chosen any other value for these numbers. In this section, we will study the impact of choosing different values for the grid size and for the number of gray values. We will study the impact on the picture quality of choosing a large size grid and making more subtle distinctions among gray values. We will also study what this option implies for the number of bits needed to represent the picture. Then we will turn our attention to movies, which correspond to a sequence of pictures shown at a given rate. We will study the impact of changing that rate on the quality of the movies and the number of bits needed to represent the movies. Finally we will discuss the issue of representing color.

4.2.1 How Many Bits per Pixel?

Quantization: The process of taking a set of continuous values and mapping them into a finite number of discrete steps represented by integers.

We defined a pixel as the smallest image element. In a black and white image, we can measure the level of brightness (or "grayness") within each pixel. Then we can assign a numerical value to it and represent that numerical value via bits. In an analog world, the gray values will be continuous and we will simply be making a qualitative statement in words as to where the gray value of a pixel stands in relation to the continuum (e.g., darkest, somewhat dark, on the lighter side). Since we want to use numbers to represent an image, we decide to break up the gray values into a finite number of discrete steps. This process is called **quantization**. You have encountered this term before in the context of pressure values as function of time (audio signals). Here we use it in the context of images, which are light intensity values as a function of space (image signals). Figure 4.13 shows a simple strip image that contains equally spaced gray values between complete dark (0) and complete white. Each strip quantizes the gray value to different number of steps. That in turn changes the number of bits required to represent the gray value. From our earlier discussion we know that as the number of bits increase by one, the number of gray values will double. So increasing the number of bits from 4 to 5 will increase number of gray values from 16 to 32. The figure also indicates the number of gray values that can be distinguished by eye. So beyond a certain number of bits (for example 8, corresponding to 256 gray values), our eye is unable to resolve differences between more closely spaced values.

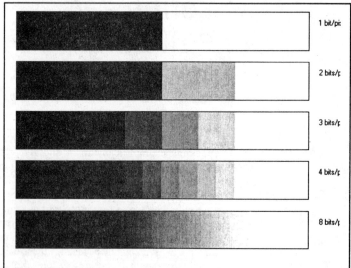

Figure 4.13: Number of gray values for different number of bits in binary representation.

The relation between the number of gray value steps and the number of bits is expressed as below in equation 4.1:

$$m \ = \ \log_2 (M) \qquad\qquad 4.1$$

$$M \ = \ 2^m \qquad\qquad 4.1(a)$$

where "m" is the number of bits, "M" is the number of gray value steps and the logarithm is with respect to base 2. This implies that for a small extra price in number of bits, the **gray value resolution** can be increased significantly. We noted earlier that for visual appearance, 256 gray value steps is more than a human eye could resolve. However, in scientific and industrial applications, a computer processes the digitized images. In that case, it is not uncommon to use 10 to 12 bit binary numbers to represent gray values corresponding to 1024 to 4096 gray value steps.

VAB Experiment 4.1:   Image Quantization

- Evaluate the visual impact of reducing the number of bits per pixel on the given set of images

- Determine the optimum quantization for each image.

16

Exercise 4.4

1. A Pentium microprocessor uses 32 bits to represent a number. If we use 32 bits to represent number of gray steps for a pixel, how many gray steps will we have? Is this number realistic?

2. How many bits are needed to represent a pixel with 32 gray levels? How many bits will be needed for 16,384 gray levels?

4.2.2 How Many Pixels in an Image?

We divide an image into regularly spaced arrays of cells called pixels. It is clear that as we increase the number of cells, the size of the cell will decrease. Suppose we have a picture that is 32 mm on a side. If we divide that picture into a grid of 32 rows and 32 columns, each cell (pixel) will be a square, 1 mm on the side. The **spatial sampling rate** is 1 pixel per mm. Now if we change the grid size to 64 rows and 64 columns, the pixel size will decrease to ½mm on the side and the spatial sampling rate is 2 pixels per mm. If we reduce the grid size to 16 rows and 16 columns, the pixel will increase to 2 mm on the side and the spatial sampling rate reduces to 0.5 pixels per mm. So the size of the pixel will be inversely proportional to the number of pixels each row or column of an image is divided into. In determining the gray value within a pixel, we ignore any variations within the pixel and assign a number that corresponds to the average gray value within the pixel. If, for example, half of a pixel area is black (corresponding to number "0") and half is completely white (corresponding to number, say, "15"), we will assign a value of "8" (average of 0 and 15, rounded to the nearest integer). As a result, some detail in the image is lost. This process of assigning an average value to a pixel is termed **sampling**. If we want to preserve fine details in an image, we need to make the individual pixels small (or make the sampling rate high) and therefore increase the total number of pixels required to represent an image.

Let us consider an image that is first sampled using a grid with 8 rows and 8 columns as shown below: This can be immediately recognized as an image of alternating black-and-white vertical bars.

> **Sampling:**
> Measuring a continuously varying image at uniformly separated points in space and assigning a value that corresponds to an average light intensity within a box surrounding that point.

> **Sampling Rate:**
> A quantitative description of the process of sampling. Sampling rate is often quoted in terms of number of samples that taken per mm. Conversely, sampling size represents the physical size of a cell within which intensity measurement is made.
> Sampling size (mm) =
> 1 / (sampling rate)

0	15	0	15	0	15	0	15
0	15	0	15	0	15	0	15
0	15	0	15	0	15	0	15
0	15	0	15	0	15	0	15
0	15	0	15	0	15	0	15
0	15	0	15	0	15	0	15
0	15	0	15	0	15	0	15
0	15	0	15	0	15	0	15

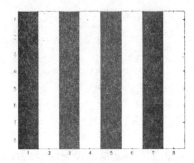

Now if we decide to represent this image by a grid with 4 rows and 4 columns, we will need to take 4 adjacent pixels from the original image and find the average gray value to assign to the new pixel in a 4 row and 4 column grid as shown below. The new image with 16 pixels is shown below:

8	8	8	8
8	8	8	8
8	8	8	8
8	8	8	8

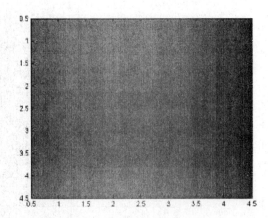

This new image contains 16 pixels, all having the same intermediate gray value of 8. The vertical bars in the original image have now completely disappeared. This kind of a loss of detail is due to inadequate sampling of an image. A simple way of describing it is as follows: if the gray value in an image is varying significantly over a small region of the original image, we need to ensure that the small region contains many pixels. In order for this to happen, the pixels themselves must be small. Since in our digital representation all pixels are of the same size, we must ensure that the whole image is covered with pixels that are sufficiently small to detect gray value variations in any regions of the image. This rule of sampling has been discussed more formally as "sampling theorem" in pervious chapters on audio signals. When we choose pixel sizes that are not small enough, the original image appears distorted. These distortions are called **sampling artifacts**. Figure 4.14 visually shows such distortions due to use of inadequate sampling rate (or large pixels)

Each images in the right hand column takes a part of the original image and increases the pixel size by a factor of two along both the row and column. So each image contains ½ the number of rows and columns as the previous image. The technical term for this operation is "down sampling by a factor of 2". These sampling artifacts can be seen when a person appearing on a TV show wears a jacket or a tie with fine design such as stripes or a checkered pattern.

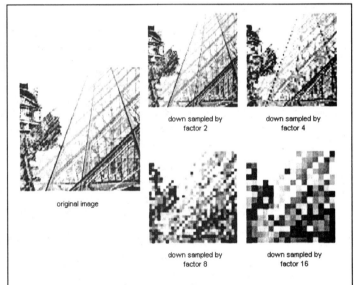

Figure 4.14 : Spatial aliasing effect – sharp features in image are distorted when sampled below a threshold spatial sampling frequency. The original image on the left is 128x128 in dimension.

Exercise 4.5

1. Consider the chessboard as an image where black squares correspond to a gray value of "0" and the white squares correspond to a gray value of "15". Write this image as a grid of numbers. How many total pixels are there in the image? How many bits are needed to represent the gray value of the pixel? How many total bits are needed to represent the chessboard?

2. If we reduce the sampling rate for the chessboard image by a fact of two along rows and columns, what will be the size of the new image? Write this image as a grid of numbers. Comment on what you observe.

VAB Experiment 4.2:   Sampling and Aliasing

- Determine the lowest spatial sampling rate for the given set of images

- Reduce the sampling rate lower than the minimum and observe the distortions produced. Write down a description of the distortions.

4.2.3 Sampling in Time for Movies

We noted earlier that a changing scene consists of images that are varying in time. When we record a changing scene as a movie, we end up recording discrete images (called frames) in time separated by a fixed interval. If we are recording an image every second, we get a temporal rate of one frame per second. Recording images every tenth of a second we get a temporal rate of 10 frames per second. This can be seen to be equivalent to spatial sampling we referred to earlier. There, a pixel was a unit in space. In temporal sampling a frame is a unit that is acquired in time. If the images are changed rapidly enough, our eye fails to perceive the individual discrete frames and instead gets the sensation of a smoothly moving object in the scene. For video, this rate is 60 frames per second. So a new frame is acquired and displayed every $1/60^{th}$ of a second.

Aliasing:
Literally meaning, "also appearing under a different name". In this case it means that the motion takes on a very different appearance in the temporally sampled movie than the original scene.

We discussed earlier the loss of fine spatial details in an image if the pixel size is not small enough. An equivalent phenomenon occurs in time if the objects in the scene are moving or changing substantially in a time much less than $1/60^{th}$ of a second. We will then fail to notice the change smoothly and instead get a smeared image. This is similar to the vertical bars disappearing in the example on spatial sampling. Furthermore, inadequate frame rate can also create more interesting artifacts called **aliasing**. The example most familiar to everybody is the wagon wheels in a western movie appear to rotate backwards as compared to the motion of the wagon. This phenomenon can be explained with the help of Figure 4.15.

Suppose each successive image in this figure is recorded one second apart giving a temporal sampling rate of 1 frame per second. In column (a), the wagon wheel is turning through 45 degrees in one second. So it will turn through 360 degrees in 8 seconds (one complete rotation every 8 seconds). This rotation rate is small enough compared to temporal sampling rate so the wheel appears to be rotating smoothly. In the second column, the wheel is turning through 180 degrees in one second or one rotation every 2 seconds. This rate is fast enough compared to the temporal sampling rate that the wheel appears to flip between the two positions instead of rotating smoothly. In the third column, the wheel is turning through 315 degree in one second. This gives a rotation rate of 1 1/7th second per full rotation. In this case, if we look at successive frames, the wheel appears to rotate counterclockwise at the rate of 8 seconds per revolution. Now the term aliasing can be understood – the wagon wheel actually rotating at 1 1/7th second per rotation in clockwise direction is now appearing as a wagon wheel rotating 8 seconds per rotation in a counterclockwise direction. If such distracting artifacts are to be avoided, the temporal sampling rate must be high enough as compared to the rate of change of the objects in the scene.

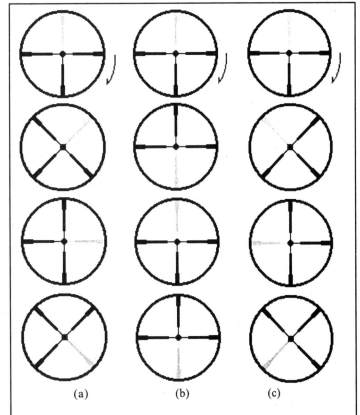

(a)   (b)   (c)

Figure 4.15 : Temporal aliasing effect – the wagon wheel effect. The wheel is turning clockwise. In column (c), it appears as if the wheel is turning counter-clockwise.

Exercise 4.6

1. Draw 4 successive frames of a movie of a wheel with one spoke in it. Select the frame rate of 2 frames per second. Draw the diagrams for the following rotation rates of the wheel:
   - One rotation every 4 seconds
   - One rotation every second
   - Two rotations per second

- Two and a half rotations per second
- Three rotations per second
- Four rotations per second.

2. Explain the effects observed.

3. A wheel is rotating at 15 revolutions per second. At what frame rates greater than 60 frames per second but less than 100 frames second will the wheel appear to be stationary?

4.2.4 How to Add Color to Our Images?

All the examples we discussed thus far involved black and white images that contained only intensity numbers per pixel. Black and white pictures can now be found only in art museums and hospitals (X-rays) and black and white TV sets can be found only in security systems. Full color is what is required for all other applications of digital imaging systems. How do we handle color? In converting then gray level information into a number, we used a strip like that shown in previous figures for comparison. If we were to take the same approach, we will need a reference strip with tens of thousands of different colors, each at several levels of brightness. This is clearly unrealistic. Fortunately for us the properties of the human visual system suggest an easy way out.

What we call "color" is in reality a complex interplay between the physics of the light waves, the design and construction of our eye and the further processing that takes place in our brain. The net result of this interplay is that we can create a very large number of colors shades by simply adding light of three different colors: red, green and blue. By varying the brightness of each of these colors, we can create a sensation of an almost infinite variety of colors. This particular phenomenon gives us an easy extension of our representation of black and white images. We create three separate images, one encoding the brightness of the red component of a scene, another with green component and the third one with blue component. By superposing these three colored images, we can effectively represent an arbitrary colored image. So now the problem of going from black-and-white to color image representation is only three times as difficult. Each color image can be spatially sampled and quantized to discrete brightness levels as was shown for black and white images. The number of brightness levels in each color image will determine how many different color combinations can be created. For example, if each color image is quantized to 16 levels (4 bits), then the full combination will have 4096 (=16*16*16) different color combinations. If each color image is quantized to 8 bit binary representation, then a full color image will have an astoundingly large number (approximately 16 million!) of color combinations. Figure 4.16 shows the three **color planes** and the superposed full color image for a picture of common vegetables.

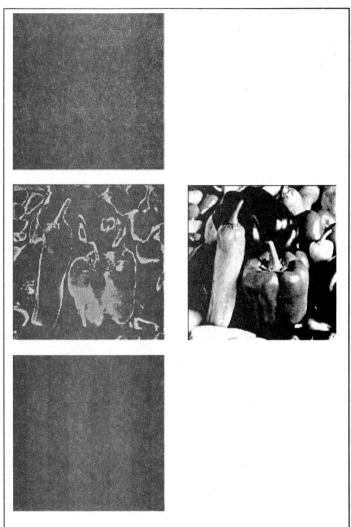

Figure 4.16 : RGB color decomposition – true color image synthesis from RGB color planes.

- Change the number of bits assigned to each color plane in the set of color images provided.

- Remove one color plane completely and observe the effect on the color

- Change the numeric values associated with the red, green and blue component of a region to produce the desired color effect.

Exercise 4.7

1. If each color image is represented by 16 bits per pixel, how many total number of colors can be obtained?

4.2.5 Summary

In this section we detailed our technique for converting still images or movies, black and white or color into a sequence of 1's and 0's. This process, called digitization, involves two distinct steps:

1. Sampling in space (creating an array of pixels) and sampling in time (creating a discrete sequence of frames)

2. Quantization in amplitude (creating discrete amplitude values per pixel) of a single pixel array for black and white images or of three pixel arrays for each of the three components (red, green and blue) for full color images.

Once we have converted an image or a movie into a sequence of numbers, we can further convert those numbers into binary representation (bits). Then we can use the full capabilities of modern computing technology (processing, storage, display) to handle the images. We saw representative operations that can be performed by a computer on digitized images. These operations were sufficient to correct the kind of defects we discussed earlier. Furthermore, binary numbers are easy to communicate since they can tolerate a large amount of distortion and noise while still retaining the essential information.

We have now identified an engineering solution to our two problems discussed at the beginning of the chapter.

Section 4.3:
Putting it
Together

So far we developed methods for representing full-color, full-motion images as binary bits. Now we will try to put everything together to get a feel for how many pixels are needed for what kinds of images, how many distinct intensity levels do we need in our quantizer and what kind of frame rate is adequate. All this translates into how many total bits are needed to store a digitized image and if we want to transmit a digitized movie, what kind of data rates are needed to accommodate the number of bits generated per second in the digitized movie. The storage capacity of our system (given in number of bits) and the data rate given in bits / second that we can transmit represent the finite resources that are available to us. Any engineering design problem involves making an optimum use of available resources. After all, a good engineering design for a car leads to a product that delivers maximum performance for least cost or weight or size or gas consumption. We will take a similar approach in analyzing our potential designs for a digital imaging system.

4.3.1 Sampling and Quantization

Let us consider the simplest case first – representing a single black and white image. The two choices we need to make are the number of pixels in a row (say, **I**) and in a column (say, **J**) of the image. Given an array size of **I** rows and **J** columns, total number of pixels in the array, **N**, will be given by

$$N = I * J. \qquad\qquad 4.2$$

A very low quality generated by a web camera will have **240** rows and **320** columns. Such an image will have **76,800** pixels. On the other hand, the next higher quality digital image will have twice the number of rows **(480)** and columns **(640)**, giving total number of pixels to be **307,200**.

We will use **M** to indicate the number of discrete steps for quantizing the gray values of each pixel. According to equation 4.1, the number of bits (m) needed to represent M gray values is related by the function "logarithm

$$m = \log_2 (M) \qquad\qquad \textbf{4.1}$$

So if we want a very crude image, we could use **16** gray value steps per pixel or **4** bits per pixel. On the other hand, a high quality image will use **256** gray value steps per pixel corresponding to **8** bits per pixel. A pure binary image will only allow **2** gray values per pixel (black or white and hence will require **1** bit per pixel.

Putting these two equations together, we get the following equation for total number of bits required to represent a black and white digitized image.

$$B = N\ m \qquad\qquad \textbf{4.3}$$

$$B = I * J * \log_2 (M) \qquad\qquad \textbf{4.3(a)}$$

Now we will calculate the number of bits required to represent digital images of varying quality. We considered two different sizes of pixel arrays **(320 x 240 and 640 x 480, respectively)** and three different values for number of steps of gray values **(2 - 1 bit, 16 - 4 bits and 256 - 8 bits).** Furthermore, we note that in computer language, bits are stored in groups of **8 bits**, called a **BYTE**. So we will calculate the number of bytes needed to represent the image. The table below gives the values for the choices described above.

Columns	Rows	Number of Pixels	Gray Levels	Bits / pixel	Total Bits	Total Bytes
320	240	76,800	2	1	76,800	9,600
320	240	76,800	16	4	307,200	38,400
320	240	76,800	256	8	614,400	76,800
640	480	307,200	2	1	307,200	38,400
640	480	307,200	16	4	1,228,800	153,600
640	480	307,200	256	8	2,457,600	307,200

4.3.2 Which is a Better Picture?

As we examine the table the range of total number of bytes per image becomes quite striking - from 9,600 bytes to 307,200 bytes. We also notice something curious - a 320 x 240 pixel image with 16 gray values per pixel requires the same number of bytes as a 640 x 480 pixel image with 2 gray levels per pixel. The question now arises:

If we have 38,400 bytes of storage at our disposal, should we use it to store a more detailed binary image or a less detailed image with more gray values per pixel?

The answer to this question is - "It depends on what the image contains!" If the original object is naturally binary (say a line drawing) then obviously we need the more detailed image. On the other hand, if the original object is the photograph of a face, then we would benefit more by having 16 gray values per pixel than by having more pixels.

This point is illustrated in Figure 4.17. The top half of the figure (a) shows a photograph of the Statue of Liberty while the bottom half of the figure (b) shows the line drawing of an apple. The images have different number of pixels (128 x 128, 64 x 64, 32 x 32, and 16 x 16) and different number of gray levels per pixel (256 - 8 bits, 16 - 4 bits, 4 - 2 bits, and 2 - 1 bit). If we perform the bit calculations for each of these images, it will be apparent that a 128 x 128 pixel image with 2 gray levels per pixel (first row, last column images in both (a) and (b))) requires the same number of bits as a 64 x 64 pixel image with 16 gray levels (second row second column images in both (a) and (b)) per pixel. Similarly a 32 x 32 pixel image with 256 gray values per pixel will need the same number of bits as a 64 x 64 pixel image with 4 gray values per pixel. Students should similarly identify pairs of images that require the same number of bits. Now we can compare the quality of these two choices to see which is better.

If we compare the 128x128 pixel image with two gray values with 64x64 pixel image with 16 gray values for the Statue of Liberty photograph, we clearly prefer to have more gray values than more pixels. On the other hand if we compare an image with 32 x 32 pixel with 4 gray values per pixel with a 16 x 16 pixel image with 256 gray values per pixel for the drawing of an apple, we clearly prefer the more detailed image with less gray values.

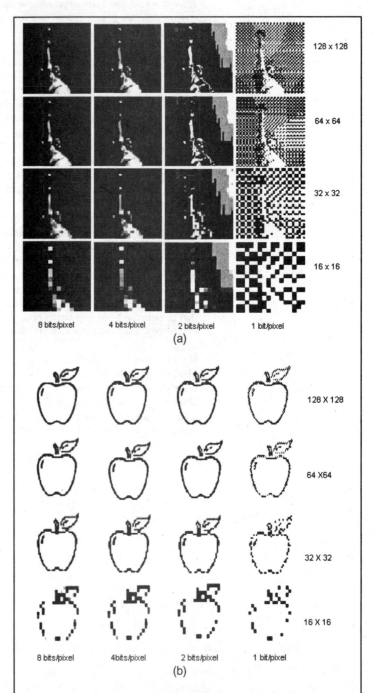

128 x 128

64 x 64

32 x 32

16 x 16

8 bits/pixel   4 bits/pixel   2 bits/pixel   1 bit/pixel

(a)

128 X 128

64 X 64

32 X 32

16 X 16

8 bits/pixel   4bits/pixel   2 bits/pixel   1 bit/pixel

(b)

Figure 4.17 : Spatial and bit quantization for two different image types – trade off.

Now let us consider the use of color in images. We saw earlier that adding color to images is easy since we can create a sensation of full range of colors by simply superposing three primary color images (red, green and blue) each quantized to specific number of intensity levels. So the complexity and storage requirement increases by a factor of three over simple black and white image. Previously we studied the trade-off between increasing numbers of pixels versus increasing number of gray levels per pixel in order to achieve the "best" image for the same number of total bits. Same analysis can be carried out with the use of color. Is it better to use three times more bits with a black and white image (giving 8 times more gray values per pixel) or is it better to add color to the image? And again, the answer is "It depends on what the image is and what you want to do with the image." Let us take the color image of a platter of fruits we studied in Figure 4.16 and see what happens when we create a black and white image of the same. Figure 4.18 shows the results.

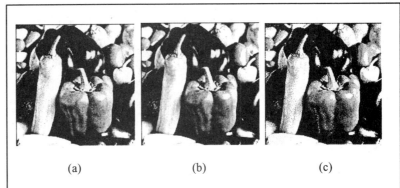

(a)   (b)   (c)

Figure 4.18: Use of color in images
(a) Color image of fruits with 256 distinct colors per pixel
(b) Black and white image of fruits with 256 distinct gray values per pixel
(c) Color image of fruits with 16 distinct colors per pixel.

In this particular case, color carries very important information about objects in the image. For example, the difference between green bell pepper and red bell pepper is (trivially) in the color with no difference in size, shape or texture. So the color image with only 16 colors per pixel is much more useful in identifying and separating different vegetables in the plate than a high quality black and white image of the same object. On the other hand, if we take a different set of objects, the role that color plays will change dramatically.

Figure 4.19 shows the image of three common objects, a spray bottle, a coffee mug and a tall drinking cup. We show images with full 256 colors per pixel and 256 gray values per pixel along with reduced color values per pixel and finally only two gray values per pixel. We can see that their shapes primarily distinguish the objects in this image with color playing very little role. So in this case, the black and white image with only two gray values per pixel is quite adequate for distinguishing among objects and the use of color will not justify the cost in terms of additional storage capacity needed to represent the color image

Figure 4.19: Image of different objects
(a) Image with 256 distinct colors per pixel
(b) Image with 16 distinct color per pixel
(c) Image with 256 distinct gray values per pixel
(d) Image with two distinct gray values per pixel

Exercise 4.8

- Show the trade-off between pixel resolution for an image and the number of steps in the gray level values per pixel when the total number of bits is fixed. Plot in a graph.

- Give examples of situations where the spatial resolution is much more important than gray level values.

- Give examples where gray level values are not more important than image resolution

- For different images, study the importance of color. You can convert a color image into black and white image by simply copying it. Make general observations on what kind of images (medical, satellite, artistic, advertisement) use the color in a critical fashion.

4.3.3 Taking Digital Pictures

Let us return to our original problem of proving to your friends that you were indeed close enough to your rock idol to see the whites of his / her eyes. In previous sections we talked technical terms like "sampling, pixels, **gray scale**. Let us put these terms in a specific context.

Let us begin with some very basic terms associated with taking pictures. First term is **field of view**. To put simply, it means the part of the scene that will be captured in your photograph. The field of view typically depends on the type of camera lens one is using, how far one is from the scene and the format of the film (or electronic sensor) the camera is using. There are three broad types of camera lenses with different fields of view (scene coverage):

- Normal lens: this is a lens that covers approximately 5 meters wide area of the scene when placed at a distance of 10 meters from the object.

- Wide-angle lens: this is a lens that covers approximately 8 meters wide area of the scene when placed at a distance of 10 meters from the object.

- Telephoto lens: this is a lens that covers approximately 1.5 meters wide area of the scene when placed at a distance of 10 meters from the object.

When we take a picture, the objects within the field of view are imaged on the sensor in the camera, whether a photographic film or an electronic sensor which contains an array of discrete pixels. A digital camera is usually specified in terms of how many pixels the electronic sensor in the camera contains. For example, a medium quality digital camera will contain an image sensor that has 1500 columns and 1200 rows of pixels. Now if we use a telephoto lens with the camera and if we are at a distance of 10 meters it means that the 1.5 meters wide area of the scene is sampled by 1,500 pixels in the digital camera. So the size of the object that is covered by one pixel in the digital image is given by:

Pixel Size = Field of View / No. of Pixels

Pixel size = 1.5 meters / 1,500   **4.4**

Pixel size = 1 mm.

We know that the normal size of the white of a human eye is approximately 3.5 cm wide and 1.5 cm tall. That means that if you take a digital picture of the performer under the conditions listed above, each eye of the star will get approximately 35 pixels horizontally (columns) and 15 pixels vertically (rows). This is enough to get sufficient details of the facial features. Figure 4.20 shows the digital photograph of the face where the section around the eye is magnified to show the details. The quality of the picture is therefore sufficient to convince even the most skeptical how close you actually were to the performer.

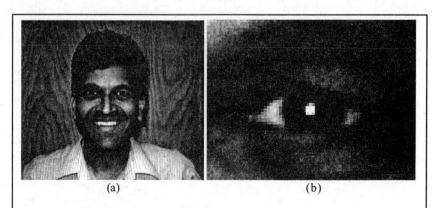

Figure 4.20: Digital picture of a face
(a) A 630 x 520 section of the full picture showing the face
(b) A blow up of a section showing 35 x 15 pixel array covering the eye.

4.3.4 Epilogue

Your job is done. You now have an understanding of the technology that will allow you to take pictures and correct for any mistakes, fix any defects AFTER they occur. You have also learned about the fundamental mathematical notions that are behind the current revolution in "digital imaging", which really turns out to be a fancy name for "imaging by the numbers". You know how to figure out what kind of digital sensor you need and what kind of camera lens you need in order to take highly detailed pictures of your object.

Then you realize one thing - this business of converting pictures into numbers can have great potential. You already saw some examples of digital animation and digital special effects. You wondered how far you can push it. You start wondering if there is a way you can convince your friends that you actually **MET** the rock star and even joined on the stage for one of the numbers. Well, you indeed can by using the color in a very special way and by using a digital computer to manipulate the images. Here is how you can do it.

Ideally what you would like to do is to take your picture standing by yourself and add it to a picture of the rock star. Now you have seen some cheap, "analog" versions of that on street corners during your visit to Washington DC. There were cardboard cutouts of the President and other celebrities. For a small amount of money the roadside photographer will let you pose with the cardboard cutout and take your picture with the "President". The quality of such pictures were so poor that they looked as phony as a three dollar bill. Then you thought that may be if you can take your picture, cut it out very precisely and simply paste it next to the picture of the rock star and then snapped another picture of the composite picture, you may be able to create a more passable forgery. But then it seemed like too much work and too much skill was involved. So you start looking for a digital approach to the problem of cutting and pasting.

In cutting out your picture, you are looking to remove all the parts of the picture that are not "you". So the background and you must not have anything in common at all. The best and the easiest way to separate the background from the main object are to use color. We hang a cloth in the background of a color that is not to be found anywhere on you. Most typically this is done with a deep blue color. So now in your picture in front of the deep blue background (say, picture **A**), wherever the computer finds the blue color, it can be sure that it is not "you". Select another picture (say, **B**) of a scene of exactly the same number of rows and columns as picture **A**. You would like to add yourself to the scene shown in picture B (say a group picture of a rock band). The computer can merge the two pictures according to the following recipe:

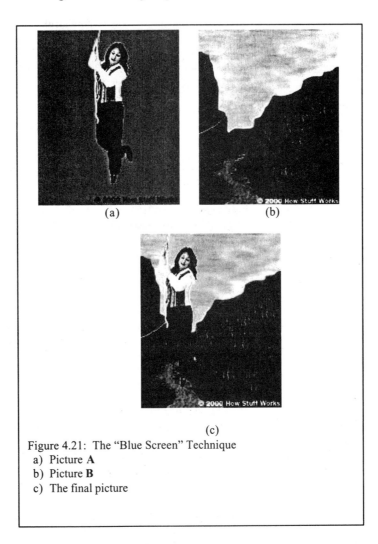

(a) (b)

(c)

Figure 4.21: The "Blue Screen" Technique
  a) Picture **A**
  b) Picture **B**
  c) The final picture

- Examine corresponding pixels (same row and same column) from each of the two pictures

- Wherever the pixel from picture **A** is deep blue, exactly matching the color of the background that you had picked, the computer will select the exact corresponding pixel from picture **B**.

- Wherever the pixel from picture **A** is **NOT** the deep blue color of the background, the computer selects that pixel as the corresponding pixel of the new composite picture. Figure 4.21 shows these operations on the picture of an actress dangling from a rope taken in the studio (picture **A**) and the picture of a river gorge as the background (picture **B**). The composite picture **C** shows the actress dangling in the river gorge.

When the computer finishes examining all the pixels, the new composite picture will contain your picture seamlessly blended into the scene. By creative selection of the scene picture and your picture (location, size, pose etc.) you can depict yourself shaking hands with the star or even giving a high-five!

We all know that computers are becoming more and more powerful and cheap everyday. It should take a relatively cheap computer to do this pixel sorting very fast. If the computer can do this within one frame time of a video (which we identified as $1/60^{th}$ of a second), then instead of doing the operation on a single frame, the computer can take two separate videos, one with the rock band doing their number and the other with you doing your part of the number in front of a blue background and merge them. The result will be a composite video showing you joining the rock group on the stage in one of their numbers. Now **THAT** will definitely impress your friends…at least the ones who haven't taken this course!

Exercise 4.9

1. Take the previous exercise in determining how many pixels are covering the eye of a portrait. Apply the same considerations in determining what kind of camera lens and what kind of image sensor is needed if you want to take the picture of a blue jay from a distance of 50 meters while having at least 5 x 5 pixel array covering its eyes. Are the numbers reasonable? Find out which of the commercially available digital cameras can satisfy your requirements.
2. Find out as many examples as you can where the blue screen technique is used to create amazing special effects.
3. Calculate the storage needed in number of BYTES for storing 100 images with the following properties:
   o 320x240 black and white images (1 bit / pixel)
   o 320x240 color images with 64 gray levels for red, 32 for blue and 4 for green images.
   o Medical X-rays with 1500 x 2000 pixel spatial resolution and 1024 gray levels per pixel.
4. You are taking the photograph of a butterfly that is 5 cm across. If your field of view exactly matches the size of the butterfly and the

photograph has 2048x2048 pixels, what is the smallest detail on the butterfly wing that you can resolve?

5. The previous picture has been quantized to a total of 2,097,152 distinct colors. If each color has been quantized to the same number of gray levels, how many bits are required for a pixel in each color image? How many total bytes are needed to represent this image?

Section 4.4:
Glossary

aliasing
Literally means, "also appearing under a different name". In this case it means that the motion in a sampled movie takes on a completely different appearance than the original scene

animation
Creation of moving images by creating individual frames that when shown in quick succession produces the illusion of smooth motion

color plane
each color image is made of three single color (red, green and blue) images placed on top of each other; these single-color components of the color image form the three color planes

digital imaging
the technique of representing images (or movies) as a matrix of numbers

field of view
When applied to an imaging system, this term describes the region of the scene that will be captured by the camera

frame
An individual image in a sequence that when displayed in time constitutes a movie. If we push the "pause" button on a VCR (or a DVD) the stationary image we see on the screen is a *frame*

gray scale
levels of gray tones covering ranges from all-white to all-black

image resolution
number of pixels along the horizontal and vertical dimension in an image

image sampling
measuring a continuously varying image at uniformly separated points in space and assigning a value that corresponds to an average light intensity within a box surrounding that point

multimedia
we mean an integration of text, numbers, sound and still and moving pictures

pixel
Contraction of the phrase "Picture Element". A pixel corresponds to the smallest detail in the image that one wants to preserve

quantization
the process of taking a set of continuous values and mapping them into a finite number of discrete steps represented by integers

sampling
Measuring a continuously varying image at uniformly separated points in space and assigning a value that corresponds to the average light intensity value within a box surrounding that point

sampling artifact
The distortion arising in a sampled image when the sample rate is too low to adequately capture the finest details in the original continuous image

sampling rate
number of samples of the image taken per unit physical size of the image

temporal aliasing
distortion observed in movies when frames are not captured fast enough by the camera

The Infinity Project

Chapter 5: Math You Can See

How can the images in digital form be manipulated to improve their usefulness?

Approximate Length: 2 Weeks

In the previous chapter we studied the motivation and techniques for representing images with numbers. The primary motivation was ease of acquiring, storing and distributing images and the possibility of correcting any errors in the image. We also mentioned other consequences of a digital representation of the image, namely building digital imaging systems for complex applications. That is the focus of this chapter. Again, we will introduce the mathematical preliminaries and follow up with applications of those mathematical techniques in digital image processing and finally systems design.

Section 5.1: Do I *HAVE* to do that?

It was been a rather stressful week. What with all the AP exams, final reports and all that coming due in the same week, you were completely swamped. Now it is Friday and you are looking forward to going to a movie with your friends and just hang out. Then you get a call from your mother. They are expecting company that evening and you are asked (ordered?) to clean up your room, tidy up the house, do the laundry. You see your plans for the movie go up in smoke! As you start with the chores, you start wondering - "Isn't there a better of doing these things?" After all everything that you are asked to do seems rather simple. Shouldn't we be able to *BUY* a robot or something that can do grunt work like this? Stacking the papers and books according to size, sorting the clothes by color (light or dark), recognizing small pieces of toys and other sundry junk lying on the carpet; all seem to be rather simple jobs. As a student enrolled in the **Multimedia and Information Engineering** class, you recognize this as a very complex engineering problem with many dimensions. Part of it is pure mechanical engineering. How does one make a robot with mechanical hands and legs that have the flexibility and strength required for the job of picking up small and large objects, handling heavy as well as delicate items? But then there is also the question of giving the robot eyesight that sees and recognizes the world around it. Having studied digital imaging in the previous chapter, you feel more comfortable thinking about this problem. Surely others have faced similar problems before and have thought of solutions to these kinds of problems.

There indeed are several situations - far too many, as a matter of fact - where problems of similar nature are encountered. Here is a small list of problems which are similar to the one you face:

- Finding a good landing site for a Mars landing pod

- Determining when to send police to foil a robbery attempt

- Destroying enemy tanks with smart missiles

- Distinguishing between normal and abnormal X-rays.

The problems vary greatly in difficulty. You should know that some of these problems are currently being solved using existing technology while others require more futuristic technologies. Nonetheless, all applications involve some common operations. These are:

Image processing, analysis and decision-making.

5.1.1 Preliminary Considerations

In the previous chapter we studied a new way of representing images (single frames and movies). This representation, which we call "**Digital Representation**", reduced the images and movies to a set of binary numbers – a *VERY LARGE* set of numbers, but just binary numbers none-the-less. Once we represent images by numbers, image manipulations can be described in terms of a very powerful collection of numeric operations that have been developed by mathematicians over several *thousand* years. It can then be implemented on the electronic digital technology that has seen explosive growth over past 50 years. In this chapter we will study various image manipulation applications in terms of simple numeric operations performed on digital images.

Let us first review the idea behind digital representation of images. We described in the previous chapter how we take a picture and superpose a grid on it. The grid consists of unit cells (which we called "pixels") of identical size arranged in a regularly spaced array along rows and columns. The average gray value within each pixel is converted into a discrete number. This results in the familiar mathematical construct called a "**matrix**". The matrix elements can also be converted to a binary number representation giving us a sequence of bits for each pixel. A movie consists of a sequence of individual images shown at a regular interval over time. Each image can be digitized as described before. So a movie can also be converted to a time sequence of bits.

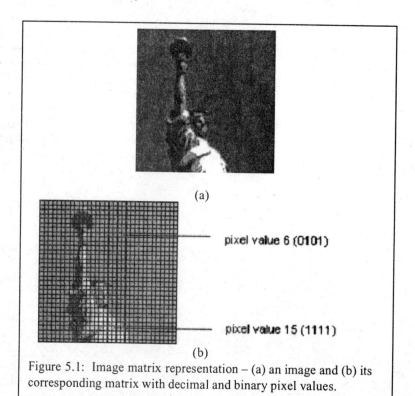

(a)

pixel value 6 (0101)

pixel value 15 (1111)

(b)

Figure 5.1: Image matrix representation – (a) an image and (b) its corresponding matrix with decimal and binary pixel values.

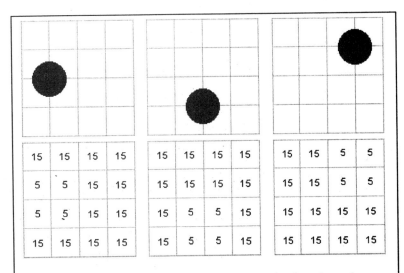

Figure 5.2: Three frames in a movie sequence simple animated movie of a ball bouncing and corresponding matrices. (4x4 pixels per frame, 4bits/pixel (0-15) gray scale resolution.)

Once we have this set of numbers, then there are a number of common operations we can perform on these numbers:

- addition/subtraction,

- multiplication/division,

- comparison,

- mapping one number to another using some function or rule (for example, square, logarithm or an arbitrary look-up table).

We can also process a group of numbers using these basic operations: average, median, maximum or minimum value. A number of interesting image manipulation applications can be implemented using basic operations of matrix algebra. Before we go deep into the mathematics, we will show some of the results of the image processing operations.

5.1.2 Simple Image Processing Operations

Image Segmentation:

A scene usually consists of a number of objects. If we want to design a system that automatically analyzes these objects, it is necessary to separate the objects from the background. This operation is called "segmentation". Figure 5.3 shows an example of a segmentation operation. It should be noted that the separation between the object and the background can be performed based on a number of properties, such as intensity, color, texture. The image manipulation operation required depend on which of these attributes (or a combination of them) separates the object from the background.

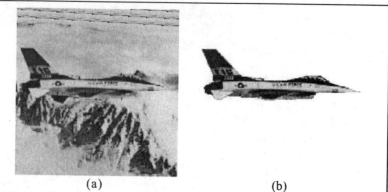

(a)   (b)

Figure 5.3 : Image segmentation, (a) a composite image containing
background and object and (b) only object present, background
suppressed.

Image Morphing

In the recent years the operation of "morphing" has started appearing in
movies and in advertisements. Morphing literally means "changing from
one form to another". Most famous examples of morphing can be found in
the movie **"Terminator – 2"** in the commercials for Exxon gas where a car
morphs (changes) in a smooth manner into a tiger. Figure 5.4 shows five
successive frames in a morphing sequence. The main idea behind morphing
is that any two successive frames only differ by a small amount but the last
image of the sequence is completely different than the original image. This
operation has become possible because of the digital representation of the
image and the easy availability of large computational resources. One can
imagine making physical models of all the intermediate stages of the
morphing operation and shooting each stage as a frame in a movie (analog
solution). But the time, skills and money involved in such an operation will
be clearly overwhelming.

Figure 5.4: Example of image morphing – digital processing to
morph a soccer ball into a football.

Noise Reduction

In the process of image acquisition, the hardware often has non-ideal behavior. One of the problems that are often encountered is noise. While noise is a hard term to define, most of us intuitively understand the difference between noise and what we are trying to observe. In analog systems, one has to work very hard in order to minimize noise in image acquisition. In digital imaging, we have the flexibility of removing the noise after the image is acquired and digitized. Figure 5.5 shows one such example of a noisy image that is cleaned up using a simple numeric recipe.

(a)   (b)

Figure 5.5: Noise removal – (a) a noisy image and (b) its cleaned version.

Image Sharpening

Images, which contain large sections of uniform intensity, are usually considered "boring" or lacking in interesting details. Most interesting images contain many transitions between light and dark areas. These transitions could be very gradual or rather abrupt (corresponding to a step change). Qualitatively human observers prefer images where the transitions between light and dark areas are more abrupt. The common phrase used is that the images look "sharp". An image could lose its sharpness due to many reasons – a bad lens, wrong focus, moving the camera while taking pictures. With the film-based analog imaging technology, once the image was fuzzy, when snapped, all was lost. Not so with digital images. The fuzzy images can be processed to make them sharper. Figure 5.6(b) shows a picture that was processed by a "sharpening filter" to bring out the details in the image. Sometimes it is desirable to make the images look soft (usually for an artistic effect). The figure also shows a softening operation being performed on the same image.

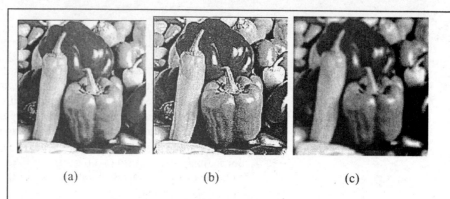

(a)   (b)   (c)

Figure 5.6 (a) Original image of a plate of vegetables, (b) same image which was sharpened by digital processing, (c) original image processed by a softening operation.

Edge Detection

The image sharpening operation kept most of the gradually varying areas in the image intact. For some applications, these areas are not of interest at all. It is only where the intensity changes abruptly between different regions of an image, that an object boundary is assumed to occur. So for those applications, only those pixels in an image that are on the boundary (edge) between dark and bright areas are made bright. All other parts of the image are suppressed to a dark value. An example of edge detection is shown in Figure 5.7. One can see that even though pixel-by-pixel, the image that contains only the edges in an image is very dissimilar to the original image, we can recognize the basic objects in the original image from its edges. There is considerable evidence to believe that human eye performs the operation of edge detection before sending the image signals deeper in the brain for further processing. That is why presenting with just the edges of an object is not that detrimental to our recognizing the object.

(a)   (b)

Figure 5.7(a) Original image, (b) original image processed only to show edges.

8

Change Detection

It is our everyday experience that if we see the same thing in the same place day in and day out that object ceases to register in our mind. But if somebody were to remove that object, we will notice that something is missing. Same considerations apply to analyzing images of earth from space - we are often more interested in determining what has changed in the scene. Below are shown examples of two images that are identical except for removing one object (a coffee mug) in the scene. In the third image only the change is seen while the parts that were common to both images are suppressed. One can readily see the similarity between the previous operation of edge detection, where a change in light intensity as a function of spatial position was considered and this operation where the change in intensity at the same spatial location in two frames acquired at two different times is considered. Gain, the comparison with human vision is worth making. It has been proven that our eyes are sensitive only to light intensity that changes in time and that if we are able to immobilize an image on the retina (light sensitive part of human eye – our "film"), we don't "see" the image.

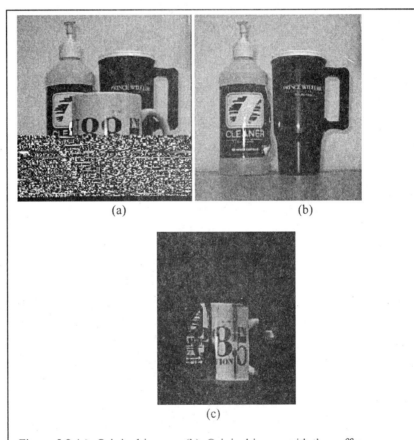

(a)   (b)

(c)

Figure 5.8 (a) Original image, (b) Original image with the coffee mug removed, (c) Difference between the two images. The difference shows the missing coffee mug as well as the part of the cleaner squeeze bottle that was previously hidden by the mug.

5.1.3 Systems Applications

We remarked earlier that the need to develop computer systems that can process images and analyze them is found in many walks of life. In this section we will present a few examples of systems that currently exist and are meeting real needs. The first example of such a system is shown in Figure 5.9. It shows a precision guided missile for destroying tanks or ships when launched from air by a fighter plane.

Figure 5.9: AGM -65 "Maverick" air-to-ground precision guided missile.

The other application that is found commonly is security applications. Security cameras are found everywhere from convenience stores to bank to parking garages. However, it is expensive to monitor them manually all the time. It is relatively simple to develop a system that only triggers a response or notifies a human monitor when motion is detected in the image. In Figure 5.10 we show one such commercially available system that employs a black-and-white camera with a wireless transmitter and a Video Cassette Recorder. The recorder is triggered only when the camera and associated processor detects motion.

Figure 5.10: Motion detector video security system made by Access Control International.

The latest toy craze has been a robotic dog from Sony Corp named "Aibo". This mechanical toy has modest amount of artificial vision that enables the dog to move around without running into things. Figure 5.11 shows the second generation Aibo.

Figure 5.11: Aibo robot dog with simple vision capability.

One spectacular example of digital techniques in image processing is the movie Matrix. It introduced a special effect called "bullet time" where an action sequence is frozen in time and different views of the three dimensional scene are presented. Same technique has now been used in Superbowl 2001 and NCAA finals. This technique requires a large number of cameras arranged in a circular fashion around the scene, very precise synchronization between the cameras and some very sophisticated digital image processing to create the feeling of smooth motions around the actors frozen in time. Figure 5.12 shows one screen from the Matrix.

Figure 5.12: One frame from the Matrix showing action frozen in time while the camera pans around the actors.

These examples are just the tip of the iceberg. There are now a large number of systems where the image processing is buried deep in the structure of the system so it is not apparent to a casual observer. The increasing miniaturization of various components (imaging lenses, electronic camera, associated processing electronics) means that digital imaging will be integrated into more and more aspects of our daily lives. Videophone, which has been touted for over 30 years, may finally find its place in our homes. A home may come fully wired with built-in cameras. School security can be enhanced enormously by increasing use of video cameras. With this increased presence of imaging systems come the requirement of every more sophisticated automated processing. It should, however, be remembered that even very sophisticated applications are built on some very simple and basic notions. The main purpose of this chapter is to introduce these elementary operations and then build interesting applications based on them.

Exercise 5.1:

1. Identify other applications for digital imaging systems from entertainment, defense, and medical fields.

2. Describe the applications in a block diagram form and identify the operations performed within each block.

3. Give examples of types of "noise" commonly encountered in TV images.

4. What property of an image can we use to segment an image that contains green apples, bananas, oranges, wicker basket and light blue Formica counter top?

Section 5.2: Matrix Formulation of Digital Image Processing

A digitized image is essentially a two dimensional (2-D) array of numbers as discussed in Chapter 4. This 2-D array of numbers can be represented mathematically as a matrix. The numbers in a matrix are arranged in forms of rows and columns. For example, a matrix **A** that contains 4 rows and 3 columns is shown in Equation 5.1 below:

$$\mathbf{A} = \begin{vmatrix} 12 & 9 & 0 \\ 5 & 6 & 23 \\ 8 & 16 & -9 \\ -13 & -12 & 0 \end{vmatrix} \quad \text{Columns} \downarrow \text{Rows} \quad 5.1$$

An element of the matrix is identified by its row index and its column index. For example, the element that is in third row and second column of matrix **A** is represented by **A**(3,2). In this particular case that matrix element will have a value of 16. The size of the matrix is indicated by the number of rows and number of columns that matrix contains. Matrix **A** is called a 4x3 matrix, indicating that it has four rows and three columns. This matrix will have a total of 12 elements. Generalizing this to an MxN matrix, we see that it will have M*N elements arranged in M rows and N columns. A matrix that has the same number of rows and columns is called a square matrix. One of the simplest matrices contains only one row and column and therefore has only one element. Such a single number is often called a "scalar". It should be noted that in our discussion we, are representing matrix elements in a decimal number system though they could just as well be represented in binary number system as discussed in Chapter 4.

The rows and columns of a matrix can be identified with the two spatial directions in an image. The X-axis (horizontal direction) of an image is along the rows of a matrix and the Y-axis (vertical direction) along the column direction. So an image that has a size of 16 pixels along the X-axis (number of columns) and 24 pixels along Y-axis (number of rows) can be represented by a 24x16 matrix and will have a total of 384 pixels (384 = 24*16). An image is a map of the light intensity and hence can never be negative. So a matrix that represents an image has only positive elements.

In computer displays, images have been set to standard sizes so once a standard configuration is defined, multiple companies can make products (graphics cards, monitors, computers) that can work with each other in a seamless manner. The first standard was termed Color Graphics Adapter (CGA) and it set the image size to be 240 rows by 320 columns. In matrix notation, we will denote this as a 240x320 matrix. However, industry practices have adopted giving the number of columns first and then the number of rows and hence it is stated as 320x240 display. The standard was set to display only 16 unique colors and hence the images were extremely crude by current standards. But as the first introduction of colors to Personal Computers, CGA images delighted a generation of young students with color video games. After CGA, the next generation of computer graphics developed the Video Graphics Adapter (VGA) standard, which doubled the number of rows and columns (640x480, or a 480x640 matrix). The number of colors was now increased to 256 (8-bit). Most current video standards (Super VGA and XGA) image sizes are even larger (800x600 – 600x800 matrix- for SVGA and 1024x768 – 768x1024 matrix – for XGA). The number colors are now increased to 16 bits for "high color" and 32 bits for "true color". It should be noted that number of columns are always larger than the number of rows, making the image rectangular, wider than taller. This is done to accommodate the human visual system, which has a wider field of view (see definition of Field of View in Chapter 4) side-to-side than up-and-down. This is why the movie screens are also wider than taller with a height-to-width ratio of 4:5. In any case, it is clear that very large matrices are needed to represent current digital images. The total number pixels range from 76,800 for CGA to 307,200 for VGA to 480,000 for SVGA and finally 786,432 pixels for XGA.

5.2.1 Elementary Matrix Operations

Matrix representation is a convenient way to manipulate images. One of the most basic operations that one can perform with a matrix is to multiply all of its elements by the same number. When a single number (called scalar) multiplies all elements of a matrix, a new matrix results. As one can imagine, the new matrix resembles the old matrix in its structure (the relation of different elements with each other) though the individual values may be different. Equation 5.2 describes a matrix **A** being multiplied by a scalar "p".

$$B = p * A \qquad\qquad 5.2$$

$$B(i,j) = p * A(i,j)$$

As a matter of notation, matrices are represented by bold upper letters while scalars are represented by lower case letters. If we us the 4x3 matrix "**A**" given in Equation 5.1 and assign a value 2 to scalar "p", the resultant matrix **B**" will be as follows:

$$\mathbf{B} = \begin{vmatrix} 24 & 18 & 0 \\ 10 & 12 & 46 \\ 16 & 32 & -18 \\ -26 & -24 & 0 \end{vmatrix}$$

If the scalar multiplying the matrix is less than one, the resulting matrix values are all lower than the original matrix. If the scalar is greater than one, the new matrix values are greater than the original matrix values. When the matrix corresponds to an image, the scalar-matrix operation is used to brighten or darken an image. Figure 5.13 shows how images are modified as result of the scalar-matrix multiplication.

Figure 5.13: Image matrix scalar multiplication – scalar > 1 results in intensity amplification, scalar < 1 results in intensity attenuation.

The second basic operation one can perform with matrices is adding two matrices of the same size (containing equal number of rows and columns). Elements that are in the same row and column of the two input matrices are added together to produce the corresponding element of the output matrix. For example if **C** is the output matrix that results from the addition of two matrices **A** and **B**, then the element in the 3rd row and 5th column of matrix can be calculated as follows:

$$\mathbf{C}(3,5) = \mathbf{A}(3,5) + \mathbf{B}(3,5) \qquad\qquad \mathbf{5.3}$$

A more general way of stating the same relation involves using letters "i" and "j" for row and column index, respectively as shown below:

$$C(i,j) \; = \; A(i,j) \; + \; B(i,j) \qquad\qquad 5.4$$

The operations of scalar multiplication and matrix addition can also be easily extended to scalar division and matrix subtraction. Care must be exercised when performing the subtraction operation with image matrices. As noted earlier, the elements of an image matrix can only be positive. This problem can be solved by ignoring the sign of the output matrix and only retaining the magnitude of the results of subtraction or alternatively converting all negative numbers to zero, representing a dark pixel in an image.

These two operations can be combined as shown in equation 5.5

$$C \; = \; A \; + \; p * B$$

$$C(i,j) = \; A(i,j) \; + \; p*B(i,j) \qquad\qquad 5.5$$

Figure 5.14 shows images resulting from these operations. Subtracting two matrices corresponding to two images that are only slightly different results in a image that only retains the difference. Image addition can be used to create special effects by essentially superposing two images creating a composite image as shown in the figure.

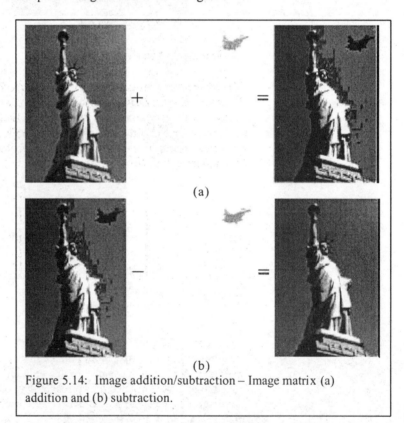

(a)

(b)

Figure 5.14: Image addition/subtraction – Image matrix (a) addition and (b) subtraction.

A more complex operation involves multiplication of two matrices resulting in a third matrix. However, in order for the matrix multiplication to be defined, the two input matrices must satisfy certain conditions. If **A** is one matrix with **M** rows and **N** columns and **B** is the other matrix with **P** rows and **Q** columns, then the matrix multiplication operation **A*B** is defined **IF**

AND ONLY IF the number of columns of matrix **A (N)** is equal to number of rows of matrix **B (P)**. Unlike the multiplication of two numbers (scalars), **A*B** is **NOT** the same as **B*A**. The matrix multiplication operation **B*A** is defined **IF AND ONLY IF** the columns of **B (Q)** and the rows of **A (M)** have to be equal. We will now define the operation of matrix multiplication in terms of matrix elements.

$$C = A * B \qquad\qquad 5.6(a)$$

$$C(i,j) = A(i,1)*B(1,j) + A(i,2)*B(2,j)+....$$

$$A(i,N)*B(N,j)$$

Where i = 1 to M, j = 1 to Q

We can see that the output matrix **C** will have **M** rows (same as input matrix **A**) and **Q** columns (same as input matrix **B**).

The matrix multiplication **B*A** is defined as follows:

$$D = B * A \qquad\qquad 5.6(b)$$

$$D(i,j) = B(i,1)*A(1,j) + B(i,2)*A(2,j)+....$$

$$B(i,M)*A(M,j)$$

Where i = 1 to P, j = 1 to N

We can see that the output matrix **D** will have **P** rows (same as input matrix **B**) and **N** columns (same as input matrix **A**). In general, each element of the output matrix is generated by the multiplication of appropriate elements of the input matrices and then summing the products. We show below a numerical example of multiplying the 4x3 matrix A shown in equation 5.1 by a 3x3 matrix **B** shown below.

$$B = \begin{vmatrix} 2 & 4 & 1 \\ 0 & 9 & 7 \\ -2 & 8 & 2 \end{vmatrix}$$

$$A = \begin{vmatrix} 12 & 9 & 0 \\ 5 & 6 & 23 \\ 8 & 16 & -9 \\ -13 & -12 & 0 \end{vmatrix}$$

$$C = A*B \qquad\qquad 5.7$$

$$C(1,2) = A(1,1)*B(1,2) + A(1,2)*B(2,2) + A(1,3)*B(3,2)$$

$$= 12*4 + 9*9 + 0*8$$
$$= 48 + 81$$

$$= \quad 129$$

The full 4x3 matrix C is shown below:

$$
C \quad = \quad
\begin{vmatrix}
24 & 129 & 75 \\
-26 & 258 & 93 \\
34 & 104 & 102 \\
-26 & -160 & -97
\end{vmatrix}
$$

The operation we defined above is the matrix multiplication. Another simpler operation can be defined between two matrices, which can be termed "matrix masking". The requirements for matrix operation are same as those for matrix addition – both matrices must have the same number of rows and columns. When this condition is satisfied, then matrix masking between matrix **A** and matrix **B** is defined as follows:

C = A ** B
C(i,j) = A(i,j) ** B(i,j)   5.8

In other words, the output matrix elements are direct multiplication between corresponding elements of the two input matrices. The term masking becomes relevant when one of the matrices has elements that are only 0 or 1. In that case those elements of an arbitrary matrix that coincide with zeros in the "mask" matrix will be blocked from being passed on to the output. In Figure 5.15 we show an example of the input matrix, the mask matrix and the output matrix for 8x8 size matrices and corresponding gray scale images

0	0	1	2	2	1	0	9
1	5	8	0	9	15	9	1
15	8	9	10	11	6	10	11
12	13	9	9	8	0	10	12
18	8	8	15	0	0	0	0
8	8	12	0	0	0	0	0
0	0	12	12	12	0	0	0
9	9	9	9	9	9	9	9

(a)

0	0	0	0	0	0	0	0
0	0	0	0	0	0	0	0
0	0	1	1	1	1	0	0
0	1	1	1	1	0	0	0
1	1	1	1	0	0	0	0
0	0	0	0	0	0	0	0
0	0	0	0	0	0	0	0
0	0	0	0	0	0	0	0

(b)

0	0	0	0	0	0	0	0
0	0	0	0	0	0	0	0
0	0	9	10	11	6	0	0
0	13	9	9	8	0	0	0
18	8	8	15	0	0	0	0
0	0	0	0	0	0	0	0
0	0	0	0	0	0	0	0
0	0	0	0	0	0	0	0

(c)

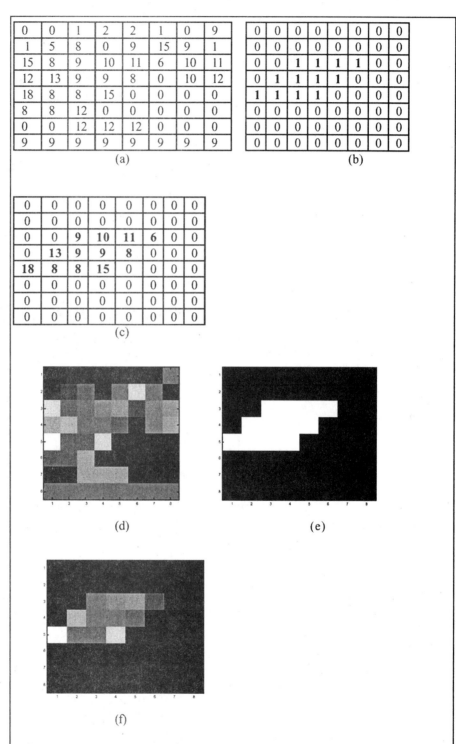

(d)

(e)

(f)

Figure 5.15(a) Input 8x8 matrix, (b) Binary 8x8 mask matrix, (c) Output matrix after the masking operation . (d) Gray scale image corresponding to the input matrix (e) Gray scale image corresponding to the mask matrix and (f) Gray scale image corresponding to the output matrix after the masking operation.

Exercise 5.2

1. Construct a 4x4 matrix where the elements are given by the sum of its indices. An element in second row and third column will be 5.

2. Construct another 4x4 where the elements are given by a product of its indices. The same element in second row and third column will be 6

3. Multiply the second matrix by Matrix by 2 and add to the first matrix.

4. A particular display has 4:5 aspect ratio between number of rows and number of columns. If the display has 1,000 rows, how many columns will it have? What will be the total number of elements in this display?

5. With the same set of matrices, now evaluate the matrix product B*A = D. Comment on the resultant matrix.

6. Take the previously calculated matrices C and D and evaluate C*A = E, and A*D = F. Compare matrices F and E. How are the matrices E and F related to the original matrix B? Is there a formal name for this operations? What are the informal names for these operations?

7. Construct a 4x4 matrix (A) with only the following elements "1" and all others "0": A(1,2), A(2,3), A(3,4). Construct another arbitrary 4x4 matrix (B) with elements between 0 and 9. Evaluate product A * B = C. Comment on the nature of C.

8. Take the same two matrices A and B and now evaluate the product B * A = D. Comment on the nature of D.

9. Modify the element of matrix A to make A(4,1) also "1" in addition to the three elements given in the previous problem. Call this matrix A'. Evaluate A' * B = C'. Compare C with C'. Comment.

10. Take the new matrix A' and evaluate B * A' = D'. Compare D' with D. Comment.

11. Construct a 4x4 matrix **P** that has all zero elements except for the following elements, which are "1": **P(1,2), P(2,3), P(3,4), P(4,1).** Multiply an arbitray 4x4 matrix by P. Comment on the output.

12. Construct an 8x8 "mask" matrix that passes the central 4x4 section of an arbitrary 8x8 input matrix. Verify the operation of masking.

13. A particular display has 4:5 aspect ratio between number of rows and number of columns. If the display has 1,000 rows, how many columns will it have? What will be the total number of elements in this display?

5.2.2 Changing Values of the Matrix Elements

The elements of a matrix can be transformed by other simple operations as well. For example, each element of a matrix can be squared to calculate a new matrix. Other functions (for example, logarithm) can also be used to calculate elements of the output matrix from the input matrix. The elements of the input matrix can be compared to a fixed number and be converted into a "0" if the element is less than the fixed number or into of greater than the fixed number. This particular operation is called "thresholding" and is commonly used to convert an image with multiple shades of gray values (using many bits per pixel) into a binary image, which requires only one bit per pixel. In its most general form, the new matrix value may be calculated by a method known as "look-up table". The table simply consists of two columns – the first column lists the input value and second column lists the corresponding output value. Below we give example of such a look up table where the inputs take on values from 0 to 7.

Table5.1: Examples of three different operations performed on the elements of an image matrix and the resulting output images.

Input	(Input)^2	Threshold at 5	Arbitrary
0	0	0	0
1	1	0	1
2	4	0	3
3	9	0	5
4	16	0	7
5	25	1	2
6	36	1	4
7	49	1	6

Such a mapping can also be represented by a graph that plots output intensity on Y-axis and input intensity on the X-axis.

Input-Output Relation

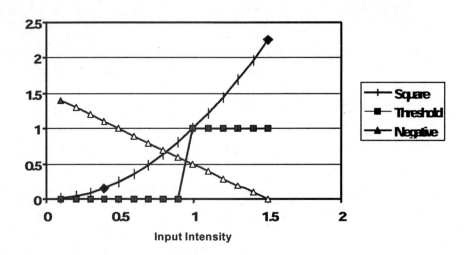

Figure 5.14(a): Input-out intensity mapping plotted as graph.

Figure 5.17 shows the input image in the left column, the input-output mapping graph in the central column and the output image in the right column. The effect of different nonlinear mappings can be seen quite clearly.

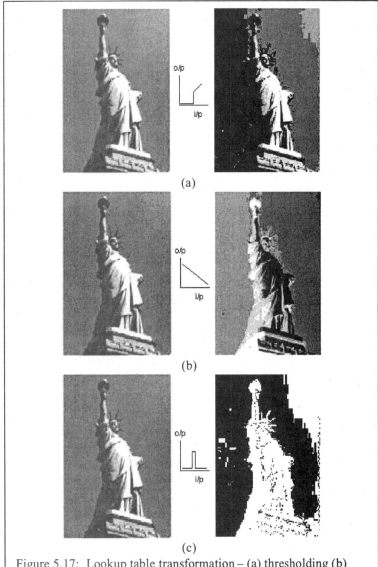

Figure 5.17: Lookup table transformation – (a) thresholding (b) contrast reversal, creating a negative image and (c)level slicing.

Exercise 5.3

1. Construct a look up table for input elements varying from 0 to 7 that will lead to a "negative" image (bright => dark, dark => bright)

2. Construct an arbitrary 4x4 matrix with elements ranging from 0 to 7. Apply the look up table created above to calculate a new matrix.

3. Add the two matrices. What do you get? Explain.

4. Construct a nonlinear mapping table where the input intensity varies from 0-9 in integer steps (10 levels) the output is square root of the input intensity. Plot this in a graph. Compare this graph to the graph where the output intensity was equal to the square of the input intensity. Comment.

5.2.3 Neighborhood Operations

Thus far the operations we introduced were one-to-one, meaning that one matrix element in the input matrix was converted according to a given rule into the corresponding element of the output matrix. Now we will describe operations where a set of elements in the input matrix decides the value of a single element in the output matrix. The set of elements in the input matrix are usually adjacent to each other, hence the name "neighborhood operation".

One of the most common operations performed on a set of numbers is to find a mean (average) value of the set. The mathematical operation of calculating an average value for a set of numbers is used to find out the central tendency of a data set. For example, the average grade for the class gives an idea of the overall performance of the students. The averaging operation is defined in equation 5.9(a) below:

$$\text{Data set} = (5, 9, 10, 7, 8, 4, 4, 3, 0, 9)$$

$$\text{Average} = (5 + 9 + 10 + 7 + 8 + 4 + 4 + 3 + 0 + 9) / 10$$

$$= 5.9 \qquad\qquad \text{5.5a}$$

Now let us consider the general case of a data set consists of a sequence of N numbers indicated by "x_i" where "i" is the running index for an element in the sequence. We now introduce the following notation that indicates a summation over all elements:

$$Sum = \sum_{i=1}^{N} x_i$$
$$Sum = x_1 + x_2 + x_3 + \ldots + x_N \qquad\qquad (5.9b)$$

$$Average = \frac{1}{N} \sum_{i=1}^{N} x_i$$

We can perform this operation on elements of a matrix that represent a digital image. The most common neighborhood utilized is that containing a 3x3 sub array surrounding the center pixel. This is illustrated in an example below:

0	1	2
6	15	7
9	2	3

The average value of all nine elements in this matrix is:

Average = 1/9 * (0 + 1 + 2 + 6 + 15 + 7 + 9 + 2 + 3)

= 5

In a neighborhood operation, the center element (which in this case is **15**) is replaced by the average value of the nine elements that surround the central element, i.e., **5**. The size of the neighborhood is somewhat arbitrary and it would be 5x5 or 7x7 or 9x9. Having a sub-array that is odd rows and columns allows us to find a central element. It can be seen how this operation in fundamentally different than the earlier point operations on an image where the new value of an element depended on that element only.

The average value is obtained by performing simple operations of addition and division. Another common function that is used to finding central, tendency of a group is the "median". A median of a group is the value where half of the elements in the group have a lower value and half of them have a higher value. In order to find the median of a group, we need to arrange the elements in an increasing order. Then one can find the value of the point that is exactly in the middle of the data set. In case the data set is even, then the middle value of the two points on either side of the center is set to be the median. By and large, a median value is one of the numbers in the data set whereas in most situations, none of the elements will have the same value as the mean. A specific example of a median operation on the same data set with 9 elements is given below. First we arrange the elements in an increasing order and then we pick the element in the middle.

0   1   2   2   **3**   6   7   9   15   5.10

In this case, the median of the group has a value "3". The central element of the neighborhood (with a value of 15) is now replaced by "3".

Finding the minimum or the maximum number form a data set is another operation that involves neighborhoods of a pixel. In the same 3x3 neighborhood example we have been discussing, we can see that the maximum value is 15 and minimum value is 0. Sometimes we may want to replace the central pixel by a minimum (or maximum) value in the neighborhood.

Figure 5.18 summarizes these neighborhood operations on a 3x3 sub-array within an image. Besides the numbers associated with the array, a gray level representation is also provided for reference.

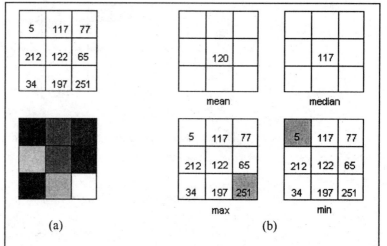

Figure 5.18: Neighborhood operations – (a) matrix and pixel shades and (b) various neighborhood operations on the image matrix.

Exercise 5.4

1. Create a data set consisting of the age (in years) of all students in the class. Calculate the mean. Calculate the median.

2. In the example discussed above, the mean and the median were significantly different. Construct two data sets of 9 numbers each (between 0 and 20).

   a. Construct one such that the mean and the median are almost equal when rounded to the nearest integer.

   b. Construct another set such that the median and mean are far apart. Comment on the distribution of values within each data set.

3. Construct a nonlinear mapping table where the input intensity varies from 0-9 in integer steps (10 levels) the output is square root of the input intensity. Plot this in a graph. Compare this graph to the graph where the output intensity was equal to the square of the input intensity. Comment.

Section 5.3: Preprocessing of Digital Images

In Section 5.1 we discussed several examples of Digital Image Processing Applications. In this section we will describe in some detail a subset of applications from there. These applications are rather simple and consist of operations that make the image "better". We put the word "better" in quotes to indicate that is a highly subjective term and will strongly depend on what we want to do with the image. These operations are often called "preprocessing" since the output of such operations is still a digital image. In contrast, the applications discussed in the next section contain outputs that are quite different from the digital images they started with.

5.3.1 Noise Removal

Noise is a term that is very hard to define. In a broad sense we can define it as "a signal that is not desired by a specific user at that time". This leaves open the possibility that a different user may consider different parts of the image as "undesirable" or that the same user at a different time and for different use may consider different parts of the image as "undesirable". In most of the situations, we have only a general idea about the nature of the objects in an image (after all if we knew exactly what the image was going to look like, why bother recording it?). So given a particular feature in an image, we never know with certainty whether that feature is an inherent part of the image or is an undesirable artifact and hence should be considered as noise. In this chapter, we will simply define certain properties of an image as undesirable under most circumstances while allowing that under special circumstances they may be considered as a part of the image to be recorded.

One common type of noise that is encountered in imaging systems is termed impulsive noise. It consists bright spots that are randomly distributed throughout the image. We can model this type of noise as "additive noise" and is illustrated in Figure 5.5.

We just noted in the previous paragraph that we don't know the precise nature of the image we are viewing. In addition to the uncertainty in the desired image, there is also an uncertainty in the additive noise (that's why we call it "random noise"!). But we do know that the expected intensity distribution of the desired image and the noise is significantly different. If it were not to be the case, the desired image and the noise would appear substantially similar and the operation of noise reduction would be difficult at best and meaningless at worst. So we may not know exactly what constitutes noise and what constitutes signal but we know which pixel (or group of pixels) does not fit in with the overall structure of the desired image. For example, we want to remove sand and dirt that is mixed in with a bag of kidney beans. We know that the grains of sand may vary in size and so would the kidney beans. Nonetheless, we know that the smallest kidney beans are still larger than the largest particles of sand. So we design a sieve where the holes are smaller than the smallest kidney beans but larger than the largest sand particles. When we pass the bag of beans through the sieve, the sand will go through leaving kidney beans above. Similarly, we know that in most images, the intensity varies smoothly at a pixel level and that there are no abrupt changes. We also know that a noise image contains isolated pixels or groups of pixels of same intensity and hence contain sharp and random variation in intensity. We would like to design an ideal "image sieve" that will filter out the noise without affecting the desired image in any way. In other words, we would like to smooth out the random and sharp variations produced by the noise signal.

In the previous section we had introduced the idea of replacing the central pixel of a sub-array of pixels by the average value of all the pixels in the sub-array. We first generalize this to the operation of "moving average" of a one-dimensional sequence of numbers. Such a sequence is typically generated when a one-dimensional time signal (such as sound) is digitized into a sequence of numbers. Here, instead of performing the average over the entire sequence, we replace each number in the sequence by an average of a few numbers in the immediate vicinity of that number. For example, the 15th number is a sequence will be replaced by an average of the 14th, 15th and 16th number in the sequence (one number before, the number, and one number after). One can imagine the sequence being written down on a strip of paper and then placing a "window" cut in a cardboard on top of the strip. Only a few numbers in the sequence will be seen through the window. We average all the numbers visible through the window and replace the central number with the average. In the previous example the window was large enough to show three numbers in the sequence. Now if we move the center of the window along the original sequence one number at a time, we will create a new sequence that will be a "filtered" version of the original sequence. This operation is indicated by equation 5.11 given below:

$$x'_i = (x_{i-1} + x_i + x_{i+1}) / 3 \qquad\qquad 5.11$$

x'_i is the ith element of the filtered output sequence which replaces the element x_i of the input sequence.

This operation runs into special situation at the end of the sequences. Since there is no element before the first element and no element after the last element, it is difficult to use a 3-element window for averaging at the beginning and the end of the sequence. This is handled as follows: the first element is replaced by the average of that element and the second element (when the window size is 3) and the last element is replaced by an average of the last element and the element before it. This can be achieved by assuming that a series of zeros are added to the sequence before and after. This operation is called zero-padding and serves to achieve the output sequence the same length as the input sequence. Equation 5.12 shows the output sequence when input sequence is filtered through a 3-element window (the average is rounded to nearest integer).

input = (8 8 8 15 8 2 2 15 2)

output = (5 8 10 10 8 4 6 6 6)   5.12

One can see qualitatively that the new data set has lot less variability than the original one. For example, the 4th element that was 15 in the input is now 10. Similarly the 8th element that was 15 is now 6.

Digitized images are represented by matrices, which are nothing but a 2-D array of numbers. Extending the concept of a moving average filter to digital images, we need to extend to two dimensions, the neighborhood over which the averaging operation takes place. So a 3x3 window will include an element in its center and 8 elements surrounding it that are immediately adjacent to the central element. Now we replace the central element by an average of 9 elements within the window. This operation can be cast in terms of the matrix multiplication between the matrix representing the input image and a special matrix that performs the operation of averaging.

It turns out that a matrix can be designed to perform averaging operation separately along the rows columns of a matrix. Let us consider an 8x8 matrix **A** shown below:

1/2	1/2	0	0	0	0	0	0
1/3	1/3	1/3	0	0	0	0	0
0	1/3	1/3	1/3	0	0	0	0
0	0	1/3	1/3	1/3	0	0	0
0	0	0	1/3	1/3	1/3	0	0
0	0	0	0	1/3	1/3	1/3	0
0	0	0	0	0	1/3	1/3	1/3
0	0	0	0	0	0	1/2	1/2

Now consider an 8x8 input matrix **B** that we considered before.

0	0	1	2	2	1	0	9
1	5	8	0	9	15	9	1
15	8	9	10	11	6	10	11
12	13	9	9	8	0	10	12
18	8	8	15	0	0	0	0
8	8	12	0	0	0	0	0
0	0	12	12	12	0	0	0
9	9	9	9	9	9	9	9

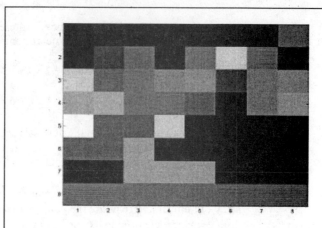

Figure 5.19: Gray scale image corresponding to the matrix **B** above

Since both matrices are 8x8, the condition for matrix multiplication **A*B** is satisfied and the resultant matrix **C** will also be 8x8. Since most of the elements of matrix **A** are zero, the summation terms for each element of matrix **C** will be very few. Below, we will show sample calculations for three matrix elements of **C**.

C(1,1) =

A(1,1)*B(1,1) + A(1,2)*B(2,1) + A(1,3)*B(3,1) + A(1,4)*B(4,1) +
A(1,5)*B(5,1) + A(1,6)*B(6,1) + A(1,7)*B(7,1) + A(1,8)*B(8,1)

5.13(a)

= 1/2*0 +1/2*1 = 1/2

C(2,5) =

A(2,1)*B(1,5) + A(2,2)*B(2,5) + A(2,3)*B(3,5) + A(2,4)*B(4,5) +
A(2,5)*B(5,5) + A(2,6)*B(6,5) + A(2,7)*B(7,5) + A(2,8)*B(8,5)

5.13(b)

= 1/3*2 + 1/3*9 + 1/3*11 = 7 1/3

C(4,6) =

A(4,1)*B(1,6) + A(4,2)*B(2,6) + A(4,3)*B(3,6) + A(4,4)*B(4,6) +
A(4,5)*B(5,6) + A(4,6)*B(6,6) + A(4,7)*B(7,6) + A(4,8)*B(8,6)

5.13(c)

= 1/3*6 + 1/3*0 + 1/3*0 = 2

The full 8x8 output matrix **C** and its gray scale representation is shown in Figure 5.20. It is clear from the previous equations that with the exception

30

of the elements of **C** along the top and the bottom row, the rest of the elements correspond to an average value of three adjacent elements along the column.

0.5	2.5	4.5	1	5.5	8	4.5	5
5.33	4.33	6	4	7.33	7.33	6.33	7
9.33	8.66	8.66	6.33	9.33	7	9.66	8
15	9.66	8.66	11.33	6.33	2	6.66	7.66
12.66	9.66	9.66	8	2.66	0	3.33	4
8.66	5.33	10.66	9	4	0	0	0
5.66	5.66	11	7	7	3	3	3
4.5	4.5	105	10.5	10.5	4.5	4.5	4.5

(a)

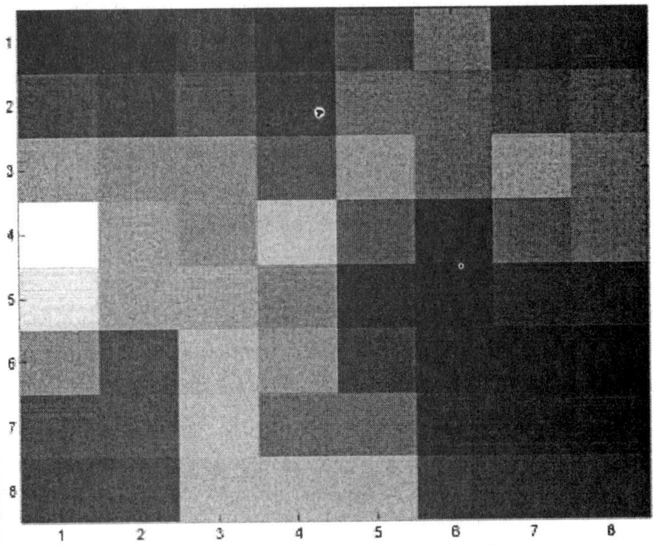

(b)

Figure 5.20: (a) Matrix C and (b) Gray scale image corresponding to matrix C

The operation of averaging along rows can be similarly carried out using another matrix, **A'**, that is shown below. We note that the matrix is almost identical except the elements in the corner. The element **A(1,2)** is exchanged with element **A(2,1)** and similarly element **A(8,7)** is exchanged with element **A(7,8)** to create the new matrix **A'**. Rest of the elements of the two matrices are identical to each other. The output matrix D is calculated by performing the matrix multiplication **B*A'**. The calculation of the elements of matrix D is carried out in a similar manner as before.

D(1,1) =

B(1,1)*A'(1,1) + B(1,2)*A'(2,1) + B(1,3)*A'(3,1) + B(1,4)*A'(4,1) +
B(1,5)*A'(5,1) + B(1,6)*A'(6,1) + B(1,7)*A'(7,1) + B(1,8)*A'(8,1)

5.14(a)

= 1/2*0 +1/2*0 = 0

D(2,5) =

B(2,1)*A'(1,5) + B(2,2)*A'(2,5) + B(2,3)*A'(3,5) + B(2,4)*A'(4,5) +
B(2,5)*A'(5,5) + B(2,6)*A'(6,5) + B(2,7)*A'(7,5) + B(2,8)*A'(8,5)

5.14(b)

= 1/3*0 + 1/3*9 + 1/3*15 = 8

D(4,6) =

B(4,1)*A'(1,6) + B(4,2)*A'(2,6) + B(4,3)*A'(3,6) + B(4,4)*A'(4,6) +
B(4,5)*A'(5,6) + B(4,6)*A'(6,6) + B(4,7)*A'(7,6) + B(4,8)*A'(8,6)

5.14(c)

= 1/3*8 + 1/3*0 + 1/3*10 = 6

Figure 5.21 shows the output matrix D and its gray scale representation as an 8x8 image.

1/2	1/3	0	0	0	0	0	0
1/2	1/3	1/3	0	0	0	0	0
0	1/3	1/3	1/3	0	0	0	0
0	0	1/3	1/3	1/3	0	0	0
0	0	0	1/3	1/3	1/3	0	0
0	0	0	0	1/3	1/3	1/3	0
0	0	0	0	0	1/3	1/3	1/2
0	0	0	0	0	0	1/3	1/2

(a)

0	0.33	1	1.66	1.66	1	1.33	4.5
3	4.66	4.33	5.66	8	11	8.33	5
11.5	10.66	9	10	9	9	9	10.5
12.5	11.33	10.33	8.66	5.66	6	7.33	11
13	11.33	10.33	7.66	5	0	0	0
8	9.33	6.66	4	0	0	0	0
0	4	8	12	8	4	0	0
9	9	9	9	9	9	9	9

(b)

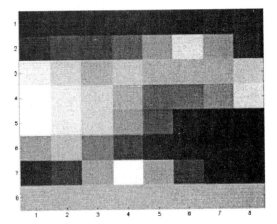

(c)

Figure 5.21: (a) Matrix A' (b) Output matrix D and (c) The gray scale image corresponding to matrix D

Figure 5.22 shows the use of such a moving average filtering in removing impulsive noise. Figure 5.22(a) shows the original image, 5.22(b) shows the noisy image and 5.22(c) shows the output image after being filtered by a 3x3 neighborhood average filter. The noise reduction is obvious. But it is also seen that the image has been degraded with respect to the original image. The sharp edges are blurred and some small features are lost. This artifact is evident in the simple 1-D example we discussed in equation 5.8. The original signal consists of a step of 5 elements with value 8, followed by 5 elements with value of 2. On top of that are added noise elements with value 15 in position 4 and 8. The output sequence has a much reduced noise part, but then it also has lost the abrupt step between values 8 and 2. This trade-off between the noise reduction and edge blurring is an unavoidable feature of the linear filtering operation. The extent of blurring will depend on the size of the neighborhood. But then the reduction of noise will also improve with increase of neighborhood size, since the extraneously large value now contributes over a much larger set of pixels instead of to a single pixel (in the noisy image case).

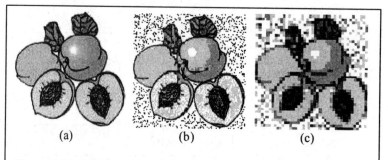

(a)　　　　　(b)　　　　　(c)

Figure 5.22 : Noise removal by a linear filter – (a) original image, (b) noisy version and (c)a cleaned version using the linear averaging filter.

The idea behind such filtering is to ensure that the image is smoothly varying by taking any "odd" elements (large spikes) and distributing their effects over a larger area. Each pixel is replaced by a value that represents the central tendency of the group of pixels surrounding the central pixel. In the previous section, we discussed another function that determines the central tendency of a set – the median operation. We take the previous formulation of the "moving average" filter and replace it with a "moving median" filter. We start with the same input sequence as shown in equation 5.8 and now perform a median operation on three elements (one before, the element, and the one after) and replace the middle element with the median of the group. the results are as described below:

input = (8 8 8 15 8 2 2 15 2)

output = (8 8 8 8 8 2 2 2 8)   **5.15**

One can see in this example that with the exception of the last element, the ideal sequence of five elements with a value of 8 followed by five elements with a value of 2 is recreated exactly. The spikes of value 15 are suppressed without losing the sharpness of the edge. This is a rather simplistic example. However, in practice we do find some loss of details in the image, though not as severe as in the case of the moving average filter. Figure 5.23 shows the operation of median filter in reducing impulsive noise.

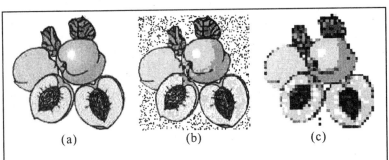

Figure 5.23: The same noise removal example but with a median filter – (a) original image, (b) noisy version and (c) a cleaned image using a median filter.

In both operations, the size of the window affects both the distortions in the output image and the ability to reduce noise. Obviously the window should be larger that the contiguous group of pixels that correspond to noise. On the other hand, any detail finer than the size of the window will be lost in the final image, thus making the degradation worse for larger size windows.

VAB Experiment 5.3: Median Filtering for Noise Reduction

• Implement the mean filter operation on images with varying amount of noise. Observe the detailed structure of the output image. Verify the statements in the section about the effects of increasing window size.

• Repeat the above experiment for median filter.

5.3.2 Image Segmentation

The image of a typical scene contains many different objects. For example, a picture taken of a classroom will contain several students, desks, books, and chairs. In addition, there is the wall in the background. Many times in automated image analysis, it is desirable to study the image one object at a time. It is also desirable to perform this operation without any manual intervention. In the previous section on elementary operations, we made a reference to using thresholding operation to isolate bright objects from somewhat darker background. In this section we will discuss it a little further. In particular, we will discuss two different types of thresholding operations – analog thresholding and binary thresholding. In the analog thresholding operation, all pixels in the image that have a value less than the threshold are set to 0. On the other hand, once the value of a pixel exceeds the pre-selected threshold, it is left unchanged. As a result, the gray value variations within the bright objects are preserved. In binary threshold, the pixels with values less than the threshold are similarly set to 0 but the pixels that exceed the threshold value are all set to 1 regardless of their actual gray value. This creates a binary image (1 bit per pixel) which does not preserve the internal structure of the object that is selected for analysis. Such a binary thresholding operation is desirable when the selected object is primarily analyzed for its shape and size. The internal details of the object are at best extraneous and at worst distracting. Figure 5.24 shows these two operations and the difference in their outputs.

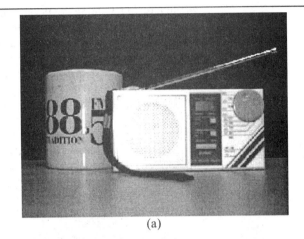

(a)

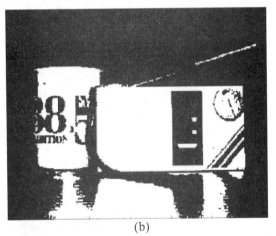

(b)

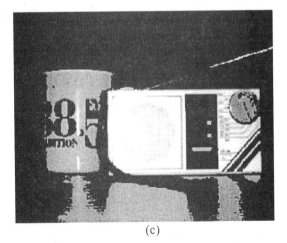

(c)

Figure 5.24: Analog and binary thresholding – (a) Original
Image (b) Image after binary thresholding and (b) Image after
analog thresholding.

37

Additional operation that was discussed earlier was called "level slicing" where only pixels that had the pre-selected gray value were set to an output value of 1 while the rest of the pixels were set to 0. The resultant image is also binary and hence this operation is useful for shape and size analysis. Figure 5.23 shows examples of such a level slicing operation. In some cases a band of gray values adjacent to the desired value are chosen instead of a single gray value. The pixel gray value and the width of the band are two parameters that are under designer control.

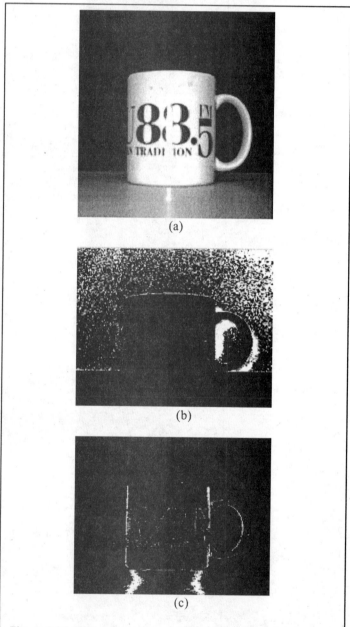

(a)

(b)

(c)

Figure 5.25: Examples of level slicing operation where the band is set to two different values – (a) Original image (b) Original image after level slicing for a higher pixel level and (c) Original image after level slicing for lower pixel value.

5.3.3 Image Blurring and Deblurring

In section 5.3.1 we discussed the impulsive noise that is added to an ideal image corrupting it. We also have other ways in which an ideal image can be affected. One that is familiar to all of us is the "out-of-focus" picture. The universal availability of auto-focus on cameras has largely eliminated the main object being out of focus. But the surrounding objects that are at a different distance do appear blurry. In this section we will study the mathematical formulation of the blurring process and then discuss some potential ways pictures can be sharpened digitally.

Image blurring and sharpening both are based on a central concept called the "Point Spread Function" (or PSF). Imagine that we have an object that when imaged, will ideally correspond to a single pixel. Such an object is (quite naturally) called a point object. When the image acquisition system has errors, the point object does not image to a single pixel but is spread over many pixels. This particular image of a point object is called the "Point Spread Function". One simplifying assumption we will make is that when the point object is shifted in the scene, the PSF simply shifts in the image in a corresponding manner. With this simplifying assumption, we can predict the output of a non-ideal imaging system for *any input* once we know its PSF. For simplicity, we will confine our discussions to black-and-white images, though the treatment extends in a straightforward manner to color images as well.

An arbitrary input can be considered to be a collection of point objects at various locations in the input, each having a specific value for its intensity. In ideal imaging case, these point objects with a given intensity are mapped onto image pixels with corresponding gray value. In non-ideal imaging system, each point object is mapped onto a PSF that is located at corresponding location in the image and that is multiplied by the intensity value of the input point object. The whole image is constructed as a weighted and shifted sum of the PSF. Let us consider a simple 8x8 input and a 3x3 PSF for the non-ideal imaging system. The input has just 4 point objects as shown. The output will consist of a summation of 4 PSFs that are centered at the locations corresponding to the locations of the point objects in the input, each weighted by the intensity value of the point object.

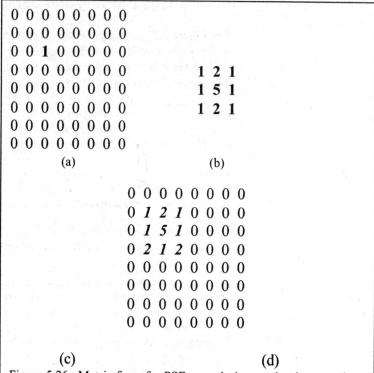

Figure 5.26 : Matrix form for PSF convolution , a simple example - (a) the original image matrix, (b) the PSF mask, (c) the output image where the (3,3) matrix element in the input is replaces by the 3x3 PSF mask in (b)

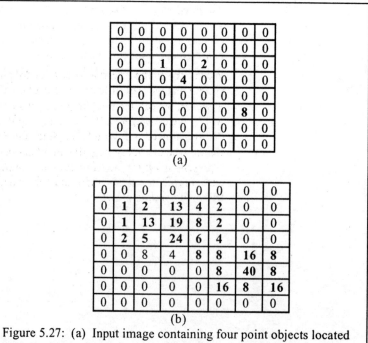

Figure 5.27: (a) Input image containing four point objects located at (3,3), (3,5), (4,4) and (6,7) respectively. Their brightness is as indicated. (b) Output image when the PSF shown in Figure 5.26(a) is applied to the image in (a).

The Figure 5.27 contained input image with very few points and was mostly 0. That was done for the sake of simplicity. But the same principle applies when a more complex image is involved. Figure 5.28 shows an image that was blurred by the PSF shown in Figure 5.26(a).

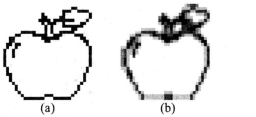

(a)   (b)

Figure 5.28: Blurring from the PSF in previous figure – (a) original 32x32 pixel image and (b) the blurred version.

```
... 0 0 0 0 0 0 ...        0  0  0  0  0  0
... 0 0 0 0 0 0 ...        0  0  1  3  4  3
... 0 0 0 1 1 1 ...        0  1  3  7  7  6
... 0 0 1 0 0 0 ...        1  3  7  4  4  3
... 0 1 0 0 0 0 ...        2  8  4  1  0  0
... 0 1 0 0 0 0 ...        2  7  2  0  0  0
          (a)                        (b)
```

Figure 5.29: Matrix form of PSF convolution for Figure 5.26. The same PSF mask as in Figure 5.22 is used – (a) a part of the image matrix and (b) its PSF.

We discussed the PSF that arises due to imperfections in the imaging system that we had little or no control over. Once the image is digitized, however, we can manipulate it in a much more flexible manner. In other words, we can *define* a PSF in the computer and perform the same operation that we showed in Figure 5.26. Since the PSF is defined in the computer, we are free to choose negative numbers in the PSF since it does not represent any light intensity but just numbers in our computer. This particular freedom has important consequences, as illustrated with the following simple example.

We will consider a simple one-dimensional sequence of numbers to introduce the concept of linear filtering. Equation 5.16 shows a sequence of numbers that we wish to process:

$$0\ 0\ 0\ 0\ 1\ 1\ 1\ 1\ 1\ 1\ 0\ 0 \qquad\qquad 5.16$$

The PSF for processing this consists of three elements as shown below:

$$-1\ 3\ -1 \qquad\qquad 5.16(a)$$

Now we perform the processing in the same manner as described before: we replace each point in the original input by the three-point pattern shown above and then sum the results. Another way to describe this operation is as follows:

- Take the original sequence, multiply it by the first number (-1)

- Take the original sequence, multiply it by the second number (3) and shift it one element to the right

- Take the original sequence, multiply it by the third number (-1) and shift it two elements to the right.

- Add all the three sequences together to create a new sequence that is now 14 elements long (two elements longer than the input sequence).

- Select the central 12 elements of the sequence and create the output sequence.

Let us now apply the algorithm to the specific example.

First sequence	0	0	0	0	-1	-1	-1	-1	-1	-1	0	0	
Second Sequence		0	0	0	0	3	3	3	3	3	3	0	0
Third sequence			0	0	0	0	-1	-1	-1	-1	-1	-1	0	0
Sum sequence	0	0	0	0	-1	2	1	1	1	1	2	-1	0	0
Output sequence		0	0	0	-1	2	1	1	1	1	2	-1	0	

The output sequence has negative numbers. If we simply suppress the negative numbers by setting them to zero, we get a new sequence where the edges of the pattern are made brighter (value of 2). The input and output sequences are shown plotted together in Figure 5.30.

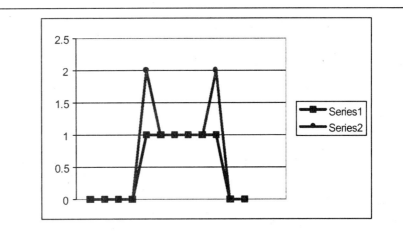

Figure 5.30: Input and output sequences plotted as curves. The edges of the pattern are made brighter by the processing operation.

This type of processing can be used with images, which are after all simply two-dimensional array of numbers represented by matrices. We can use a bipolar PSF (one containing negative numbers) to process the blurry input image like the one shown in Figure 5.27 and make it sharper. Figure 5.31 shows the blurry image when processed using this kind of center-positive-surround negative PSF.

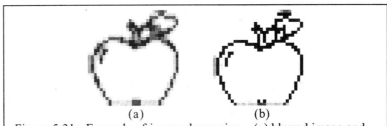
(a)   (b)

Figure 5.31: Example of image sharpening – (a) blurred image and (b) sharpened image.

The previous PSFs discussed are but two examples of a very rich area of research and applications. A clever selection of PSF sizes, shapes and values can implement very interesting image processing operations with very simple hardware. Indeed such processing forms the heart of many digital image processing systems.

Exercise 5.5

1. Create an 8x8 matrix **A** that contains a border of zeros 2 rows on top and bottom and two columns on left and right end. The middle 4x4 sub matrix is uniformly bright with a value of 15. Now create another matrix by shifting the 4x4 sub matrix by one column to the right within the 8x8 matrix. Call this new matrix **B**. Subtract B from A. Make all

elements of the new matrix **C** positive by keeping only the magnitude of any negative elements. Interpret the matrix **C**.

2. Shift the 4x4 sub matrix by one row up and repeat the same operation calling the new matrix **D**. Interpret matrix **D**.

3. Shift the 4x4 sub matrix by one column to the right and one row up and repeat the same operation calling the new matrix **E**. Interpret matrix **E**.

4. Add the two matrices **C** and **D** together. Compare the new matrix **F** with **E**. Interpret the results.

5. Consider the following 1-D sequence: 0 0 1 2 3 4 3 2 1 0 0. Create another sequence by shifting the original sequence one element to the right. Make sure and add one element "0" to the left of the new sequence and drop the right most element "0" in the shifted sequence to maintain the same length for both sequences. Now subtract the two sequences. Plot the original and the new sequence. What do you get? Comment on the output sequence.

6. Take the output sequence from the previous problem and apply the same procedure as discussed in that problem. Plot all the three sequences on the same graph. Comment on the relation between the three sequences.

7. **CHALLENGE ASSIGNMENT:** Consider the fourth assignment in Exercise 5.2. Using that assignment as a basis, create an edge detection operator via elementary operations like matrix multiplication and addition / subtraction. Show the results on the 8x8 matrix A given above.

Section 5.4:
Putting the
Pieces Together

We started this chapter with a problem - designing the vision system for a robot that can clean up the house, do the laundry and allow you to go see a movie with your friends. In the previous section we described various operations that one can perform on a digital image. Now we will put the pieces together.

The robot we want to design must be capable of capturing a scene, say the living room, and identify objects laying on the carpet that need picking up. It should be smart enough so it does not try to "pick up" the design on the throw rug but should pick up the Lego blocks your kid brother left on the floor. It should have the capability of separating light clothes from dark clothes before putting them in the laundry. It should know better than to try and "gather" your cat that is scurrying around the room.

Let us try to describe our requirements in a more formal language that will allow a generalization of our design to other problems. Our needs are:

- Object identification and classification

- Detection and tracking of moving objects

We will immediately recognize that other problems such as designing smart missiles and designing a video security system will also benefit from developing approaches to solving these two problems.

5.4.1 Object Identification and Classification

A scene captured by a digital camera is usually very rich containing a complex background and many objects. The first task we need to perform is to separate the background from the objects. In the previous section we called this operation "segmentation". This operation assumes that we have some notion about the nature of the background and the nature of the objects we are trying to identify. We can use several different attributes individually or in combination to achieve this operation. Intensity, color and texture can be the image parameters that will allow us to separate background from the objects.

Once we isolate the objects from the background, the next task is to classify the objects into pre-determined groups (for example, toys, clothes, food). The classification can bee performed based on the following properties of the isolated objects:

- Brightness

- Size

- Shape

- Color

- Texture

An object may be completely characterized by one or many of these properties. Measuring these properties of an object is made possible by the digital representation of the image. Furthermore, the object properties we chose to focus on will depend on the other objects in the scene. For example consider the problem of trying to locate your friend on a hiking trail where the background consists of tress, earth, and sky and other objects include wild animals and markers. There you will try to look for an object with smooth texture corresponding to the windbreaker or a hat. On the other hand, locating the same friend in a crowded cafeteria will involve focusing on what color clothes the friend was wearing, the color of heir, height and finally facial features (when the "object" is close enough).

Coin Counter

We will now consider a very simple problem of capturing an image containing several coins, identifying quarters among them and calculating total number of quarters in the scene. A block diagram of such a system is shown in Figure 5.32

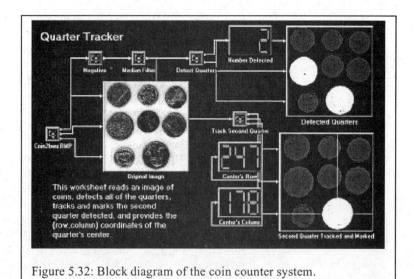

Figure 5.32: Block diagram of the coin counter system.

An image of the scene is first captured by the camera, and is then sampled and digitized by the computer. The image is found to be dark objects on a bright background. So a contrast reversal operation is performed on the image as discussed earlier in this chapter. Now the coins appear as bright objects on a dark background. The image is processed using median filter discussed earlier in the chapter to reduce noise in the background. Then it is subject to binary threshold. This generates a binary image where only the coins contain bright pixels with highest value (typically 255, for an 8-bit number) and everything else has a gray value of 0. The scene is now separated into disconnected objects that are identified with individual coins. Next we find that what distinguishes a quarter from a dime or a nickel is the size. One way to measure size is to count the number of pixels that are bright (value 1) in a given object. Once this operation is done, we will get a number for each coin that indicates the size. The largest number is seen to correspond to the largest coin, which in this case is a quarter. In the output only the largest coins are set to highest intensity value and other coins are assigned an intermediate intensity value. The number of objects that have the highest value now corresponds to the number of quarters in the scene. In addition, the center of the second quarter is determined and a cross hair is added to the center and the row and column index of the center of the second quarter is shown as an output.

How flexible is this system? Can we use this system to identify dimes instead of quarters? The answer is obviously "yes". The three coins have very distinct sizes and hence can be distinguished on the basis of size alone. Since the dime is the smallest coin, we can now set the size value to the lowest among the group and select only those objects that are the smallest. The system thus does provide the flexibility to adjust the size value to identify the particular coin.

The size will also vary depending on the distance between the camera and the scene. It is well known that as an object is moved farther away, its image becomes smaller. Since all the coins are moved away, all of them will get smaller and their relative sizes will still remain the same. We can handle the variability in distance by simply comparing the sizes of the coins or by first calibrating the system by putting only quarters in the scene and measuring their sizes. By taking an average of all the different quarters, we will have a reliable value for the size of a quarter from that distance. This value can then be used to compare the size of an unknown coin to this value to determine if it is a quarter.

The angle of the camera to the scene could change introducing distortions in the image. Instead of perfect circles, the coins will appear to be elongated oval shaped. In spite of this distortion, the relative size of the coins will not change. Since our object identification algorithm does not depend on the shape of the object but the size, the algorithm will continue to work.

If we change the background from light to dark, we will have bright objects. We can handle this situation by simply eliminating the contrast reversal block in our system and proceed as usual with the rest of the operations.

One possible variation that the system will *not* be able to accommodate is when two coins partially overlap. The system will then consider the two coins together as a single object and will come up with an erroneous answer for the size of the object.

Size is just one parameter of objects that can be used in classifying them. Other parameters are: color, shape, texture. Of these, color is easiest to use and you will be asked to design a system that uses color to provide further discrimination between different objects in the **VAB** laboratory that follows.

Sock Matcher:

A common problem encountered by *EVERYBODY* regardless of age or sex is to make sure that the two socks we wear match in color. Given that we perform this operation early in the morning when we are not fully awake and when the light in the room may be dim makes this task particularly challenging. One (obvious) solution is to buy socks all of the same color (white, black, dark blue). But we are learning to be engineers! We don't go for the "obvious" solution. So we apply our knowledge of digital image processing to design a sock matcher. Figure 5.33 shows a block diagram of the Sock Color Matcher system.

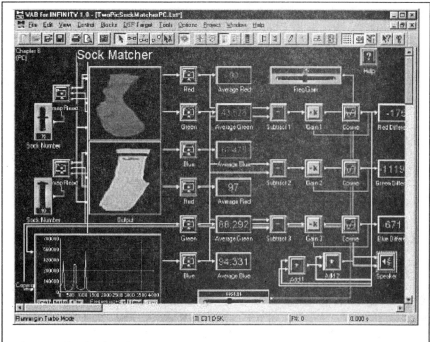

Figure 5.33: Sock color matcher system block diagram

The first step is, always, to acquire the image, sample it and digitize it. Since we are considering color objects, we will get three images corresponding to the red, green and blue component of the full color image, respectively. We saw in Chapter 4 that any arbitrary color can be fully characterized by the intensity of its red, green and blue components. So for the colors of two different socks to match, the values of red, green and blue components of the pixels must also match.

A major issue in engineering design is to account for the variability in the data we deal with. For example, if we have a very detailed image of the sock, the colors of different pixels in a sock image will vary slightly due to fading, cloth weave and general non uniformity of the dyeing process. So it will be desirable for us to take several pixels and calculate the average value of the color components of these pixels.

Similarly in any engineering system, we need to set performance objectives. In other words, we need to know when something is "good enough". This problem is no exception. We need to specify ahead of time how close the colors have to be for us to declare a match.

We perform the following computation to determine the color match:

$$M = ABS(IR_1-IR_2) + ABS(IG_1 - IG_2) + ABS(IB_1-IB_2) \qquad 5.18$$

Where IR, IG and IB correspond to the intensities of red, green and blue components of a pixel and the subscripts indicate whether the pixels were from image 1 or image 2. We take the absolute value of the intensity difference since we care more about the mismatch rather than the sign of the mismatch. The parameter M indicates the degree of mismatch. In our system design, we set a maximum value for M, below which we declare that there is a close enough match between the colors of the two socks. This is

49

done for two reasons: (i) It will be practically impossible to find two socks which match in color *EXACTLY* (corresponding to M = 0), (ii) It is impossible for most people to tell the difference between two colors when the mismatch value is below a certain number. This minimum number for the mismatch can be determined by performing an experiment with human observers to determine when most people will notice two colors as being different. The maximum value for M can then be set just below that.

VAB Experiment 5.5: Sock Matcher

- Determine when two socks match in color. Calculate the mismatch for different socks
- Try the experiment with live inputs from video cameras.

5.4.2 Moving Object Tracker

One way to distinguish between objects is to note the movement of individual objects. In many cases (such as with video security system) the goal is imply to detect any movement in the scene and identify the object that is moving. The block diagram for a system that can perform this operation is shown in Figure 5.34.

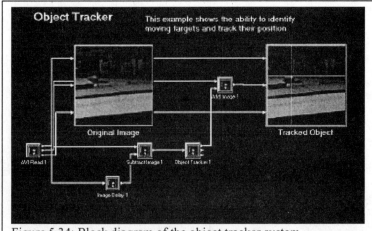

Figure 5.34: Block diagram of the object tracker system.

One major difference between this system and all other systems discussed thus far is the use of time dimension. Previous applications worked with a still, single frame images. The object tracker works with and indeed requires movies, which consist of individual images acquired sequentially in time. The principle of the object tracker operation is very simple. In a scene that contains fixed objects and moving objects, images that are acquired at adjacent time intervals are compared. The objects that don't move will occupy the same place in the image and will be mostly identical (other than some random noise). The objects that move between acquisitions of frames in adjacent time intervals would have shifted their position in the image. Now we employ the operation of image subtraction that we discussed earlier. The fixed objects will cancel each other leaving behind only those objects that have moved. Again, we need to handle negative numbers that will inevitably result from the subtraction operation. Different procedures for handling and displaying negative numbers were discussed earlier. The result of the subtraction was processed by a thresholding operation to suppress extraneous signals. The brightest parts of this image will indicate the presence of moving objects. The center of such bright regions is calculated and a cross hair is added to indicate the position of the moving object. This cross hair is now superposed to the original scene where it will indicate the object that has moved.

The speed of the moving object will determine the extent to which the object would have moved between adjacent time frames. If the object is moving very slowly, then it may not move an appreciable amount between adjacent frames and hence will not be tracked. One solution to overcome this problem is to subtract frames that are separated by a longer amount of time (frames separated by greater than one unit time interval). In the system shown, this parameter is under operator control and can be adjusted.

This is where another important aspect of engineering comes into play. Most of the times there are two or more competing goals that we want to achieve. However, these two goals often conflict. Decision has to be made on achieving a balance between these tow goals depending on the requirements. In case of the motion detector, we would like the system to be sensitive to objects that move very slowly. At the same time, we would like the system to be robust and not to generate extraneous (unintended) signals. In a security context, a system that generates a high level of extraneous signals will lead to false alarms, which makes the law enforcement officers rather unhappy. Achieving the balance between sensitivity and low false alarm rate is one of the most common design principles in many detection systems. Motion detector system is no exception to that. In the following VAB exercise, the trade off between sensitivity and accuracy can be fully explored by changing the parameters in the program.

VAB Experiment 5.6: Moving Object Tracker
• Change the amount of time delay between the frames to be subtracted. Study the correlation between the time delay and the minimum speed of objects that can be detected.   • Change the brightness of the moving object. Does this affect the performance of the system? Try to fool the system. What variations succeeded in creating errors?

5.4.3 Epilogue

We started this chapter thinking about the kind of robot vision needed to perform the simplest and most mundane household chores. The feeling was that it should be relatively easy to perform the kind of object identification and classification tasks that we need to accomplish to do clean up around the house. Indeed, we did formulate the problem in terms of simple mathematical operations on matrices, and then identified the necessary image processing operations that can be accomplished and finally did some preliminary system design study. You of course realize that the job could not be that simple. Otherwise we will all be having robot maids cleaning our house. The challenge lies in the fact that the world around us comes in almost infinite variations where no single set of rules can be applied to all situations. A sleeping cat should not be treated like a rug and vacuumed and simple change of hairstyle should not keep us from recognizing our best friend. These are the things that we accomplish effortlessly and without receiving formal education. On the other hand, the jobs like playing chess, making medical diagnosis is treated as a very challenging job for which we need to undergo years of training. Surprisingly, computers can be programmed to play chess and Grand Master level and to make accurate medical diagnosis. But making a computer vision system as powerful and flexible as a human being is a goal that has eluded engineers for the past 50 years. One area in which processing digital images has achieved enormous success is the area of compressing the amount of data needed to represent a picture or a movie. Some of the image and video compression algorithms indeed utilize the simple operations described in this chapter.

Section 5.6:
Glossary

digital image processing
field of digital signal processing dealing with manipulation of digital images and movies

edge detection
a special kind of image filter which retains only the object edges in the image

image filter
a numeric manipulation of an image to get desired results, usually used for enhancing image quality

image morphing
a process that changes one image to another in a series of smooth transitions

image segmentation
the operation of separating (picking out) the objects from the background or any other object in an image from its surroundings

impulsive noise
it is a form of noise that results in bright spots that are randomly distributed throughout the image

matrix
an array of numbers laid out in rows and columns

neighborhood
a set of matrix elements that are adjacent to a given element in rows or columns

neighborhood operations
image manipulations which act on a group of neighboring pixels in an image

noise
any unwanted information within our image, generally used to denote irrelevant or meaningless information

point operations
elementary matrix manipulations on each pixel of the image, one at a time

point spread function
when the image acquisition system has errors, the point object does not form a single pixel in the image, but is spread over many pixels. This particular image of a point object is called the PSF

random noise
uncertainty in the additive noise

scalar
the simplest type of matrix that contains only one number, i.e., only one row and one column in the array

threshold
a benchmark value chosen to make decisions on image pixels

The INFINITY Project

Chapter 6: Digitizing The World

How do we represent and store information digitally?

Approximate Length: 3 Weeks

Section 6.1:
Digital Yearbook

As suggested by the title of this chapter, we are now about to study digitization of information. Digitization is the representation of information in a form that can be stored on a computer hard drive, or in RAM, or on a CD. To provide motivation for this study and to add some flavor of engineering design, we will develop the topics in this chapter within the context of a hypothetical design scenario.

Perhaps you are one of the students working on your school's yearbook this year. If not, you may be one of the many students who will be purchasing a copy. Have you ever thought about the possibility of improving the format of your yearbook? Is it necessary to continue with the same hard-bound paper copy as last year? Why not go digital? A digital yearbook could offer many possible advantages, including:

a) In addition to the usual text and images, you could include speech, music, and video to more completely document the school year.

b) The yearbook would be more portable. It could be accessed over the web or perhaps it could be made available on a CD or DVD ROM.

c) Perhaps a digital version could be produced at lower cost, especially if you wished for the yearbook to contain a great quantity of information.

d) There would be no limited printing of a digital yearbook. More copies, including hard copies, could be made at any time.

e) A digital yearbook could be more than a yearbook. It could be a "living document." That is, material could be added in future years, as classmates went on to college, took jobs, married, had children, or had other significant events occur in their lives. In fact, the yearbook could form the core of a larger web site that the graduating class would update throughout their lifetimes.

Suppose we wish to explore, in some detail, the feasibility of a digital yearbook. What process should we follow? First, we must define the specific objective. That is, exactly, what is it that we wish to design? Second, we should sketch out some of the details of the engineering design process that we think will lead us to a satisfactory design. After performing the design, we then must evaluate it. How successful were we in meeting the original objective? We also should think about the numerous consequences of the design. After all, engineering technology is implemented in the real world with real people! Will all of the consequences be good? Might there be some negative consequences that we had not intended? Once we have a feel for the design process and some of the considerations in the design, we must ask what we need to learn in this chapter to carry out the design. The resulting list of topics will form the basis for our study in this chapter.

6.1.1 Specific Objective

So, we want to design a digital yearbook. Can we be more specific? Clearly, we are talking about storing a lot of data digitally in some sort of computer memory device. Deciding exactly what to store and arranging it in a convenient and artistic way would fall under the duties of the yearbook editor and his or her staff. Although this would be a fun part of the design process, let's not worry about this aspect. Let us suppose that our job is more on the technological side. We are to determine how much storage or computer memory will be required to store some appropriate mix of text and photos from a traditional-style hard-copy yearbook, plus any speech, music, and video that we might wish to incorporate. This amount of storage will be measured in bits, which is a measure that we will be talking about in great detail later in this chapter. Furthermore, we will want to determine if our proposed digital yearbook will fit on a CD or DVD ROM.

6.1.2 Engineering Design

To determine the amount of storage required for our digital yearbook, we can divide study of the storage problem into multiple tasks, according to the various types of information that the yearbook will contain. An engineer might follow these steps:

1) Text: Specify the amount of text to be stored, measured in number of pages, lines, or words. Estimate the number of letters in all words, combined. Assume that each letter will be stored in the ASCII binary code. (We will discuss ASCII in this chapter.) Further, assume that the size of the resulting computer file size can be reduced by a factor of 2 via compression. (We will discuss compression briefly later in this chapter. This subject is studied in detail in Chapter 9.)

2) Speech: Specify the total number of seconds of speech that we wish to store. Determine the necessary sampling rate, measured in samples per second, and the number of bits per sample needed to faithfully represent the speech. Assume that the size of the computer file containing the speech information can be reduced by a factor of 5 via compression, without significant loss of fidelity.

3) Music: Same considerations as for speech, but assume a compression ratio of 3.

4) Images: Specify the number and sizes of images to be stored. The size is measured in number of pixels, or picture elements. Determine the number of bits per pixel needed to faithfully represent an image. Assume that the size of the computer file can be reduced by a factor of 10 via compression.

5) Video: Specify the number of seconds of video to be stored. Specify the number of pixels for each frame (a frame is a single image in the video sequence). Determine the necessary frame rate (images per second) and number of bits per pixel needed for faithful representation. Assume a compression factor of 100.

6) Overhead: In addition to storing the raw text, images, etc., we will need to allow for storage of other information that will permit a nice appearance or layout of the information and that will allow us to index all of the information in the yearbook so that we can easily search for items of interest. We shall allow 10% for overhead, but this could be either high or low, depending on the complexity of the page layouts and indexing schemes.

7) Total the amount of required storage from 1) – 5) and add an extra 10% to account for 6).

6.1.3 Evaluation of the Design

One of our goals is to determine whether the digital yearbook can fit onto a CD ROM. If not, how about on several CDs or onto a DVD? The answers to these questions will determine whether we can store the yearbook on some form of portable media, or whether we must rely on access of the contents via the web. Of course, if the storage requirements are too vast, we can modify our design. We can decide to store less information in the yearbook (e.g., fewer seconds of video) or we can store the information with less fidelity (e.g., fewer bits per sample, or smaller images and video, or with higher compression factors). If one type of information (e.g., video) requires much more storage than other types, then this category of information will be an obvious candidate for reduction.

If we change the specifications for the original design and perform a second design based on the new specifications, we will be *iterating* on the design process. In the engineering of complex systems, it is common to iterate multiple times in an attempt to *optimize* the design (i.e., to make it as good as it can be). In each iteration, or design cycle, there are tradeoffs that must be decided. For example, we can store an image with high fidelity, or we can reduce the amount of necessary memory by storing smaller images, or by using fewer bits per pixel, or by using a higher compression factor. What choices should we make? The answers are best given by persons who are experts in the technology and who have experience working on the application under consideration. Engineers who work in multimedia have a good feel for what specifications are necessary to store information at a fidelity that will be found acceptable. After you finish this course, you will be a good judge of this, too!

6.1.4 Consequences of the Design

Before accepting a final design, we should at some point consider the consequences of the design. We are expecting that a digital yearbook will cost less than a hard-bound copy. Is this true? Will the images look as good as or better than those in the hard-copy version? Do virtually all of your classmates have computers at home, or would some of them need to purchase hard copies of the yearbook? Will some students and families prefer a nicely bound copy that can be placed on a coffee table? How would hard copies be produced from the digital version? Of course, none of the speech, music, or video would be preserved in a hard copy. Is your yearbook staff equipped with the digital still-frame and video cameras, digital audio recording devices, and computerized editing and page-layout facilities that will be needed? If not, can you contract the work to a company? Can your staff complete a digital yearbook on schedule? Might there be any unforeseen risks in undertaking a digital yearbook? After all, the hard-copy route that has been used for many years will work for sure. In this chapter we are going to assume that all of these potential problems can be resolved. We shall march ahead with a digital design!

6.1.5 Topics Required for the Digital Yearbook Design

To make further progress in designing a digital yearbook, there are many topics that we will need to learn about in considerable depth. We will address the following topics and questions in this chapter:

• Representation of information as numbers: As we have seen in earlier chapters, waveforms and images can be represented by their uniformly-spaced samples. The samples are simply sets of numbers. How do we represent all types of information, including text, as numbers?

• Sample spacing: How closely spaced should samples of a waveform be? Similarly, if we wish to scan a photograph, how small should the pixels be?

• Bits: What is a bit? How does computer memory store bits? How many bits can be stored on a CD or on a DVD ROM?

• Binary numbers: What are binary numbers and how are they related to regular numbers?

• ASCII Code: The ASCII code lets us represent text and punctuation in bits. What is this code?

• Quantization: So far, samples of a waveform or values of pixels have been real numbers that can take on any value within some range. How are these real numbers converted into binary numbers or bits for storage? (This process is called quantization.)

• Noise: Quantization introduces noise into the storage process. How does the amount of noise depend on the number of bits?

Exercise 6.1

1. A digital yearbook would contain what types of information?

2. Name five possible advantages of a digital versus hard-copy yearbook.

3. Name three possible difficulties in producing a digital yearbook.

4. Why do engineers often iterate the design process?

Section 6.2:
Representation
of Information
as Numbers

In earlier chapters we saw that waveforms, such as from music or speech, can be represented by the numerical values of samples of those waveforms. An analog-to-digital converter collects uniformly-spaced samples of a waveform. The values of these samples are simply a list of numbers. If the samples are densely enough spaced, then the values of the samples provide a faithful representation of the original waveform.

The situation is similar with images. A scanner samples a printed photograph or image, except that the sampling is performed at different points in space (typically on a Cartesian or rectangular grid), instead of time. The gray-scale value (lightness or darkness) at any point in a black-and-white image can be represented as a nonnegative number. A set of samples of an image can be thought of as an array or matrix of nonnegative numbers. A digital camera collects such an array of numbers directly. A digital camera contains a CCD array, which is a set of tiny light sensors, arranged on a rectangular grid. The light sensors sample the light field at all points on the grid and store the intensity as numbers.

A digital video camera adds an extra dimension – time. A video camera collects a new image, called a frame, every 1/30 of a second. Each video frame is a rectangular array of numbers. Together, the frames are a stack of such arrays of numbers, which represent a sequence of digital images. Each individual number represents the brightness of the scene at a specific spatial location at a specific time.

Of course, most scanners and cameras record color imagery. In this case, each pixel is represented by more than one number. For example, three numbers per pixel might be used to record the amounts of red, blue, and green contained in the light field at a pixel location. Together, these three numbers allow the representation of any color, as a combination of red, blue and green. The important point is that waveforms, images, and video all can be stored as numbers.

What about text? Can we store text as numbers? Yes! We can assign a specific number to represent each letter or punctuation mark on a computer keyboard. We can then store the numbers corresponding to the letters and punctuation that make up a section of text. In Section 6.6 we will discuss a special code, called ASCII, for storing text as numbers. As we shall see, the concept of storing information as numbers underlies the process of digitization, whether we are talking about text, speech, music, images, or video.

Exercise 6.2

1. A speech waveform is sampled at a rate of 8,000 samples per second for a duration of 2 minutes. How many numbers are needed to represent the speech samples?

2. A music waveform is sampled at 40,000 samples per second for a duration of 30 minutes. How many numbers are needed to represent the music samples?

3. One-hundred digital, color images are collected, each of size 1200 by 900 pixels. Assuming that three numbers are needed to represent each color pixel, how many numbers are needed to represent all of the images?

4. Digital video is collected at a rate of 30 frames per second for a period of one hour. Each frame is 800 by 600 pixels. Assuming that three numbers are needed to represent each color pixel, how many numbers are needed to represent the video?

Section 6.3:
Sampling Rate
for Waveforms,
Images & Video

Before we discuss how numbers are stored in a computer memory, it is worth examining the sampling process a bit further. In particular, we must determine how closely spaced the samples of a waveform must be. Likewise, what should be the spacing of the pixels in an image or video? To minimize storage requirements, we hope that samples can be spaced far apart, so that not too many numbers will be needed to represent the speech, music, image or video. On the other hand, if a waveform varies rapidly with time, or the intensity of a scene changes suddenly from one region to the next, then closely spaced samples or pixels will be needed.

In this section we explain that for faithful representation, waveforms must be sampled at a rate greater than twice the highest frequency contained in the waveform. This rate is called the **Nyquist rate** and it governs the necessary sampling rate in practical systems. We will see that sampling below the Nyquist rate produces a sequence $s[n]$ from which the analog waveform $s(t)$ cannot be exactly recovered. Sampling at a rate that is too low leads to a bad effect called aliasing, where high frequencies can masquerade as low frequencies.

> **Nyquist Rate**: The slowest sampling rate that permits recovery of the signal $s(t)$ from its samples $s[n]$.

6.3.1 Sampling Theorem

To help gain some intuition about the question of minimum sampling rate, let us consider the sampling of a sinusoid. After all, since complicated signals can be constructed as sums of sinusoids, we reason that if we can figure out the necessary sampling rate for a sinusoid, we can also determine the necessary rate for a more general signal. Consider Figure 6.1, which shows a sinusoid sampled at a moderate rate (several samples per period) and Figures 6.2 – 6.4, which show the same sinusoid sampled at lower rates (2 samples per period, 3/2 samples per period, and 1 sample per period, respectively).

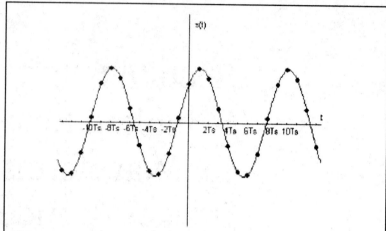

Figure 6.1: Sinusoid sampled at a moderate rate, several samples per period.

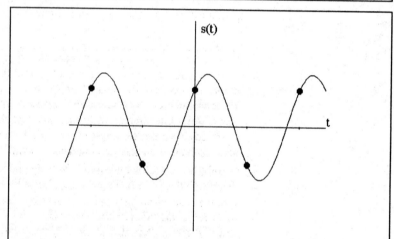

Figure 6.2: Sinusoid sampled at a lower rate, 2 samples per period.

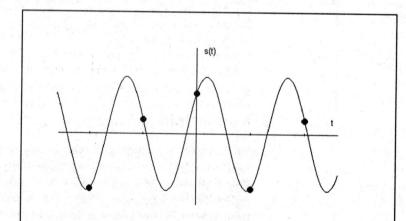

Figure 6.3: Sinusoid sampled at a lower rate, 3 samples per every 2 periods.

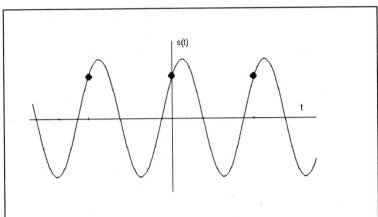

Figure 6.4: Sinusoid sampled at a lower rate, 1 sample per period.

If we were given the samples in Figure 6.1 and told that they came from a signal having frequency no higher than that shown, we would have little trouble recovering the analog waveform s(t). In Figure 6.2, the sampling rate has been reduced to 2 samples per period, i.e. to $f_s = 2f$ samples per second, where f is the frequency of the sinusoid. Here it looks like we could still barely recover the sinusoid from its samples. Notice, however, that if the sinusoid had been shifted to the right so that a zero-crossing occurred at the origin, then all sample times would occur at zero-crossings and, therefore, all samples would be zero. In this case, we would not be able to tell the difference between a sinusoid and a signal that was identically zero. Thus, we conclude that sampling at the rate, $f_s = 2f$, as in Figure 6.2, seems to be on the borderline of the minimum that is necessary. In Figure 6.3, we see that the sampling rate seems to be insufficient. It would be hard to guess from the given samples that the original waveform was a sinusoid of the frequency shown. Finally, in Figure 6.4 the sampling rate $f_s = f$ is clearly insufficient. Indeed, the samples are constant, so that we would be unable to tell the difference between the sinusoid shown and a signal that was constant! The underlying signal would not appear to be sinusoidal at all!

From our intuitive analysis of these figures, we suspect that the minimum sampling rate must be greater than two samples per period, or equivalently more than twice the frequency of the sinusoid, if we hope to recover the analog waveform s(t) from its samples s[n]. Mathematically speaking, if T_s is the spacing between samples (called the sampling period and measured in seconds) and f_s is the sampling frequency (measured in samples per second), then we are suggesting

$$f_s = \frac{1}{T_s} > 2f.$$

Nyquist Rate : The Nyquist rate tells us that we must sample the signal s(t) at a rate that is at least twice as high as the highest frequency in the signal. We can determine the highest frequency in the signal by using the Fourier transform or the spectrogram.

Indeed, it turns out that this is correct and that this same idea generalizes to signals other than sinusoids. As we discussed in Chapter 2, a waveform s(t) with an arbitrary shape can be written as a sum of sinusoids having appropriate frequencies, amplitudes, and phases. For this general case we must sample at a rate that is larger than twice the *highest* frequency contained in s(t). The general sampling rule is

$$f_s > 2 f_{max}$$

where f_{max} is the highest frequency contained in s(t). This is a major result in the theory of signal processing and communications. $2f_{max}$ is called the **Nyquist rate**. One will often hear in the engineering world that "you must sample above the Nyquist rate."

Signals whose maximum frequency f_{max} is finite are called **bandlimited**, because their frequency content lies within the finite band

$$0 \leq f \leq f_{max} .$$

Figure 6.5 shows an example spectrum for a bandlimited signal. In this case, the highest frequency contained in s(t) is 10,000 Hz which means that we must sample at a rate of at least 20,000 times per second for the set of samples to provide a faithful representation of the analog waveform s(t).

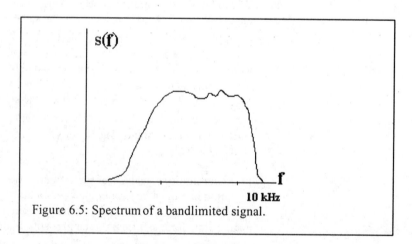

Figure 6.5: Spectrum of a bandlimited signal.

The considerations in sampling an image are very similar to the sampling of waveforms. If we represent an image as a two-dimensional signal s(x,y), where x and y are spatial variables, it is possible to write s(x,y) as a sum of two-dimensional sinusoids where the sinusoids extend throughout space rather than time. We won't discuss the details here, but images with lots of edges tend to have considerable high frequency content, whereas images with a smooth appearance have less high frequency content. Images with a smoother appearance can be sampled at lower rates. The sampling rate for an image should be twice the highest *spatial frequency* contained in the image. Thus, the sampling criteria for waveforms and images are basically the same.

Returning to one-dimensional signals, the **sampling theorem** states that if a signal s(t) is bandlimited and if it is sampled above the Nyquist rate $2f_{max}$, then it can be exactly reconstructed from its samples. Furthermore, there is a formula for the recovery:

$$s(t) = \sum_{n=-\infty}^{\infty} s[n] \frac{\sin(\frac{\pi}{T_s}(t - nT_s))}{\frac{\pi}{T_s}(t - nT_s)}.$$

We refer to this as the ideal reconstruction formula, since it permits exact recovery of s(t). (This formula is for the case where s(t) is non-periodic; the formula for the periodic case is slightly different.) We will not dwell on this formula. The important thing is that there is a formula and that it can be approximately implemented in a very practical way. In fact, CD players contain digital-to-analog (D/A) converters that implement a form of this type of reconstruction. The implications of all this sampling theory are nothing short of remarkable. The fact that we can go back and forth between waveforms or images and their samples suggests that the samples contain all of the information contained in the original analog waveform of image! Thus, as we shall see in Chapter 12, any time we wish to process an analog waveform or image, say to remove noise or emphasize one frequency band over another, we can accomplish this operation digitally inside a computer. We can sample the signal, manipulate the resulting numbers inside the computer to form a new sequence, and then reconstruct a new analog waveform that passes through the computed samples. This is called **digital signal processing (DSP)**!

Exercise 6.3

1. What is the definition of *Nyquist rate*?

2. What is the definition of *bandlimited*?

3. State the sampling theorem.

4. Suppose $s(t) = \cos(4\pi t)$ and $T_s = 0.1$. Find and plot the values of $s[n] = s(nT_s)$, n = 0, 1, 2, ..., 10.

5. Given the waveform, $s(t) = \sin(4\pi t)$, plot its samples, assuming $T_s = 0.3$. What has happened? Have you sampled above the Nyquist rate?

6. Telephone speech is bandlimited to approximately 3,500 Hz. What is the minimum sampling rate that will avoid information loss for this signal?

7. In a recording studio, music is intentionally bandlimited to 20 KHz, since that is the upper range of human hearing. What is the minimum sampling rate for digital music? (As a point of interest, the sampling rate used for music CDs is 44.1 KHz.)

8. The bandwidth of a demodulated AM radio broadcast is 5 KHz. Commercial AM radio is currently analog. If AM radio were to "go would be the minimum sampling rate needed?

9. The bandwidth of a demodulated FM radio broadcast is 15 KHz. Commercial FM radio is currently analog. If FM radio were to "go digital," what would be the minimum sampling rate needed?

10. What does *DSP* stand for? What is DSP?

6.3.2 Effects of Sampling Below the Nyquist Rate

We have seen that bandlimited signals can be perfectly recovered from their samples, so long as the sampling rate is greater than twice the highest frequency contained in the signal to be sampled. Some signals, however, are not bandlimited; that is, they have energy at frequencies extending to infinity. For example, in Chapter 2 we saw that triangle waves are not bandlimited. What if we try to recover one of these signals from its samples? What happens? Likewise, suppose we sample a signal that is bandlimited, but that we do not sample above the Nyquist rate? Again, what happens?

This question can be answered in two different ways. First, from a mathematical point of view, the difficulty is that sampling below the Nyquist rate means that the samples s[n] will no longer uniquely specify s(t). This is illustrated in Figures 6.6 – 6.8. Figure 6.6 shows an analog waveform to be sampled and then reconstructed. Figure 6.7 illustrates that if the signal is sampled above the Nyquist rate, then it can be correctly recovered from its samples. Figure 6.8 illustrates that if the sample spacing is too large, there will be an infinite number of signals s(t), having bandwidths greater than or equal to $f_s/2$, that will pass through (agree with) the collected samples s[n]. Unlike in Figure 6.7, these signals s(t) are all possible candidates for the true s(t) because we have not sampled above the Nyquist rate and so we do not have $f_{max} < f_s/2$. The ideal reconstruction formula will produce just one of these signals – the one that is bandlimited to precisely $f_s/2$. This s(t) will be incorrect. Sampling below the Nyquist rate is called **undersampling**. With undersampling we have no way of telling which of the s(t) in Figure 6.8 is the correct reconstruction, because the samples are spaced too far apart to uniquely characterize the analog waveform. In a DSP system, composed of a sampler, processor, and digital-to-analog converter, undersampling at the sampler will create error in the system output signal.

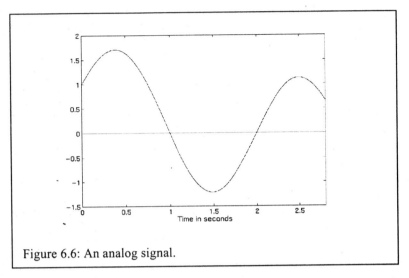

Figure 6.6: An analog signal.

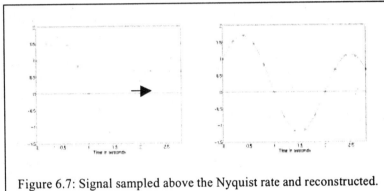

Figure 6.7: Signal sampled above the Nyquist rate and reconstructed.

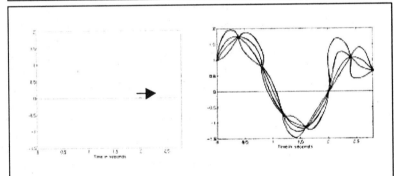

Figure 6.8: Signal sampled below the Nyquist rate. Many bandlimited signals with $f_{max} > f_s/2$ can be reconstructed that pass through the samples.

There is a second way to think about sampling below the Nyquist rate, by examining the effects on sinusoids. Let us consider a particular sinusoid that is sampled below the Nyquist rate. Figure 6.9 shows 18 periods of a 720 Hz sine. The Nyquist rate for this signal is 1440 Hz. Suppose that we sample this signal at 660 samples per second – far below the Nyquist rate. The resulting samples are denoted by the dots in Figure 6.9. Quite surprisingly, these samples take on the shape of a sinusoid, but the frequency represented by the samples is far different from the original analog frequency of 720 Hz! A careful examination of the figure shows that the samples correspond to a frequency of 60 Hz, which is the difference between the true frequency and the sampling frequency. This phenomenon is indicative of what happens when a signal is undersampled. Frequency components of the signal that are insufficiently sampled are changed to incorrect frequencies! This effect is called **aliasing** because one frequency can masquerade as another.

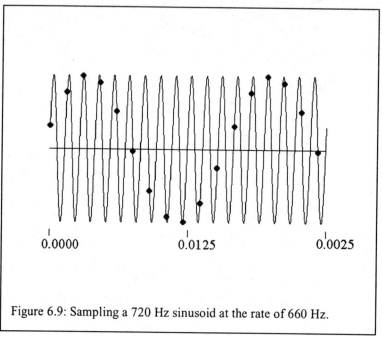

Figure 6.9: Sampling a 720 Hz sinusoid at the rate of 660 Hz.

A similar effect occurs in undersampling of images, except that the effect in that case is the appearance of distorting patterns or artifacts. These distorting patterns are called **Moire patterns**. In video, aliasing can occur both within individual image frames and also from one frame to the next. If the scene contains objects that are moving, an insufficient frame rate (sampling in time) can lead to quite amazing effects. A visual demonstration of such aliasing is present in many "cowboy films" when you see a wagon either speeding up or slowing down. A movie camera captures only 24 image frames per second, so it cannot sample a quickly rotating wagon wheel, without aliasing. This effect is complicated by the fact that a wagon wheel is symmetric, with many spokes. Thus, the necessary camera frame rate f_s needed to avoid aliasing is multiplied by the number of spokes. For example, if a wheel with 12 spokes is rotating at 1.5 revolutions per second, then a camera frame rate of 24 Hz is already insufficient to avoid aliasing (the frame rate would need to be greater than $2(12)(1.5) = 36$ frames per sec). If a wagon wheel maintains a constant rotational speed, then the effect of aliasing is to make the wagon wheel appear to rotate more slowly than is actually the case, or possibly to rotate backwards. If the wagon changes speed, a more interesting effect occurs. As a wagon wheel slows down it may at times appear to speed up or to suddenly rotate backwards!

To see a demonstration of such aliasing, go to the project web site. You will see a bicycle wheel with four reflectors placed on it at evenly spaced angles. As the wheel speeds up and slows down you will see that it sometimes appears to rotate clockwise and other times counterclockwise. (The wheel is actually always rotating clockwise.) There are also brief times when the reflectors appear to be stationary. Can you explain why? This movie file is reasonably large: 9.7 megabytes. If your web connection is slow, try the same URL except with Small substituted for Med. This will allow you to access a much smaller version of the same movie (but the movie will also be small on your screen).

The bottom line is that aliasing is a serious effect that must be avoided if we are to represent information faithfully. Aliasing causes frequencies to change, whether the signals are in 1-D, 2-D, or 3-D. For typical signals, images, and time-varying scenes that are composed of many frequencies, the effects are very complex. If s(t) is a bandlimited 1-D signal, then we can avoid aliasing by simply sampling above the Nyquist rate. If s(t) is not bandlimited, then we must first pass s(t) through a low-pass filter (composed of resistors, capacitors, and transistors in the analog world) to reduce its frequency range to some f_{max}. Such a filter is called an **antialiasing filter**. This filtering operation will cause error (we will be throwing away the high frequencies in s(t)), but it can be shown that this error is always less than that which would have been caused by aliasing. Thus, antialiasing filtering is commonly performed prior to sampling to ensure that the signal to be sampled is bandlimited, so that the Nyquist criterion can be satisfied. Digital cameras use a similar concept, except that the filtering is a spatial blur introduced by a lens, which acts to smooth the image (which removes high frequencies) prior to sampling on the CCD sensor.

Appendix A briefly examines the mathematical aspects of aliasing for those students who might wish to investigate this subject further. Completion of the reading and homework exercises in this appendix would constitute a nice project for such students.

Experiment 6.1: Aliasing Wheel Movies
Student watch some movies that demonstrate aliasing.

- The following signals have been sampled and then reconstructed from their samples, assuming the sampling rates indicated. The results were recorded onto the VAB platform. Listen to these signals to discover what the results of aliasing sound like in the analog world!

500 Hz tone sampled at f_s = 1200 Hz (no aliasing)

500 Hz tone sampled at f_s = 800 Hz (aliasing)

500 Hz tone sampled at f_s = 400 Hz (aliasing)

Speech bandlimited to 3,500 Hz and sampled at 8 KHz (no aliasing)

Speech bandlimited to 3,500 Hz and sampled at 4 KHz (aliasing)

Speech bandlimited to 3,500 Hz and sampled at 1 KHz (aliasing)

Music bandlimited to 7 KHz and sampled at 15 KHz (no aliasing)

Music bandlimited to 7 KHz and sampled at 8 KHz (aliasing)

Music bandlimited to 7 KHz and sampled at 2 KHz (aliasing)

- Can you hear why we need to avoid aliasing?

Exercise 6.4

1. What is meant when we say that a signal is undersampled?

2. For a sinusoidal signal, what is the effect of aliasing?

3. For a 1-D signal that is not bandlimited, how can aliasing be avoided?

4. In a digital camera, how is aliasing avoided?

5. In the wheel movie referred to in this section, 4 reflectors are placed at evenly spaced angles around the wheel. If the video camera uses a frame rate of 30 frames per second, the wheel will appear to be stationary in the video for what speeds of the wheel (revolutions per second)?

Section 6.4:
Storage Devices
and Bits

Computers and other digital devices represent information in a binary
number code where the smallest unit is called a **bit**. A single bit represents
one of two states – we can think of them as 0 and 1. Many bits are grouped
together to represent numbers that are larger than 0 or 1. A group of 8 bits
is called a **byte**. Why do computers and other digital devices represent
information in terms of bits or bytes? The short answer is that memory
devices are built from transistors or physical materials that can represent
only two different states. Chapter 17 gives comprehensive coverage of this
topic; here we briefly introduce the main idea.

Let us first consider semiconductor memory, sometimes called random
access memory (RAM) because data can be accessed in any order at the
same speed. RAM is constructed of millions of small transistor circuits on
a chip (integrated circuit) where each small circuit can maintain either a
high voltage or a low voltage, depending on an applied control signal. The
key is that there are only two possible voltage levels: either high, about 3
Volts, or low, less than 1 Volt (assuming a 3 Volt power supply). Thus, the
state of each circuit (high voltage versus low voltage) can represent only
two different numbers.

Both the hard drive and floppy drive in your computer share this same
characteristic, although they work on a magnetic principle. Each designated
location on the disc can be magnetized in one of two directions, depending
on the value of an applied control signal. This magnetization is sensed by
the disc head, which rides over the spinning disc. Once again, only two
states are possible at each physical location on the storage device.

Finally, let us consider the CD or DVD ROM. Each designated spot on the
CD or DVD surface (these spots are microscopic in size) has either a pit or
no pit. Laser light is reflected off the surface, and depending on the amount
of light returned to a sensor, the presence or absence of a pit is easily
detected. As is the case with RAM and magnetic drives, a CD or DVD
ROM stores only one of two possible states at each physical location on the
device. To simplify discussion of all these devices, we refer to the two
states as 0 and 1. We could equally well call them A and B or by any other
two identifiers, but we will use 0 and 1, which is standard nomenclature.

Since computer memory devices can store only 0's and 1's, the computer must use binary number representation and binary arithmetic. We study binary number representation in the next section.

Exercise 6.5

1. What is a *bit*?

2. What is a *byte*?

3. Explain why computer storage devices store bits.

4. Explain how a set of blue cards and red cards could be used to represent bits.

5. Explain how a handful of pennies could be used to represent bits.

6. Explain how a set of apples and oranges could be used to represent bits.

7. Explain how a set of right-handed and left-handed people could be used to represent bits.

8. Explain three other ways that bits could be represented, other than those mentioned so far.

Section 6.5:
Binary Number Representation

In our coverage of sampling in Section 6.3, we assumed that the values of a sampled speech waveform or image can be stored as numbers in a computer. For example, a given speech sample may take on the value 2.5 or –0.75. However, in the last section, we just learned that computers and digital memories store bits. How can numbers such as 2.5 be represented by bits? The answer is that computers and digital storage devices use the **binary number system**, rather than the decimal number system that we are more familiar with. A binary number groups together the values of many consecutive bits, each either a 0 or 1, to represent a number such as 2.5. Representing numbers as bits affects how numbers are stored and how arithmetic operations (e.g. addition, multiplication) are performed. In this section, we shall focus on how numbers are stored in binary format as bits.

The binary number system has only two symbols, unlike the decimal system, which has ten symbols. The binary number system provides a way to represent a wide range of numerical values using devices that can exhibit only two states at any specific location, e.g. high voltage or low voltage, left or right magnetization, or pit versus no pit. A single bit can have values 0 or 1. Two bits together can represent (0,0), (0,1), (1,0), or (1,1), i.e. four different possibilities. Thus, two bits can represent four different states or values. Likewise, three bits together can represent (0,0,0), (0,0,1), (0,1,0), (0,1,1), (1,0,0), (1,0,1), (1,1,0), or (1,1,1), i.e. eight different possibilities. Thus, three bits can represent eight different states or values. In general, B bits, where B is a positive integer, can represent 2^B states or values. Therefore, 4 bits can represent 16 values, 5 bits can represent 32 values, and so forth. To see how this is accomplished, in the next subsection we begin the development of binary numbers by discussing the representation of positive decimal integers as binary numbers. In later sections we talk about the binary representation of fractions and negative numbers.

Exercise 6.6

1. How many different values or states can be represented by (a) 6 bits, (b) 7 bits, (c) 8 bits, (d) 10 bits?

2. You are given a dozen fireplace matches and you line them up, side by side. Each match head can be either up or down. The number of states or values that can be represented by this set of match heads is the number of possible orientations of the match heads. How many states can these matches represent?

3. Driving down a long street you pass under 20 streetlights. Most are working, but some are burned out. How many possible combinations are there of burned out and working street lights?

4. Suppose that a tiny super-fast memory contains 100 bits. How many states or values can this memory represent? Your answer will be a very, very large number. Approximate it by a number written in the form 10^M for some integer value of M, by using the fact that 2^{10} is about equal to 1000.

5. How does the answer to Problem 4 compare to Avogadro's number from chemistry class? How does the number of different states that the 100-bit memory represents compare with the number of atoms in 1,000 kilograms of hydrogen?

6. A new memory technology is developed where each bit can represent three states, instead of just two. How many different values or states can be represented by (a) 6 bits, (b) 7 bits, (c) 8 bits, (d) 10 bits? How do these answers compare with those in Problem 1?

6.5.1 Positive Binary Integers

In the decimal number system, the number 10 is the base or radix. The integer 238 takes the value 2(100) + 3(10) + 8(1) where 100, 10, and 1 represent the values of the positions of 2, 3, and 8, respectively. Here, the value of each position is 10 times the value of the position to its right, with the rightmost position having value one. This is illustrated pictorially in Figure 6.10.

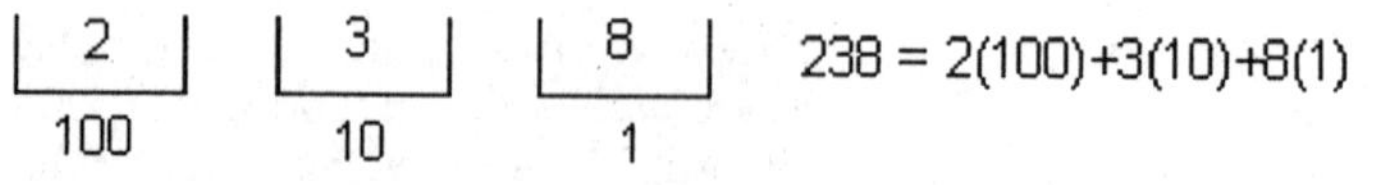

$$\boxed{2} \quad \boxed{3} \quad \boxed{8} \quad 238 = 2(100)+3(10)+8(1)$$
$$100 \qquad 10 \qquad 1$$

Figure 6.10. Decimal representation of 238.

In the binary number system, the number 2 is the base and each binary number is a string of 0 and 1 bit values. The value of each bit position in the string is 2 times the value of the position to its right, with the rightmost position having value one. Thus, the 4-bit number 1011_2 takes the value $1(8) + 0(4) + 1(2) + 1(1) = 11$. Here, we have used the subscript 2 to indicate that we are writing a binary number, so that the values of the positions are 8, 4, 2, and 1. Figure 6.11 shows the binary representation of the decimal number 11.

Figure 6.11. Binary representation of the decimal number 11.

How would we represent the decimal number 238 in binary form? This will require quite a few binary digits! Each binary digit represents a power of two. We can write 238 as some combination of 128, 64, 32, 16, 8, 4, 2, and 1, which are each a power of two. The particular combination we need is $1(128) + 1(64) + 1(32) + 0(16) + 1(8) + 1(4) + 1(2) + 0(1) = 238$, which is 11101110_2 in binary form. Figure 6.12 illustrates this binary representation.

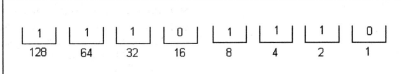

Figure 6.12. Binary representation of 238.

To help cement these ideas, study the binary numbers and their decimal equivalents in Tables 6.1 and 6.2. For the 8-bit numbers in Table 6.2, the value of the leftmost bit position is, $2^{(8-1)} = 128$, so that the bit values are (128, 64, 32, 16, 8, 4, 2, 1). For the 10-bit numbers, the value of the leftmost bit position is $2^{(10-1)} = 512$, so that the bit values are (512, 256, 128, 64, 32, 16, 8, 4, 2, 1). The largest integer that can be represented by a B-bit binary number is $2^B - 1$.

Table 6.1. The First 32 Integers Expressed in Binary and Decimal Form

Binary	Decimal	Binary	Decimal
00000	0	10000	16
00001	1	10001	17
00010	2	10010	18
00011	3	10011	19
00100	4	10100	20
00101	5	10101	21
00110	6	10110	22
00111	7	10111	23
01000	8	11000	24

01001	9	11001	25
01010	10	11010	26
01011	11	11011	27
01100	12	11100	28
01101	13	11101	29
01110	14	11110	30
01111	15	11111	31

Table 6.2: Some Larger Integers Expressed in Binary and Decimal Form.

Binary	Decimal	Binary	Decimal
01011011	91	1100101011	811
10101100	172	0111001100	920
11010110	214	1011001010	714
11111111	255	1111111111	1023

Exercise 6.7

1. Convert the following positive numbers from binary to decimal: 010_2, 101_2, 011_2, 111_2.

2. Convert the following positive numbers from binary to decimal: 1000_2, 1010_2, 0111_2, 1101_2.

3. Convert the following positive numbers from binary to decimal: 10000_2, 10101_2, 01111_2, 11011_2.

4. Convert the following positive numbers from binary to decimal: 100001_2, 110001_2, 011110_2, 101101_2.

5. Convert the following positive numbers from binary to decimal: 10101010_2, 11011011_2, 11100001_2, 01110111_2.

6. Convert the following positive numbers from binary to decimal: 11111111_2, 111111111_2, 1111111111_2, 111111111111_2.

6.5.2 Conversion from Decimal to Binary

In the last section, we concentrated mainly on conversion from binary to decimal numbers. What about the other way around? There are efficient recipes for making the conversion from decimal to binary representation, but some intuition is lost in these approaches. Here, we shall discuss a straightforward method for converting from decimal integers to binary integers. Instead of specifying a formal procedure, we will illustrate by example.

Suppose we wish to convert 213 to binary. 213 lies between $128 = 2^7$ and $256 = 2^8$, so we will need 8 bits to represent this number. Our job is to find the bit values B_1 through B_8 satisfying

$$
B_1(128) + B_2(64) + B_3(32) + B_4(16)
$$
$$
+ B_5(8) + B_6(4) + B_7(2) + B_8 = 213
$$

We refer to B_1 as the **most significant bit**, since it carries the most weight in the binary representation, and we call B_8 the **least significant bit**. The position of the most significant bit, B_1, has value 128. Set this bit equal to 1 and then subtract 128 from 213 to give the remainder 85. The remainder is larger than 64, which is the value of the position of B_2. Thus, set B_2 to 1. Subtract 64 from 85 to give the next remainder, 21. This remainder is smaller than 32, which is the value of the position of B_3. Thus, set B_3 to 0. The position of B_4 has value 16, which is smaller than the remainder 21. Thus, set B_4 to 1. Subtract 16 from 21 to give 5. The position of B_5 has value 8, which is larger than 5. Thus, set B_5 to 0. The position of B_6 has value 4 which is smaller than 5. Set B_6 to 1 and subtract 4 from 5 to give 1. The position of B_7 represents 2, which is larger than the remainder 1. Set B_7 to zero. Finally, set B_8 to 1. The result is $213 = 11010101_2$. The overall procedure used in this example is illustrated in Figure 6.13.

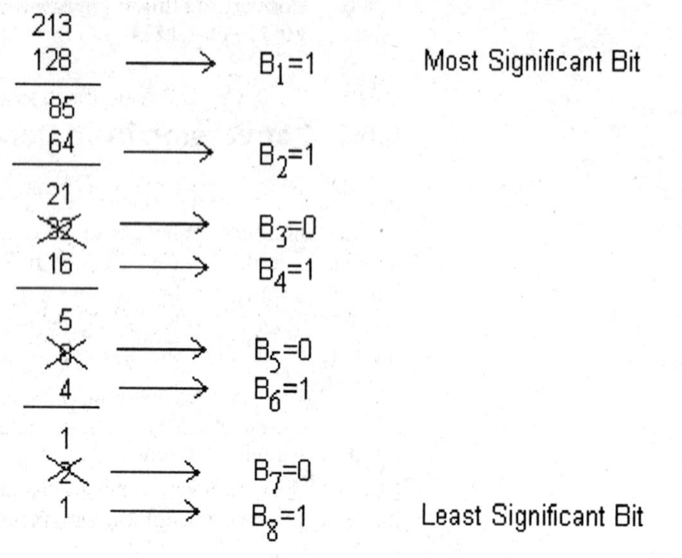

```
213
128  ────────▶  B₁=1        Most Significant Bit
────
 85
 64  ────────▶  B₂=1
────
 21
 ✗✗  ────────▶  B₃=0
 16  ────────▶  B₄=1
────
  5
  ✗  ────────▶  B₅=0
  4  ────────▶  B₆=1
────
  1
  ✗  ────────▶  B₇=0
  1  ────────▶  B₈=1        Least Significant Bit
```

Figure 6.13: Conversion of the decimal number 213 to the binary number

11010101_2.

You may wonder how computers carry out arithmetic operations such as addition and multiplication with the numbers represented in binary form. There is no need (or desire!) to convert numbers back to decimal form prior to undertaking computations. The procedures for binary arithmetic are completely analogous to those for decimal arithmetic. The main difference is that when adding a column of binary numbers, as soon as the total is 2 or greater, there must be a carry because the only binary digits are 0 and 1. With decimal numbers there is no carry until the total reaches 10 or greater.

Exercise 6.8

1. What is meant by *most significant bit* and *least significant bit?*

2. Write the following positive decimal integers in binary form: 4, 6, 8, 12, 13, 15.

3. Write the following positive decimal integers in binary form: 16, 20, 25, 31.

4. Write the following positive decimal integers in binary form: 32, 39, 57, 63.

5. Write the following positive decimal integers in binary form: 64, 70, 111, 127.

6. Write the following positive decimal integers in binary form: 128, 138, 195, 255.

7. Write the following positive decimal integers in binary form: 256, 1024, 2048, 4096.

8. Write the following positive decimal integers in binary form: 743, 1500, 3001, 4104.

9. Write the following positive decimal integers in binary form: 849, 1800, 3307, 4608.

10. Write the year of your birth in binary form.

11. Write the year of Thomas Edison's birth in binary form.

12. Write the year in binary that Alexander Graham Bell first demonstrated the telephone.

13. Write the year in binary that Bardeen, Brittain, and Shockley received the Nobel Prize for invention of the transistor.

14. Write the year in binary that Jack Kilby received the Nobel Prize for invention of the integrated circuit.

6.5.3 Binary Fractions

The binary representation of positive fractions, or positive integers plus fractions, is similar to the representation of integers. In the decimal world, we have a decimal point that separates the fractional part from the integer part. In the binary world we use the same point notation but we call it the **binary point**. The binary number 1101.101_2 represents the decimal number $1(8) + 1(4) + 0(2) + 1(1) + 1(1/2) + 0(1/4) + 1(1/8) = 13.625$. This is illustrated in Figure 6.14.

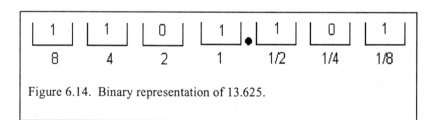

Figure 6.14. Binary representation of 13.625.

When binary fractions are involved, the value of the position to the left of the binary point is 1 and the values of the positions to the right then decline by factors of two giving position values 1/2, 1/4, 1/8 and so forth. Movement of the binary point P places to the right corresponds to multiplying the number by 2^P. Likewise, movement of the binary point P places to the left corresponds to division by 2^P. See Table 6.3. for several binary fractions and their decimal equivalents. Notice that the numbers in the right-most columns are 1/2 those in the leftmost columns because the binary point has been shifted one place to the left.

Table 6.3: Some Fractional Numbers Expressed in Binary and Decimal Form.

Binary	Decimal	Binary	Decimal
100.01	4.25	10.001	2.125
011.10	3.5	01.110	1.75
110.00	6.0	11.000	3.0
111.11	7.75	11.111	3.875

Exercise 6.9

1. Convert the following positive fractions from binary to decimal: 10.1_2, 10.0_2, 01.0_2, 11.1_2.

2. Convert the following positive fractions from binary to decimal: 1.01_2, 1.00_2, 0.10_2, 1.11_2. How are your answers related to the answers to Problem 1? Do you know why?

3. Convert the following positive fractions from binary to decimal: 100.01_2, 111.11_2, 011.10_2, 101.101_2.

4. Convert the following positive fractions from binary to decimal: 1000.1_2, 1111.1_2, 0111.0_2, 1011.01_2. How are your answers related to the answers to Problem 3? Do you know why?

5. Convert the following positive fractions from binary to decimal: 10101.010_2, 11011.011_2, 1110.0001_2, 0111.0111_2

6. Convert the following positive fractions from binary to decimal: 111.111_2, 1111.1111_2, 11111.11111_2, 111111.111111_2.

7. Write the following decimal fractions in binary form: 5.5, 7.25, 9.125, 12.0625.

8. Write the following decimal fractions in binary form: 8.375, 14.625, 15.875, 19.3125.

6.5.4 Negative Binary Numbers

There are numerous ways in which to represent negative quantities in the binary number system. The simplest is to add a single leftmost bit to the representation, where a 0 value for that bit indicates a positive number and a 1 indicates a negative number. Thus, if 101_2 is two bits plus a sign (leftmost) bit, it represents the decimal number -1, whereas 001_2 represents $+1$. Likewise, if 11010_2 is four bits plus a sign bit, it represents the decimal number -10, whereas 01010_2 represents $+10$. This combined representation of negative as well as positive numbers is called **sign-magnitude form**.

Exercise 6.10

1. Convert the following integers, in sign-magnitude form, from binary to decimal: 110_2, 010_2, 111_2, 011_2.

2. Convert the following integers, in sign-magnitude form, from binary to decimal: 1100_2, 0100_2, 1111_2, 0111_2.

3. Convert the following integers, in sign-magnitude form, from binary to decimal: 11000_2, 01000_2, 1111_2, 01111_2.

4. Convert the following integers, in sign-magnitude form, from binary to decimal: 11011_2, 10101_2, 01101_2, 10011_2.

5. Convert the following fractions, in sign-magnitude form, from binary to decimal: 110.01_2, 010.10_2, 111.11_2, 011.00_2

6. Convert the following fractions, in sign-magnitude form, from binary to decimal: 1100.001_2, 0100.010_2, 1111.100_2, 0111.111_2.

7. Write the following decimal fractions in sign-magnitude binary form: -5.25, +7.5, -9.375, +12.625.

8. Write the following decimal fractions in sign-magnitude binary form: -8.625, +14.875, -15.625, -19.5625.

Section 6.6: The ASCII Code

Our digital yearbook will certainly need to contain some text. How can text be represented using binary numbers? Engineers and computer scientists have developed a common code to represent letters of the alphabet in binary form. This code is called the **ASCII code** or ASCII table. Table 6.4 shows the entries of the ASCII code for the letters a-z, the capital letters A-Z, the numbers 0-9, and some important punctuation marks. It is almost a certainty that all the digital devices that you use—from your CD player to an LCD watch to a computer—use the ASCII code representation to store letters. From this table, one sees that the ASCII table uses 8 bits or 1 byte to store its values. Hence, there are 2^8=256 possible entries in the ASCII code table. The entries not shown include certain control characters and symbols that are used less often.

Table 6.4 Entries of the ASCII Table for few commonly used characters

Character	ASCII Code in Decimal	ASCII Code in Binary
A	65	01000001
B	66	01000010
C	67	01000011
D	68	01000100
E	69	01000101
F	70	01000110
G	71	01000111
H	72	01001000
I	73	01001001
J	74	01001010
K	75	01001011
L	76	01001100
M	77	01001101
N	78	01001110
O	79	01001111
P	80	01010000
Q	81	01010001
R	82	01010010
S	83	01010011
T	84	01010100
U	85	01010101
V	86	01010110

W	87	01010111
X	88	01011000
Y	89	01011001
Z	90	01011010
a	97	01100001
b	98	01100010
c	99	01100011
d	100	01100100
e	101	01100101
f	102	01100110
g	103	01100111
h	104	01101000
i	105	01101001
j	106	01101010
k	107	01101011
l	108	01101100
m	109	01101101
n	110	01101110
o	111	01101111
p	112	01110000
q	113	01110001
r	114	01110010
s	115	01110011
t	116	01110100
u	117	01110101
v	118	01110110
w	119	01110111
x	120	01111000
y	121	01111001
z	122	01111010
0	48	00110000
1	49	00110001
2	50	00110010
3	51	00110011
4	52	00110100
5	53	00110101
6	54	00110110
7	55	00110111
8	56	00111000
9	57	00111001
@	64	01000000
;	59	00111011
.	46	00101110
,	44	00101100
~	126	01111110
\	92	01011100
"	34	00100010
!	33	00011011
?	63	00111111
`	96	01100000

From this table, you to can learn to code text in ASCII and to read ASCII like a computer.

Exercise 6.11

1. Write the ASCII sequence representing the word *Hi*.

2. Write the ASCII sequence representing the term *MP3*.

3. Write the ASCII sequence representing the word *digital*.

4. Write your first name in ASCII.

5. What does the following ASCII sequence spell? 01001101, 01100001, 01110100 , 01101000

6. What does the following ASCII sequence spell? 01001001, 01101110 , 01100110 , 01101001, 01101110, 01101001, 01110100 01111001. Imagine trying to read an entire book this way!

Section 6.7: Quantization of Signal Samples

In our discussion of sampling we overlooked a very important point that we shall now consider. We implicitly assumed that computers can store numbers with infinite precision. That is, we assumed that the values of samples of a waveform, or pixels of an image, can be stored *exactly*. Real-world digital memory devices assign a fixed number of bits for the storage of each number to be stored, which implies fixed (and finite) precision. In this section, we shall focus on how numbers are coded in finite-precision binary format.

The accuracy achieved in storing a sequence s[n] depends on how many bits are allocated to the storage of each sample. More bits means greater precision. The following example will illustrate. Figure 6.15 shows a sampled analog signal. The computer cannot store every possible

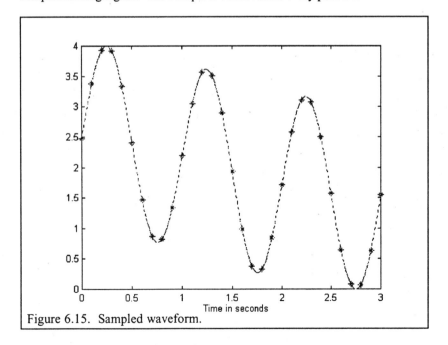

Figure 6.15. Sampled waveform.

level that this this waveform might take, because there are an infinite number of them! For example, suppose that one of the samples had the value of $\pi = 3.14159......$. This number cannot be represented exactly as a binary (or decimal!) number with a fixed number of digits. Using binary numbers with a fixed number of zeros and ones, we can represent only a finite number of amplitude levels. The signal in Figure 6.15 ranges between 0 and 4 in value. Suppose we decide to represent each signal sample by just two bits, which corresponds to only four levels. Let us set the levels at 0, 1, 2, 3. (The levels could be set differently if you like.) For each signal value s[n], we will then round the true value to the nearest of the four levels that we have agreed to represent. This process of signal **rounding**, shown in Figure 6.16, is called **quantization**. In Figure 6.16 the dashed lines indicate whether a sample is rounded up or down. Figure 6.17 shows the quantized version of the s[n] that would actually be stored.

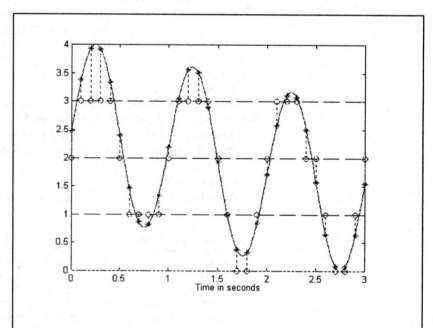

Figure 6.16. Quantization process. Plus signs and circles represent original and quantized values, respectively.

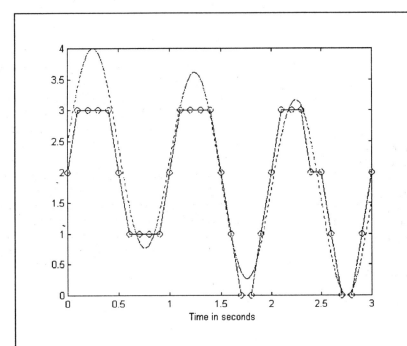

Figure 6.17: Signal quantized to 4 levels (2 bits). Original waveform is shown in dashed line.

The values of the quantized samples are listed in Table 6.5. Since there are only four quantization levels, these can be represented by two binary digits. We represent the levels 0, 1, 2, and 3 by the binary values 00, 01, 10, and 11, respectively. The binary representation for the quantized values is shown in the third and sixth columns of Table 6.5. Using this scheme, the storage of the number 3, represented by 11, on a CD ROM would require two successive pits, assuming that a 1 is represented by a pit and a zero by the absence of a pit. Storage of the number 2, represented by 10, would require a pit followed by an absence of a pit, and so forth.

True Value	Quantized Value	Binary Number	True Value	Quantized Value	Binary Number
2.0000	1.9750	10	-0.7078	0.0000	01
3.1674	3.9500	11	-1.5621	-1.9750	00
3.8694	3.9500	11	-1.6835	-1.9750	00
3.8289	3.9500	11	-1.0707	-1.9750	00
3.0465	3.9500	11	0.0000	0.0000	01
1.8006	1.9750	10	1.0807	1.9750	10
0.5412	0.0000	01	1.7233	1.9750	10
-0.2814	0.0000	01	1.6508	1.9750	10
-0.3885	0.0000	01	0.8637	0.0000	01
0.2217	0.0000	01	-0.3601	0.0000	01
1.2732	1.9750	10	-1.5718	-1.9750	00
2.3188	1.9750	10	-2.3223	-1.9750	00
2.9112	1.9750	10	-2.3346	-1.9750	00
2.7748	1.9750	10	-1.6092	-1.9750	00
1.9113	1.9750	10	-0.4244	0.0000	01
0.8002	0.0000	01			

Table 6.5: Sampled Values are quantized and Stored as Bits.

The approximation of the samples in Figure 6.16 is a very crude one, because we used only four quantization levels, or equivalently 2 bits. We could greatly improve the accuracy of our representation by using more bits per sample. Using B bits per sample gives 2^B quantization levels. Thus, 2 bits gives 4 levels, 3 bits gives 8 levels, 8 bits gives 256 levels (often used to represent image pixels), and 16 bits (used for music CDs) gives 65,536 levels. Clearly, as the number of levels increases, the error associated with the quantization operation in Figure 6.16 goes down. This is illustrated in Figure 6.18 where we show the quantized signal assuming 3 bits (8 levels) per sample, and in Figure 6.19, which assumes 4 bits (16 levels) per sample. As you can imagine, the 16 bits per sample used for recording music CDs gives an exceedingly fine approximation to the original analog sample values!

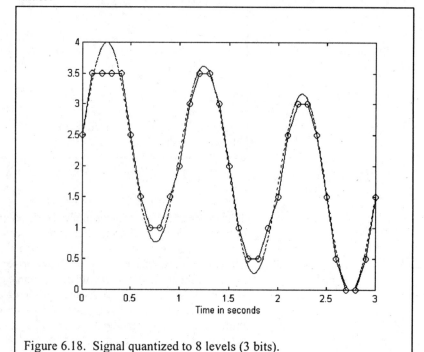

Figure 6.18. Signal quantized to 8 levels (3 bits).

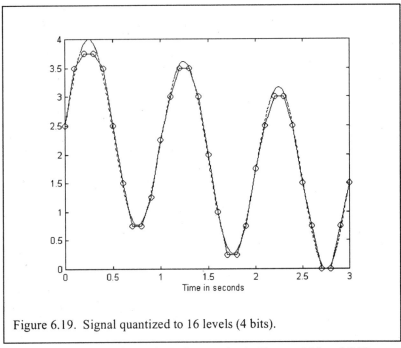

Figure 6.19. Signal quantized to 16 levels (4 bits).

Notice that unlike in Figure 6.17, the samples in Figures 6.18 and 6.19 can have fractional values. This is not a problem, because fractions can be represented in binary form as discussed in Section 6.5.3. Similarly, signals may go negative as well as positive. Both negative and positive quantized values can be represented in sign-magnitude binary form as discussed in Section 6.5.4. Usually, the quantization interval is not equal to 1 and the signal can be both positive and negative. In general, a signal may have a so-called **dynamic range** extending from +A to -A, in which case the total interval of size 2A would be divided up into 2^B amplitude levels for B-bit binary representation. The levels must be set to cover the full dynamic range of the signal; otherwise there will be amplitude **clipping**, whereby large amplitudes will be lowered in the coding process to the largest quantization level available.

Before leaving the topic of quantization, we raise one additional point. Quantization levels do not have to be spaced uniformly. Suppose that the signal to be quantized has a wide dynamic range, but is usually small and very seldom large. In this case, it would make sense to use more quantization levels for low amplitudes (where the signal is likely to have values) and fewer levels for higher amplitudes. Your computer uses floating point number representation, which is a particular form of nonuniform quantization intended for numbers that may have a huge dynamic range. In floating point, a number is represented as

$$M \, 10^E$$

where M is called the mantissa and E is the exponent. M and E are separately coded in binary and stored as a pair of numbers. This is equivalent to nonuniform quantization that devotes the same number of levels to each decade (ratio of 10) of amplitudes.

No matter what type of quantization we are talking about, the number of bits per sample determines the fidelity of the representation. This form of

Digital Storage
refers to the storage of digital signals, i.e. storage of quantized values represented in binary form as strings of bits.

storage, where numbers are quantized and approximated by fixed-precision binary values is called **digital**. The combined processes of sampling and quantization is called **digitization** or **analog-to-digital (A/D) conversion**. Digital systems are systems that manipulate or process digital signals. Furthermore, digital signal processing (DSP) refers to the manipulation of digital signals.

Exercise 6.12

1. $s[n] = 1.5 + 2\cos(0.6n)$ is quantized to the four levels 0, 1, 2, 3. Find and plot the original samples $s[n]$ and their quantized values for $n = 0, 1, 2, \ldots$, 12. Assign 2-bit binary values to the quantized samples.

2. $s[n] = 1.75 + 2\cos(0.6n)$ is quantized to the eight levels 0, 0.5, 1, 1.5, 2, 2.5, 3, 3.5. Find and plot the original samples $s[n]$ and their quantized values for $n = 0, 1, 2, \ldots$, 12. Assign 3-bit binary values to the quantized samples.

3. $s[n] = 0.035 \, n^2$ is quantized to the four levels 0, 1, 2, 3. Find and plot the original samples $s[n]$ and their quantized values for $n = 0, 1, 2, \ldots$, 10. Assign 2-bit binary values to the quantized samples.

4. Same as Problem 3, except with 3 bits and the levels 0, 0.5, 1, 1.5, 2, 2.5, 3, 3.5

5. Same as Problem 3, except with 4 bits and the levels 0, 0.25, 0.5, 0.75, 1, 1.25, 1.5, 1.75, 2, 2.25, 2.5, 2.75, 3, 3.25, 3.5, 3.75.

6. Define, in words, the operation of *quantization*.

7. What is the *dynamic range* of the signal in Problem 1?

8. What is meant by *clipping*?

9. There is some clipping in the quantization of the signal in Figure 6.16. The clipping can be reduced by resetting the quantization levels. How should the four levels be set to minmimize the maximum quantization error? Each pair of adjacent levels should be the same distance apart.

10. What is a *digital signal*?

11. What is *A/D conversion*?

Section 6.8: Quantization Noise

As we saw in the last section, the quantization operation inserts error into the storage process. We refer to this error as **quantization noise** because the effect of quantization error sounds like noise in a musical signal and looks like noise in an image. The amount of noise decreases as the number of bits B and the corresponding number of quantization levels increase. The amount of noise also depends on the dynamic range of the signal, since a wider dynamic range implies that the quantization intervals must be spaced more widely. Given these considerations, the most useful measure of the impact of quantization noise is **signal-to-noise ratio (SNR)**. There are many different, but similar, definitions of SNR. In this book, we use the simplest, which is the ratio of the largest signal amplitude to the largest noise amplitude:

$$SNR = \frac{\max \langle |signal| \rangle}{\max \langle |noise| \rangle} .$$

Here, the vertical bars indicate magnitude or absolute value and max means maximum over all time.

Suppose that the quantizer rounds sample values to B bits. In general, a signal can be negative as well as positive and so we use sign-magnitude binary representation. Thus, reserving the leftmost bit as the sign bit, we have B-1 bits remaining to represent the magnitude. These B-1 bits can represent 2^{B-1} levels. Suppose we set the highest level, 2^{B-1}, to represent the maximum signal amplitude. Assuming a rounding quantizer, the quantization error can be no larger than half a quantization interval. Thus, the SNR formula above reduces to

$$SNR = \frac{2^{B-1}}{\frac{1}{2}} = 2^{B} .$$

In typical digital storage applications, B ranges from about 6 bits to 24 bits. Thus, the SNR take on a huge range of values from one application to the next.

When engineers deal with quantities that take on a wide dynamic range, they sometimes express the quantity on a logarithmic scale called decibels (dB). A number x expressed in dB is defined to be $20 \log_{10}(x)$. Perhaps you have seen the log function in class. $\log_{10}(x)$ increases very slowly as x increases. Table 6.6 lists some values of $\log_{10}(x)$ for different x.

x	Log10(x)
1	0
10	1
100	2
1000	3
10000	4
100000	5
1000000	6

Table 6.6 Table of Values of $\log_{10}(x)$

Expressing SNR on a dB scale, we have

$$SNR = 20\log_{10}(2^{B}) = 20B\log_{10}(2)$$
$$= 6.02B.$$

Thus, on a dB scale, the SNR goes up linearly with the number of bits, B. Each additional bit raises the SNR by 6 dB. A music CD quantizes the sound samples with 16 bits, resulting in a SNR of about 96 dB. To put this number in perspective, an increase in audio level by 3 dB is perceptible by the human ear, and the difference in sound level between a quiet conversation and a jet aircraft taking off is about 100 dB.

> Sound quality will increase by 6dB for every additional bit we use to store signal samples.

VAB Experiment 6.3: Quantization and Clipping

- Listen to:
  1. Speech coded at 7 bits, 5 bits, 3 bits, and 1 bit per sample. The corresponding SNRs are 42, 30, 18, and 6 dB, respectively.
  2. Music coded at 16, 13, 10, 7, 4, and 1 bit per sample. The corresponding SNRs are 96, 78, 60, 42, 24, and 6 dB, respectively.
  3. Speech that is clipped in amplitude because the quantization levels do not cover the full dynamic range of the signal.
  4. Music that is clipped.

Exercise 6.13

1. What is *quantization noise*?

2. What is the definition of SNR, on a linear (not logarithmic) scale?

3. Suppose that the SNR in a particular application can range from 1 to 1,000,000 when measured on a regular linear scale. What would be the range of SNR when measured in dB?

4. Suppose a speech signal is quantized with 8 bits per sample. What is the SNR, measured in dB, of the speech signal? A music signal to be stored on a DVD is quantized at 22 bits per sample. How much higher is the SNR (again measured in dB) for the music signal?

Section 6.9:
Design of the
Digital
Yearbook

Given what we have learned about sampling and quantization, let us now return to the design of the digital yearbook. We will study separately the storage of text, speech, music, images, and video. Our design will be for a high school with about 1500 students and teachers. We will assume that the "conventional" part of the yearbook (text and images) occupies about 250 pages. After initial binary coding, we will assume that text can be compressed by a factor of 2, speech by 5, music by 3, images by 10, and video by 100. Chapter 9 discusses compression. Briefly, the idea underlying compression is that signals and images contain a great deal of redundancy. By removing this redundancy, the bit rate can be reduced. As an example, in an image, adjacent pixels are likely to have similar values. Thus, by coding the difference between adjacent pixels, which is likely to be a small number, fewer bits can be used. The decoding or decompression process needs to know how the signal or image was compressed so that the effect of the compression later can be undone. See Chapter 9 for details. After summing the required storage for the compressed text, speech, music, images, and video, we will add in an extra 10% for "overhead" to account for page layouts, background graphics, means of indexing the digital material, etc.

6.9.1 Text

We need to decide how much text our yearbook will contain. Let's assume that each page contains 25 lines of text on average (most of the space will be used by images), with 10 words per line and 6 letters per word on average. We will ignore punctuation. These assumptions give $(250)(25)(10)(6) = 375,000$ letters. Coding a single letter in ASCII requires 8 bits, or 1 byte. Thus, we project that the text in our yearbook will require 375 kilobytes (Kbytes) of storage, prior to compression. Assuming a compression factor of 2, the required storage for text would be 187.5 Kbytes.

6.9.2 Speech

We will assume that 1,600 sound clips are collected, with an average length of 1 minute. Speech can be bandlimited to 3,500-4,000 cycles per second without significant loss in sound quality. Thus, let's assume that the speech is sampled at 8,000 samples per second (Nyquist rate). Furthermore, assume that each sample is represented in 8 bits. This gives a total of $(1,600)(60)(8,000)(8) = 6,144,000,000$ bits = 768 megabytes (Mbytes), prior to compression. A compression factor of 5 would reduce this to 153.6 Mbytes. We see that the speech will require nearly 1,000 times as much storage as the text!

6.9.3 Music

The music to be stored might be recordings of the high school orchestra or jazz band, or might be from musical theater productions. We will also lump any other audio, such as nonmusical theater productions into this category if they are to be recorded with high fidelity. Let us assume that 12 hours of high-fidelity stereo will be recorded. The human ear can hear frequencies up to about 20,000 cycles per second (20 KHz). Thus, let's assume that the musical signal is bandlimited to 20 KHz prior to sampling and then is sampled at a rate of 40,000 samples per second (Nyquist rate). Each of the two stereo channels will be coded with 16 bits per sample. This gives a total of $(12)(60)(60)(40,000)(2)(16) = 55,296,000,000$ bits = 6,912,000,000 bytes, prior to compression. A compression factor of 3 would reduce this to 2,304 Mbytes. The music will require about 15 times as much storage as the speech.

6.9.4 Images

Assume that each of the 1,500 students and faculty will have a medium-sized portrait photo of size 800 by 600 pixels. Each pixel will consist of 3 colors represented in 8 bits per color. Furthermore, let's assume that there will be 500 other photos of all sorts of student activities. Some of these photos will be larger than the student/staff portraits, so we will assume image sizes of 1,200 by 900 to enhance image quality. The total amount of storage required for images will be $(1,500)(800)(600)(3)(8) +$ $(500)(1,200)(900)(3)(8) = 30,240,000,000$ bits = 3,780 Mbytes, prior to compression. A compression factor of 10 would reduce this to 378 Mbytes. The images will require more storage than the speech, but far less than the music.

6.9.5 Video

We may wish to incorporate long video segments from athletic contests or theater productions, as well as many, much shorter segments. Video will require lots of storage, though, so we will try to limit the amount of video in our yearbook. We will plan on storing 100 short video clips, with an average duration of 1 minute each, plus 10 longer video segments having an average duration of 15 minutes each. The total number of minutes of video will be 250. We will assume a frame rate of 30 frames per second, and a frame size of 640 by 480 pixels. As with the still-frame images, each pixel will be represented by 3 colors with 8 bits per color. This leads to (250)(60)(30)(640)(480)(3)(8) = 3,317,760,000,000 bits = 414,720 Mbytes, prior to compression. A compression factor of 100 would reduce this to 4,147.2 Mbytes. Despite the high compression factor, this amount of video requires more storage than any other component of our yearbook – about twice the storage required for the 12 hours of music.

6.9.6 Total Storage and Iteration on the Design

The sum of the storages required for the compressed text, speech, music, images, and video is 0.1875 + 153.6 + 2,304 + 378 + 4,147.2 = 6,983 Mbytes = 6.983 Gbytes. Adding in an extra 10% overhead (multiplying 6.983 Gbytes by 1.1) gives a total storage figure of 7.68 Gbytes. How does this compare with the amount of storage available on a CD? The answer is not too well! Chapter 9 tells us that a CD can hold only 650 Mbytes of information. Thus, our digital yearbook will not fit onto a single CD. Twelve CDs would be required. The problem is that both the music and the video require a tremendous amount of storage. A single DVD could do the job, though, depending on the type of DVD. Chapter 17 states that a one-sided, single-layer DVD can hold 4.7 Gbytes. A double-sided single-layer DVD can hold 9 Gbytes. Thus, we are in luck – one double-sided single-layer DVD will easily hold our digital yearbook. Of course, if we had hoped to store many hours of video, even the double-sided DVD would not do the job.

Suppose we wish to iterate on (revise) our above yearbook design so that the digital yearbook can fit onto one single-sided DVD. What information from the above yearbook should we cut out? Since the music and video dominate the storage requirement, let's reduce those. Reducing the music from 12 hours to 10 hours will cut the required music storage from 2,304 Mbytes to 1,920 Mbytes, giving a savings of 2,304 − 1,920 = 384 Mbytes. Completely eliminating the longer video segments will reduce the required video storage from 4,147.2 Mbytes to 1,658.9 Mbytes, giving a savings of 4,147.2 − 1,658.9 = 2,488.3 Mbytes. These two reductions, plus the corresponding overhead savings will be (1.1)(384 + 2,488.3) = 3.16 Gbytes. Subtracting this savings from the original total required storage gives 7.68 − 3.16 = 4.52 Gbytes, which is just about right for a single-sided DVD!

Exercise 6.14

1. Suppose an average page in a book contains 500 words and that words average 6 letters in length. How many pages of text can fit onto a CD? Assume that no compression is used.

2. Assuming that speech is sampled as outlined in Section 6.9.2, how many seconds of speech could fit onto a CD? Assume that compression by a factor of 5 is used.

3. Assuming that music is sampled as outlined in Section 6.9.3, how many minutes of music could fit onto a CD? Assume that a compression factor of 3 is used.

4. How many color images of size 1200 by 800 pixels can fit onto a DVD having 4.5 Gbyte capacity? Assume 8 bits per color and compression by a factor of 10.

5. How many hours of video can fit onto a DVD having 4.5 Gbyte capacity? Assume 800 by 600 pixel image frames, 30 frames per second, 8 bits per color, and compression by 100. How many hours of video can fit with no compression?

6. In the above yearbook design, suppose we decide to store only the text and images to make a very basic form of the digital yearbook. Will they fit onto a CD? If so, how much could the image compression factor be reduced such that the text and images would still fit onto a CD? Don't forget the 10% overhead factor!

7. In the above digital yearbook design, suppose that we wished to increase the number of hours of music from 12 to 20. How much would the compression factor have to be increased so that the entire digital yearbook would still fit onto a single double-sided DVD having capacity 9 Gbytes? Don't forget the 10% overhead factor!

8. In the above digital yearbook design, suppose we increase the number of minutes of video from 250 to 350. How much would the compression factor have to be increased so that the entire digital yearbook would still fit onto one double-sided DVD having capacity 9 Gbytes? Don't forget the 10% overhead factor!

Appendix A: Analysis of Aliasing

This appendix is included for those students who may wish to undertake a project on the subject of aliasing. A more thorough study of the aliasing phenomenon boils down to asking the following question. Suppose we sample

$$s_1(t) = \cos(2\pi f_1 t)$$

with period T_s to produce the sequence

$$s_1[n] = \cos(2\pi f_1 n T_s).$$

Then, sampling analog sinusoids of what other frequencies f_2 will produce exactly this same sequence $s_1[n]$ (in which case these sinusoids will be indistinguishable in the sampled world)? We claim that there are two sets of analog frequencies that will be aliased to the same set of samples $s_1[n]$, namely

$$f_2 = f_1 + k f_s, \quad k = \text{any integer}, \qquad (1)$$

and

$$f_2 = -f_1 + k f_s, \quad k = \text{any integer}, \qquad (2)$$

where $f_s = 1/T_s$ is the sampling frequency. Here, we are saying that there are an infinite number of frequencies that all alias to exactly the same set of samples! For example, if $f_1 = 200$ Hz and $f_s = 500$ Hz, then the following frequencies would all masquerade as 200 Hz in the sampled world: 700, 1200, 1700, etc., from aliasing condition (1), and 300, 800, 1300, etc., from aliasing condition (2). Amazing! Figure 6.20 shows the first several frequencies that will alias to 200 Hz if the sampling rate is $f_s = 500$ Hz.

46

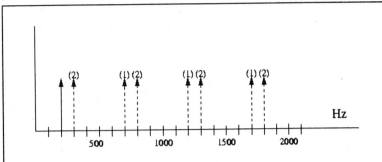

Figure 6.20: Frequencies that will alias to 200 Hz when sampled at 500 Hz. Parenthesized numbers indicate whether the alias results from condition (1) or condition (2).

Going back to an earlier example from Section 6.3.2 and Figure 6.9, letting $f_1 = 720$, the 60 Hz alias that we observed is obtained from alias condition (1) above with $k = -1$ as

$$f_2 = 720 + (-1)660 = 60.$$

In Figure 6.9 there are an infinite number of other higher-frequency sinusoids (corresponding to larger k) that will produce the same samples shown by the dots. Visually, however, we pick out the 60 Hz sinusoid since it is the simplest in the sense that it is the lowest frequency sinusoid that passes through the given samples.

Let us now prove the first aliasing condition. We will leave proof of the second condition as a homework problem. Suppose we have $s_1(t)$ at frequency f_1 and its samples $s_1[n]$ as given above. Consider a second sinusoid

$$s_2(t) = \cos(2\pi f_2 t)$$

where we choose $f_2 = f_1 + kf_s$, and $f_s = 1/T_s$ is the sampling rate. The samples of $s_2(t)$ would then be

$$\begin{aligned}
s_2[n] &= \cos(2\pi(f_1 + kf_s)nT_s) \\
&= \cos(2\pi f_1 nT_s + 2\pi kn) \\
&= \cos(2\pi f_1 nT_s) \\
&= s_1[n]
\end{aligned}$$

which is exactly what we set out to show. Notice that this proof holds for any integer value k. Thus, as stated earlier, there are an infinite number of frequencies that can alias to the same sequence. The saving grace is that all of these other frequencies violate the Nyquist condition. That is, if s(t) is

47

bandlimited to f_{max} and we choose $f_s > 2f_{max}$, then all potential alias frequencies will lie above f_{max} and therefore cannot be part of s(t). In the example above with sampling frequency $f_s = 500$, the implication of the Nyquist assumption is that there is no frequency being sampled having value greater than 500/2 = 250 Hz. Notice that all of the potential aliases we listed (700, 1200, 1700 etc. and 300, 800, 1300 etc.) are greater than this number.

Web Site Demo

Remember the web site demo of the rotating bicycle wheel from Section 6.3.2? See the movie demo in VAB. We noted that if the wheel rotates at a constant speed, then it sometimes can appear to be rotating backwards (see Problem 5 below). If the wheel changes speed, an even more interesting effect occurs. Basically the integer k in the above analysis can change with time so that as the wheel slows down it may at times appear to speed up or to suddenly rotate backwards.

Exercise 6.15

1. Suppose a sinusoid of frequency 1 KHz is sampled at a rate of 2,200 samples per second to produce a sequence s[n]. What other analog frequencies would produce this same sequence?

2. Same as Problem 1, except for a sinusoid of 500 Hz sampled at 2,200 samples per second.

3. For the conditions in Figure 6.9, what other (higher) frequencies would produce the same samples shown?

4. Prove aliasing condition (2), i.e., if a sinusoid of frequency f_1 is sampled at a rate f_s, then show the same sequence would be created by sampling a sinusoid of frequency $-f_1 + kf_s$ where k is any integer. Hint: The cosine is an even function (cos(t) = cos(-t)).

5. Suppose a circular piece of cardboard is cut out and a dot is painted near its edge. Now, suppose the cardboard disc is spun on the end of a drill and photographed with a video camera operating at 30 frames per second. If the disc completes one rotation each 1/30 of a second, explain what will you see in the video. If the disc completes slightly more than one rotation each 1/30 of a second, explain what you will see in the video. If the disc completes slightly less than one rotation each 1/30 of a second, explain what you will see in the video. Justify your explanations by showing the location of the dot on the disc at various times in the video sequence. Now, explain what you will see if the disc completes two, slightly more than two, and slightly fewer than two rotations every 1/30 of a second. Does your explanation generalize to k rotations where k is any integer? How does your analysis relate to the aliasing phenomenon?

Section 6.10: Glossary

aliasing
An undesired effect due to undersampling where one frequency can masquerade as another

analog-to-digital conversion
Sampling, of an analog signal followed by quantization of the samples to binary numbers

antialiasing filter
An analog lowpass filter preceding the sampler to assure that the analog signal is bandlimited prior to sampling

ASCII Code
Binary representation of the letters and other characters on a keyboard

bandlimited
A signal whose highest frequency is limited to a fixed, finite value

binary representation
An arithmetic system for representing numbers as a series of bits

bit
One of two physical states, typically thought of as a zero or a one.

clipping
Result of the signal amplitude exceeding the range of available quantization levels

digital signal processing (DSP)
The manipulation or processing of digital signals

digital signals
Sampled values represented in binary form as bits

digital systems
Systems that store, manipulate, or communicate digital signals

digitization or Analog-to-Digital (A/D) Conversion
Combined operations of sampling and quantization

dynamic range
The difference between the largest and smallest values that a signal takes on

Nyquist rate
Minimum necessary samp ling rate = twice the highest frequency contained in the analog signal

quantization
Rounding of numbers to one of a finite number of levels

quantization noise
The error or noise introduced into signal samples through the process of quantization

sampling frequency (f_s)
Number of samples per second = 1/(sampling period)

sampling period (T_s)
Spacing in time between two adjacent samples

Sampling Theorem
Provides a formula for exact reconstruction of an analog signal from its samples, provided that the signal is bandlimited and sampled above the Nyquist rate

sign-magnitude Form
A binary representation for negative, as well as positive, numbers where the leftmost bit indicates the sign of the number

Signal-to-Noise Ratio (SNR)
Ratio of maximu m signal level to maximum noise level, usually expressed in a logarithmic form in decibels (dB). For quantization noise, SNR = 6B where B is the number of bits per sample

undersampled
A signal sampled below the Nyquist rate

The Infinity Project

Chapter 7:
Communicating with
1s and 0s

Approximate Length: 3 Weeks

Section 7.1:
Introduction

7.1.1 What is Communication?

In earlier chapters we found that voices, music, images, and other types of signals and media can be represented digitally as a list of symbols or bits. We have determined that each of these media can be converted into symbols, and we have determined that these symbols can be converted back into a form that humans can easily recognize and interpret. In addition, we have learned that those symbols can be stored and recovered from storage, and that they might be manipulated or "processed" to improve some aspect of how we interpret them.

In this chapter, we examine the **communication** of digital information. We define communication as the movement of digital or symbolic representations from one location to another physically separate location, which may be millimeters or millions of miles away. At the new location the information might be stored, reconstructed for human use, or retransmitted to yet another destination. A wide variety of communications systems such as the telephone, electronic mail, and broadcast radio are part of everyday life.

7.1.2 Design Objectives for Communications Systems

Before we begin our investigation of communications systems, we will first define, from the perspective of engineering design, exactly what our objectives are. We do this by asking and answering the following questions.

- **What problem are we trying to solve**? We desire to move multimedia information from one location to another. In principle we'd like to move it as fast as possible, receive it as accurately as possible, and to do it as cheaply as possible.

- **How do we formulate the underlying engineering design problem**? The potential user of a communications system will usually specify his or her requirements, that is, the set of characteristics the system must have to make it worth paying for. These characteristics will typically include the ones just mentioned, speed, accuracy, and cost, but may also include others, such as the delay between transmission and reception, the security of the communications system, and its ease of use. A common additional requirement is that the communications system work in accordance with government regulations or industry standards.

Interesting fact: Sound travels at 331 meters per second at sea level. As the altitude increases, the atmospheric pressure drops and so does the velocity of sound.

- **What will the consequences be if the communications system satisfies the user's need?** Engineering work is often driven by an economic objective. When this is true, the consequence of producing a communications system that meets the customer's need is being paid for the work and having the opportunity to do even more challenging, interesting, anf profitable work for the same happy customer or for new customers. More generally, the consequence of developing better technology for multimedia communications is that the whole world will be able to work together on a far broader scale than was ever possible before.

- **How will you test your design?** As we just discussed, most potential users of a communications system will provide a list of specifications. An important part of the engineering design process is to turn those specifications into a list of tests. Both the designer and the customer agree that successful completion of the tests means that the system works as desired. In the case of multimedia communications systems, these tests will typically include some of the following:

  - Peak data rate

  - Average data rate

  - Number of errors as a fraction of the total number of bits transmitted

  - Number of seconds in which no data is received

  - Maximum and minimum delay between transmission and successful reception

  A carefully written "test plan" is typically part of the initial contract between the supplier of a communications system or service and the prospective customer. This plan describes exactly how each of the measurements will be made so that both have confidence in how the system well will work.

- **What do we need to learn in order to understand multimedia communications systems and then to design them?** In this chapter we will explore the common elements of all communications systems. We will explore the design goals of each of these systems and then consider different methods of achieving those goals. We will start with simple solutions using basic technology and then watch them evolve into higher performance systems that employ state-of-the-art technology.

7.1.3 Basic Concepts and Definitions

Although there are many different ways to communicate information, all of them rely on the same basic principles and use the same three basic components to successfully transfer information from an originating source to a destination. This process is shown in Figure 7.1. On the left side of the figure a user provides information to the **transmitter**. This transmitter accepts the information from the user and turns it into a signal that can be conveyed over the **communications channel**, which is the physical medium that carries the signal. This medium might be an optical fiber, a pair of telephone wires, or the air itself. The channel carries the signal from the transmitter to the **receiver** shown on the right side of the figure. At the destination the signal is captured with an appropriate device, such as an antenna, and supplied to the receiver. The receiver attempts to recover the transmitted data as accurately as possible from the signal, and then deliver this recovered data to the user at the distant location.

<table>
<tr><td>
Communication channel:
It is the physical medium that carries the signal from the transmitter to the receiver, for example, telephone wires, optical fiber, or air.
</td></tr>
</table>

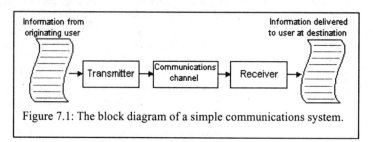

Figure 7.1: The block diagram of a simple communications system.

Spoken communication between two people in a room, as illustrated in Figure 7.2, can be interpreted in terms of the block diagram in Figure 7.1. The information to be communicated exists in the mind of the talker. The transmitter consists of the talker's vocal cords and the parts of the brain which convert mental information into mechanical motion of the vocal cords and mouth to generate speech sounds. The communications channel is the air between the two people. The speech sounds travel through the air from the mouth of the speaker to the ear of the listener. The receiver consists of the listener's ears, which capture and respond to the speech sounds, and the mental processes which convert that response into meaningful ideas in the mind of the listener.

Figure 7.2: Spoken communication between two people.

7.1.4 A Bit of History

The timeline on the left highlights some important events in the evolution of digital communications. While people have been communicating since before recorded history, the invention of the first practical electrical telegraph by Samuel Morse in the 1830s marks the beginning of modern digital communications. The telegraph was the first commercially successful application of electricity. It made it possible to govern large countries, to

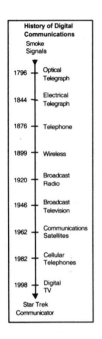

History of Digital Communications

Smoke Signals

1796	Optical Telegraph
1844	Electrical Telegraph
1876	Telephone
1899	Wireless
1920	Broadcast Radio
1946	Broadcast Television
1962	Communications Satellites
1982	Cellular Telephones
1998	Digital TV

Star Trek Communicator

conduct business on an international scale, and to operate military forces across the continents and oceans. With it also came the field of electrical engineering, and the concept of venture capital, private investment in technology as a way of creating financial wealth. As the timeline shows, innovations have continued since the advent of the electrical telegraph, and they can be expected to continue into the future. All of them are based on the fundamental ideas that made the telegraph work.

Exercises 7.1

1. For the following communications systems, identify and describe the transmitter, the channel, and the receiver:

    a cellular telephone, broadcast radio, cable television, a facsimile (fax) system, a television remote control

2. Determine the rate of communication for talking by having several students read a selected paragraph out loud and measuring the time it takes them to say the words. Count the number of words in the paragraph and the number of characters in the paragraph.

    a.) Compute the communication rate in units of words per minute.

    b.) Compute the communication rate in units of characters per second.

    c.) Which units, words per minute or characters per second, are the most natural measure of spoken language?

    d.) Which units, words per minute or characters per second, are the most natural measure of typing on a keyboard?

3. Repeat parts (a) and (b) of Exercise 2 by measuring the rates at which students can type the same paragraph on a keyboard.

Section 7.2:
A Simple
Communications
System

The example of two people speaking to each other seems "simple" in that it is intuitive, commonly practiced, and easily accomplished even by young children. However, the cognitive and physiological processes of transmitting and receiving are so complex that many important components of these activities are currently not fully understood. These processes are still areas of active research for psychologists and biologists. In contrast, manmade communications systems, in which the receiver is not a human listener, rely on simpler, well-understood transmitters and receivers. These systems use signals with simple mathematical representations that are less intuitive for human perception than speech. However, the decision to use simpler signals allows us to design and build inexpensive, yet reliable, communications systems that can operate over great distances at rates much faster than human speech.

7.2.1 Design of a Simple System

A simple example of a digital communications system can demonstrate the functions shown in the block diagram of Figure 7.1. The objective of this system is to send text information wirelessly over a short distance such as across a room. Consider a method of sending textual information from one side of the room to the other by turning the individual letters of the text into "music" instead of using speech. The text characters will be represented by a set of audio tones that are transmitted using an audio loudspeaker as the transmitting device instead of vocal cords. The communications channel will be the air between the transmitting and receiving sites. The receiver will capture the audio tones transmitted through the air by listening with a microphone instead of with ears. If this system works correctly, the receiver will recognize the transmitted tones, and will be able to turn them back into the original transmitted text. Although there are many possible ways to transmit information wirelessly, we consider an acoustical system first because it is most similar to our direct experience of talking and listening.

Tone: a signal of constant frequency and amplitude, for example, a sinusoid.

There are many possible ways to represent letters with audio tones. One simple approach would be to associate each letter with a unique **tone**, a signal of constant amplitude and frequency. Figure 7.3 shows how this system could function. Note that this figure closely resembles Figure 7.1, except that each block in Figure 7.3 has more specific detail than the corresponding general block in Figure 7.1. The receiver in Figure 7.3 must attempt to determine what characters were transmitted based on the sound patterns it receives. Then the receiver must make a list of the identified characters and display them to the recipient of the message. The successful functioning of the receiver requires that we identify each character by a unique sound pattern, so that the receiver can distinguish each one from the other possibilities.

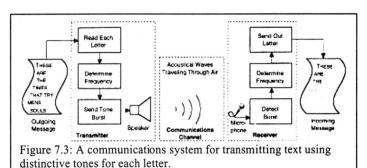

Figure 7.3: A communications system for transmitting text using distinctive tones for each letter.

Other possible sound patterns for this system might also be considered. A person could simply choose to spell the word, saying each letter in turn so that the signal associated with each letter would be the sound of someone saying that letter. However, human speech creates a very complex signal that usually communicates far more information than just the text of the message. From speech a listener might also learn the identity of the speaker or the mood of the speaker or the regional dialect used by the speaker. Because of these factors and the similarity of many letter sounds such as "b" and "p", speech signals would be difficult for a simple electronic receiver to interpret correctly. To make it easier for the electronic receiver, simpler signals, in this case unique sinusoidal signals, are chosen to represent the letters. Although these sinusoids might be perceptually unfamiliar to people as a way of communicating information, they are far easier for an electronic system to interpret correctly. When we analyze this method later, we will also find that it can transmit data much faster than a human speaker and listener.

7.2.2 Operation of a Simple System

Assume that the information to be transmitted is a list of capital letters chosen from the Roman alphabet. Examples of messages might be "MEET ME AT FOUR OCLOCK" or "HOUSTON WE HAVE A PROBLEM." Designing the system shown in Figure 7.3 requires the selection of a different tone to represent each character. How this might best be done is discussed later, but for now we'll use the "mapping" shown in Table 7.1. Using this table, sending the letter A requires the creation of a sinusoid with frequency 300 Hz. The letter B would be indicated by a sinusoid with frequency 400 Hz. The other letters would be generated in the same way with 100 Hz difference between each adjacent pair of character tones. Thus with 27 tones we can send all 26 capital letters and the space character needed to separate words. If punctuation or numbers were also needed, more tones would have to be added.

The "air modem" is used to demonstrate that textual messages can be sent through the air from a transmitter to a receiver one letter at a time using short bursts of audio tones. The tone burst for each letter has a different frequency than the frequency for every other letter.

Letter	Frequency (Hz)	Letter	Frequency (Hz)	Letter	Frequency (Hz)
A	300	J	1200	S	2100
B	400	K	1300	T	2200
C	500	L	1400	U	2300
D	600	M	1500	V	2400
E	700	N	1600	W	2500
F	800	O	1700	X	2600
G	900	P	1800	Y	2700
H	1000	Q	1900	Z	2800
I	1100	R	2000	space	2900

Table 7.1: A mapping from the capital letters to the frequencies in Hertz of audible tones with 100-Hz separation between adjacent tones.

The general mathematical representation of a tone is given by Equation 7.1,

$$s(t) = a * \cos(2\pi f t + \phi) \tag{7.1}$$

where *s(t)* represents the strength of the signal as a function of the time variable *t*. The tone is completely specified by three parameters. The frequency of the sinusoid is represented by *f*, the maximum amplitude by *a*, and the phase by ϕ. The specific signals for the letters A and B are given by $s_A(t)$ and $s_B(t)$, respectively, in Equation 7.2.

$$s_A(t) = a * \cos(600\pi t + \phi)$$
$$s_B(t) = a * \cos(800\pi t + \phi) \tag{7.2}$$

Operation of the Transmitter: How then would this communications system send the word **MEET**? Upon receiving the list of the characters in the message from the originating user, the transmitter would send a sequence of four tone bursts for the four-character word **MEET**. A **burst signal** is a signal that is present for only a short time interval rather than forever. A tone burst is a burst signal that has a constant frequency and amplitude during the short time interval. For the message **MEET**, the first tone burst, corresponding to the letter **M**, would have a frequency of 1500 Hz. The next two tone bursts would be at 700 Hz, and the last would be at 2200 Hz. There would be some quiet time between each burst so the two **E**s would be distinguishable from each other at the receiver. After completing the set of four tone bursts, the transmitter would then wait for another message to transmit. Figure 7.4 shows a waveform for this signal in which the tone bursts last 4 milliseconds (ms) and the quiet time between the tone bursts is 2 ms. Note that the frequency for the **T** signal is about three times the frequency for the **E** signal, so the number of cycles in the 4-ms tone burst for the **T** is about three times the number of cycles for the tone bursts for the **E** signal.

Filter: A simple system (digital or analog) which only allows prespecified frequencies to pass from the input to the output.

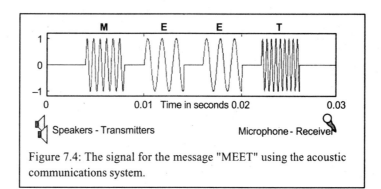

Figure 7.4: The signal for the message "MEET" using the acoustic communications system.

Operation of the Receiver: How would the receiver function? If the receiver is close enough to "hear" the transmitted signal clearly, its job is to separate the signal into individual tone bursts and then determine which of the 27 possible frequencies was used for each tone burst. If it can accurately estimate the frequency for each tone burst, it can look up the frequencies in its own copy of Table 7.1 to determine which characters were sent. Presuming that the receiver successfully detected bursts at frequencies of 1500, 700, 700, and then 2200 Hz, it would display the letters of the word **MEET** to the recipient as intended by the sender.

A closer look at the receiver is shown in Figure 7.5. The incoming signal is provided as the input to 27 separate **bandpass filters.** Each filter is tuned to respond strongly to a narrow range or band of frequencies around the specified frequency for one of the characters. Frequencies outside this range would create a very weak response. The segment of the message signal representing the letter **E** is shown as the input to all of the filters. The output of the filter tuned to the frequency for the letter **E** has a strong output signal. A good receiver will also generate a strong response to frequencies very close to the desired frequency so that small errors in tuning will not cause complete operational failure. Because the filters will not be "perfect," the filters for other letters will have a low-level output rather than a zero-level output. As shown in the figure, the two adjacent filters for **F** and **D** have very weak output signals and all other filters have a flat output. In this case, the decision logic has a relatively easy task to determine which character was sent.

The receiver structure, shown in Figure 7.5, is similar in some ways to the anatomical structures in the human ear, shown in Figure 7.6. The sound travels into the ear and causes vibrations in the fluid of a coiled tube inside the ear. This tube, which looks like a snail shell, is called the **cochlea** Due to the shape of this tube with its varying stiffness and diameter, different locations along the tube respond best to different frequency ranges. The length of the tube in the cochlea might be about 30 mm [1]. The maximum response for the lowest frequency that humans can hear, approximately 20 Hz, occurs at the most distant end of the tube which is the center of the coil. A 500-Hz signal would cause a maximum response about 80% of the distance to the end of the tube while a 1500-Hz signal would cause a maximum response at about 60% of the distance along the tube [1]. Nerve receptors at sites all along this tube pick up these responses and communicate them to the brain where decision analysis is far more complex than the simple logic needed for Figure 7.5.

Bandpass Filter: A filter which only "passes" or responds strongly to a specific range or "band" of frequencies.

Interesting fact: The human auditory system can hear sounds in the range from 15 Hz to 20,000 Hz, but it is most sensitive to sounds in the range from 1000 to 4000 Hz.

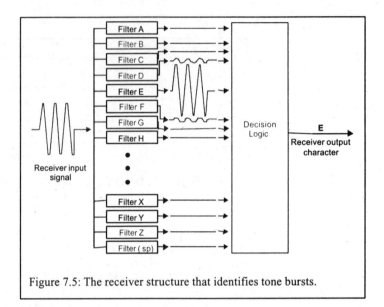

Figure 7.5: The receiver structure that identifies tone bursts.

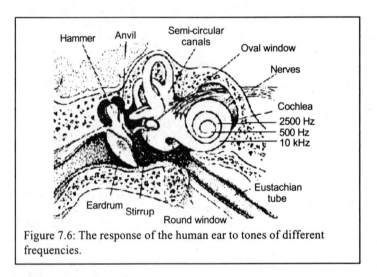

Figure 7.6: The response of the human ear to tones of different frequencies.

7.2.3 Key Concepts

A message of any length can be sent using the simple system that we have designed as long as the message only uses the 26 capital letters and the word space. One measure of performance of a communications system is how long it takes to send a message. Clearly, for this system, the transmission time depends directly on the number of characters to be transmitted, the durations of the tone bursts, and the quiet time between tone bursts. When the duration of the tone bursts is increased, the characters are more easily and confidently distinguished by the receiver, but the overall transmission rate is reduced. If a system can successfully receive ten bursts per second, then the transmission rate is ten characters per second, or ten symbols per second.

The fundamental idea that makes this system work is that the transmitter sends an acoustic signal for each character that is designed to be **sufficiently different** from the acoustic signals it sends for all other characters so that the receiver can reasonably be expected to distinguish that character's signal from the signals of other characters. This fundamental concept applies to all communications systems no matter what channel, message coding, or transmission is used. The specifications for what will make the signals sufficiently different will depend on the resources available for the receiver design, the characteristics of the communications channel, and the environment in which the communications system is expected to operate. If the signals associated with all possible transmitted characters are sufficiently different, then the receiver should be able to accurately distinguish and interpret all of them. Therefore, at the signal's destination the receiver should be able to accurately report the whole message to the user without error.

Interesting application:
Using tones to help the visually impaired read printed text.

There are some communications systems that do use audio tones to communicate directly to a human listener rather than to an electronic receiver. A device developed to assist a blind person in reading printed text, called a Stereotoner, converts printed character images into combinations of ten tones [2,3]. Horizontal strips of printed text are sliced vertically as shown in Figure 7.6A(a). Each vertical strip is divided into ten squares corresponding to the frequencies of the ten tones. If a square is mostly black, then the corresponding tone is turned on, and the pattern of the vertical strip creates a particular multitone sound. The vertical strips are scanned from left to right, creating a sequence of multitone sounds that can communicate the image information to a trained listener. A similar concept is used in navigation aids which convert the distance of objects into sound patterns.

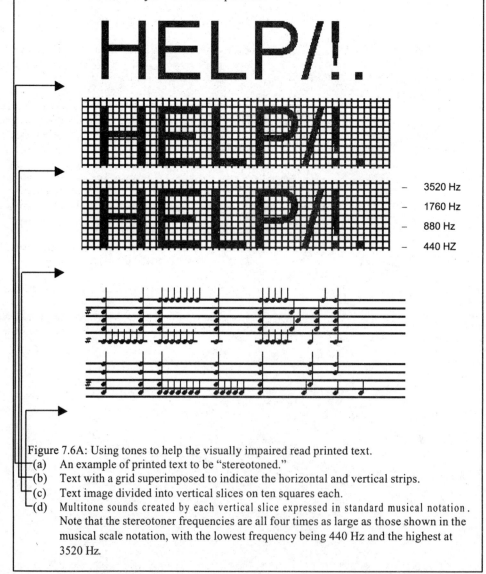

Figure 7.6A: Using tones to help the visually impaired read printed text.
- (a)   An example of printed text to be "stereotoned."
- (b)   Text with a grid superimposed to indicate the horizontal and vertical strips.
- (c)   Text image divided into vertical slices on ten squares each.
- (d)   Multitone sounds created by each vertical slice expressed in standard musical notation. Note that the stereotoner frequencies are all four times as large as those shown in the musical scale notation, with the lowest frequency being 440 Hz and the highest at 3520 Hz.

Exercises 7.2

1. Assume that the receiver in your communications system is capable of distinguishing tone bursts that are only 25 Hz apart instead of 100 Hz apart, and the letter **A** is still assigned the frequency 300 Hz. Make a mapping of letters to frequencies similar to Tables 7.1 and 7.2 with a 25-Hz spacing between adjacent pairs. What frequency range would be needed for the character set if your table were used?

2. You are designing an expanded communications system that still uses a single frequency to represent each character, but the characters will include both upper- and lowercase letters, all 10 decimal digits, and a space.

    a.) How many different tones are needed?

    b.) If the lowest acceptable frequency is 300 Hz and the highest acceptable frequency is 3400 Hz, what frequency separation between the tones should be used?

3. If each tone burst lasts 4 msec and the bursts are separated by 1-msec quiet intervals, how long would it take to transmit the message "MEET ME AT FOUR OCLOCK FOR COFFEE?" (Hint: Don't forget that the space between a word is represented by a tone burst.)

4. One way to estimate the frequency of a signal is to count the number of times the signal changes sign from positive to negative. This is often called counting "zero crossings." A sinusoidal signal should have two zero crossings for every cycle.

    (a) How many zero crossings would be counted for a 300-Hz sinusoidal tone burst of 10 msec? Of 5 msec?

    (b) Repeat part (a) for a 400-Hz sinusoidal tone burst.

    (c) How many "zero crossings" would distinguish an A at 300 Hz from a B at 400 Hz if tone bursts last 10 ms? If tone bursts last 5 mec?

Section 7.3:
Sources of Error in
a Communications
System

7.3.1 Causes of Errors in Communications Links

Suppose the receiver cannot distinguish between a tone burst at 700 Hz and one at 800 Hz. In this case the receiver will not be able to accurately reconstruct the transmitted message if it contains either **E** or **F**. Similarly, if the receiver is unable to "hear" bursts with frequencies above 2500 Hz, then it will simply not realize the letters **X**, **Y**, and **Z**, and the space have been transmitted. Clearly, both of these cases must be avoided since they degrade the accuracy of the received message. The system's designer would be responsible for making sure that the receiver can distinguish all of the possible tones and that all of the tones are within the receiver's "hearing range." There are practical circumstances, however, that can limit the system's performance in spite of the care used by the designer. Three common causes of degraded performance in a communications system are weak signals, noise, and interference. Figure 7.7 shows examples of these conditions using the **MEET** signal shown in Figure 7.4.

Weak signals: Suppose that the transmitter is far enough away from the receiver that the acoustical signal becomes inaudible. At some point it will become too weak to be accurately detected. This is illustrated in Figure 7.7(a). If this weak signal is the input to the bank of filters as shown in the block diagram in Figure 7.5, then even the strongest outputs will be very weak and the decision logic will not be able to determine with confidence that it is much stronger than the outputs of all other filters.

Experiment 7.2: Effect of Weak Signals, Noise and Interference

For the air modem's receiver to work properly, it must be able to detect the occurrence of each tone burst. Then it must determine the burst's audio frequency, and finally determine the letter corresponding to that frequency. Inability of the receiver to "hear" the transmitted signal clearly will degrade its ability to detect the tone and accurately determine its frequency.

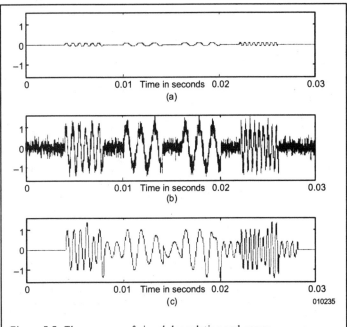

Figure 7.7: Three causes of signal degradation and errors demonstrated with the signal for the message MEET – (a) weak signal, (b) noisy signal, and (c) interference from weaker message HELP.

Noise: In many circumstances the transmitted signal is received in the presence of noise. This noise may mask the transmitted signal enough to make it difficult to detect the presence of each tone burst or to accurately determine its frequency. If either of these problems exist, then characters will be missed altogether by the receiver or they may be interpreted as the wrong character. Figure 7.7(b) demonstrates this problem.

> **Experiment 7.2 contd: Effects of Weak signals, Noise and Interference**
>
> We observed in Experiment 7.2 that when the signal arriving at the receiver is weak, the ability of the receiver to properly recover the message that was sent is degraded. Even if the transmitted signal is received loudly enough at the receiver, the presence of noise at the receiver can also limit how well it recovers the message.

Interference: In some cases the receiver may be able to "hear" from two or more transmitters simultaneously. If the frequencies used by the various transmitters overlap to the extent that they are not separable by the receiver, then, just as with the noisy case, the receiver may make errors and even miss the desired tone bursts altogether. In Figure 7.7(c) the **MEET** signal and a weaker **HELP** signal, which begins 2 ms later, are received at the same time. This problem is similar to trying to listen to one person talk while another is also talking.

From these examples we can see that a number of problems can arise in the communications channel which make a communications system prone to make errors or miss signals altogether. We will find that much of the work associated with designing a communications system deals with the problem of making it as resistant as possible to noise, interference, and some other conditions that distort or impair the transmitter's signal.

Experiment 7.2 contd: Effect of weak signals, Noise and Interference

The ability of the receiver to "hear" the transmitted signal is crucial to properly recovering the message that was sent. The presence of interference at the receiver can also degrade how well it recovers the message. While noise is technically "interference" as well, the term usually refers to signals that are similar to the one that is of interest to the user. This similarity can often confuse the receiver more than an unrelated random noise signal with the same power level.

Experiment 7.3: Impact of Transmission Speed on the Accuracy of Reception

When receiving a signal in the presence of noise or interference, more confident detection and more accurate determination of frequency will be possible if the burst carrying each letter is allowed to remain on for a longer time. While this will usually improve the accuracy of transmission, it also slows it down. This tradeoff between speed of transmission and accuracy of reception is a fundamental one in communications systems.

7.3.2 Coordination Between Sender and Receiver

Although it might seem obvious, reliable reception of messages also requires that the transmitter and receiver have a common agreement about how the characters are to be represented by the tones. In particular, for the one-tone-per-character system to work there must be a **prearranged agreement** on:

(1) What tone frequencies are to be used for each character

(2) How long each tone burst will last

(3) How much quiet time there will be between the end of one tone burst and the beginning of the next tone burst

Using this information the receiver can "listen" at the right times and at the right frequencies. This will allow it to be less sensitive to noise and interference as well.

The importance of the requirement for "prearrangement" can be illustrated by again assuming that the transmitter sends the letters **MEET** using the tone frequencies specified in Table 7.1 and the signal shown in Figure 7.4. Suppose, for some reason, that the receiver uses Table 7.2 instead of Table 7.1 to interpret the meaning of the received tone bursts. Even if the frequencies of the tone bursts are received and identified perfectly, the receiver will decide that the letters **NVVG** have been sent, which is obviously not the desired result. Another receiver using Table 7.3 would interpret the signal as **YII**. This is analogous to a case of human oral communication in which a German speaker is trying to communicate with one listener who only understands Chinese and another listener who only understands Arabic.

Letter	Frequency (Hz)	Letter	Frequency (Hz)	Letter	Frequency (Hz)
A	2800	J	1900	S	1000
B	2700	K	1800	T	900
C	2600	L	1700	U	800
D	2500	M	1600	V	700
E	2400	N	1500	W	600
F	2300	O	1400	X	500
G	2200	P	1300	Y	400
H	2100	Q	1200	Z	300
I	2000	R	1100	space	2900

Table 7.2: A second mapping from the capital letters to the frequencies in Hertz of audible tones with 100-Hz separation between adjacent tones.

Letter	Frequency (Hz)	Letter	Frequency (Hz)	Letter	Frequency (Hz)
A	300	J	750	S	1200
B	350	K	800	T	1250
C	400	L	850	U	1300
D	450	M	900	V	1350
E	500	N	950	W	1400
F	550	O	1000	X	1450
G	600	P	1050	Y	1500
H	650	Q	1100	Z	1550
I	700	R	1150	space	1600

Table 7.3: A mapping from the capital letters to the frequencies in Hertz of audible tones with 50-Hz separation between adjacent tones.

For the single tone communications system, one might assume that this type of error is not a big problem. Since the received letters **NVVG** or **YII** are clearly not English words, it would seem clear that a mistake had been made somewhere. Now consider instead the case where a third receiver is erroneously using the mapping table shown in Table 7.4. When the letters **MEET** are sent as tone bursts with frequencies of 1500, 700, 700, and 2200 Hz respectively, the receiver will, in this case, decode them as **FOOL.** This is an English word and might convey the wrong message to the recipient!

Letter	Frequency (Hz)	Letter	Frequency (Hz)	Letter	Frequency (Hz)
A	300	J	1200	S	1800
B	400	K	1300	T	1900
C	500	L	2200	U	2000
D	600	M	2300	V	2100
E	1400	N	2400	W	2500
F	1500	O	700	X	2600
G	1600	P	800	Y	2700
H	1000	Q	900	Z	2800
I	1100	R	1700	space	2900

Table 7.4: Another mapping from the capital letters to the frequencies in Hertz of audible tones with 100-Hz separation between adjacent tones.

One way to guarantee that the sender and the receiver are using the same lookup table and the same timing of the communications signals is to design the system to allow only one option. However, if a communications system must function correctly in different environments of signal level, noise, and interference, it would be wise to include some options so that performance can be optimized when the environment changes. In this case either the receiver must know the current characteristics of the transmitter before communications start, or the receiver must be designed to be clever enough to figure out the transmitter characteristics for itself based on signals that it receives.

Experiment 7.4: Different Codes/Music Receiver

The simple air modem works by assigning each letter in the alphabet to a different audio tone frequency. The association of a letter to a tone frequency is called the "mapping." This map must be used at the transmitter to select the tone frequencies to be transmitted, and at the receiver to convert a measured tone frequency back into a letter. If the same maps are not used at both ends, then the wrong message will result.

Exercises 7.3

1. Assume that the maximum frequency your receiver can "hear" is 3000 Hz. What potential advantage does the system using Table 7.3 have compared to a system using Table 7.1? What potential disadvantages does it have?

2. It might initially seem that the number of different symbols transmitted could be increased indefinitely by reducing the frequency spacing of the symbols. (For example, see Tables 7.1 and 7.3.) This would mean that the data transmission rate could also be increased indefinitely. However, the duration, T_d, of a burst must be equal to $C/\Delta f$ for accurate reliable detection of tones, where Δf is the minimum frequency separation of the symbols and C is a constant that depends on the particular filter implementation. Assume that $C = 4$. In addition, assume that the quiet

17

time between bursts is 1/Δf, that the lowest frequency that can be used is 300 Hz, and that the highest frequency that can be used is 3480 Hz.

a.) For Δf = 100 Hz, how many different symbols can be used? How many different messages can be sent in 100 msec.?

b.) Repeat part (a) for Δf = 50 Hz.

c.) Repeat part (a) for Δf = 200 Hz.

3. If the zero crossing method described in Exercise 4 of Section 7.2 is used to estimate frequency, explain how the following signal degradations might cause problems.

(a) weak signal

(b) added noise

(c) interference

4. Using Table 7.1, the letter **A** is represented by a cosine signal at 300 Hz and the letter **B** is represented by a cosine signal at 400 Hz. A receiver could distinguish bursts of these two signals by aligning the first peaks of the signals in time and then subtracting one from the other.

a.) Use a calculator to plot the <u>difference</u> of a cosine at 300 Hz and a cosine of 400 Hz over the interval from time = 0 to time = 0.5 msec. What is the maximum value of the difference?

b.) Use a calculator to plot the <u>difference</u> of a cosine at 300 Hz and a cosine of 400 Hz over the interval from time = 0 to time = 10.0 msec. What is the maximum value of the difference?

c.) If you were a communications system designer, which time interval would you choose for a tone burst? Why?

5. Repeat Exercise 4 for the letters **Y** and **Z**.

6. How would you modify the single tone system described in Section 7.2.2 to allow the comma and period to be included in messages (without having to spell out the words "comma" and "period" in the text of the message)?

7. Suppose that the single-tone-per-character system were to be used in a room with a significant amount of acoustical reverberation. How might that affect the receiver's performance? What characteristic of the transmitted signal might be changed to reduce the effect of reverberation and echo? How would changing this parameter affect the transmission rate which can be achieved by the system?

8. You are in charge of designing a receiver such as the one shown in Figure 7.5 for an application in which Table 7.1 is used. The main cause of potential errors is assumed to be interfering tone bursts from other sources that can occur at any frequency. Suppose further that the transmitter may wander in frequency so that all its transmitted

frequencies may be as much as 20 Hz too high or 20 Hz too low. Discuss the following tradeoffs for the design of the 27 bandpass filters.

a) If each bandpass filter gives a strong response over a range of 40 Hz, for example, from 380 Hz to 420 Hz for the letter **B**, then the full range of the transmitted signals can be received correctly. What percentage of the interfering tone bursts will also be interpreted as valid characters?

b) If each bandpass filter is designed to give a strong response over a narrow frequency of 10 Hz, for example, from 395 Hz to 405 Hz for the letter **B**, then what percentage of the interfering tone bursts will be interpreted as valid characters? What impact would the narrower filters have on the reception of messages from the desired transmitter? What would you need to know about the transmitter to estimate the percentage of character signals that would be incorrectly detected?

Section 7.4:
The Craft of Engineering – Improving the Design

The previous section demonstrated that it is possible to convey information from one location to another and defined the basic components of a communications system that would accomplish this information transfer. A specific example demonstrated the basic concepts by sending text messages for short distances through the air using a sequence of audio tones to represent a sequence of characters. The range of this communications system is limited by the ability of the microphone receiver to "hear" the transmitting speaker as the distance between them grows. There are many other ways to represent the characters of a message and there are many more ways to transmit a message. In this section we will consider two alternative methods of data representation, and we will develop performance criteria to compare these new methods.

The performance and cost comparisons we will do are part of the engineering design process. Engineering, by definition, is the application of scientific knowledge to practical uses which satisfy human needs. In choosing the best design for a practical use, engineers must compare alternative implementations and make judgements about the cost and technical performance of each alternative within the context of a specific application.

7.4.1 Improving the Design of a Digital Communications System

The communication method developed in section 7.2 and illustrated in Figure 7.3 can be called a **one-tone-per-character** method since each of the 26 letters and the space is assigned to a unique frequency or tone. The receiver implementation for this system with the best performance has 27 narrow band filters and was shown in Figure 7.5. Each filter is **tuned** to respond to exactly one of the frequencies used to represent a character so that each filter had a maximum response at the frequency associated with the character it was designed to detect. The outputs of these filters are examined constantly by the receiver's decision logic. When the energy from one filter substantially exceeds the energy from any of the other filters for a sufficiently long time, then the receiver declares that a tone burst has been transmitted. The character that was transmitted is determined to be the character cor responding to the filter which responds most strongly to the tone burst.

While the 27-filter receiver design works very well, it can be quite expensive compared to other approaches. The overall cost and performance of a receiver will depend on several factors, including the number of filters and the performance and cost of each individual filter. Factors affecting the cost of individual filters will include frequency selectivity and the minimum time duration needed to determine which tone was transmitted. In many implementations the filters are the most expensive components used. If we wanted to build a cheaper receiver, we would explore strategies to reduce both the number of filters needed and the cost of each filter.

We must first ask if it is possible to design the transmitter and receiver pair so that fewer tone frequencies, and hence fewer expensive receiver filters, are needed. In fact this is possible. But when we redesign in this way to reduce cost, we will see that other attributes of the communications system must also change. The engineer's objective of finding the "best" design will come down to comparing the alternative designs and their attributes, and finding the one best suited to both the customer's desires and financial resources.

The next section presents two alternatives to the communications system defined in Section 7.2. The second one is cheaper than the first one in the sense that fewer tone frequencies are used, and hence fewer expensive filters are needed. After each of these alternatives is described, it is analyzed in terms of **implementation cost**, **transmission speed**, and **accuracy.**

Cost

When we talk in this section about costs, we focus principally on the cost of implementing a particular design. Engineers must also be concerned with the cost of maintenance, the cost of training personnel to use a system or device, and the cost to keep it in operation. All must be in proper balance once the design is complete.

7.4.2 Binary Representation for Each Character

The design from section 7.2 can be viewed as extravagant because only one of the 27 receiving filters is being used at any time. Specifically, the transmitter sends one tone burst for each character in the message and one of the 27 filters at the receiver is expected to detect it. The other 26 are said to "reject" the tone and have no significant energy output. We consider here the possibility of sending more than one tone at the same time. We believe that this might improve the implementation cost of the receiver by requiring fewer of the costly filters.

Using More Than One Tone

Consider a system using a small number of tones to represent characters so that any combination of any number of the tones will represent some valid character. Let P be the number of tones. This would allow 2^P combinations of tones since each of the P tones could independently be on or off. Actually, the P tones will be used to represent only $(2^P - 1)$ characters since we must exclude the case where all tones are off. (Why?)

.If P = 5, then $2^P - 1 = 31$, and we can send our 26 Roman letters, a space, and a few control characters using a combination of up to five tone frequencies for each letter. The method would work as shown in Figure 7.8. Each character from the originator's list would be converted into a burst, just as before, but the multitone burst would consist of the sum of between one and five sinusoids. The choice of the number of sinusoids, and their frequencies, would be determined by a table of the type shown in Table 7.5 where an "x" in a column indicates that a tone is on for the letter specified. For example, the transmission of the letter **A** would be sent via a burst consisting of two sinusoids, one at frequency F1 and the other at F2. Similarly the letter **B** is represented by a burst containing three sinusoids, while the space character is a burst with only one sinusoid at frequency F3.

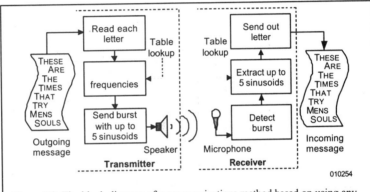

Figure 7.8: The block diagram of a communications method based on using any combination of five possible tone frequencies.

	A	B	C	D	E	F	G	H	I	J	K	L	M	N	O	P	Q	R	S	T	U	V	W	X	Y	Z	s
F1	X	X		X	X	X			X	X					X		X		X		X	X	X	X		X	
F2	X		X				X	X	X	X				X	X	X			X	X	X			X	X		
F3		X			X		X	X		X		X	X		X	X		X	X		X	X			X		X
F4		X	X	X		X	X			X	X		X	X	X			X				X		X			
F5		X				X	X			X	X		X	X	X	X			X		X	X	X	X	X		
	A	B	C	D	E	F	G	H	I	J	K	L	M	N	O	P	Q	R	S	T	U	V	W	X	Y	Z	s

Table 7.5: A possible mapping table for sending characters as the combination of up to five sinusoids.

An example of such a composite burst is shown in Figure 7.9. Suppose that we desire to send the letter **Z** from Table 7.5 and that our choice for frequency F1 were 300 Hz and our choice for frequency F5 were 2300 Hz. The top two traces of Figure 7.9 show bursts of 300 and 2300 Hz, while the bottom trace shows the composite burst containing the sum of the upper two.

Experiment 7.5: Multitone Audio Communication of Text

The simple air modem used in the earlier experiments uses a single burst consisting of a single sinusoidal signal to carry a single letter of text. It is possible to use more complicated bursts, consisting of the sum of one or more sinusoids, to carry textual traffic as well. These more complicated methods usually bring the benefit of improving the rate at which messages can be sent or decreasing the implementation cost of the transmitter or receiver.

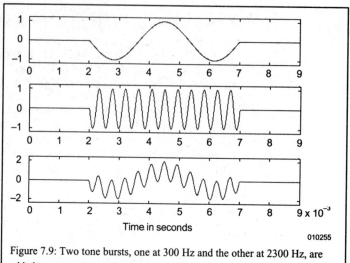

Figure 7.9: Two tone bursts, one at 300 Hz and the other at 2300 Hz, are added to represent one character.

As shown in Figure 7.8, the receiver is based around a set of five filters, each one tuned to one of the five frequencies F1 to F5. The receiver watches the outputs of all five filters. The receipt of a transmitted character is indicated by the presence of sufficient energy in any of the five filters. The receiver then examines all five filter outputs to determine which are responding to a transmitted sinusoid and which are not. By referring to Table 7.5, the receiver can determine which of the characters were sent.

The expected cost and performance of the new alternative we have just developed can be evaluated in terms of four important factors.

- **Implementation Cost**: We now need only five filters, making this method substantially cheaper to implement than the 27-filter single tone system. However, since from one to five sinusoids may be transmitted simultaneously, the microphone input system must be designed to respond correctly over a wider range of input powers than a system which expects a fixed number of sinusoidal inputs for each character.

- **Transmission speed**: We still send one character with each multitone burst, leaving the transmission rate the same.

- **Complexity**: The logic needed at the receiver to detect the appearance of one or more frequencies in each burst is more complicated than the logic to detect a single tone, but not prohibitively complex.

- **Accuracy**: This new scheme is less reliable than the single-tone-per-character scheme since any combination of possible tones represents a valid output. The first method allows the receiver to confirm the presence of exactly one frequency before declaring that a character has been received.

Binary Interpretation of Character Codes

We can interpret this five-tone method as a method of sending characters using a 5-bit binary code. Each letter and control character is represented by a unique set of five bits. Each of the bits controls one of the tone frequencies which might be present in the burst. If only 1 bit is '1' for the character, such as the space, the E or the T, then the burst carrying the character to the receiver consists of only one sinusoid at the designated frequency. If two of the bits are '1,' then the corresponding two frequencies are sent. Other characters may have three or four bits equal to '1' and will have three or four frequencies used for transmission. This is perhaps more apparent when Table 7.5 is reorganized as Table 7.6. The contents of the table are the same, but by reordering the letters we can see that there is a clear binary counting pattern present. Only 27 of the 31 possible patterns are identified, allowing four additional control characters to be assigned to the four unnamed patterns if desired.

```
F1  X   X   X   X   X   X   X   X   X   X  X   X   X   X  X  X
F2    X X     X X     X X     X X     X X     X X     X X     X X
F3      X X X X       X X X X       X X X X       X X X X
F4          X X X X X X X X             X X X X X X X X
F5                        X X X X X X X X X X X X X X X
    E    A sp S I U   D R J M F C K T Z L W H Y P Q O B G   M X V
```

Table 7.6: A reordering of Table 7.5 to reveal the binary counting progression of the transmission patterns.

When viewed this way we see that we have assigned a binary pattern to each character which is different from all other patterns. We then used the values of the bits to determine whether a sinusoid at the frequency corresponding to each bit is a part of the character's multitone burst. The receiver observes the burst using five filters, each one looking for one of the five sinusoidal frequencies. When a sinusoid is detected at the filter's output, a "1" is reported for that bit. If no sinusoid is detected by a filter, a zero is reported for that bit. The pattern of five ones and zeroes is then compared to Table 7.5 or 7.6 to determine which character was transmitted. Since the bits are sent together in a single burst, and therefore in "parallel," we call this the **parallel binary** method.

One can show mathematically that if our transmitter is limited to simply turning sinusoidal bursts on and off for each character, then the parallel binary method we just discussed is the best method assessed in terms of minimizing the number of expensive filters in our receiver. However, if that limitation is not present and a single character can be represented by a sequence of tones, then there are ways of using even fewer filters. The next method we examine will reduce the number of filters to two. There is a price to pay, however, for the reduced number of filters, and the engineer must evaluate these performance and cost factors to determine the best choice for a particular application.

7.4.3 Serial Binary Representation for Each Character

Suppose now that our objective is to find a method of conveying the capital letters and a few control characters with just two frequencies and two corresponding filters at the receiver. A common procedure is a variation on the parallel binary scheme we just examined. In that method Table 7.5 was used to convert each character we desire to transmit into a binary pattern of ones and zeros. The bit position was identified by one of five frequencies and the bit value was identified by whether or not the frequency was present in a single burst.

To transmit "serial binary," however, we use a sequence of five separate tone bursts for each character, with each burst containing one of two sinusoids. The sinusoid at frequency F_{one} is used if a one bit is to be sent and the frequency F_{zero} is used if a zero bit is to be sent. In this case the bit value is identified by one of two frequencies, and the bit position is identified by the tone's position in the sequence of five tones. We can rewrite Table 7.5 again as Table 7.7 to show how this is done.

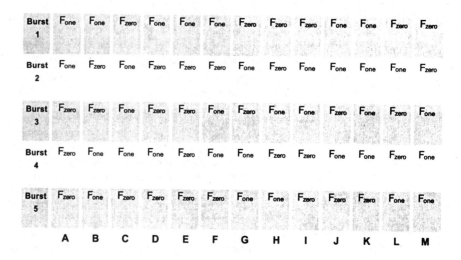

	A	B	C	D	E	F	G	H	I	J	K	L	M
Burst 1	F_{one}	F_{one}	F_{zero}	F_{one}	F_{one}	F_{one}	F_{zero}	F_{zero}	F_{zero}	F_{one}	F_{one}	F_{zero}	F_{zero}
Burst 2	F_{one}	F_{zero}	F_{one}	F_{zero}	F_{zero}	F_{zero}	F_{one}	F_{zero}	F_{one}	F_{one}	F_{one}	F_{one}	F_{zero}
Burst 3	F_{zero}	F_{zero}	F_{one}	F_{zero}	F_{zero}	F_{one}	F_{zero}	F_{one}	F_{one}	F_{zero}	F_{one}	F_{zero}	F_{one}
Burst 4	F_{zero}	F_{one}	F_{one}	F_{one}	F_{zero}	F_{one}	F_{one}	F_{zero}	F_{zero}	F_{one}	F_{one}	F_{zero}	F_{one}
Burst 5	F_{zero}	F_{one}	F_{zero}	F_{zero}	F_{zero}	F_{zero}	F_{one}	F_{one}	F_{zero}	F_{zero}	F_{zero}	F_{one}	F_{one}

Note : Table continued on the next page

	N	O	P	Q	R	S	T	U	V	W	X	Y	Z
Burst 1	F_{zero}	F_{zero}	F_{zero}	F_{one}	F_{zero}	F_{one}	F_{zero}	F_{one}	F_{zero}	F_{one}	F_{one}	F_{one}	F_{zero}
Burst 2	F_{zero}	F_{zero}	F_{one}	F_{one}	F_{one}	F_{zero}	F_{zero}	F_{one}	F_{one}	F_{one}	F_{zero}	F_{zero}	F_{zero}
Burst 3	F_{one}	F_{zero}	F_{one}	F_{one}	F_{zero}	F_{one}	F_{zero}	F_{one}	F_{one}	F_{zero}	F_{one}	F_{zero}	F_{one}
Burst 4	F_{one}	F_{one}	F_{zero}	F_{zero}	F_{one}	F_{zero}	F_{zero}	F_{zero}	F_{one}	F_{zero}	F_{one}	F_{zero}	F_{zero}
Burst 5	F_{zero}	F_{one}	F_{one}	F_{one}	F_{zero}	F_{zero}	F_{one}	F_{zero}	F_{one}	F_{one}	F_{one}	F_{one}	F_{zero}

Table 7.7: A table for mapping characters into a sequence of bursts, each of which contains a sinusoid at one of only two possible frequencies.

We can call this technique the **serial binary** method, since binary representations are still used for each character, but we send the bits out serially, with the ones and zeros represented by bursts of one or the other of the two frequencies. This method is also historically called **frequency shift keying (FSK)**, since the transmitter's frequency is shifted from one frequency to another to indicate the shift from a zero to a one, and vice versa.

The difference between the serial and parallel binary transmission methods can be visualized by looking at the patterns of the transmitted signals. Figure 7.10 shows the waveforms for the word **IF** when transmitted in both ways. We assume that Table 7.5 is used for the parallel method and Table 7.7 is used for the serial method. Recall that the mapping from each character to the binary pattern is the same in both cases, but the way in which the bit positions and bit values are represented is different. In the upper left part of Figure 7.10 we see a single multitone burst for the parallel binary representation of the letter **I**. It is formed by adding two sinusoids, one whose frequency is F2 and the other whose frequency is F3. Below it we see a train of five pulses, each of which consists of only a single sinusoid for the serial binary representation. In this case the letter **I** is sent with F_{zero} as the frequency of the first, fourth, and fifth pulses in the train, and F_{one} being the frequency of the second and third. In both cases, the same 5 bits of information is sent. On the upper right side of the figure we see the multitone burst of three frequencies for the letter **F**, and below that we see its serial representation as a train of five pulses.

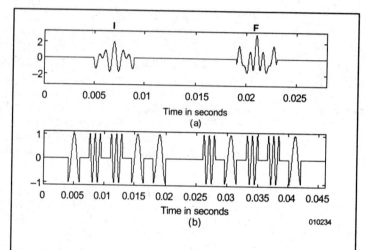

Figure 7.10: A view of the waveforms used to carry the letters **I** and **F** – (a) a single multitone burst for each character using the "parallel binary" method with frequencies of 500, 900, 1300, 1700, and 2100 Hz and (b) a sequence of 5 tone bursts for each character using the "serial binary" method with 500 Hz representing a zero and 1200 Hz representing a one.

The cost and performance of the serial binary method can now be evaluated.

- **Implementation Cost**: This method requires only two sinusoidal frequencies at the transmitter and only two filters at the receiver, making it substantially cheaper, when measured in "filters," than either of the two previously discussed methods.

- **Transmission speed**: It takes longer to transmit a character with this method if the tone burst is of the same duration as the tone bursts of the other methods. The two previous methods needed only one burst of tones to send a single character, while this method uses a sequence of five tone bursts for a single character.

- **Complexity**: The decision logic to determine whether a bit is '1' or '0' is very simple. However, synchronization and timing logic must be added to identify the start and end of the bit sequences for each character and to put the bit values in the correct order to make the desired binary pattern of the character that was sent.

- **Accuracy**: This method is as robust as either of the methods examined so far. In fact, the serial binary method is one of the most reliable transmission methods that there is.

In comparing the serial binary method to other methods, it is clear that the specific application requirements and priorities will determine if it is a better or worse choice. If the customer's primary objective is fast transmission, then perhaps this is not the best method to use. In contrast, if the primary objective is reliable, inexpensive communications, then this just might be the best method.

Experiment 7.6: Serial Binary Audio Communication of Text

The use of multiple sinusoids in each burst increases the transmission rate at the possible cost of making the receiver more vulnerable to making errors in the presence of noise and interference. An alternative method is to use only two tone frequencies, one denoting a binary 1 and the other 0, and sending sequences of five bits to indicate each letter.

7.4.4 A Comparison of the Three Transmission Methods

The search for less expensive alternatives to the single-tone-per-character communication method led to two possible alternatives. (See Exercise 7.4 for yet another.) We have seen that there are many ways to accomplish this communication task, even if we confine ourselves to the acoustic channel and the use of audio-range frequencies. Although each new method could be refined with some variations, the basic methods can be compared to show the cost and performance characteristics of each. These general characteristics are listed in Table 7.8 for a system designed to transmit 30 distinct codes.

We can observe that both of the alternatives to the single-tone method use fewer filters, and therefore are presumably cheaper than the single-tone-per-character method. We also note, however, that in trade for less implementation cost, other difficulties grow and the speed of transmission may decline. These different methods have different attributes, some of which might be more attractive to a user than others. Some might want speed, while others want the lowest cost, while others want the best possible accuracy when attempting to communicate through noise and interference. There is no absolute best choice among these methods. The most appropriate method for a particular use will depend on the application's communication needs and priorities.

Method	Number of filters	Relative Speed	Reliability	Automatic Detection
Single-tone-per-character	30	5	Medium	Very easy
Parallel Binary	5	5	Medium-low	Medium
Serial Binary (FSK)	2	1	Good	Easy

Table 7.8: A comparison of the basic attributes of three alternatives for designing an acoustical communications system for textual messages using 30 characters.

7.4.5 Using Pairs of Tones to Communicate

The touchtone telephone shown in Figure 7.11 uses a representation method that is different from the three discussed so far in this section. A set of seven possible tone frequencies is used to send any of twelve possible numbers and symbols on the touchtone keypad. A different frequency is used for each of the three columns of buttons and four more frequencies are used to indicate each row. Pressing a button sends a burst containing exactly two frequencies, the appropriate one for the row and the appropriate one for the column for the button that is pressed. (To verify this for yourself, simultaneously press two buttons on a column and then two buttons on the same row. Doing this produces the single tone associated with that row or column.)

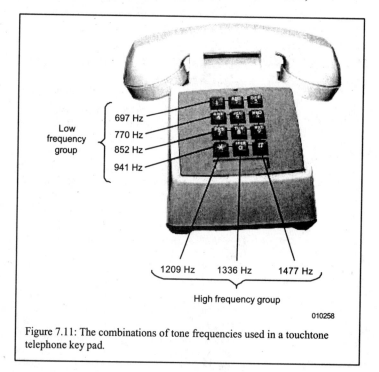

Figure 7.11: The combinations of tone frequencies used in a touchtone telephone key pad.

Consider now a scheme for sending textual messages using an extended version of the touchtone concept (which is technically called dual tone multifrequency (DTMF). Suppose that we use the sets of frequencies shown in Figure 7.12 to transmit each of the capital Roman letters. For example, the letter Z would be carried by a burst consisting of tones at 1100 and 1700 Hz. (This method has actually been used to transmit messages over shortwave radio communications systems.)

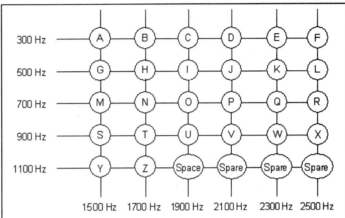

Figure 7.12: A method for transmitting characters using the combination of two tones.

Exercises 7.4

Questions:

1. Characterize this acoustical transmission system just as we have done the others, in terms of (1) transmission speed, (2) implementation cost, (3) complexity, and (4) expected accuracy.

2. Fill in the appropriate entries in Table 7.8 for this scheme.

3. Provide a general mathematical rule for determining the number of tones necessary to transmit N different characters.

4. Determine the minimum number of different tone frequencies needed to convey any of 64 characters. What is the minimum number for a character set with 78 members?

Experiment 7.7: Dual Tone Audio Communication of Text: Sending and Receiving Touch Tone

The simple air modem uses tone bursts containing a single sinusoid to send the letters in the alphabet. The "dual tone" method is the next step in complexity. Each letter is still sent using a single burst, but in this case, the burst is the sum of two sinusoids. The receiver determines the frequencies of the two sinusoids, and, from those, the letter that was sent. The dual tone method can be more reliable and cheaper to implement than the single tone method.

Section 7.5:
Extending Our
Reach

7.5.1 Larger Character Sets – From Morse Code to ASCII and Beyond

The methods of data representation described in this chapter have been demonstrated with a limited character set of 27 codes to emphasize the differences in the methods and the design process of selecting a method. However, in modern communications systems a much larger set of characters is typically used. In order to represent upper- and lowercase Latin alphabet characters and the ten digits, 62 character codes would be needed. Adding punctuation and control codes further increases this number. There are two basic approaches to extending the set of characters that can be represented. One is to simply add more character codes. The second uses sequences of codes from a small set to make a larger number of codes. These two approaches may also be used in combination.

Experiment 7.8: Morse Code Audio Communication of Text / Communicating with More Characters

The first important method of electrical communications was the telegraph. It used a code devised by Samuel Morse. It employed neither tones nor ones and zeros. It used "dots" and "dashs," short and long electrical pulses, in a code devised to carry the English language as fast as possible.

The implications of adding new characters to the character set are different for the methods already discussed. For all methods except the serial binary method, adding new codes requires fundamental changes in the transmitter and receiver capabilities. For example, with the one-tone-per-character method, the addition of new character codes requires the addition of a new tone for each new character. If we assume that we want to keep all of the tones within the same frequency range, the addition of new tones means that the frequency separation between tones must be reduced and the number of filters must be increased. Both of these factors increase the implementation cost of the system. The same considerations apply to the parallel binary method, but fewer new tones would be needed. The number of additional tones would increase with the base 2 logarithm of the number of characters for the parallel binary method. If the character set contained 127 characters, the single tone system would require 127 tones while the parallel binary system would require seven tones. Only the serial binary FSK method continues to use the same two tones as the number of characters increases. If the size of the character set is increased from 31 to 128, then the serial binary method simply increases the number of tone bursts for each character from 5 to 7. This causes the transmission time for each character to increase in proportion to the number of bits representing the character, but no new filters are needed by the receiver.

Because the sender and receiver must be using the same method and the same character codes for successful communication, standard coding methods are extremely important for communications systems. The International **Morse Code** standard evolved from the code first used by Morse in the United States in the early 1840s [4]. It consists of 50 codes for uppercase Latin alphabet characters, numbers, punctuation, and control codes [5]. The 5-bit **Baudot code** used in Tables 7.5, 7.6, and 7.7 was developed in 1875 [6] for international communications and was widely used for more than 100 years.

In 1963 a new 7-bit code was defined for computer communications systems. It increased the number of codes from 31 to 127 and explicitly represented lowercase characters. The American Standard Code for Information Interchange (**ASCII**), which is currently used on most computer systems, is shown in Table 7.9. From this table the 7-bit code for the letter **A** is 1000001 and the code for **a** is 1100001. Codes below 0100000 are used for printing and device control, and only selected codes are shown. For example, BS is backspace, FF is form feed or new page, and EOT is end of transmission.

Although the Latin character set was historically important in the development of computers, the 127 character codes are insufficient to represent the characters used in the many languages of the world. For effective global communication a new standard was needed. The **Unicode** is a standard 16-bit code adopted by many communications and computer equipment manufacturers. In version 3.0, 57,709 of the 65,535 possible codes have been assigned, which includes 49,194 codes representing characters and symbols from languages worldwide. Additional information about this standard can be obtained from www.unicode.org.

	000	001	010	011	100	101	110	111
0000---	NUL				EOT			BEL
0001---	BS	HT	LF	VT	FF	CR		
0010---								
0011---				ESC				
0100---	SP	!	"	#	$	%	&	'
0101---	(	)	*	+	,	-	.	/
0110---	0	1	2	3	4	5	6	7
0111---	8	9	:	;	<	=	>	?
1000---	@	A	B	C	D	E	F	G
1001---	H	I	J	K	L	M	N	O
1010---	P	Q	R	S	T	U	V	W
1011---	X	Y	Z	[	\	]	^	_
1100---	`	a	b	c	d	e	f	g
1101---	h	i	j	k	l	m	n	o
1110---	p	q	r	s	t	u	v	w
1111---	x	y	z	{	|	}	~	DEL

Table 7.9: 7-bit ASCII character codes.

Exercises 7.5

1. Access the unicode web site and find the codes for two languages that do not use the same characters as English. How many codes are used for each character set?

2. For the ASCII character set, what is the difference between an "A" and n a "B" and a "b"? Find a rule to tell whether a code represents an upper case character or a lower case character.

3. Translate the phrase QUICK BROWN FOX into binary form using both the Baudot alphabet (Table 7.5) and the ASCII alphabet (Table 7.9).

7.5.2 Binary Data Streams

How would we modify a communications system if the objective were not the transmission of a single text message from one human to another, but rather the continuous transmission of a very long stream of binary data representing images or sound or other nontext data?

A simple, and perhaps obvious, answer is to use one of the communication methods already described but with the minor modification that the incoming bits are grouped together into sets, and then each set of bits is sent as a "character." The character sequence would have no meaning to us as text, but it would represent the complete bitstream for the audio or video. At the receiver each transmitted character is determined and then the bits defining that character are added to the data stream that is passed on to the consumer of the data at the destination. For example, Table 7.10 shows a typical table for the single-tone-per-character method. At the receiver the presence of each pulse is detected, the tone frequency is determined, and the corresponding set of bits is sent on to the user. If the binary sequence 0101000100 were to be transmitted, it would first be divided into the two 5-bit sequences of 01010 and 00100. From Table 7.10 the frequency for the tone burst for the first sequence is determined to be 1300 Hz. After the 1300 Hz signal is transmitted, the frequency for the second 5-bit sequence is found to be 700 Hz, and then sent. At the receiver the 1300 Hz signal is interpreted as 01010. When the second tone burst at 700 Hz arrives it is interpreted as 00100, and this set of 5 bits is appended to the first set of 5 bits to recreate the original 10-bit data binary sequence. This scheme can clearly be extended to handle longer and longer streams of bits, carried 5 bits at a time on each single-tone symbol.

Binary Pattern	Frequency in Hz	Binary Pattern	Frequency in Hz	Binary Pattern	Frequency in Hz
00000	300	01011	1400	10110	2500
00001	400	01100	1500	10111	2600
00010	500	01101	1600	11000	2700
00011	600	01110	1700	11001	2800
00100	700	01111	1800	11010	2900
00101	800	10000	1900	11011	3000
00110	900	10001	2000	11100	3100
00111	1000	10010	2100	11101	3200
01000	1100	10011	2200	11110	3300
01001	1200	10100	2300	11111	3400
01010	1300	10101	2400		

Table 7.10: A table for translating sets of binary bits into a tone burst's frequency.

If all methods use the same time duration for tone bursts, then the single-tone-per-character method will transmit the binary data five times faster than the serial binary method, since 5 bits are carried with each tone burst. However, the serial binary method is a more natural match for a stream of binary data because, as discussed in the previous section, that method can send bitstreams longer than five without significant modifications.

Now, faced with the objective of transmitting large streams of data using already existing character-oriented communications systems, we need to divide our bit stream into character-sized patterns and transmit the data as if it were a text message regardless of the original source of the bit stream. In this case we are converting binary data into equivalent text messages in order to use readily available transmitters and receivers. This is the basis of modern high-speed data communications. It works because the communications systems are designed to accurately transmit one of a fixed number of symbols during a specified time interval, but it does not matter to the communications system whether we interpret that bit pattern as a textual character or a number or part of an image.

Figure 7.13 shows this process graphically. On the left are two different types of inputs to a data communications system, and on the right are two different ways that the data is commonly conveyed over the transmission channel. On the left, the data to be transmitted is often provided as a file of characters, the letters in the English language for example. In other cases it is presented as pure binary data, a long list of ones and zeros. On the right, some communications channels permit the transmission of symbols that can have many values, such as tone bursts with many different frequencies, while others can only reliably convey ones and zeroes. An example of the latter is the FSK transmission scheme discussed in Section 7.4.3. Another important one is an optical fiber which carries only the presence or absence of light.

Experiment 7.9: Transmitting a FAX

The first electrical and optical communications systems were designed to carry text messages. The most reliable ones actually broke the text into binary ones and zeros. From this basis, it was easy to extend these communications systems to carry black and white image pixels as well.

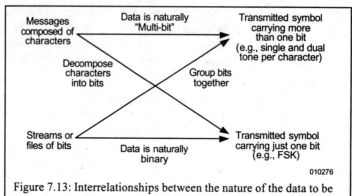

Figure 7.13: Interrelationships between the nature of the data to be transmitted and the type of transmissions used.

Figure 7.13 illustrates that either type of source on the left can be used with either type of transmission system on the right, if the designer makes the proper adjustments. Character-oriented data, which is destined to be carried over a binary transmission system, like the Internet, must be decomposed into bits before transmission. The "serial binary" method could do this. Conversely, a transmission system capable of carrying multi-valued symbols, such as tone bursts with multiple frequencies, can be used to carry natively binary data by grouping bits together before transmission. The big incentive to do this is that all the bits in each group are carried on a single symbol, making this scheme much faster than when only 1 bit is carried on each symbol. For example, if a symbol can have 64 possible values, then it can "carry" 6 bits, and transport those bits six times faster than it could have if each required its own two-valued binary pulse.

Exercises 7.6

1. Consider the binary sequence { 01101 01011 00101 10110 }. How would it be transmitted using

    a.) a series of tones using Table 7.10

    b.) parallel binary using Table 7.5

    c.) serial binary using Table 7.7

2. An old saying states that "a picture is worth a thousand words." Suppose that a thousand words requires 6000 characters, each of which is represented by an 8-bit pattern.

    a.) If an image is transmitted as a data steam with three 8-bit patterns specifying the red, green, and blue component values of each pixel, what is the size of a square image that takes the same transmission time as the thousand words?

    b.) Suppose that digitized speech is transmitted as a data stream with 8-bit values representing the amplitude of each data sample. If the speech has been sampled at 8000 samples per second, what is the duration of a speech segment that takes the same transmission time as the thousand words?

Section 7.6: Other Transmission Channels

The communications system discussed in this chapter sends the characters of a message as discrete tone bursts through the air (the "acoustic channel") to the intended receiver. There are many variations on this idea, all of which conform to Figure 7.1. Our first example used a single frequency to represent an individual letter, but then we branched out to try other approaches. In practice many combinations of frequencies, amplitudes, and phases are used in the quest to transmit multimedia data faster and faster.

Similarly, practical communications systems use other channels for carrying their signals. So far we have focused on sending our signals as sound waves through the air. This has the attraction that it resembles the way that humans communicate, but in reality it is not the fastest way to send information. Most practical communications systems send their signals as electrical waves through wire, as electromagnetic (radio) waves through the air or space, or as even higher frequency electromagnetic waves which we call **light**. These light waves are sometimes conveyed through the air or space (like the **infrared** signal from your TV remote control), and sometimes they are guided through **optical fiber** from the transmitting laser to the receiving detector located miles away. In later chapters we will examine these different kinds of transmission channels in more detail. We will find, however, that in all cases the fundamental concepts are the same. The input data is applied to the transmitter, which is designed to best convey that data over the transmission channel that has been chosen by the design engineer. The receiver collects the signal and recovers the transmitted data, as best it can in the presence of noise and interference.

Experiment: Text Transmission over a Wired Circuit

All of the communications experiments done so far in the chapter employ bursts of audio frequency tones carried over the air to a microphone and a receiver. These tones can be carried electrically instead, by connecting a pair of wires between the transmitter and receiver.

Section 7.7: Summary

In this chapter we have defined the three basic components of any communications system: the transmitter, the channel, and the receiver. As we developed a simple acoustic system for text communication to illustrate the operation of each component, we learned the following.

- It is possible to reliably move text information, that is, communicate the information, between two different locations.

- The transmitter is designed to put the data into a form so that the electronic receiver can distinguish it from all other patterns of data.

- The receiver must know the characteristics of the transmitted signal so that it can do its job properly.

- Noise, interference, weak signals, and possibly other factors can degrade the received signal to the point that it can no longer be accurately recovered by the receiver.

With these basic points established, we extended them in three ways:

- We found alternatives to the **single-tone-per-character** scheme, and determined that some carry data faster than others do, some are more robust to noise and interference than others, and some are cheaper than others. Engineers evaluate alternative schemes and their capabilities to find the best one for each particular application.

- Many communications systems need to transmit very long streams of binary data, such as digitized voice and images, rather than just short messages using the Roman alphabet. We found, however, that this binary data can be sent as a sequence of "characters" or **symbols** by grouping the bits together into bundles and then mapping those bundles into characters that the transmitter is capable of sending and the receiver is capable of accurately detecting.

- Our final observation was that most communications systems use electrical or electromagnetic waves to convey their signals, rather than the acoustic waves assumed in this chapter. Chapters 13 and 14 discuss these systems in much more detail. Chapter 15 describes ways in which the best performance can be obtained from a communications system.

In the next chapter we turn our attention to using the communications systems we've examined here to build **networks** which can efficiently provide multimedia telecommunications services to many users.

Section 7.8: Glossary

antenna (7.1.2)
a device used for radio communications that converts electrical signals into electromagnetic (e.g., radio) waves and vice versa

ASCII code (7.5.1)
American Standard Code for Information Interchange, which is currently used on most computer systems and shown in Table 7.9; can represent 128 different symbols using 7 bits of information

bandpass filter (7.2.2)
a filter which has a strong response to frequencies in a specific range or band and has minimal response to all other frequencies; an "ideal" bandpass filter would have zero response to all other frequencies

Baudot code (7.5.1)
a 5-bit code used in Tables 7.5, 7.6, 7.7 used with a keyboard for international communications until the mid 1970s; it represents 32 different symbols using 5 bits of information per symbol or character

buffer
a place to temporarily accumulate characters in a message while it is being transmitted or received

burst signal (7.2.2)
a signal that only exists for a short time

cochlea (15.2.2)
a fluid-filled tube in the ear that responds to different frequencies at different points along its length

communication (7.1.1)
the movement of information, which may have digital or symbolic representation, from one location to another physically separate location

communication channel (7.1.2)
the physical medium that carries a communications signal from the transmitter to the receiver; for example, telephone wires, optical fiber, air

FSK (frequency shift keying) (7.4.3)
a method of representing a character by a series of bits which can be 0 or 1 and transmitting it by sending the bits one at a time using one frequency to represent a 0 and a different frequency to represent a 1

interference (7.3.1)
signals or disturbances unrelated to the signal you wish to receive that can make it more difficult for the receiver to operate correctly

Morse code (7.5.1)
a code devised by Samuel F. B. Morse [4] to send characters by telegraph, the first commercially successful electrical communication system; it uses combinations of short intervals (dots) and long intervals (dashes) to represent characters and can be used with electrical signals, lights, or sounds

noise (7.3.1)
disturbances in signals that can make it more difficult for the receiver to operate correctly

parallel transmission (7.4.2)
sending communication units such as bits or tones all at the same time

receiver (7.1.2)
a receiver recovers the transmitted information from the transmitted signal and converts it into a form that the recipient can use

serial transmission (7.4.3)
sending communication units such as bits or tones one at a time

tone (7.2.2)
a signal that has a constant frequency and amplitude, i.e. a sinusoidal signal

tone burst (7.2.2)
a sinusoidal signal that only exists for a short time, in contrast to the mathematical definition of an infinite duration sinusoidal signal

transmission speed (7.4.1)
the rate at which information can be communicated which is measured in communication units per time interval such as words per minute or characters per second or bits per second

transmitter (7.1.2)
a device or circuit which converts a communication signal into a form that can be conveyed to a distant physical location; for example, a signal might be converted into sound vibrations in the air or an electrical signal on a wire

Unicode (7.5.1)
a 16-bit code that can represent 65,535 different symbols from international alphabets

Section 7.9:
References

[1] Margaret W. Matlin and Hugh J. Foley, *Sensation and Perception*, Allyn and Bacon, Needham Heights, Massachusetts, 1997, p. 306.

[2] G. C. Smith and H.A. Mauch, *Summary report on the Development of a Reading Machine for the Blind*, Mauch Laboratories, Inc., Dayton, Ohio, 1975.

[3] Robert A. Weisgerber, Bruce Everett, and Claudette A. Smith, *Evaluation of an Ink Print Reading Aid for the Blind: The Stereotoner*, Final Report AIR 39600-12/75 FR, American Institutes for Research, Palo Alto, California, submitted to Veterans Administration Res. Ctr. for Prosthetics.

[4] Lebow, *Information Highways and Byways*, IEEE Press, 1995.

[5] Leon W. Couch, *Modern Communications Systems*, Prentice Hall, Englewood Cliffs, NJ, 1995, p. 536.

[6] Leon W. Couch, *Modern Communications Systems*, Prentice Hall, Englewood Cliffs, NJ, 1995, p. 537.

[7] Tom Standage, *The Victorian Internet*, Walker and Company, New York, 1998.

Figure References:

[7.6] *** unknown**

[7.11] Courtesy of Rich Taylor, Applied Signal Technology, Inc

[7.14] Courtesy of Applied Signal Technology, Inc.

The Infinity Project

Chapter 8: Networks And The Internet

From the Telegraph to the Internet

Approximate Length: 3 Weeks

Section 8.1: Combining Communications Links to Build a Network

In Chapter 7, we learned to build reliable signal transmission systems that can carry multimedia data from one location to another. We would now like to figure out the best way to use transmission systems such as these to interconnect many users at the same time and to do it as inexpensively as possible. We will find in this chapter that to do this, we need to abandon our simple broadcast and point-to-point concepts. In place of them we will design an interconnected set of transmission links called a **network**. Once we do this, we will find that there are different types of network architectures and that none of them satisfies all customers' objectives equally well.

Before we begin our investigation of networks we will once again define, from the perspective of engineering design, exactly what our objectives are.

- **What problem are we trying to solve?** Just as with simple communications systems, we desire to move multimedia information from one location to another. As before, we'd like to move it as fast as possible, receive it as accurately as possible, and do it as cheaply as possible. The introduction of the concept of a network is intended to reduce the cost as much as possible when trying to offer communications services to many users instead of just a few.

- **How do we formulate the underlying engineering design problem?** The potential user of a communications network will have virtually the same objectives as he or she would for a simple communications link. These objectives will typically include speed, accuracy, and cost, but may also include others, such as the delay between transmission and reception, the security of the communications system, and its ease of use.

- **What will the consequences be if the communications system satisfies the user's need?** Engineering work is often driven by an economic objective. In the development of communications networks this is even more true, since the whole purpose of a network is to minimize the cost to each of the connected users.

- **How will you test your design?** From the user's perspective, the services offered by a network are the same as those that would be provided by separate transmission links, except that it is cheaper and has a broader reach to other potential users. As a result, the list of specifications for a network closely resembles those of a simple communications system. If those specifications are met, then the customer should pay. In the case of multimedia communications systems, whether they be simple or a network, the tests needed to verify that the specifications have been met will typically include some of the following:

- Peak data rate

- Average data rate

- Number of errors as a fraction of the total number of bits transmitted

- Number of seconds in which no data is received

- Maximum and minimum delay between transmission and successful reception

In a network, it is also common to add requirements that specify how many other users can be reached and with what difficulty and cost.

- **What do we need to learn in order to understand multimedia communications networks and then to design them?** In this chapter, we first examine the economic rationale for developing a network and then develop several approaches for doing the sharing of transmission resources on which a network depends. Based on these ideas, we'll explore the first important data communications system, the telegraph, and then move right on up to the modern Internet.

8.1.1 What is a Network?

A **network**, by its definition, is a group of interconnected individual elements. When used in the communications business, the word means an arrangement which permits any of many users to communicate with any of the other users. For example, the telephone network for a small town might have 5000 telephone subscribers and might permit any 500 of them to be speaking simultaneously with any of the other 4500. How might a network be constructed? How might the design be different if the objective were to carry messages rather than voice? Further, how would we minimize the cost of service to each of the users of the network? To answer these questions, we need to examine the types of communications systems we have developed so far and then determine how they need to be adapted for operation as a network.

8.1.2 Background

The communications systems that we have examined so far fall into one of two categories. They are either broadcast or point-to-point. **Broadcast** systems use one transmitter to send information to many receivers, as shown in Figure 8.1. These systems have traditionally used radio, rather than wire transmission, and, in most cases, the users who receive the signal do not send any information back to the transmitter. Obvious examples are television and FM radio broadcast systems, but there are less obvious ones as well, such as the Global Positioning System (GPS) whose satellites transmit digital information constantly to any and all GPS receivers so that they can determine their locations. Cable television, as originally implemented, also falls into this category although it uses wired connections.

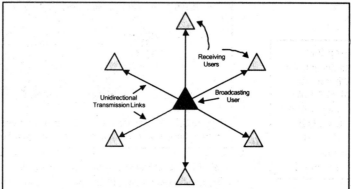

Figure 8.1: A Simple Graphical Description of a Communications Network in which one user "broadcasts" to six others using transmission links that send in only one direction.

The second category is **point-to-point** systems, in which two users are connected directly to one another using a radio or wire transmission system. This type of system is shown in simple form in Figure 8.2. The presence of arrows on both ends of the line connecting the two users indicates the ability to transmit in both directions. While there are exceptions, point-to-point systems usually permit communication in both directions, allowing the two users to respond to each other's messages and to start communications whenever they are ready.

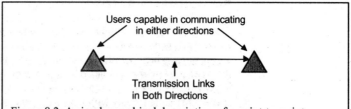

Figure 8.2: A simple graphical description of a point-to-point communications system in which both users are able to send information to each other.

How would these two approaches be used to build a communications network? The broadcast approach is very difficult to adapt, since it implicitly assumes that one user does the talking and all others do the listening. We will assume that for the design of our network we would like all of the users to be able to both talk and listen. With this assumption, the point-to-point approach can be used to meet our objective. We simply construct point-to-point transmission links between every pair of users. Such an arrangement is shown in Figure 8.3 for a network of six users. Fifteen bidirectional links are needed to make all the pairwise connections.

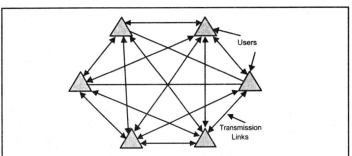

Figure 8.3: A first attempt in the development of a communications network – direct interconnection between all pairs of users.

With this network design we have achieved our system specification goal. Any user can send data to any other user by simply picking the transmission link that connects the two and then transmitting the bits comprising the data. However, as we shall see shortly, this is an expensive way to meet our objective, and we might want to refine our specifications and consider other alternatives. For this approach we must build two transmission links between each pair of users because we need one for each direction of transmission. Since each link is used only for traffic between the connected users, most of the links can be idle for much of the time. Our new objective is to figure out how we can still connect any pair of users, but do it less expensively by making more efficient use of transmission resources.

8.1.3 Reducing the Cost of a Network

As an alternative to the network design shown in Figure 8.3, consider the approach shown in Figure 8.4. In this case, there is only one two-way link to each user. Instead of connecting to some other user, however, the link from each user connects to a central **relay** point. The relay's role is to accept messages or bits from any user and "relay" them on to the intended destination. If the relay is capable of performing this function, and if it can perform it at the rate required by all of the users, then this network should still be able to meet its primary objective, which is the delivery of information from any user to any other user. For this design six bidirectional links are needed for the six users.

What would make this approach attractive compared with the network of direct user-to-user links shown in Figure 8.3? The simple answer is that, in most practical situations, it costs less. To understand this, assume for the moment that the relay point costs nothing, and that the cost of each transmission link is proportional to its length. By simply counting the number of links and the total length of the links in both Figures 8.3 and 8.4, we find that the network using the single relay is significantly less expensive to build than the one that relies completely on point-to-point links.

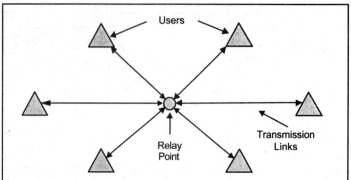

Figure 8.4: The second attempt in designing a network – two-way links from users, joining at a single central relay point, reduces the number and length of transmission links.

The relative advantage of the network with a relay point increases as the number of users grows. To find the number of links needed for a fully connected network with N users, we need to count the combinations of N things taken two at a time. A little bit of mathematics shows this number to be N(N-1)/2, so that is the number of two-way links that are needed. In contrast, if the central relay-based scheme is used for N users, then only N links and a relay are required. As N grows to be very large, the relay scheme's advantage grows with N/2. The ratio of costs, which depends on the total length of the links, goes up similarly. Consider, for example, the relative advantage for a telephone network where there are roughly 1 billion telephones in the world. However, even for modest sized networks we can reasonably conclude that the relay scheme is a cheaper way to build a network than the fully connected (or "fully meshed") method.

Although the relay method will be less expensive, we should still explore ways to improve it. Consider the network shown in Figure 8.5 where we permit the use of more than one relay. Specifically, each user is connected to a relay point, and those relays are connected to others. Is this cheaper? If the total length of the transmission links is still the indicator of total network cost, as it is in many cases, and if the relays can be trained to deliver the users' messages properly to their intended recipients, then this network is clearly cheaper than either of the previous two we examined.

Although this multi-relay network is cheaper in terms of transmission links, we achieved this savings by requiring that the relay points have additional performance capability. This can be illustrated with an example. Our objective is still to be able to send messages from any user to any other user. Suppose first that we would like to send a message from user A to user B in Figure 8.5. The two users, A and B, use the leftmost relay, marked R1, just as we described before. A message from A and intended for B is sent to relay R1. The relay then forwards the message to user B. In this case the relay does not need more capability than it did in the "one-relay" design.

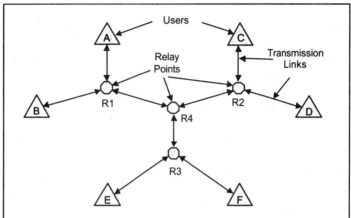

Figure 8.5: Using more than one relay point to further reduce the cost of a network.

Suppose now that we want to send a message from user A to user D. User D is not directly connected to relay point R1, so the operation of relay R1 must become more complicated. In this case the relay must decide which link it should use to forward the message to another relay point rather than directly to the intended recipient. In this example, the answer seems clear. User D is served by relay point R2. If A's relay point, R1, sends the message over its link toward relay R2, then R2 can deliver the message to user D once it receives it. If this "forwarding" mechanism is available at all of the relays, then it should be clear that any user can send messages to any other user. In the case of this example, R1 must relay the message to R4, which then sends it on to R2 for final delivery.

Since the multi-relay method reduces the total cost of the expensive transmission links, we should consider other system costs and determine if it will always be the least expensive network solution. If transmission costs dominate, then it probably will be, but in general there is no simple answer because the best design depends on many factors. Specific applications may have great differences in the number and location of the users, the amount of message traffic to be sent between each pair, the price of the transmission systems, the cost of the relays, and more.

Despite these many variable factors, several general observations can be made. In geographically diverse networks with many users, it is common to use many relay points to minimize the cost of the system, since that cost tends to be dominated by the costs of transmission. In contrast, in geographically confined networks such as **local area networks (LANs)**, the tendency is to use only one relay, as shown in Figure 8.4, since in this case the cost of the relay might be more than the transmission system. In both the "fully-meshed" network of Figure 8.3 and the central relay network of Figure 8.4, the links to each user are used only for traffic to or from that user. When more relays are used, the traffic for more than one user will be present on the links connecting the relays. This sharing of the links makes the network more efficient and less expensive for each user, but it also makes it more prone to congestion.

The financial bottom line is that modern telecommunications networks are designed to minimize their cost. This leads the designer away from simple mesh and single-relay designs and toward more complicated ones with many relays and interconnecting links that carry the traffic of many users. These principles apply to multimedia communications networks, like the telephone network and the Internet, but also to highways, railroads, and airlines.

Exercises 8.1

1. Compute the number of two-way transmission links needed to build a fully-meshed communications network connecting 4, 10, 100, and 1000 users. How many links are needed to connect the same number of users when using the **star** configuration shown in Figure 8.4?

2. Consider the construction of a network to serve six users who are distributed evenly about a circle which has a radius of 10 miles. Assume that cable containing a pair of wires is used to make the connections.

    (a) How much cable, measured in miles, is needed to build a fully meshed network?

    (b) How much is required to build the network if we use the star configuration seen in Figure 8.4 and assume that the relay point is located at the exact center of the circle?

    (c) Now assume that we use the multi-relay design shown in Figure 8.5. Assume further that the center relay is located at the center of the circle and that the three other relays are located on a concentric circle with a radius of five miles. How many miles of cable are needed for this network design?

    (d) Assume that the wire and its installation costs $10,000 per mile and that a relay costs $100,000 to build and install. What is the total cost for each system? Which is the cheapest to build?

    (e) Assume that the networks and the price of a relay stay the same as in (d), but that the price of the cable falls to $100 per mile. How does this change the costs of the three approaches? Which scheme is now the cheapest?

3. Suppose that a network is built in the form of the multi-relay approach shown in Figure 8.5. Suppose further that every user desires to send a total of 100,000 bits/second to the other five destinations, and, for the moment, that this "load" is split evenly, that is, each user sends at the same rate, 20,000 bits/second, to each of the other five users.

    (a) Compute the number of bits that must be carried on each of the transmission systems used in the network.

    (b) Compute the number of bits/second that each of the relays must be able to handle to allow complete flow of the user's data.

    (c) Suppose the central relay is only capable of handling 300,000 bits/second. How will this affect the users?

(d) Suppose that the transmission links connecting users with their associated relay can only carry 150,000 bits/second in each direction? How will this affect the network's operation?

(e) Suppose instead that the links connecting the central relay with the outer relays can carry only 150,000 bits/second in each direction? What effect does this have?

(f) So far we've assumed that the "load," that is, the amount of data transmitted, is the same from and to all users. How do the answers found above change (qualitatively) if the load is not the same for all users?

Section 8.2:
The Relay and
its Basic
Operation

In the previous section we found that the ability to relay a message from one communications link to another was the key to being able to minimize the cost of building a network. We learned that the relay must be able to accept a message, and then send it on toward its intended destination. In this section we examine how this might be done, and how we need to modify our communications procedures to make it work.

There are many types of relays used in telecommunications systems and we won't explore more than two of them in this chapter. We start with a simple but very important case, the design of relays to move a message from one user, the originator, to other user at the destination. The relay must be able to accept a message, and then send it on toward its intended destination. In order to meet that objective, several requirements must be met.

When a relay receives a message, it must first determine to whom the message is directed. Then the relay must find a way to send the message on a path that will lead to the intended recipient. If the relay is directly connected to the destination, the relay should send the message there. However, if the relay is not directly connected to the destination, the relay must determine the best path available toward that destination and then send the message along that path.

These basic requirements presume that there is no other traffic being carried in the network. Since consideration of competing traffic is almost always necessary, additional requirements are needed. When a relay has received a message destined for a user, but the link toward that user is already occupied with another message, the relay should hold onto that message by storing it until the link to the destination is available. When the link does become available, the relay should send or "forward" the stored message.

In order to make the network as reliable as possible in the face of equipment failures and times of heavy traffic load, we will add one more requirement. When the relay has received a message and ascertained the best path for it, but finds that route to be unavailable, the relay might consider the use of another route, even though the alternate route would normally not be considered to be the best one.

The first step to making this all happen does not occur in the relay device, but rather back at the origination point. In addition to sending the message, the originator must also send information that identifies the destination of the message. All relays involved in the transmission of the message will use this address information to route the message to its ultimate destination. There are several ways to send the address, but we will describe the one that is used most frequently. When using this method, the message itself is extended in length by adding two pieces of information. The first is an identifier or address for the originator. The second is an identifier for the destination. These pieces of information are usually prepended to the message by placing them in front of the message text. An example is shown in Figure 8.6.

Destination Address	Source Address	Message to Be conveyed

Example:

BILL%FRED!I NEED TWO TICKETS TO THE GAME#

Figure 8.6: A common method of adding address information to a message – "prepending" the source and destination addresses to the beginning of the message.

When each relay receives this new extended message, it proceeds as previously described. It extracts the destination identifier from the front of the message, **BILL**, and uses it to determine which way the message should be sent. When it sends out the message, it sends the full extended message so that any succeeding relays will have all of the address information that is needed to deliver the message.

It is obvious that the destination address must be included with the message, but why send the address of the source? The answer is that once the message winds its way through a network that uses even one relay, the destination user has no way to know who the originator was unless he or s he is told. Without this information it is not possible to respond. The source address information attached to the message provides this information.

Experiment: Relaying a signal

Networks consist of interconnected communications links. When a user's data signal arrives at a "node", a junction of two or more links, it must be relayed from the incoming link to one of the outgoing links. Which direction it is sent might be decided before the data is sent, a technique called **switching**, or when the data actually arrives at the node, a technique called **routing**.

8.2.1 Store and Forward Networks

The network structure described in the preceding section is commonly termed a **store and forward** network, since each relay point stores each message until it is ready to forward it to either its ultimate destination or to another relay point. This method of constructing a communications network is a famous one. The original telegraph networks of the 1800s were built this way. What might be more surprising is that this is also an accurate description of the modern Internet as well. Figure 8.7 shows a simple example of what such a network might look like on a broad geographical scale. The solid dots show terminals points where traffic can be originated or "terminated." The open dots are relay points where traffic may be forwarded as well as originated or terminated. The lines connecting the relay and terminal points are the transmission links. In the 1800s these links were telegraph wires carrying messages at about 5 bits/s. In today's Internet, they are likely to be fiber optic lines operating at billions of bits/second. The speed and scale have changed, but the concepts have not.

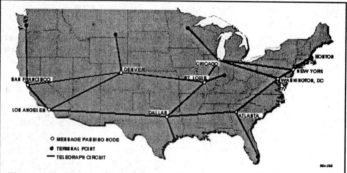

Figure 8.7: The configuration of a simple "store and forward" network.

Figure 8.8. Construction of the transcontinental telegraph along the route of the Pony Express in 1861. Once the telegraph line was complete, the Pony Express was rendered obsolete [7].

8.2.2 How is the Relay Implemented?

In the days of the telegraph, the relay points were implemented with human beings. Because it is relatively intuitive to understand the strategies and procedures these workers developed on a human scale, it is instructive to examine them here. Modern networks have automated these processes and procedures so that human action at a relay point is no longer needed. Although the rate at which these automated dispatchers make decisions is much faster, the issues and decisions they face are very much the same as those faced by the human operators in the earlier systems.

At a telegraph relay point, all but one of the workers operated the telegraph lines, sending and receiving messages using telegraph keys and "sounders" of the type shown in Figure 8.9. The remaining person was the dispatcher. This person was constantly making decisions regarding how each received message was to be handled.

Definition:

We use here the modern definition of a **relay**, meaning a person or device capable of receiving, directing, and then transmitting a signal. The original definition of a relay presumed that it served only two links, and that it had no ability to route or schedule a retransmission. They were used only to boost, amplify, or "clean up" a signal traveling from one user to another. When performing this simpler function, they were called **repeaters** rather than relays.

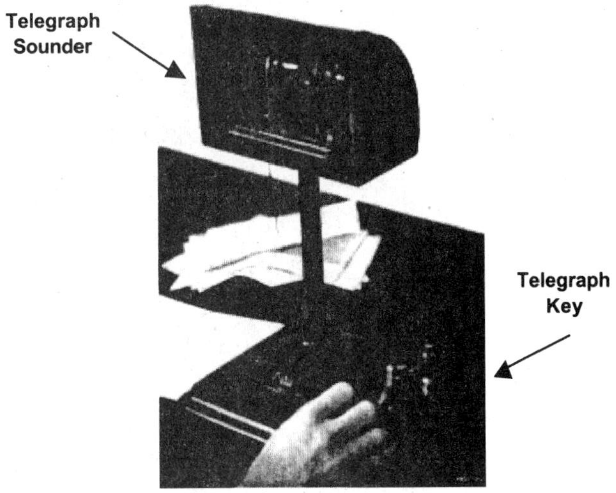

Telegraph Sounder

Telegraph Key

Figure 8.9: A telegraph key and sounder of late 1800s vintage.

For each incoming message the dispatcher first determined whether the message terminated at that office or whether it should be relayed to another office. If the destination of the message was not the local office, it needed to be routed to another relay point. The dispatcher had to decide where it should be sent next. Usually there was a preferred path to reach a particular destination, but that line might be down or the queue of messages waiting for that line might be long. Either of these conditions might lead the dispatcher to select a less usual path to the destination. Finally, the dispatcher had to make a decision about the relative priority of each incoming message. Among all the messages leaving the relay point on a particular transmission link, which one should go first? Late messages should go before ones that were not late, and ones for which priority treatment had been paid should go earlier than messages without priority. The human dispatcher used his or her knowledge of how the network was configured in order to make these decisions. Since the length of time needed to send out each message was a few minutes, decisions had to be made at a rate of one every minute or so.

The modern Internet uses all of the same concepts, but operates at much faster rates. Fiber optics, as noted earlier (and in Chapter 13), provide the medium to support higher data transmission rates. The human dispatcher is replaced by a special purpose computer called a **router**, so named because it chooses the route for each of the outgoing messages. A photograph of a modern router from 1999 is shown in Figure 8.10. It is built by Cisco and accepts messages at data rates of up to 40 billion bits/second. The dispatcher function, that of choosing routes for the messages, must be done at the rate of almost 1 million decisions per second.

Figure 8.10: A Cisco 12016 Gigabit Switch Router, capable of accepting messages from up to 16 streams each operating at 2.5 Gb/s.

8.2.3 Characteristics of "store and forward" networks – the telegraph and the Internet

We can summarize the characteristics of the store-and-forward network structures. First, each message must have the destination address attached to it so that each dispatcher/router can intelligently choose the next transmission link for it. Although the message has the originator's address attached, the originator has no idea whether or not a message reaches the intended destination unless a separate message is explicitly sent back to acknowledge the receipt. The general strategy applied by the dispatcher/router is to find the path for each message that is likely to result in the least delay in reaching its destination. The time it takes to send a message between the same origination and destination points can be different when attempted at different times of the day. The time it takes will depend on the specific path the message takes and on the amount of time it spends in queues awaiting transmission on each link that has been chosen for it.

From an economic perspective the store-and-forward method of operating a network is very efficient. By allowing messages to be stored, or **queued**, at each relay point, the network's operator can keep the transmission links, or "data pipes," full much of the time. Since these transmission systems have historically been the most expensive part of the system, using them

14

efficiently means that the whole system operates at its best efficiency. This interest in cost efficiency is why both the Internet and the U.S. Postal System are designed in the same way.

8.2.4 Another Way of Relaying Data

One might conclude from the previous discussion that the store-and-forward method is the only practical way to build a communications network. In fact it is not true. Further, in some important cases, it is not even the best way to do it. To see why this might be, consider a different problem. Instead of sending messages from one user to another, suppose the objective is to send streams of digitized voice and to use the network to carry telephone calls. When we send messages through a network, our expectation of service is that we will receive the message with reasonable reliability and within a reasonable amount of time. As a service user, we are unsatisfied if we do not receive the message at all, but we accept the fact that the amount of time to get the message to us varies widely. For a telephone, however, our expectations are very different. When we speak, we expect to be heard and heard immediately. When the person we are talking to speaks, we expect to hear that person immediately and we expect to hear everything that is said. Delays destroy the rhythm of an interactive conversation and make effective communication extremely difficult for human speakers. In short, our expectation for **quality of service (QoS)** is different for voice than it is for messages, files, broadcast television, or other types of data.

How then should a network be designed to handle real-time interactive media like voice and video teleconferencing? The answer will often not be the store-and-forward method, because of the variable delay it introduces as well as some other quality problems. The most common approach in these cases is the use of a **switch** at the relay points, instead of a dispatcher or a router. The switch reserves a certain amount of data-transferring capacity for each "connection" between a pair of users and moves their data through the switch expeditiously when it arrives. In this way it can guarantee that each pair of users obtains the quality of service that they desire. To do this, the switch must be able to handle data at the peak rate at which is applied, since delaying it is not acceptable.

Although networks based on switches seem best for real-time interactive media, it is not the best structure for all networks. Switched networks are not as efficient as routed networks when measured in terms of how full the transmission pipes are kept. Reserving capacity for each pair of talkers so that they can get through the switches immediately means that the links will be idle for some portion of the time, typically 50%. An analogous situation is waiting in line for an ATM. The banks like it when people line up for the ATMs, keeping the ATMs efficiently occupied. The ATM users like it when there are extra ATMs so that one is always idle when the user arrives. The bank sees this condition as less efficient, although the customers like it better.

8.2.5 Comparison of Routing and Switching

Now that two kinds of networks have been defined, we can explore which of the two methods, routing or switching, is best The answer, as in most engineering problems, is that the way we measure "best" depends on the problem to be solved. The Internet may not be the best way to carry voice and other interactive multimedia, where best is measured in low delay, but neither is the telephone the best way to carry delayable message-based services like e-mail, where best is measured in terms of low cost. However, some general observations apply to both type of networks.

- The goal of a communications network is to permit any of its users to communicate with any other user.

- If cost were no object, the best network would be fully meshed, consisting of a direct link between every pair of users. This approach is usually cost-prohibitive, however, driving us to different designs. To make the network cheaper, we can introduce the concept of relays, which accept data from one link and place it on another. To make this possible, the data must be augmented with information that indicates the destination of the data or message.

- The relays can be implemented in two important ways. When low delay and reliable transmission are important, switches are often used. These switches guarantee the allocation of bandwidth from one link onto the next, and introduce very little delay. They are commonly used for telephony and interactive multimedia.

- The other important way of implementing the relay point is with a router, a special purpose computer which stores incoming data, and uses the attached address to determine which output path to send the message on, once that path is available. Networks using routers make very efficient use of the transmission bandwidth, and are often the most cost effective solution when carrying traffic, such as e-mail and data files, that can tolerate variable delay in its delivery. The original telegraph, the Internet, and the U. S. Postal System are all examples of "routed networks."

Exercises 8.2

1. Most routers use a **routing table** to determine how to handle each incoming message. Such a table consists of a list of all possible destinations for the message and the most appropriate transmission path for each. The table shown below would be appropriate for router R1 in Figure 8.5.

Ultimate Destination	Next Hop
User A	User A
User B	User B
User C	Router R4
User D	Router R4
User E	Router R4
User F	Router R4

The information in this table can be summarized in the following way. A message arriving at router R1 and addressed to User A or B will be sent directly there since R1 is connected directly to both of them. If the destination is not A or B, the router R1 will send the message on to router R4, which will then, hopefully, pass them on in the right direction.

Now build the routing table for router R2 and then R4.

2. Redraw the network shown in Figure 8.5 to include a two-way transmission link directly connecting relays R1 and R2.

   (a) Assuming the availability of this new link, what is the best way to send messages from User A to User D? What is the best way to send traffic from User A to User F?

   (b) There are now two ways to send traffic from R1 to R2, directly and via relay point R4. Revise the routing table for R1 shown in exercise 8.2.1 to include alternative paths where they exist. The table should now have columns labeled:

   Ultimate Destination Preferred Next Hop Alternative Next Hop

   (c) When might the alternative path be used instead of the preferred direct path to send traffic between R1 and R2? Who would make the decision to do this?

3. Look at the simple network in Figure 8.7 and list seven different ways a message could travel from Boston to Houston. Are there more? Which of these paths would you actually consider using?

4. How would you modify the message format in Figure 8.6 to include information about the priority of a message? You might consider two types of priority information. One would indicate a class of service paid for by the originator, analogous to indicating the difference between third class and first class mail. The second type of priority information might be added by the system based on how the network is doing in meeting its delivery time commitment.

5. Suppose that the traffic arriving at a relay point is "bursty" in the sense that many messages arrive for a certain time interval and fewer arrive at other times. Specifically let's assume that a data transmission link entering a relay brings 50 messages per minute for two minutes, none for a minute, 200 message per minute during the next minute, and then none for two minutes. Assume further that this pattern repeats continuously.

   (a) How many messages must a store-and-forward processor (that is, a router) be able to process per minute to keep up with the traffic load brought by this transmission link?

   (b) When operating at the rate barely able to keep up with traffic, what is the maximum delay encountered by any message, assuming that they are handling on a "first-come, first-served" basis by the router? How big must its message buffer be to avoid losing any messages?

   (c) How fast must the router be able to process messages in order to ensure the very smallest amount of delay between the arrival and successful dispatching of a message?

   (d) When the router operates fast enough to reduce the delay to its minimum, how much time does the router spend idling, waiting for the next burst of traffic?

   (e) Suppose that we use a switch at the relay point instead of a router. How many messages per second must it be able to handle to avoid losing messages? What percentage of the time over a six-minute cycle is it idle?

   (f) Compare the results from (a) and (e) and comment on the probable difference in implementation cost between the barely adequate router and the barely adequate switch.

   (g) How do all of these comparisons change if the traffic flow on the incoming transmission link is constant at 50 messages per second?

Section 8.3: The Internet

In this section we explore the Internet in more detail. It is not the most widely accessible data network in the world (That's the telephone network!) but it is growing rapidly, and more and more people are gai ning access to it daily.

8.3.1 The Origin of the Internet

In the late 1960s the Advanced Research Projects Agency (ARPA) of the U.S. government was confronted with a vexing problem. It was running a very successful program that had as its goal the development of supercomputers, computers capable of operating at hundreds or thousands times faster than those commercially available at the time. A few of these were built but they were very expensive. More researchers wanted to be involved in the project, but they either had to move to where the supercomputers were located, or ARPA needed to buy more supercomputers. The researchers didn't want to leave their home laboratories and universities, and ARPA didn't have the money to buy everyone a supercomputer.

A potential solution was suggested. Engineers and computer scientists were already developing something new called "data networks" for both military and commercial applications. ARPA decided that the same concepts could be used to build a communications network that would let many physically dispersed researchers "reach out and touch" the agency's small number of supercomputers. Figure 8.12(a) shows the actual situation in the late 1960s and Figure 8.12(b) shows the approach to a solution. ARPA started a research program to design and build such a network. It was termed the ARPANet and ended up having much more of an effect on how modern data networks are designed than anyone ever imagined.

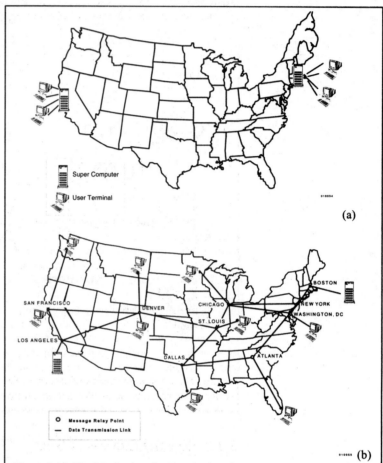

Figure 8.12. The Motivation for Developing the ARPANet. (a) The few expensive supercomputers could only be used by those researchers physically close them. (b) The development of a data network permitted researchers at other universities and labs to "log on" at a distance to the scarce computers.

The distant users were attached to the ARPANet using the smaller computers located in their own labs and universities. Special routers were built to relay the messages sent to and from the supercomputers. The transmission links were rented from the Bell System, America's telephone company at the time. **Protocols**, which are the procedures by which the network operates, were developed, tested, and improved by a large number of engineers and computer scientists who were encouraged by ARPA, and, later, by the National Science Foundation (NSF).

Another view of the ARPANet's architecture is shown in Figure 8.13. On the left are the users, in the middle is the communications network itself, and on the right are the supercomputers. In modern nomenclature, the users **access** the network, which then connects them to the **servers,** those devices that provide the services desired by the users. In the case of the ARPANet, the servers were the supercomputers themselves and the service that the users desired was the ability to send commands to the supercomputers (and get results back) as if they were sitting next to them. The modern Internet takes this same idea and extends it to a much wider range of access and services.

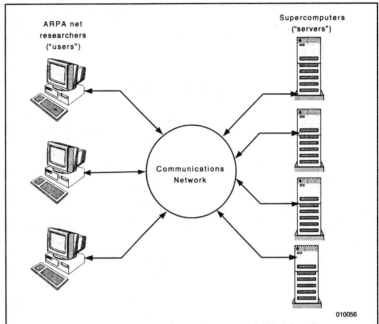

Figure 8.13: A simple version of the ARPANet, highlighting the separate roles of the users, the network itself, and the supercomputers.

8.3.2 How the Internet is Built

As computers became cheaper and data transmission systems got faster, the concepts used to build the ARPANet were commercialized and spread rapidly. We call it the Internet today cause it was specially designed to permit networks owned and operated by many universities and laboratories to smoothly interconnect with each other. This concept of "internetworking" was shortened to become simply the **Internet.**

Design of a Local Internet Service Provider

Figure 8.14 shows the block diagram of what we'll call a local **internet service provider (ISP)**. (Caution: We'll find out as we proceed that the nternet service provider" can mean many things!) As we dissect this diagram we find, just as with the ARPANet, this ISP can be broken into three basic segments: (1) access, (2) connectivity, and (3) service. The ISP's users gain access to the ISP through telephone modems, digital subscriber loop (DSL) modems, cable modems, and, now, via their cellular phones and personal digital assistants (PDAs). Once they've gained access, a router sends their messages to whichever service they are seeking. Thus the router provides the connectivity between the users and the services they desire. Examples of these services are e-mail, file storage, video games, and more that we'll discuss shortly.

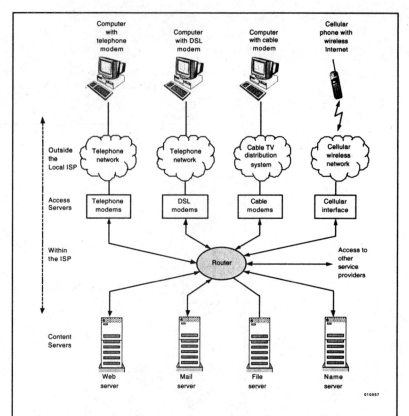

Figure 8.14: An Internet Service Provider (ISP) with a broad range of access technologies and content services

Following the diagram in Figure 8.14 from the top, we see that computers equipped with telephone dialup modems (such as the V.90 modems found in most modern personal computers) connect to modems at the ISP through the telephone network. The bank of modems at the ISP is termed an **access server** since it is accessible by, or "serves," many users. Similarly, the next computer to the right, equipped with a DSL modem, connects to the ISP, again via the telephone network, into a DSL access server. Clearly then, the entry or access into the ISP from any user is via an access server of the appropriate type. (Can you think of other methods of internet access beyond just these four?)

The **content servers** are shown along the bottom row of Figure 8.14. These servers are computers that are augmented with special software, and sometimes hardware, to perform their tasks.

Some of the content servers that one might commonly find at an ISP include the following:

- A mail server – a computer equipped with software to store and retrieve e-mail for the ISP's clients

- A web server – a computer equipped with software (and, possibly, extra hardware) to send web pages to anyone who asks

- A multimedia server – a computer with the software and hardware to provide audio or movies to customers

- A game server – a computer capable of providing customers with access to interactive games

- And a **name server** – a computer whose function is to "find names." This function will be described in Section 8.3.4.

As new services are added, new servers will be added to provide them. In principle, the access servers and the router do not need to change to handle new services. Sometimes, however, the increase in data flow caused by the new services requires that they be upgraded to handle it. For example, simple dialup modems and a low-capacity router can handle all of the data needed to send e-mail between users. As soon as the ISP's customers want to obtain streaming video (for example, movies), faster access and faster routing will be required.

Interconnecting Two Local ISPs

We have assumed, so far, that all of the services that a user needs can be provided by its own ISP. What if this is not true? A simple example is the one in which a user desires to send e-mail to a user of another ISP. Another is the case in which a user desires to surf World Wide Web pages held on a remotely located server.

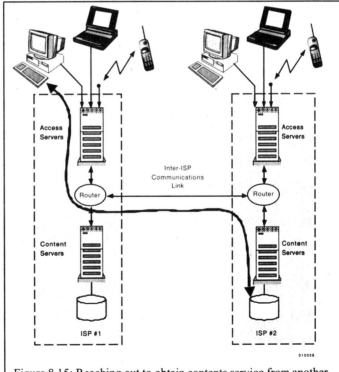

Figure 8.15: Reaching out to obtain contents service from another local ISP.

The first step in solving this problem is shown in Figure 8.15. Here we consider two ISPs and desire to make any of the services (such as e-mail) which are available on either of them now available to the users on both of them. This can be accomplished by setting up a two-way (**full-duplex**) data communications link between the route used in two ISPs. If the **routing tables** in these two routers are set up properly, then a user who gains access to the ISP on the left side of Figure 8.15 can send its e-mail to the **mail server** located in the ISP on the right side, and vice versa. This happens as follows:

- The user on the left ISP gains access through an **access server**. He or she then writes the e-mail message and asks that it be sent to the **mail server** that holds incoming mail for the intended receipt. We assume here that the recipient is served by the mail server in ISP #2 in Figure 8.15.

- When the user's computer sends the mail message, it adds to it the address of the recipient and the user's own address, as shown in Figure 8.6 and discussed in Section 8.2. Once the addresses are added, it sends the message on to its router.

- The router serving the originating user examines the address and consults its **routing table.** It finds that the message is to be sent not to its own mail server, but rather to the one in ISP #2. The routing table instructs the router to send the message on to its counterpart in ISP #2. That router again examines the address in the message, determines that ISP #2 is the proper destination, and then sends the message on to its own mail server.

- At some later time, the intended recipient of the e-mail gains access to ISP #2 and sends a message to ISP #2's mail server asking if there is "new mail." A message would be returned informing the recipient of the waiting message.

We've focused here on the transmission of electronic mail, but it should be clear that this same scheme will work with any of the users of either of the two ISPs and with any of the services offered by either. In particular, a Web page available on the **web server** on ISP #1 can be browsed by any user of either of the two ISPs as long as a good communications link exists between the two routers and the routers are properly instructed, through their routing tables, about where to find the desired services.

Using a Network of Links and Routers to Connect the Two Local ISPs

In Figure 8.15, we used a dedicated transmission link between the two local ISPs in order to let the users of the two ISPs have access to any of the services offered on either of them. From Sections 8.1 and 8.2, it should be clear that it is often more economical to use a network to connect the two, rather than a dedicated link. This concept is shown in Figure 8.16. On the left and right are the two local ISPs, as before. Instead of connecting their two routers directly, however, Figure 8.16 shows them connecting through a network of digital communications links and routers. If we assume the network can provide the same transmission rate, accuracy, and response time of the dedicated link, then the users of the two ISPs can retain all of the services and quality they had, while reducing the cost of the whole operation.

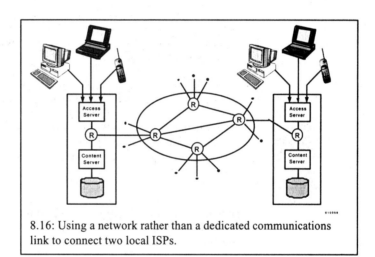

8.16: Using a network rather than a dedicated communications link to connect two local ISPs.

Before we go on, we should observe that we have built an **internetwork.** Both of the ISPs are networks in their right, and we have now connected them to each other using yet another network (often called a **backbone** network). To do this kind of interconnection requires that all of them use the same rules for addressing data and sending it through routers, even though all three of them are typically owned and operated by different people or organizations. Again, this ability to "internetwork" is where the **Internet** gets its name.

Interconnecting Many Local ISPs – The Internet

It is a simple but important step to go from connecting two local ISPs to connecting many of them. This is shown in Figure 8.17. Each of the little triangle symbols is all or part of an ISP of the type shown in Figure 8.14. Each serves a local area or perhaps a region of the United States. All of them are connected to each other with a network of high-speed routers and digital communications links often called a **backbone** or **transport** network. It is called this to emphasize that its principal function is moving data from ISP to ISP, rather than providing access to individual users or providing **content services**.

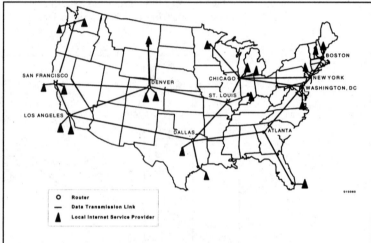

Figure 8.17: A geographical view of local and regional ISPs interconnected by a backbone transport network.

Revisiting the Definition of an ISP

One of the things that confuses people new to the Internet is the fact that almost anyone and anything is called an **internet service provider**. A key reason for this confusion is that some companies and organizations have "specialized." Rather than offering all aspects of what we consider to be an ISP, they have focussed on just one or two. We have already seen one example, the so-called **backbone** ISP, an ISP that simply provides data transport services between other ISPs. This concept of specialization can be understood using the diagram in Figure 8.18.

The ISPs in Figure 8.18 marked ISP #1 and #4 are the "full-service" ISPs of the type shown in Figure 8.14. They provide access to their customers, some of the services needed by their customers, and the connectivity needed to provide their services to any "Internet citizen." ISP #5 is a **transport** or **backbone ISP**, connecting all of the others. ISP #2 provides access but no content. Its customers must obtain any content services they desire by "reaching out" to other ISPs through the backbone network. ISP #3 is just the converse. It has no customers who directly access it, but rather it "serves content" over the network instead. Most industrial-grade Web servers take the form of ISP #3. Many large computers equipped with large amounts of semiconductor and disk memory (that is, the content servers) are connected through very fast routers to one or more transport ISPs. These "server farms" can provide Web access to thousands of users at the same time. The same concept is used for video and game servers as well.

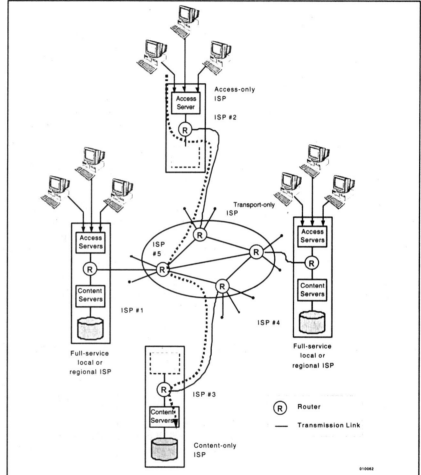

Figure 8.18: Illustrating the role of specialized ISPs – ISPs #1 and #4 provide full service, while ISP #2 provides only access to users, and ISP #3 only "serves content." The role of ISP #5 is to transport data packets between the other four.

A Quick Review – Building the Internet in Big Steps

It is useful to step back and re-examine how we reached this design for the modern Internet.

- In Section 8.1 we stated the objective of sending messages between any pair of many possible users who are distributed over a wide geographical area. We examined several approaches and decided that a network of nodes and short communications links (Figures 8.4 and 8.5) is often less expensive than a "full mesh" (Figure 8.3), the scheme that provides direct links connecting all possible users.

- The first national telegraph system, deployed in the mid-1800s, used the "store-and-forward" concept to pass messages through a network which spread over the whole country (Figure 8.7).

- The ARPAnet, developed in the early 1970s, was developed to cheaply provide access to a few very expensive servers (specifically supercomputers) by researchers all over the country (Figure 8.12(b)). The ARPAnet used the same store-and-forward idea as the telegraph but at much faster transmission rates.

- Once built, the ARPAnet's capabilities grew by offering many more services than just access to supercomputers. Examples were sending messages between users (e-mail) and transferring files and documents between researchers.

- By developing a set of rules or **protocols** that everyone could use, it became possible to interconnect many local and regional networks. This came to be called the **Internet**.

- Since their first appearance, internet service providers have changed from nonprofit organizations supporting government-funded research into commercial organizations driven by sales and profit. With this has come specialization, where ISPs focus on only the portion of the network where they "add value" and can make money. For some this means only providing user access. Some only transport messages, and, for some it means only "serving content," such as Web pages and movies.

8.3.3 Protocols and Their Use in the Internet

It was mentioned in the previous section that a key to the successful development of the Internet was a general agreement to use a standard set of rules regarding how messages are built and how services will be provided. These rules are called **protocols**. In this section we examine what a protocol is, why we call it that, and why we want them. We also describe a few of the most important ones used in the Internet.

The term "protocol" comes originally from the conduct of diplomacy between two nations. Rules and procedures were developed so that "heads of state," like kings, could deal with each other without misunderstanding. When engineers began developing communications systems, it quickly became clear that the lack of clear agreement on the rules for sending messages led to chaos and confusion. Thus rules, or **protocols**, were born. They define how messages are built, when they are sent, how they are acknowledged, and all other aspects of data communications. Many engineers and computer scientists worked together to develop the protocols used on the modern Internet.

There are literally hundreds of protocols that define how the Internet operates. (You can find them at www.ietf.org/rfc). We will describe a few of the most important ones here.

TCP/IP – Making Communication Reliable and Fair

When a computer on the Internet has a message ready to send to another, the two of them frequently use a pair of protocols called **transport control protocol (TCP)** and the **internet protocol (IP)** to send the message and ensure its reliable delivery. More detail on how they work appears in Chapter 16 and in reference [2] but their operation can be summarized in three steps:

1. At the transmitter, software using TCP breaks the message to be sent into **packets** of smaller length and places an "address header" of the type seen in Figure 8.6 on each of them according to the **internet protocol (IP)**.

2. The "IP packets" are delivered from the transmitting computer to the receiving computer via one or more routers which use the address information in the header to seek out the proper destination.

3. At the receiver, software using TCP reassembles the received packets into a complete and accurate message and delivers it to the user.

There are many motivations for breaking up each message or file into packets but one of the most important is to make the network "fair" in the sense that many users are not delayed by one user who is sending a very long message. Breaking all messages into packets allows the routers to intermix traffic from many users in a relatively fair way.

SMTP – Sending and Receiving Electronic Mail (e-mail)

Simple Mail Transfer Protocol (SMTP) is the procedure by which most computers send and receive electronic mail. It works in a very straightforward fashion [2]. Its basic steps are as follows:

1. A computer (at Rice University, for example) with an e-mail to send informs the recipient's mail server (at SMU, for example) that it has mail to send. The mail server responds with a message (delivered with TCP and IP) of the form

    smu.edu SMTP service ready

2. To that, the originator at Rice University sends

    HELO rice.edu

3. The receiving mail server then responds with

    smu.edu says hello to rice.edu

4. With this "handshake" complete, the computer at Rice states its purpose and sends the following line of ASCII characters

    MAIL FROM john@rice.edu

In this same back-and-forth fashion the computer at Rice identifies the user to whom the e-mail is directed and then sends the mail message itself. The receiving computer at SMU accepts the message and then the two agree that the transmission is complete.

An e-mail composed purely of text is easily sent using this line-by-line scheme. What if the e-mail users want to send multimedia data like music or pictures? A scheme called **Multipurpose Internet Mail Extensions (MIME)** was developed for just this purpose. It works by taking strings of bits and mapping them into ASCII characters. These characters are sent using SMTP as if they were a textual message. The recipient's mail viewer recovers the bits and then displays or plays them appropriately.

HTTP – Finding and Retrieving Web Pages

The World Wide Web (WWW) was developed in the early 1990s as a way of using the Internet to reach out and obtain information held anywhere in the "web" of computers present on the network. The information is prepared for display on other people's computers by converting it into **hypertext**, a format which uses ASCII characters to describe not only the textual information but also the placement of text and figures on the display screen. Once one has found the web server containing the desired information, it is necessary to transport it to the user who wants it. For this purpose a protocol called **hypertext transfer protocol (HTTP)** was developed [2].

The software that performs the functions defined by HTTP works very much like the software that handles Simple Mail Transfer Protocol (SMTP), which was described above. The "calling" computer uses TCP and IP to establish a connection with the computer containing the information of interest to the caller. In an exchange much like that shown above, the "called" computer responds to the request by finding the data file containing the desired hypertext and then sending it, line by ASCII line, to the calling computer. The web browser at the calling computer then interprets the hypertext and displays it appropriately.

FTP – Sending and Receiving Data Files

Many activities among computers in a network rely on the ability to reliably transfer data files from one computer to another. A simple example is the case where a computer user wishes to print a document. In the early days of personal computers, each computer had its own printer. Now it is common for the computers in a laboratory or office to be "networked" together and to share the use of a single "print server," a printer which has been augmented with a built-in computer that accepts files from the client computers, stores them, and then sends them to the attached printer when it becomes available. A necessary part of this scheme is the ability for the client computers to reliably send the document to be printed to the server. To make this and many similar cases very simple, the **file transfer protocol (FTP)** was developed.

The computer originating the transfer begins by establishing a connection with its counterpart using TCP. Over this "control connection," the two computers identify the file or files to be transferred and then start and stop the transfer. The software that performs the FTP functions actually opens a second connection between the two computers in order to carry the data agreed to over the control connection. Once the data transfer is complete, and TCP confirms that it is accurate, then the two computers agree to break down the connections and continue with their tasks.

As mentioned earlier, these are but a few of the hundreds of protocols used by the Internet. There are other types of data networks as well that also have their own numerous sets of protocols. (See reference [2] for a readable description of these and more.) The details of these protocols is not important here, but the fact that they exist and why they exist is. If not for an orderly and well understood set of rules, it is impossible to reliably interconnect two computers, much less millions of them.

8.3.4 Domain Names and Finding Computers in the Internet

Before we leave our description of the Internet, there is one more protocol that needs to the described. It helps to perform a vital function, that of converting the name you know a computer or service by, and the actual binary address that the Internet Protocol (IP) uses to route and deliver the packets of any data transfer. To see why this new protocol is needed, we must first understand how this situation of having two names came about.

Internet Protocol Addresses

When the Internet Protocol was first defined, the engineers involved decided to use the scheme shown in Figure 8.6. Each data packet comprising a message from one user to another would be "prepended" with two addresses, the first being that of the destination computer and the second being that of the source computer. It was further decided that each of these addresses would be 32 bits long. Since there are about 4 billion different combinations of 32 bits, it was felt at the time (about 1982) that this was more that enough addresses to handle all of the computers that there might ever be. Even though there are efforts under way now to increase the address length to 128 bits, virtually all computers on the modern Internet still use these 32-bit addresses.

Thus each of the computers in the network has its own 32-bit address and they are all different. It is also true that all devices comprising the Internet, including routers and servers, also must have unique IP addresses.

While computers easily accommodate 32 binary numbers, the humans who administer them are not so flexible. For them the addresses are usually written in "three-dot" form. The 32 bits are broken into four blocks of eight bits and then each of the blocks is converted in to a decimal number. Those numbers are separated by dots. For example, the 32-bit address given in binary form by

11000000 00110101 11011110 00000111

would be written by humans as [192.53.222.7].

Uniform Resource Locators (URLs)

Even with the IP address simplified to a string of decimal numbers, it is inconvenient for human computer users to remember how to address a particular computer or resource. To solve this problem, an even more "human friendly" method was developed. Computers and the services that they provide were thought of as operating in "domains" and a scheme of "domain names," or, more formally, **uniform resource locators** (URLs) was developed. The universe of all Internet users was split into disjoint sets. In the original system, a computer might be part of an educational domain, a government domain, a military domain, a nonprofit organization, or a part of the network itself. These domains were designated by the letters EDU, GOV, MIL, ORG, and NET. (Recently many more of these "root domains" have been added to accommodate the rapid growth in the number of users of the Internet.) Each of these large domains contained many members. RICE and SMU, for example, are both members of the educational domain EDU. This relationship of being a member was written by placing the member's name in front of the larger domain. The URL for Rice University, following this example, is RICE.EDU.

In fact, this scheme can go on indefinitely. There might be many organizations at Rice which have separate computers and need separate addresses. This is indicated by adding these organizational or functional designators in front of the URL for Rice. For example, the Electrical and Computer Engineering Department at Rice can be reached at

ece.rice.edu

while the university's web server is addressed with the URL

www.rice.edu

This type of addressing scheme is termed "hierarchical," since there is a clear hierarchy in the names for each computer or service. These relationships can be seen easily using Figure 8.19. The internet universe is split into "root domains" that include many members. Each of those members can be composed of many submembers, and, progressively, each of them can have many parts. The URL describes the progression from the smallest member up to the root domain.

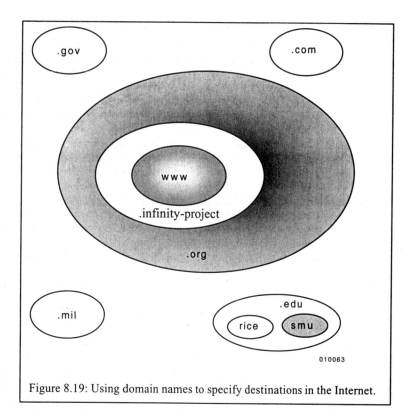

Figure 8.19: Using domain names to specify destinations in the Internet.

Relating the Two Addressing Schemes

Computers easily handle 32-bit addresses and work less well with variable length ASCII URLs. Humans are just the opposite. How can the two be used together? The answer comes in the creation of a new service and the definition of a new protocol. The new service translates a URL into a 32-bit IP address. The computer which offers this service is called a **name server**. The new protocol is called **domain name service (DNS)**. It is used by any computer that needs to have a URL translated. How this works is shown in Figure 8.20.

Suppose a user of the computer on the left of Figure 8.20 writes an e-mail message and directs it to a mail server located in the domain on the right side. This e-mail cannot be sent until the sending computer knows the IP address of the mail server. It might have the address stored in its **cache**, a temporary memory for IP addresses, but, if not, it sends a message defined by the domain name service (DNS) to its name server. This name server might be in the user's ISP or it might be a more complete one to which it has been referred. In either case the name server looks up the IP address of the computer corresponding to the URL in a look-up table and returns it to the requesting computer. Now equipped with the proper IP address, the mail transfer can proceed (using SMTP and TCP/IP).

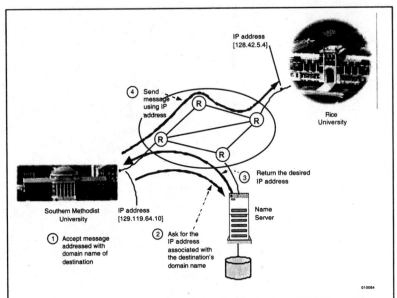

Figure 8.20: Using the Domain Name Service to obtain an IP address from a Name Server.

Just as the domain-naming convention is hierarchical so are the name servers. If a computer asks for a name unknown by its local name server, it can "go up the hierarchy" to get the answer. For example, the name server at SMU can be expected to know the IP addresses for all of the departments and services at SMU but might not know the IP address for the mail server at Rice. To find it, the name server at SMU can go ask, using DNS, the master name server for the whole EDU domain for the name server at Rice. Armed with that answer, it will send a message to the Rice name server asking for the IP address for the mail server. Using this scheme any computer in the Internet can, with successive queries using DNS, find the IP address of any computer or service on the net. This is the glue that holds the net together.

Exercises 8.3

1. Using the Web page for the Internet Engineering Task Force (www.ietf.org/rfc) find the oldest "request for comments" (RFC) that defines the Internet Protocol.

2. Convert the following bit strings into "three-dot" decimal IP address:

    (a) 11101010 00100100 11111000 00110011

    (b) 00010100 11100001 01010101 0000001

3. Convert the following "three-dot" sequences into 32-bit IP addresses:

    (a) [192.34.208.56]

    (b) [36.1.23.255]

    (c) [255.255.255.255]

4. Using a personal computer with an Internet connection, enter the DOS shell and type the command TRACERT followed by any URL that you know. (Use rice.edu if you do not know one.) The first response from TRACERT is the IP address corresponding to the URL.

    (a) Write down the IP address in both the three-dot and its binary form.

    (b) What protocol was used by TRACERT to convert the URL?

5. TRACERT, which is short for "trace route," is used by network professionals to determine how the routers in a backbone network will direct the packets in a data transfer from your computer to the destination that you have designated with the URL.

    Use TRACERT to find the path to any three universities. Determine as best you can from the cryptic responses to each TRACERT how the Internet would send messages to the respective destination.

6. Use the TRACERT command to find the path to each of the following addresses:

    cornell.edu

    ee.cornell.edu

    anise.ee.cornell.edu

    www.ee.cornell.edu

    Does the name server identify the same computer for each of these URLs? Would you expect it to? Are the routing paths different? Would you expect them to be?

7. The Multipurpose Internet Mail Extensions (MIME) provide a set of schemes for converting binary voice and image data into ASCII characters so that they can be attached to a text -based e-mail. To see how this can be done, translate the binary string.

    0000 1111 0001 1110 0010 1101 0011 1100 0100
    1011 0101 1010 0110 1001 0111 1000

    into an ASCII string using the "hexadecimal" conversion of A = 0000, B = 0001, C = 0010, and so forth.

8. Find a line of hypertext and explain what it causes a web browser to do.

Section 8.4: Issues in the Design of a New Network

After studying all of the concepts in this chapter it is reasonable to ask how a new network should be designed. It should be clear at this point that the answer to that question depends on what the network will be expected to do. Specifically, we ask questions like the following:

- How much traffic must be transmitted?
- How many users are there?
- How much delay can be tolerated?
- How much are the users willing to pay?
- How secure must the system be?
- How much data loss can be tolerated?
- What will government regulators permit?
- Must I use equipment from a specific vendor?
- Must the new system work with older "legacy" equipment and software?
- And probably many more.

What we find, of course, is that there is no single "best" design for a network. The definition of best depends on the application and the needs of the customer. The engineer's design problem is to best match the available technology with the desires and needs of that customer.

Class Activity

- What would Internet III be?
- What would be its *requirements* and what design choices might we make?
- What would be the requirements for a network optimized to carry multimedia signals like video teleconferencing?

Section 8.5:
Summary

There are already many types of communications networks in the world and the number of them will gro w as computers and microprocessors become a part of every home, business, automobile, and appliance. The designers of microprocessors are now using the concepts of data networking to design how the various parts of a single silicon chip communicate with each other. While the applications of networks will become more widespread and diverse, the fundamental principles will remain those examined in this chapter. They can be summarized in the following points.

- The objective of a communications network is to permit any of many users of the network to communicate information with any or all of the other users.

- There are many ways to design a network and the choices made depend on the requirements that the users have. These requirements might include speed of delivery, accuracy of delivery, reliability of delivery, instant availability, the ability to obtain immediate responses, and any combination of these and others.

- The requirement that most frequently drives the design of a communications network is to minimize its cost.

- We analyzed the important case in which minimizing cost is the key issue and when the network must operate over long distances, making the costs of transmission dominate the total cost of the system. In this case we found that the cheapest way to build a network is not to connect all users directly to each other with their own dedicated links, but rather to share transmission links. This was done by connecting links to each other with "relay points." When one of these relays receives data destined for a user, it selects, by some strategy, an outgoing transmission link which carries the data to, or closer to, the intended recipient.

- Two key strategies have emerged over the past two hundred years for relaying messages. The first is now termed "rout information, called the address, added to the message by the outgoing user to determine the next best step in the network toward the intended recipient. Since routers are permitted to temporarily store messages until the proper output link is available, this method makes very efficient use of expensive transmission systems. The U.S. postal system, the telegraph, and the Internet use this strategy in their relays.

- The second key strategy is termed "switching." When using this scheme, all of the switches involved in a planned data exchange first send messages to each other to determine a path through the network for the data and to ensure that there is enough network capacity for it. Once this prearrangement, termed "call setup," is done, the data can then flow with assurance and reliability, two aspects of what is commonly called "quality of service." The telephone network is an important example of a switched network.

- The technology used to implement electrical communications networks has improved markedly from its first appearance in the 1830s. It operated then over a few miles and at rates of about 5 bits per second using human transmitters and receivers. Modern versions send data at up to 10 billion bit/s and use high-speed electronic routers to send, receive, and dispatch the data. The concepts are identical to those used 150 years ago, but the implementation has improved enormously.

- Modern networks still use both approaches. The modern version of the router-based scheme is called the Internet, so called because the rules, or "protocols," developed for its use permit the interconnection of networks owned and operated by many different organizations and companies.

- The Internet's ability to "internetwork" has allowed specialization in "internet service providers (ISPs)" and the services that they offer. Some specialize in providing access by users to the Internet, some provide content like Web pages and video, and some provide the communications paths between the users and the servers that they desire.

The role of network engineers is the same as for other types of engineers. They must determine what the users really want and what constraints, such as cost and performance, that apply. They must then develop a number of alternative designs and then compare them to determine the best one. They supervise the construction of the system, test it once complete, and verify that it will operate as designed. Engineers also typically contribute to the maintenance of the network and the long-term support of its users.

Section 8.6: Glossary

access (8.3.1)
a term used to describe a user's entry point into a data network

access server (8..3.2)
a set of equipment located at an internet service provider through which users gain access to the network and its services. An access server is commonly composed of a large number of voiceband or DSL modems, a controlling computer, and a router which aggregates the data from the modems and sends it into the network.

broadcast (8.1.2)
a communications system that uses one transmitter to send information to many receivers

backbone network (8.3.2)
an internet service provider whose sole function is to connect other internet service providers to each other. A backbone network tends to use high-capacity transmission systems and very fast routers in order to handle the traffic load imposed by the numerous client ISPs.

content server (8.3.2)
a network-connected computer that has the ability to provide informational services to network users upon request

DNS (8.3.4)
Domain Name Service – a protocol used in the Internet to obtain from a **name server** the translation from a uniform resource locator (that is, a domain name) of a computer on the network and its associated 32-bit IP address

FTP (8.3.3)
File Transfer Protocol – a protocol used to reliably transfer disk files from one computer to another over the Internet

HTTP (8.3.3)
Hypertext Transfer Protocol – a protocol used to reliably transfer hypertext from a Web server to a client computer

guaranteed service (8.3.1)
an attribute of a network capable of ensuring each user that they will receive the quality of service for which they have asked. This quality might be measured in terms of bit rate, maximum error rate, delay, variation in delay, reliability, or a combination of these.

Internet (8.3.2)

the name for a networking technology that permits the interconnection of many smaller networks

Internet service provider (ISP) (8.3.2)

an organization or business that provides users with access to the Internet, connection with the various services available in the Internet, or the services themselves. Some ISPs can provide all three of these, while others specialize to just one or two.

Internetwork (8.3.2)

a set of networks that are interconnected and can act as a whole in communicating data from a user to any other user or service

mail server (8.3.3)

a network-connected computer equipped with the software to receive e-mail for its clients and send it on to them when it is requested

network (8.1.1)

a group of interconnected individuals

packet (8.3.3)

a segment of a complete message or file; messages are commonly broken into packets to permit smooth and fair sharing of a data communications network by many users

point-to-point (8.1.2)

a system in which two users are directly connected to each other by a wire or radio transmission system

protocol (8.3.2)

a set of rules that all parties in a network use to format and communicate data

quality of service (8.2.4)

a term used in data networks to describe the performance attributes of a network

queue (8.2.3)

a line in which messages wait to be transmitted

relay point (8.1.3)

a point in a network that can accept messages from one user over a transmission link and then forward them to other users on other transmission links

repeater (8.2.2)

a device that can amplify or "clean up" a communications signal but can not change the routing or scheduling

router (8.2.2)

a network relay point that operates by receiving messages, selecting an outgoing transmission link based on address information carried in the message, storing the message temporarily in a queue, if necessary, and then sending it out the prescribed link as soon as its turn comes in the queue; a router executes a strategy termed "store and forward."

routing table (Exercise 8.2, 8.3.2)
a table held by a router that instructs the router as to the best (and, possibly, alternative) transmission path to be taken out of the router to send a message toward its ultimate destination

server (8.3.2)
a network component that performs its service for many users, and, possibly, for many at the same time. A server is typically a network-accessible commuter that has been augmented with software and additional equipment to perform its function. They typically fall into two classes, those that support a user's access to the network (that is, an access server) and those that provide information at a user's request (that is, a content server).

SMTP (8.3.3)
Simple Mail Transfer Protocol – a protocol used to reliably transfer electronic mail from a user's computer to a mail server located somewhere else in the Internet

star (8.1.2)
a network design in which all of the users are directly connected to a single relay point using individual transmission links

store and forward (8.2.1)
a strategy for relaying a message in which a network relay point, by receiving messages, selects an outgoing transmission link based on address information carried in the message, stores the message temporarily in a queue, if necessary, and then sends it out the prescribed link as soon as its turn comes in the queue

switching (8.2.4)
a strategy for relaying messages in which a network relay point receives a message and immediately sends it on toward its destination. To make this possible, capacity on both the switch and the selected transmission link must be reserved before the data transmission can begin

TCP/IP (8.3.3)
Transport Control Protocol/Internet Protocol – a pair of protocols used in the Internet to fairly and reliably carry information between any two computers; TCP breaks a user's message into packets and then sends each of them to the destination computer using IP. The computer at the destination reassembles the original message after it has received all of the packets.

web server (8.3.2)
a network-connected computer equipped with the software to receive requests for World Wide Web (WWW) pages and then send them to the requestor

Section 8.7:
References

[1] Tom Standage, *The Victorian Internet*, Walker and Company, New York, 1998.

[2] Andrew J. Tanenbaum, *Computer Networks, Revision 3*, Prentice-Hall, Upper Saddle River, New Jersey, 1996.

Figure References:

[8.8]] Tom Standage, *The Victorian Internet*, Walker and Company, New York, 1988.

[8.9] Courtesy of The World Book Encyclopedia, © 1955

[8.10] Courtesy of Cisco Systems, Inc.

The INFINITY Project

Chapter 9: Compressing Information

What does it mean to code information? What is compression? Why are coding and compression important?

Approximate Length: 2 Weeks

Section 9.1: Introduction to Codes and Coding

Approximate Length: 1 Lecture

9.1.1 Codes : The Language of Digital Systems

The world today is full of different societies of people who speak and write different languages. English is predominantly used in the United States, but there are hundreds of other languages that are used by many people throughout the country as well. In some other countries, there are thousands of different languages in common use. For example, in India, 1683 different mother tongues are spoken, including Hindi, the so-called "national" language. The common set of sounds, symbols, words, and meanings that make up a language make it possible to people within a society to communicate easily. When two people who speak different languages want to communicate, generally a translator—someone who knows the meanings and structure of both languages—is needed.

Digital devices also need a common language in order to "talk" to one another. For example, a set of bits that is created on one device can only be understood by a program on another device if there is a set of rules that define what each bit means. We call the set of rules that define the meanings of bits a *code*. *Encoding* describes the act of translating symbolic information into the bit language of a digital device. Similarly, *decoding* describes the act of interpreting the bits as meaningful symbols to the user of the device. Devices and programs that encode and decode information are digital "translators" that change digital information from one useful form to another. One good example is the ASCII Code, discussed in Chapter 6.

Example: The ASCII Code

From the ASCII Code table, you to can learn to read like a computer. What does the following sequence spell?

2

01001001 01101110 01100110 01101001 01101110 01101001 01110100
01111001

There are a number of features of a code that we can describe. A few of these features are discussed below:

Codebook – The codebook is the table that describes the translation of important symbols to their binary representation. In the case of the ASCII Code, the codebook consists of all of the text, number, and punctuation symbols in the code and their corresponding binary representations.

Codeword—A codeword is any encoded version of a symbol in a codebook. For example, the letter "I" has the codeword 01101001 in the ASCII Code.

Codebook Size—The codebook size is the number of entries in the codebook. In the case of the ASCII Code, the codebook size is 256.

Codeword Length—The codeword length is the length in bits of a particular codeword in a codebook. For example, in the ASCII Code, all codewords are 8 bits long. Having all codewords of the same length is not a requirement. We will consider examples later when talking about compression.

In many cases, the codeword length and the codebook size are directly related. Suppose that C denotes the codeword length and N is the codebook size. In the case of the ASCII Code, C and N are related by

$$N = 2^C$$

In other words, the number of unique codewords corresponds to the number of different forms that the binary codewords can take. In fact, every possible 8-bit binary sequence is represented in the ASCII Code, and we could not represent even a single additional codeword uniquely. This fact immediately shows us that

$$N \le 2^C$$

for codebooks made up of binary codewords. An example shows us how to use this fact to design codebooks.

> **Recall:** If we want to store N symbols, we must have at least $\log_2 N$ bits.

Example: A Codebook for the Whole Numbers 0 to 9

In many digital systems, we only need to store the whole numbers 0 through 9 as digits of a longer number (for example, the time on a digital clock). In such cases, we can assign binary codewords to the individual numbers. Suppose we use the same number of bits in each codeword. How many bits in each codeword do we need to represent these numbers?

3

The answer to this question can be reasoned from the relationship between N and C given above. Since we need to represent ten numbers (0,1,2,3,4,5,6,7,8,9), we set $N=10$. Thus, we need a value of C such that

$$10 \leq 2^C$$

Since C must be an integer greater than one, we can simply start calculating the possibilities for the right-hand side of the inequality as

$$2^1 = 2$$
$$2^2 = 2 \times 2 = 4$$
$$2^3 = 2 \times 2 \times 2 = 8$$
$$2^4 = 2 \times 2 \times 2 \times 2 = 16$$

Since 16 is greater than 10, we have that $C=4$ gives a valid codeword length to represent the whole numbers between 0 and 9. We can even design a code as

$$"0" = 0000 \quad "1" = 0001 \quad "2" = 0010 \quad "3" = 0011 \quad "4" = 0100$$
$$"5" = 0101 \quad "6" = 0110 \quad "7" = 0111 \quad "8" = 1000 \quad "9" = 1001$$

We've used the binary-to-decimal number conversion relationship in Chapter 6 to design this code. This assignment is not the only possible choice, however. In fact, any unique assignment of four-bit codewords to numbers works. Moreover, there are 6 codewords "left over," so you could even store six additional characters or pieces of information. The *Hexadecimal Code* does this by including the assignments

$$"A" = 1010 \quad "B" = 1011 \quad "C" = 1100$$
$$"D" = 1101 \quad "E" = 1110 \quad "F" = 1111$$

along with the ten numerical codewords above.

Since the Hexadecimal code uses 4 bit codewords, we can describe 8-bit codewords using two Hexadecimal symbols. In this way, we can express ASCII letters pairs of Hexadecimal codewords. For example, the letter "N" in ASCII has the codeword 01101110, which can be represented in Hexadecimal Code as the symbol pair "6E". Can you write your first name using Hexadecimal?

Exercise: Coding the Time on A Digital Clock

Figure 9.1 shows the typical face of a digital clock. On the clock, there are four different types of information:

Time and Base-60:
As you know, time is counted out using multiples of 60—60 seconds equals one minute, 60 minutes equals one hour, and even 12 hours is a simple fraction (1/5th) of 60. The reason for this choice is due to the Ancient Babylonians, who used base-60 to count all numbers. Their counting system also is used in angle calculations (that is, there are 360 degrees in a circle).

4

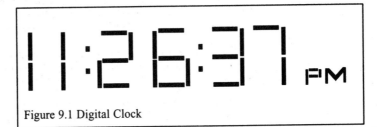

Figure 9.1 Digital Clock

1. The Hours display (from 1 to 12);

2. The Minutes display (from 00 to 59);

3. The Seconds display (from 00 to 59);

4. The AM/PM symbol.

What is the smallest number of bits are needed to store each type of information in its standard representation? If you were to build a digital clock, how many bits could you use to store the current time of day in the digital memory of the clock?

Now, look at the current time and write it down (if the clock or watch you're looking at doesn't have a seconds display, pick the day of your birthday as the time in seconds). Then, code each of the numbers (hours, minutes, seconds, and AM/PM) as bits, and write down the sequence of bits, one after the other. This list of bits is quite similar to how digital devices store the time.

Exercise: Coding the Dates of a Universal Calendar

Figure 9.2 shows the face of a universal calendar that keeps the date in its memory.

Figure 9.2 : Universal Calendar

There are three different types of information:

1. The Month display (from 1 to 12);

2. The Day display (from 1 to 31);

3. The Year display (which we'll assume only needs to store numbers between 2000 and 2099—in the 22nd century, you might want to get another universal calendar!)

What is the smallest number of bits are needed to store each type of information in its standard representation? For the Year display, you can assume that the two leading number "20" of every year don't need to be stored.)

The procedure for encoding and decoding information can be described using simple block diagrams. Figures 9.3(a) and (b) show the two generic systems of an encoder and a decoder, respectively.

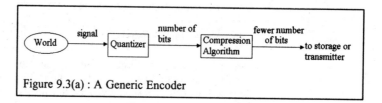

Figure 9.3(a) : A Generic Encoder

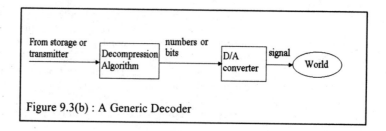

Figure 9.3(b) : A Generic Decoder

An encoder takes information, such as characters, letters, or signal samples, and turns these symbols into bits for storage or later use. Both the input to the encoder and the output of the encoder are digital signals; that is, the encoder simply changes the representation of the discrete-valued information, not its overall content. A decoder reverses the encoding process, taking the encoded bits and turning them into useful information or signals. Both of these processes are essentially identical to the procedure you took to decode the ASCII message in the example above. These procedures involve scanning the entries of a table for the proper translation from symbols to bits and vice versa. This process is very similar to a translation language dictionary; for this reason, the codebook is sometimes called the *codebook dictionary*.

An important feature of the coding process is that it is reversible; you can recover the desired information back without any loss. In other word, the correct decoder turns an encoded message back into the original message before it was encoded.

6

Uses for Coding

We can use coding for a number of purposes. Generally, there are three reasons to encode information:

1.   We can attempt to reduce the number of bits needed to store the information. This type of coding is usually called *compression*.

2.   We can add additional information to a message so that any introduced errors in the message can be detected and even corrected. This procedure is called *error correction*.

3.   We can hide the information from others, so that it remains secret. This procedure is called *encryption*.

The term "coding" is a generic term that sometimes is used to describe certain types of compression technology (such as MP3 encoding and decoding). In this chapter and subsequent chapters, we'll use the name coding to refer to any one of the terms compression, error correction, and encryption. In this chapter, we'll focus on compression.

Exercise 9.1

1.   The ASCII Code encodes English text using one byte (8 bits) per symbol (either letter, space, or punctuation mark). Take any book that you've read recently, and pick a page at random. Take any full line on that page, and count the number of symbols in the line of text. Don't forget to count spaces and punctuation marks. Next, count the number of lines on the page. Now, multiply the number of symbols per line by the number of lines, and then multiply the result by the number of pages in the book. About how many bytes would it take to store this book using ASCII? A CD-ROM holds around 650MB of information. How many books could be stored on a CD-ROM using the ASCII Code?

2.   An internet store has 45835 different items that it carries in its inventory. You are asked to design an efficient scheme for indexing all of the items using binary codewords. How many bits per codeword should you use?

3.   Encode the phrase "We the people of the United States" using the ASCII code. Then, represent this encoded version using Hexadecimal code.

4.   Design a codebook dictionary for the seven colors of the rainbow (red, orange, yellow, green, blue, indigo, violet). Suppose you wanted to then store the color white—is it possible to change the codebook without lengthening the number of bits per codeword? How would you do it?

5. Suppose you could record everything that you see out of both of your eyes using digital video cameras. Let the digital video cameras record video at 30 frames per second. Assuming that you live to your 110th birthday, how many frames of video would you record? How many bits would you need to index (not store) your stereo "life movie" using one unique index per frame? Does the answer surprise you?

6. Take a look at any car license place in your vicinity. What is the format of the license plate? (That is, what characters/symbols are allowed in each position? How many characters are there?) Using this information, figure out how many cars can be uniquely identified using the current license plate format. How many people live in your state? Does the license plate format allow for more cars than there are people currently living in your state?

Section 9.2:
Introduction to
Compression

*Approximate Length: 2
Lectures*

9.2.1 Compression of Matter vs. Information

After drinking a can of soda, it is quite common to crush or compress the aluminum can. When transporting aluminum cans to a recycling center, a crushed can saves physical space. If you were talented, you might try to "uncrush" the can or put it back to its cylindrical shape. Most of us would agree that expanding a can back to its original form is hard. We'd also agree, though, that the compressed can contains the same amount of metal as the original can; we haven't changed the physical makeup of the can in any way.

In digital systems, *compression* describes the act of reducing the amount of characters, symbols, or bits used to store any given signal. Compression of information is analogous to compression of something physical; by compressing information, the amount of physical resources (e.g. space) that we use to keep it around is reduced. Unlike crushing a soda can, however, a compressed signal can be easily reconstructed to resemble its original form. Moreover, compression does not necessarily mean that the quality of the information is compromised; in fact, there are ways to compress information without any loss of quality, as we shall see.

In the remainder of this chapter, we are going to describe the basics of compression technology—how it works, what it is useful for, and why it is so important in today's digital world.

Compression is an important technology today. There are many physical technologies to store information—optical discs such as CDs and DVDs, magnetic discs such as computer floppy and hard drives, and memory chips. As we learnt in the previous chapter, some of these technologies can store an amazing amount of information in a small physical space. If we are getting so good at storing larger and larger amounts of information in the same amount of material, why do we need compression? There are several reasons for why compression continues to be an important part of many digital solutions:

- For many applications, the "raw" data format requires too many bit locations to store easily in existing media. A good example is digital video, as will be illustrated shortly.

- As consumers, we have an ever-greater desire to store more information digitally. Digital data forms are easier to use, easier to manipulate, and easier to communicate to other digital devices. Compression gives us a way to make better use of our existing storage capabilities.

- Compression gives us the ability to transmit data more efficiently, using fewer resources (such as energy, transmission time, or communication bandwidth).

With all these advantages, it would seem that compression would and should be a natural part of every system. There are two drawbacks to compression, however.

- Since compression and decompression are both coding steps, they require additional computational effort, making the overall system more complex than if the "raw" data format were used exclusively.

- Almost all compression methods work on blocks of data. That is, they need to store several symbols (such as letters) before they can perform a compression or decompression step. For this reason, the resulting compressed or decompressed signal can only be made available after a delay of one block. In many systems, the delay is small (e.g. 24 milliseconds for digital audio that is compressed using the MP3 audio compression standard), but it can be a limiting factor in some cases.

Example: The Need for Video Compression

DVD Video discs store about 2 hours of high-quality video and multichannel surround sound. DVD Video uses clever audio and video compression technologies to get this information onto one single-layer DVD, which has a capacity of about 4.7 GB. How many DVD Video discs would we need to store the "raw" information of a 2 hour movie? For simplicity, we will consider only the video portion of the data stream, as the audio portion is of a much lower storage capacity.

A television broadcast (NTSC quality) presents a full frame of information every $1/30^{th}$ of a second. Each frame consists of 480 lines of horizontal information; these are the scan lines that we see when we look closely at a standard NTSC quality TV set. Standard TV sets have an aspect ratio of 4:3, Figure 9.4 shows how this ratio is defined.

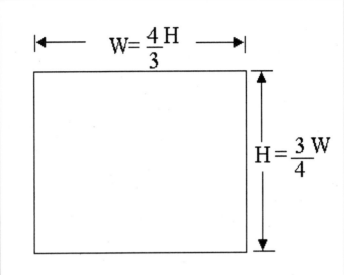

Figure 9.4 The aspect ratio for a standard NTSC quality TV set

Assuming that the vertical information is stored at the same quality, there are effectively

$$N_{Pixels} = N_{Horizontal} \times N_{Vertical}$$
$$N_{Pixels} = 480 \times 640 = 307200 \; pixels$$

Remember: A byte is eight consecutive bits.

of vertical resolution, so that the entire image size is

$$N_{Vertical} = N_{Horizontal} \times \frac{4}{3}$$

$$N_{Vertical} = (480 \; lines) \times \frac{4}{3} = 640 \; lines$$

The lowest-quality raw color format uses 24 bits per pixel to store each color pixel in a digital image. Since there are 3 bytes in every 24 bits, the total number of bytes needed to store each TV frame is

$$N_{Bytes \, / \, Frame} = N_{Pixels} \times \frac{3 bytes}{1 pixel}$$

$$N_{Bytes \, / \, Frame} = 307200 pixels \times \frac{3 bytes}{1 pixel} = 921 kB$$

All we need to do now is multiply out the number of bits per frame value above by the number of frames of video there are in a 2-hour movie. This calculation is

$$N_{Bytes/Movie} = N_{Bytes/Frame} \times \frac{30\,frames}{1\,sec} \times \frac{3600\,sec}{1\,hour} \times \frac{2\,hours}{1\,movie}$$

$$N_{Bytes/Movie} = 199\,GB$$

In other words, to store the video for a 2-hour movie in its digital "raw" format would take 199 Gigabytes, which is *over forty times* the storage capacity of one single-sided DVD! The video compression method used in DVD Video (called MPEG-2) reduces the size of this information to less than 2.5% of its original bit size, allowing a movie to be saved in a much smaller package. Isn't it amazing how well it works ?

Compression Ratio

The *compression ratio* of a compressed signal is defined as the ratio of the number of bits of the compressed version of the signal to the number of bits in the original representation of the signal, or

$$R_{Compression} = \frac{(\#\ of\ bits\ in\ compressed\ signal\,)}{(\#\ of\ bits\ in\ original\ signal\,)}$$

The compression ratio is a number between zero and one, with smaller values corresponding to higher levels of compression. For example, in digital video, a compression ratio of

$$R_{Compression} = 0.025$$

means that, on average, every 40 bits of the original video signal is reduced to one bit in the compressed video signal. A compression ratio of 40 to 1 is pretty high; typical compression ratios for images are much less (about 8 to 1).

Compression methods can also make up for bad encoding choices when assigning symbols to bits. Remember the ASCII Code? It uses 8 bits per character. It turns out that ASCII text can be compressed on average to roughly 1.35 bits per character—a compression ratio of about 6 to 1 (0.168). The way we achieve this compression shall be described shortly.

9.2.2 Lossless Compression vs. Lossy Compression

There are two general types or flavors of compression technology. While both achieve the overall goal of compression—to reduce the number of bits needed to store a set of numbers—they each do it in slightly different ways.

Lossy compression describes a class of compression techniques that throw away information about the sets of numbers or symbols being compressed.

Lossy compression literally amounts to "giving up" certain features of the set of numbers or symbols. The goal in lossy compression is to "lose" those features that wouldn't be noticed or that aren't considered important in the application. By its very nature, lossy compression is not reversible; that is, one cannot reconstruct the exact set of numbers from the compressed version.

Lossless compression describes a class of compression techniques in which nothing about the original set of numbers or symbols is lost in the compression process. Lossless compression is akin to letting the air out of a balloon; nothing about the balloon has been changed in the process except its size. Unlike a balloon, however, compressing a set of numbers requires a mathematical process or algorithm. In this case, the process is reversible, and the exact set of numbers can be recovered by undoing the compression process.

Examples of lossy and lossless compression illustrate some of their features.

Lossy Compression Example: Rounding Off a Numerical Result

There are many situations throughout life where being overly precise ends up wasting effort and resources. A good example can be found throughout this text. In most places, we have rounded all calculations to some useful set of digits, usually 3 to 4 significant digits. Numbers such as the value of pi,

$$\pi = 3.141592653589793 \quad \ldots$$

have been rounded to a suitably-smaller set of significant digits, such as

$$\pi \approx 3.1416$$

so that we can perform calculations with it. Although your calculator shows more digits of precision than that above, it too approximates the true value to some number of digits. Here, we have used the wavy equals sign to denote that the value of pi shown is not the exact value. Even so, the approximate value is remarkably useful when performing calculations, such as finding the area or circumference of a circle. We can view the approximate value as a form of lossy compression, where we need not express the value to a large number of digits.

(NOTE: Realize that the use of the symbol π to represent the value of pi is in fact a form of symbolic compression; however, when performing calculations, the symbol itself is not very useful unless certain trigonometric identities appear.)

Suppose instead of an irrational number such as π, we had a binary fraction such as 19/64=0.296875. We can represent this number as the bit sequence .010011, as this sequence represents the number as

$$0.296875 = 0(0.5) + 1(0.25) + 0(0.125) + 0(0.0625) + 1(0.03125)$$
$$+ 1(0.015625)$$

| **Irrational Numbers:** Numbers that can not be expressed in the form $\dfrac{M}{N}$ where both M and N are integers. Ex: $\pi, \sqrt{2}$. |

Thus, to store this number in binary form takes 6 bits. A 5-bit lossy compressed version of this number would be

$$0(0.5) + 1(0.25) + 0(0.125) + 0(0.0625) + 1(0.03125)$$
$$= 0.28125$$

The value of the number clearly has changed, but so have the number of bits we need to store it.

VAB Experiment 9.1: Rounding

This experiment demonstrates how computers store decimal numbers in binary format.

Samuel F. B. Morse: Born April 27, 1791 - Died April 2, 1872. The American inventor of the Electric Telegraph (1832-35) and the Morse Code (1838).

Lossless Compression Example: Morse Code

Named after Samuel Morse, inventor of the telegraph, Morse Code is a way to encode letters of the English alphabet using a four-symbol language of "dot", "dash", "space" and "break/end of sentence". Table 9.1 shows the Morse code.

The Morse Code Table

Letter	Morse	Letter	Morse	Letter	Morse
A	. -	N	- .	Ä	. - . -
B	- . . .	O	- - -	Á	. - - . -
C	- . - .	P	. - - .	Å	. - - . -
D	- . .	Q	- - . -	Ch	- - - -
E	.	R	. - .	É	. . - . .
F	. . - .	S	. . .	Ñ	- - . - -
G	- - .	T	-	Ö	- - - .
H	. . . .	U	. . -	Ü	. . - -
I	. .	V	. . . -		
J	. - - -	W	. - -		
K	- . -	X	- . . -		
L	. - . .	Y	- . - -		
M	- -	Z	- - . .		

14

Digit	Morse	Punctuation Mark	Morse
0	- - - - -	Full-stop(period)	. - . - . -
1	. - - - -	Colon	- - - . . .
2	. . - - -	Comma	- - . . - -
3	. . . - -	Question mark (query)	. . - - . .
4	. . . . -	Apostrophe	. - - - - .
5	. . . . .	Hyphen	- . . . . -
6	-	Fraction bar	- . . - .
7	- - . . .	Brackets (parentheses)	- . - - . -
8	- - - . .		
9	- - - - .		

Table 9.1 : The Morse Code Table

Binary Code using Morse Code: For Morse Code

"dot"   : 00
"dash" : 01
"space" :10
"."   :11
So "A"=.-
or 0001

 In this code, letters that are used more frequently, such as "E" and "T", use short sequences, whereas infrequently used letters, such as "Z" and "Q", have long sequences. The use of variable-length symbols helps to minimize the amount of transmission time it takes to send English text using Morse code.

Example: Compress Text Using Morse Code

To see how Morse Code can be used to compress text, take a sentence from any English language source (such as this book). Then, count the number of letters and spaces in the sentence (ignoring punctuation marks). An encoding scheme that assumed equally-likely letters would use about 5 bits per symbol, because

$$2^5 = 32$$

is the smallest binary number greater than 28 (26 letters plus the space and period). Therefore, a simple coding scheme would require 5 bits times the number of letters and spaces in the sentence.

Now, count how many bits it would take to encode this sentence using Morse Code. To count the bits, use the Morse Code table, and count every dot, every dash, every space, and every period as two bits. (It might be easier for you to write down the Morse Code table and put the number of bits per letter next to each entry in the table.) How many bits would it take to store the sentence you picked using Morse Code. What is the compression ratio of Morse Code in this case?

Lossy and lossless compression methods have their general uses in modern technology, as summarized below:

- Lossless compression methods are most useful for signals and information that we count on for accuracy. Financial and medical records, computer programs, and text documents are some examples of data that is often compressed by lossless compression methods.

- Lossy compression methods are most useful for signals and information that we perceive, such as sounds and images. The quality of what we see and hear is closely tied to the limitations of our eyes and ears, which, while being remarkable measuring devices, do have their limits in measurement ability.

Exercise 9.2

1. Name two different lossy compression methods and two different lossless compression methods.

2. How good would Morse code be at compressing random sets of letters, such as "weofijc vhkjsdhfkasudy rweruiolghs mvn"? Do you think the compressed version would require fewer total bits as compared to the ASCII version? Why or why not?

3. Rounding is often used in engineering calculations to save storage space. How many bits are required to store the mantissa of a number when rounded to four significant digits?

4. When using a cellular telephone, what part of the system most limits the ability to compress the speech signal being transmitted? Is it (a) the allowable size of the telephone, (b) the typical quality of the microphone, or (c) the amount of allowable delay in the connection?

Section 9.3:
Lossless
Compression

Approximate Length: 4 Lectures

As mentioned in the last section, one way to reduce the amount of storage we need for information is to throw unimportant information away, using lossy compression. Lossy compression makes a lot of practical sense; why bother saving what you don't need? In many situations, however, there are good reasons to store information without any degradation or loss:

- For data records that must remain error-free, lossy compression cannot be used. Imagine the uproar that would be created if banks started using lossy compression to store amounts of money in personal bank accounts—even small errors in account balances would not be tolerated by the public.

- In some situations, we might not be sure about which component of the signal is unimportant. A good example of this situation is in data collection for scientific experiments. Recently, scientific organizations from several countries have, at great expense, taken digital images of most of the surface of the Earth using cameras on orbiting satellites. This information is expected to be useful for many future studies about our planet—weather patterns, crop yields, and other phenomena. Because of its diverse applicability, it is not clear what would be "lost" if lossy compression were to be used on these images. Even a slight degradation might limit the usefulness of the data for certain as-yet-unknown studies.

- In many situations, lossless compression can be combined with lossy compression to reduce the number of bits to be stored even further. A perfect example of this situation is in JPEG still image compression, in which Huffman coding and run-length coding, both lossless compression methods, are combined with transform coding to compress image data.

In this section, we discuss lossless compression methods, and point out some important features of lossless compression methods.

9.3.1 Run-Length Coding

Run-Length coding is a lossless compression technique that we'll use to introduce the concept of lossless compression. Run-length coding is applicable to streams of bits that have several 1's or 0's that appear in succession. A simple example illustrates the technique. Suppose we have the 30-bit sequence

010000000111111000111111100000

This sequence could be described in the following way:

"A single 0, followed by a single 1, followed by seven 0's, followed by six 1's, followed by three 0's, followed by seven 1's, followed by five 0's."

We could use shorthand to represent the above sentence as

(1,0)   (1,1)   (7,0)   (6,1)   (3,0)   (7,1)   (5,0)

The first number in each pair denotes the number of bits to be represented, and the second number in each pair denotes the bit value. Since the bit values are alternating, we could shorten this list of numbers even further as

[0]   (1)   (1)   (7)   (6)   (3)   (7)   (5)

In this new representation, the first bracketed symbol is the first bit value, and the remaining numbers are the numbers of 0's and 1's that should be repeated.

We still cannot store this sequence of numbers, however, because they are not in the form of bit values. To get these numbers to bit values, we convert the decimal numbers to binary representation, noting that all numbers are between zero and 7, a 3-bit range. So, we have

[0]   [001]   [001]   [111]   [110]   [011]   [111]   [101]

This sequence of 22 bits can be used to reconstruct the original 30-bit sequence without any errors. Since 22 is less than 30, we have performed compression without any loss of information, or lossless compression. In this example, the compression ratio (22/30=0.73) is not that great, because we do not have many long sequences of successive 1's and 0's. In other situations, the number of successive zeros and ones is much greater, and thus the compression ratio can be higher. Of course, more bits would then be needed to represent each number of successive zeros (that is, 3 bits per sequence would not be enough).

The above example illustrates the general features of lossless compression:

- Lossless compression maps bits to bits; that is, a series of bits is turned into another, hopefully shorter, series of bits.

- There is a clearly identifiable encoding procedure that creates the shortened bit sequence from the original bit sequence.

- There is a clearly identifiable decoding procedure that expands the shortened bit sequence out to the original bit sequence.

Run length coding is a simple lossless compression scheme that works well for certain types of signals. It is not obvious, however, how such a method could be extended to arbitrary bit streams. Later, we shall talk about more general lossless compression methods.

First, however, we shall ask a more fundamental question: When can lossless compression be done? A simple mind experiment suggests that a fundamental limit exists on all such compression methods.

Example: A "Great" Compression Method?

The head of a large Internet company has an idea one day. He is only barely familiar with compression technologies, but he knows that, after compression, the number of bits needed to store a particular signal is always less than the number of bits in the original signal. So, why not take the compressed signal and run it through the compression routine again? There should be fewer bits remaining after compression, meaning a better compression ratio. The company head immediately calls up his corporate lawyers, who start discussing the idea of a patent.

Before long, word of this idea gets to the chief engineer in the company. She is a creative person and willing to listen to ideas, so she goes to talk to the company head. After hearing about his idea, she claims that it cannot work? What is her reasoning?

The flaw in the company head's argument is the assumption that "after compression, the number of bits needed to store a particular signal is always less than the number of bits in the original signal." If this were the case, then he could run the compressed version of the compressed signal through the compression routine again and reduce the number of bits further. If this process is repeated, then at some point, only one bit remains at the output of the compression routine. What if this one bit "signal" is sent through the compression routine? The company head's assumption means that there would be *no* bits at the output of the compression routine. Such a result makes no sense: how can you store a signal using *no* bits? In other words, the compression routine cannot always reduce the number of bits in the original signal; there must be a fundamental lower limit to the ability to compress a signal.

9.3.2 Relative Frequency and Entropy

The ability to compress any set of data in a lossless fashion is closely tied to the concept of *relative frequency*. Relative frequency is defined as

$$f_R(E) = \frac{(Number \quad of \quad Observations \quad of \quad an \quad Event \quad E \quad)}{(Number \quad of \quad Total \quad Observations \quad)}$$

where E is an event of interest.

Relative frequency is closely tied to our notion of probability and chance. For example, if the weatherperson tells us that there is an "80% chance of rain today," we understand this phrase to mean

"For days with weather conditions like today, it rains on 4 out of every 5

In the weatherperson's case, the event of interest E is daily rain. While this simple example illustrates the concept of relative frequency, the actual measurement of relative frequency in compression is much more direct than weather prediction. In fact, it is almost always the case that the data we're compressing gives us both the number of observations of the event E and the total number of observations. Could relative frequency provide the key to lossless data compression? The answer to this question can be reasoned out fairly easily:

1. Events that happen more frequently should be described or stored using as short a name or label as possible.

2. Rare events, on the other hand, can be described or stored using a long name or label, since they occur less often.

Some practical examples illustrate the use of this concept.

Long Distance Area Codes and Telephone Pulse Dialing

All modern-day telephones use *touch tone dialing*, in which a pushbutton keypad creates a series of paired tones to denote the twelve numbers and characters on the telephone keypad. Prior to the invention of touch tone dialing, telephones had a rotary dial like the one in Figure 9.5. This dial had a label with the numbers 0 to 9 on it. Attached above the dial was a clear plastic disc with finger-sized holes over each of the numbers. To dial a telephone number, one inserted her or his finger into the holed disc at the digit to be dialed, rotated the disc to the metal hook stop point at approximately the 2 o'clock position on the dial, and let go. The dial would slowly rotate back to its original position while creating a series of electrical pulses that would be sent down the telephone line. Larger numbers, such as 8 and 9, took much longer to dial than smaller numbers, such as 1 or 2.

Figure 9.5 Rotary Dial of the Telephone

As the population of the United States grows, the number of available telephone numbers across the country has to increase to support more distinct users of the telephone network. Before the invention of the long-distance area codes, the number of digits per telephone number was seven, but direct-dial long distance was not possible; all telephone calls to far-away places had to be operator-assisted. By adding three additional numbers at the front of all telephone numbers, the convenience of local calling without an operator was extended to long distance telephone calls. The question is: How should the long distance area codes be assigned?

The engineers at American Telephone and Telegraph (AT&T, also called "Ma Bell" at the time), figured that the cities with the highest population would receive the most long-distance telephone calls. To make dialing to these big cities easier, these engineers gave area codes with the fastest-dialing three-digit numbers to the largest populated cities: Area Code (212) for New York, Area Code (213) for Los Angeles, Area Code (312) for Chicago, Area Code (214) for Dallas, and so on. Less populated areas, such as Alaska, received long-distance area codes that took a long time to dial, such as Area Code (907). Choosing area codes in this way makes a lot of sense. It reduces wear and tear on average over all telephones and on the telephone user's fingers by minimizing the amount of rotation the pulse dial must move for frequently-called numbers.

With the advent of touch tone dialing, the advantage of smaller-numbered area codes went away, and so newer area codes need not be designed with this efficiency in mind. The area codes example above shows us how careful assignment of labels in a code can help save time and effort in an application. This form of convenience is used in many aspects of our everyday lives. Sometimes, we employ such conveniences without directly thinking about them.

We can use the concept of relative frequency to develop a procedure for assigning short labels or codewords to items that we have to store more often. Our labeling procedure, if done right, will save us valuable storage resources. Let's first show how one calculates relative frequencies.

Example: Relative frequencies of letters in a sentence

Consider the following popular phrase amongst children:

"I scream-you scream-we all scream-for ice cream"

This phrase contains 6 e's, 5 a's, 5 c's, 5 r's, 4 m's, 3 s's, 3 dashes, 2 l's, 2 -I, small i, f, u, w, and y. Including the 6 spaces in-between words, the phrase contains a total of 47 characters. What is the relative frequency of each of these letters and symbols?

Using the definition of relative frequency, we can calculate them easily.

$$f_R(letter \quad e) = \frac{6}{47} = 0.1277$$

$$f_R(letter \quad a) = \frac{5}{47} = 0.1064$$

$$f_R(letter \quad m) = \frac{4}{47} = 0.0851$$

$$f_R(letter \quad s) = \frac{3}{47} = 0.0638$$

$$f_R(letter \quad l) = \frac{2}{47} = 0.0426$$

$$f_R(letter \quad i) = \frac{1}{47} = 0.0213$$

To store this phrase efficiently, it would make sense to assign short labels to the most frequent symbols, which are "e" and " " in this case. Similarly, we can assign long labels to the infrequent letters, such as "i", "I", and "y."

In practice, the relative frequency of letters in any given sentence or phrase will be different, so the labels would change according to these numbers. We could use the average number of appearances in typical sentences of a language to choose the labels as an approximation if we don't want to count letters for a given sentence or document. In the English language, the letter "e" appears most often, so it would be assigned the shortest label, whereas the letters "q" and "z" are rare and would be given long codes.

Aside: The above compression strategy doesn't take into account the relationships between the letters in the phrase. In most languages, it is possible to make use of these relationships to shorten the phrase even further. For example, since the five-letter sequence "cream" appears multiple times, we could create a new symbol for it, say "*", and write out the phrase as

"I s*-you s*-we all s*-for ice *,*=cream"

This version of the phrase has only 39 characters, which is 17% fewer characters than the original message. An even shorter version of the phrase could be had if we used the phonetic sounds associated with words, letters, and numbers. Many people use a similar idea when getting customized license plates for their cars. In our case, we could store

"I#U#WeAll#4I#,#=Scream"

which only requires 22 characters—about a 53% reduction in character length.

Average Codeword Length

How do we judge the quality of a code? One way is to determine the *average codeword length*, or the average number of coded symbols it takes to store an original symbol sequence. The average codeword length can be calculated as

$$L_{Average} = \sum_{i=1}^{N} \left[f_R(S_i) \times L_i \right]$$

$$L_i = \left\{ Code \quad Length \quad of \quad ith \quad Codeword \right\}$$

$$f_R(S_i) = \left\{ \text{Re}\, lative \quad Frequency \quad of \quad ith \quad Symbol \right\}$$

In the above equation, N is the number of different symbols that appear in the signal or message.

The average codeword length depends on the way labels are assigned to particular sets of data. If shorter names are assigned to symbols that occur more frequently, then the average codeword length can be reduced.

Example: Compressing Text

Consider the phrase from a previous example:

"I scream-you scream-we all scream-for ice cream"

The ASCII Code could be used to store these letters, using 8 bits per letter in the process.

The ASCII Code gives us 256 different possibilities, which is many more than we need in this case. In the above phrase, there are 16 different characters. Thus, we could set up a code that assigns four bits to each of the 16 different characters. For example,

" "=0000, e=0001, a=0010, c=0011, r=0100, m=0101, s=0110, "-"=0111, l=1000, o=1001, I=1010, i=1011, f=1100, u=1101, w=1110, y=1111

Then, we could save the sequence of bits, as well as the codebook that translates these bits to the individual characters (which amounts to an

Entropy: The fewest number of bits per symbol required to store and recover a data set.

23

ordering of the associated characters). The average codeword length per symbol is clearly L=4 bits, since the individual codewords are each 4 bits long. Assuming that the codebook were available, the number of bits needed to store the phrase would be (47)(4)=188 bits.

We can do even better than this value, however, since certain characters such as "e" and "a" appear more frequently than others. Suppose we use the code given by

" "=100, e=101, a=000, c=001, r=010, m=1100, s=1101, "-"=0110, l=0111, o=11100, I=111010, i=111011, f=111100, u=111101, w=111110, y=111111

This codebook uses codewords with fewer bits to name characters that appear more often in the message. The average codelength for this new code is

$$L_{Average} = \sum_{i=1}^{N} \left[f_R (ith \quad letter) \times L_i \right]$$

$$L_{Average} = \frac{1}{47} [6(3) + 6(3) + 5(3) + 5(3) + 5(3) + 4(4) + 3(4) + 3(4) + 2(4)$$

$$+2(5) + 1(6) + 1(6) + 1(6) + 1(6) + 1(6) + 1(6)]$$

$$L_{Average} = \frac{175}{47} = 3.72 \, bits \, / \, symbol$$

The above number might seem strange to you – what defines a 0.72th of a bit, anyway? Remember, however, that this number is an *average* value. It only makes sense over a block of symbols. What it states is that we can encode about 372 symbols (or letters, in this case), using only 100 bits for groups of sentences having the same amount of redundancy in them as our example.

The sequence of bits for this coded version of the phrase would be 175 bits long, not including the codebook information, which is shorter than the lengths of the previously coded versions. Of course, the codebook has been designed to store just this message, so it is less likely that it would be useful for other phrases. For long text messages (say, a book), however, the size of the codebook would be much shorter than the coded message, so we would get the savings in storage using this method.

Fundamental Codeword Length Limit—Entropy

We can design several different codebooks to store a given set of symbols. How do we know which codebook is best, or even if there is perhaps a better codebook with a lower average codeword length? There is a fundamental result in coding theory that states the compressibility limits of any data set. This fundamental limit is known as the *entropy*. We first give a mathematical definition of entropy, after which we describe some of its properties.

Suppose we have a set of N unordered symbols S_i with relative frequency $f\_R(S\_i)$. Then, the entropy of the symbol set is defined as

$$H = -\sum_{i=1}^{N}\left[f_R(S_i) \times \log_2\left(f_R(S_i) \right) \right]$$

In this equation, the logarithm is log-base-2 as opposed to the natural logarithm or log-base-e. The units of entropy are bits.

The entropy of a symbol set is a measure of its randomness. A symbol set in which the numbers of each of the symbols are roughly the same has high entropy. By contrast, a symbol set in which certain symbols occur much more often than other symbols has lower entropy. An example illustrates this fact.

Example: Apples, Oranges, and Pears in a Bag

Suppose there are some apples, oranges, and pears mixed together in two different brown paper bags. The first bag contains 3 apples, 4 oranges, and 3 pears. The second bag contains 8 apples, 2 oranges, and one pear. Which bag has the contents with the higher entropy?

We can calculate the relative frequencies of each symbol set, which in this case is the content of each of the bags. The relative frequencies of fruits in the first bag are

$$f_R(apple) = \frac{3}{10},\ f_R(orange) = \frac{4}{10},\ f_R(pear) = \frac{3}{10}$$

We then calculate the entropy of the first bag's contents as

$$H_{1stBag} = -\frac{3}{10}\log_2\left(\frac{3}{10}\right) - \frac{4}{10}\log_2\left(\frac{4}{10}\right) - \frac{3}{10}\log_2\left(\frac{3}{10}\right) = 1.57\ bits$$

We repeat the calculations for the second bag's contents.

$$f_R(apple) = \frac{8}{11},\ f_R(orange) = \frac{2}{11},\ f_R(pear) = \frac{1}{11}$$

$$H_{1stBag} = -\frac{8}{11}\log_2\left(\frac{8}{11}\right) - \frac{2}{11}\log_2\left(\frac{2}{11}\right) - \frac{1}{11}\log_2\left(\frac{1}{11}\right) = 1.10\ bits$$

The second bag's content has a lower entropy. This answer makes sense. If you were to reach into the second bag without looking, the chances that you pick out one type of fruit consistently—in this case, apples—is much greater than the same experiment using the first bag's content.

The entropy of a symbol set defines a lower limit on the average codeword length that can be achieved when compressing the information in the set. This fundamental bound is

$$L_{Average} \geq H$$

In other words, we can compress a data set no further than its entropy value times the number of elements in the data set. An example shows how this limit works.

Example: Entropy of letters in a sentence

Recall the phrase from the previous example:

"I scream-you scream-we all scream-for ice cream"

If we treat each letter as a separate symbol and ignore the spatial placement of letters, we can compute the entropy of the symbol sequence quite easily. We use the formula given by

$$H = -\sum_{i=1}^{N} \left[f_R(S_i) \times \log_2 (f_R(S_i)) \right]$$

using the relative frequencies that we calculated for each of the $N=16$ different letters in the 47-letter phrase. This calculation results in the value

$$H = 3.70 \; bits$$

This value is remarkably close to the average code length of the proposed code in a previous example, which achieved a value of $L=3.72$bits per coded symbol. The entropy limit tells us that the proposed code is quite good, and near the best it can be under the assumptions we've used to compute the code.

Example: Coding general English Text

Using the ideas of coding frequently occurring letters with short sequences, and rarely occurring letters with longer sequences, we can calculate the minimum number of bits required to coded the letters of English text. Table 9.2 shows the relative frequencies of all of the letters plus the space character in typical English language text.

Letter	Relative Frequency	Letter	Relative Frequency	Letter	Relative Frequency
A	0.0642	J	0.0008	S	0.0514
B	0.0127	K	0.0049	T	0.0796

C	0.0218	L	0.0321	U	0.0228
D	0.0317	M	0.0198	V	0.0083
E	0.1031	N	0.0574	W	0.0175
F	0.0208	O	0.0632	X	0.0013
G	0.0152	P	0.0152	Y	0.0164
H	0.0467	Q	0.0008	Z	0.0005
I	0.0575	R	0.0484	space	0.1859

Table 9.2: Relative frequencies of letters in the English language.

If we calculate the entropy of English text assuming that every letter in English is completely random but still occurs with the relative frequency given in the above table, we obtain 4.08 bits per character. What does this mean? Well, we have already seen that standard ASCII coding of English text uses 8 bits, so, this means that by taking advantage of the relative frequencies in the letters in English text, we can save approximate 40% in storage.

We can go even further if we also realize that the letters are not random from one another. For example, consider the short phase "I love have a pretty good idea that the next letter should be a "r". Using this information together, the entropy of English text has been estimated to be 1.3 bits per character. This means that a book with approximately two million characters (80 characters per line, 50 lines per page, 500 pages) can be stored using only 2.6Mbits or 325KB of data. This would allow us to store about four books on a single floppy disk. Compare this with standard ASCII where we would have to store 2MB of data! We can't even fit one book on a standard floppy using ASCII.

What does all this tell us? We can use the natural redundancy in English text to efficiently store books, magazines, or anything else with lots of words in it.

9.3.3 Huffman Coding

We now describe a clever way to construct a variable-rate code that comes very close to the entropy limit—in other words, it compresses the data as much as possible. This method is called *Huffman Coding* in honor of D.A. Huffman, its inventor. Huffman Coding uses the relative frequency of the symbols directly to produce the bit sequences to assign to each symbol. The procedure for designing a Huffman Code is now described.

Procedure for Constructing a Huffman Code

1. Calculate the relative frequencies of the symbols in the signal or information to be compressed.

2. Construct a list of all of the symbols that appear in the signal or information to be compressed, where the most-frequent symbols are listed first, followed by the less-frequently-occurring symbols.

3. Starting from the bottom of the list, pair the last two symbols together using two lines meeting at a point. Write the sum of the two relative frequencies of the symbols being paired.

4. Reorder the list with the new relative frequencies, treating the new pairing as a single symbol.

5. Repeat Steps 3 and 4 until there is only one element in the list. You now should have something that looks like branches of a tree growing right to left.

6. Assign a zero or "O" to all branches that go up to the left in the tree, and assign a one or "1" to all branches that go down to the left in the tree.

7. To figure out the codeword for any symbol, start from the rightmost point in the tree and go left, writing down the zeros and ones that you come across, until you get to the desired symbol.

An example illustrates how this procedure works:

Example: Constructing a Huffman Code

Consider the sentence:

"I scream-you scream-we all scream-for ice cream"

We've already calculated the relative frequencies, so we can list the letters and their relative frequencies, as shown in Table 9.3. Here, we've only listed the numerator part of the relative frequency, because we don't need to use the fractional form to construct the code. We have also ordered the symbols so that the most-frequent symbol is at the top. In other words, we've completed Steps 1 and 2 of the above procedure.

Letter	Relative Frequency (out of 47)
space	6
e	6
a	5
c	5
r	5
m	4
s	3
dash	3
l	2
o	2
I	1
i	1
f	1
u	1
w	1
y	1

Table 9.3 Relative Frequencies

We now start generating the tree that will give us our binary code. This tree begins on the left in Figure 9.6(a), where we've copied the entries from the table. We now start pairing the most infrequent symbols by going to the bottom of this list and drawing two lines from each value to a common point and summing the two relative frequencies. This action is described in Step 3. We continue to do this pairing by drawing the lines to points and summing the frequencies until we come to pairings that would generate numbers larger than values above the pair in the list. Then, we have to reorder the symbol list as described in Step 4 (because the order would change at that point). To do this reordering, we draw dashed lines to "move" the individual symbols up or down as needed. This step is shown to the right of all of the pairings that we just made in Figure 9.6(a). Then, we go back to Step 3 and do the pairings, then the reordering, then the pairings, and so on until we come to only one point. This point must sum to 47, because there are 47 letters in our original signal.

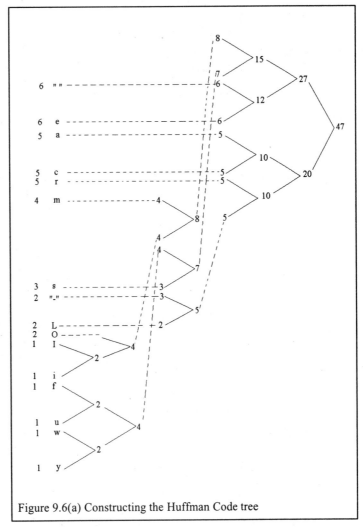

Figure 9.6(a) Constructing the Huffman Code tree

Now, to get the actual codewords from this strange picture, we redraw the picture as in Figure 9.6(b). The only difference between Figure 9.6(a) and Figure 9.6(b) is that we've physically moved the elements of the tree as opposed to using dashed lines for this move. We now perform Step 6 by assigning a 0 to every upward-going branch and a 1 to every downward going branch.

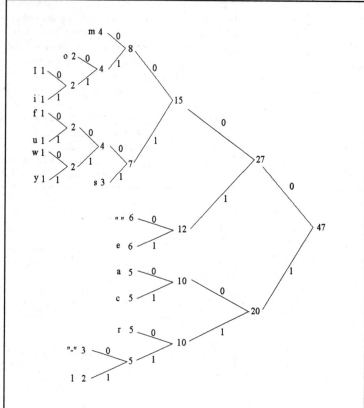

Figure 9.6(b) The finished Huffman Code tree with codewords assigned.

From this branching, we can now read the codewords by going right to left in the tree. Table 9.4 shows the codewords that we have constructed using the Huffman Code method. We've also shown the relative frequencies of the letters (out of 47). Notice how the most likely symbols (space) and "e" have shorter codewords, whereas the infrequent symbols "I", "l", "f", "u", "w", and "y" have longer codewords. In fact, this code is exactly the one we used in the average codeword length example a few pages back, which came to within a fraction of the best coding of the methods using relative frequencies. From this fact, we see that Huffman Coding has a very good coding efficiency.

Letter	Codeword	Relative Frequency (Out of 47)
space	010	6
e	011	6
a	100	5
c	101	5

r	110	5
m	0000	4
s	0011	4
dash	1110	3
l	1111	3
o	00010	2
I	000110	2
i	000111	1
f	001000	1
u	001001	1
w	001010	1
y	001011	1

Table 9.4 Letters and their Huffman codes

What would the final compressed signal look like? We use the codebook to look up the codewords for each symbol in the 47-character phrase. Thus, we have

00011001000111011100111000000111000101100010001001010001 1101 11001110000001110001010011010100111111111010 0011101110 011100 00001110001000000101100100001111101011 0101011100111000000

Advantaged and Disadvantages of Huffman Coding

The Huffman Coding technique has an important advantage:

- It can be shown that, if the codebook size is ignored (more on this later), the average codelength of the code is

$$L_{Huffman} \leq H + 1 \qquad bits$$

In other words, the average codelength of the Huffman Code comes to within a bit value of the optimal codelength for the data. This result is about the best we can hope for, as we'll always be off of the optimal codelength by a fraction anyway.

There are several disadvantages to the Huffman Code, however:

- The Huffman Code changes for each data set. Because of this fact, we have to recalculate the relative frequencies and go through the code construction each time we want to compress a signal.

- Since the codewords depend on the data, the codebook, which includes the binary tree and the locations of the symbols in the tree, must be stored along with the actual coded signal. The storage of the codebook is not included in the average codelength bound above, so in practice, the code is never that efficient. We can guess, however, that for very large data sets, the codebook is not going to be that big relative to the information that we have to save. So, Huffman Coding makes sense mainly when we have large amounts of data to store.

Huffman Coding is used in a number of compression applications, most notably as part of the JPEG still image compression standard.

Exercise 9.3

1. Which of the following three cities would take the longest time to dial using a rotary telephone? (a) Spokane, Washington (Area Code 509), (b) Santa Barbara, California (Area Code 805), or (c) El Paso, Texas (Area Code 915)?

2. Which of the following three cities would take the shortest time to dial using a rotary telephone? (a) Hackensack, New Jersey (Area Code 201), (b) Cleveland, Ohio (Area Code 216), or (c) Beverly Hills, California (Area Code 310)?

3. Run-length encoding works well for bit strings that have long runs of ones and/or zeros. Do you suspect that ASCII-encoded text would be easily compressed using run-length encoding? Why or why not?

4. In the summertime in a particular area, the weatherperson predicts that, on any given day, the probability of rain is 10%, the probability of sunny weather is 89%, and the probability of snow is 1%. What is the entropy of the weather, assuming that the weatherperson is accurate? Suppose you were to write a program to store the weather over the entire 92-day summer period. About how many bits would you need to store one summer's weather report, assuming that the weather could radically change from one day to the next?

Section 9.4: Lossy Compression

Approximate Length: 3 Lectures

9.4.1 Is Being Lost Always Bad?

The word "loss" has a bad feeling about it. Being lost means not knowing where one is. Lost money or possessions may be gone forever. Even the phrases "I'm lost" or "You lost me" usually mean that one is confused.

In compression, methods that throw away information for the sake of efficiency are called *lossy* compression methods. Like other things that are lost, the information removed during the compression step cannot be regained. The key to lossy compression is to remove the portion of the signal or information that doesn't matter to us. If done well, lossy compression can bring incredible benefits. In particular, lossy compression methods save a lot of storage space. It may surprise you to learn that MP3 audio and DVD-quality video signals are lossy-compressed versions of the raw audio and video sources, respectively. These signals still sound and look remarkably good, and we can store a lot more audio and video signals in the same bit space when these signals are compressed using lossy compression methods.

An Example of Lossy Compression: MP3

MP3 (or, more accurately, Audio Layer 3 of the MPEG1 Multimedia Coding Standard) is the name used for a group of popular Internet audio compression standards. MP3 is popular right now because it is useful. By encoding a CD-quality audio track into one of the MP3 formats, one reduces the file size required to store the track on to a computer hard disk, CD-R, or flash RAM card. Since the file size is reduced, one can store more songs onto the corresponding storage medium, allowing hours of music to be stored inside of a portable MP3 player, for example. The quality of the sound in the compressed format is not as good as that of the original CD, but the difference is not that noticeable to most people in casual listening situations. The MP3 format is also helpful when one wants to download music off of the World Wide Web, because the reduced file size effectively reduces the amount of download time needed.

9.4.2 The Role of Perception in Lossy Compression

Most lossy compression methods are closely tied to our own perceptual abilities. To really understand how lossy compression methods work, we need to understand how human sensory systems work. In this section, we briefly discuss the abilities and limitations of the human hearing system as it pertains to lossy audio coding methods. A similar description of the human visual system would help in understanding how lossy video coding works, but we have omitted a description of the human visual system for brevity.

Perceptual Characteristics of the Human Ear

As you probably are aware, human hearing abilities differ from those of other animals. A so-called dog whistle is a device that produces a loud high-frequency sound that is beyond our own hearing abilities, but it can be heard clearly by dogs. The fact that we cannot hear certain types of sounds means that we need not store all sounds when saving digital representations of audio signals. This concept is at the root of all perceptual coding methods.

Figure 9.7 shows a plot of the range of sounds that we can hear in terms of their frequency content and their intensity.

> **dB sound pressure level:** From chapter 5, we know that sound intensity can be measured on an absolute scale (where zero is a reference point). Since the ear responds to the sound intensity logarithmically, we also use the decibel (dB) scale. Table 5.2 shows some common sounds and their typical levels.

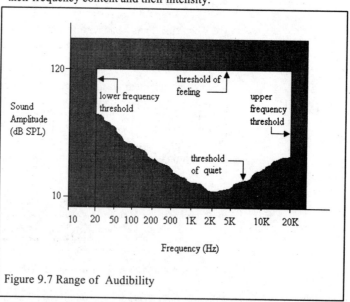

Figure 9.7 Range of Audibility

The left-to-right scale is frequency of the sounds in cycles per second or Hertz (Hz), and the vertical scale is the intensity of the sound, also known as the sound pressure level (SPL) in decibels (dB). The borders of the shaded area are the boundaries of human hearing and are worth exploring further:

- At the bottom of the graph, we find the fundamental lower limit of human perception, the *threshold of quiet*. Sounds that fall below this threshold are too soft for humans to hear. The threshold of quiet is different at different frequencies; in particular, it is higher at low frequencies and lowest at mid frequencies. Thus, it is the bass sounds in a signal that are hardest to hear when the volume of the sound is low.

- At the top of the graph, we have the fundamental upper limit of human perception, the *threshold of feeling.* The threshold of feeling is beyond the point at which one's ears can be permanently damaged. This limit is somewhat arbitrary, however. It is known that long exposures to sounds as soft as 75 dB SPL in some frequency ranges can cause permanent hearing damage. For this reason, it is always a good idea to turn down your music playing device whenever you have been listening to it for more than a few minutes.

- At the left of the graph is the *lower frequency threshold* of human hearing. Sounds below 20 Hz are no longer perceived as signals with pitch; rather, they are perceived more as vibrations in the body.

- At the right of the graph is the *upper frequency threshold* of human hearing. Sounds higher than about 20000 Hz (or 20kHz) cannot generally be heard. These sounds are the ones that your dog and other animals can hear, although they too have limits to their high-frequency listening capabilities.

There is another important aspect of the human hearing system that is useful for perceptual coding systems. When two sounds of a similar frequency but different intensity are played, the louder sound tends to dominate the perception of the listener. If the louder sound is much louder than the corresponding softer sound, one only perceives that the louder sound is playing. This psychoacoustic effect is known as *masking.* Figure 9.8 shows how masking works for narrowband signals.

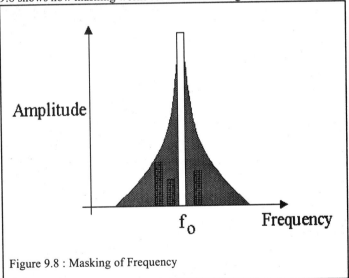

Figure 9.8 : Masking of Frequency

The clear rectangle represents a sound concentrated at one frequency, and the dark rectangles represent sounds nearby in frequency and of lower intensity. The shaded area around the clear rectangle represents the effect of masking. Any sounds that lie within the shaded area in frequency and intensity cannot be perceived. The effect is analogous to what you see when looking at a mountain range at sunset. You can only perceive the tallest mountain along the horizon; all of the other mountains are lost in the tall mountain's shadow.

The human hearing system is an amazing scientific instrument. With it, we can understand much about the world around us. Although we haven't gone into the details of how the human hearing system works, the above description will help us understand why audio systems—CD players, MP3 players, and other digital audio systems—work the way they do.

Human Hearing and CD Audio

The frequency range of human hearing is from 20Hz to 20kHz, and the *dynamic range* of human hearing (the ratio between the loudest and softest sound we can hear) is about 110dB, from Figure 9.7. As we know from our study of sampling in Chapter 2, a signal needs to be sampled at twice the frequency of the highest frequency content in the signal. Moreover, we know that every bit used to store values of the signal gives us 6dB of dynamic range.

The CD Audio playback standard uses a sampling rate of f=44.1kHz and 16 bits/sample to represent the high fidelity audio signal, which translates to an upper limit of 22.05kHz and a dynamic range of

$$16 \, bits \, / \, sample \; \times \; \frac{6 \, dB \quad SPL}{bit} = 96 \, dB \quad SPL \, / \, sample$$

Like the human perception limits, we can draw the limits of the CD Audio playback standard in terms of frequency and sound intensity. We show this plot in Figure 9.9. Note that this drawing represents a particular amplifier volume setting that places the 96dB maximum range at 96dB SPL, and it is possible to either increase or decrease the volume setting.

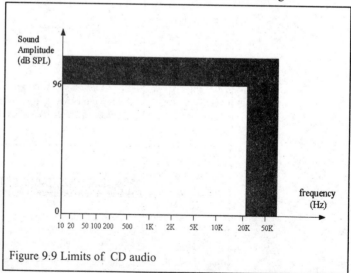

Figure 9.9 Limits of CD audio

As can be seen, the usable range of CD Audio is greater than that of the human hearing system both in terms of frequency range and loudness level at this amplifier setting. This fact is the main reason why CDs sound so good; the standard defines a quality that is *better* than even the best human ear can sense for moderate volume levels. The fact that the CD Audio standard exceeds the human hearing abilities mean s that we can reduce the amount of bits further and still get good-sounding audio. This fact is even more apparent when masking is considered, because CD Audio doesn't even use masking in its operation. Perceptual coding systems employ masking and other attributes, as we shall see next.

Example: Limits of DVD-Audio

Because DVD discs can hold more than 7 times the amount of information CDs discs can, manufacturers have been working on a standard for DVD Audio sound that surpasses CD Audio quality in every respect. While this standard is still in a state of change at the time that this book was being written, one of the standards allows for a sampling rate of $f = 96$kHz and 24 bits/sample in a stereo sound setting. What are the limits of this type of DVD Audio?

We can calculate the resulting bandwidth of the audio signal as

$$f_{Bandwidth} = \frac{f_{Sample}}{2} = \frac{96\,kHz}{2} = 48\,kHz$$

The dynamic range R of the standard is easily found as

$$R = 24\,bits/sample \times \frac{6\,dB \quad SPL}{1\,bit} = 144\,dB \quad SPL$$

The limits of this audio standard are shown in Figure 9.10. Clearly, they exceed the limits of the human hearing system in all respects. In fact, one could argue that the excess is wasteful; using more bit s in this way doesn't add to the listening experience to any great measure.

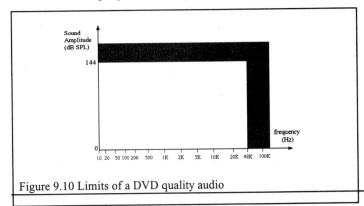

Figure 9.10 Limits of a DVD quality audio

The 24-bit/96kHz stereo audio format in the DVD Audio standard is not the only format that is allowed. It is more likely that the extra storage capacity will mainly be used for other signal types, such as high quality multichannel surround sound. This multichannel surround sound standard will likely be of higher quality than that currently provided by DVD Video

---

Transparent Compression

Transparent compression (also sometimes called *transparent coding*) describes any signal or data compression method that introduces no perceptual distortion in the encoded signal or data. Transparent compression is the goal of every lossy compression method. We don't care if the signal that we're watching or listening to is different from the original signal—we just don't want to see or hear the differences.

Lossy Compression Example: MP3 Audio Encoding

The popular MP3 coding format is actually three different formats whose long names are MPEG Audio Layers I, II, and III, respectively. The common component of the MPEG Audio Layers I and II coding methods is quite similar to the subband coding diagram (appendix B), except a separate processing module that implements a fast Fourier transform (FFT) is used to compute the frequency-selective power information needed to select the appropriate masking thresholds. These thresholds are used to select quantizer resolutions for the quantizing device. On the reconstruction side, the system performs in a similar fashion as the subband decoder above.

> **FFT** was invented by Cooley and Tukey in 1965 and set the stage for many of today's technologies.

Some of the more important details about the audio format are listed below.

- All MPEG Audio methods use 32 subbands for encoding and decoding signals.

- One of several bit rates can be selected between 32 kilobits/second for the lowest-quality monaural sounds to 384 kilobits/second stereo sounds. Note that CD-quality audio signals require approximately 1.41 megabits/second, so MP3 files offer compression ratios of about 4 for the most-bit-costly encoding method and even higher compression ratios for other levels of quality. Moreover, extensive tests with listeners indicate that MPEG Layer II encoded audio at 192 kilobits/second is similar to CD-quality audio at 1.41 megabits/second. This represents a compression ratio of about 7 to 1, a significant saving in storage needs.

Question: How much MP3 music will fit on a CD-ROM?

The MPEG Audio Level III standard is known to achieve near-CD quality at bit rates of about 128 kilobits/second. How many minutes of MP3-encoded music could be stored on one CD?

To answer this question, we simply need to set up the proper ratios for the calculation. A CD can store up to 788 MB of information. The number of bits stored on a CD is therefore

$$N_{Bits/CD} = 788 \times 10^6 \, bytes \times \frac{8 \, bits}{1 \, byte} = 6.3 \times 10^9 \, bits$$

The rate at which the MP3 music is to be read off the encoded CD is

$$R_{MP3} = 128 \times 10^3 \, bits / \sec = 1.28 \times 10^5 \, bits / \sec$$

The amount of music that can be stored can be found by dividing these two numbers, or

$$T_{MP3/CD} = N_{Bits/CD} \times \frac{1}{R_{MP3}}$$

Plugging in the values, we get

$$T_{MP3/CD} = 6.3 \times 10^9 \, bits \times \frac{1}{1.28 \times 10^5 \, bits / \sec} = 4.92 \times 10^4 \, \sec$$

Finally, converting the time units into hours, we obtain

$$T_{MP3/CD} = 49200 \, \sec \times \frac{1 \, hour}{3600 \, \sec} = 13.6 \, hours$$

In other words, a CD can hold over 13 hours of music in MP3 format at near-CD quality. If you repeat the calculation for a single-layer single-sided DVD, the amount of music that such a DVD can hold is about 82 hours!

_____

Exercise 9.4

1. Is the ability of the human ear better or worse than the sonic capabilities of a CD-quality audio system?

2. Suppose you are listening to the car radio at such a high volume that you could not hear the warning siren of a nearby ambulance. What psychoacoustic effect are you demonstrating?

3. Suppose that a computer hard disk can store 10GB of information. How many minutes of MP3 audio can be stored on this hard disk?

9.4.4 Video Compression Methods

Introduction to Video Compression

Video compression or video coding refers to a class of techniques for reducing the number of bits needed to store a sequence of digital images. Video coding is used in high-quality digital television formats such as DVD Video and high-definition TV (HDTV) as well as lower-quality video conferencing systems and Internet video delivery.

As the example in the Introduction to this chapter has shown, movies and video places the greatest demand on storage and communication resources out of all multimedia signal types. Fortunately, most moving image streams possess a lot of redundancy that can be used to reduce the number of bits that need to be stored. This redundancy can be easily seen through an example.

The 3D World of Visual Motion

All modern video and movie sources are three-dimensional data sets. The three dimensions of data include

1. The two dimensions that make up each of the viewed images, and

2. The third dimension of time.

If we consider each movie or video frame like a translucent playing card and stack the cards one after the other, the picture in Figure 9.11 is obtained. Each two-dimensional object becomes a three-dimensional object whose volume is equal to the size of the object times the amount of time that the object is in existence in the image sequence.

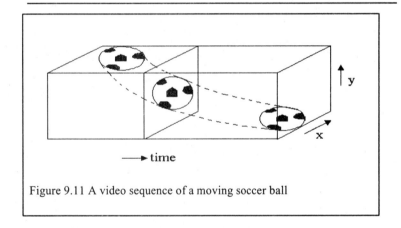

Figure 9.11 A video sequence of a moving soccer ball

A Generic Video Coder

International standardization groups such as the Motion Picture Experts Group (MPEG) and other technical groups have come up with standards for compressing digital video. While the standards for video compression differ slightly from one another, they all have similar functional parts.

Figure 9.12 shows a diagram of a typical video compression scheme.

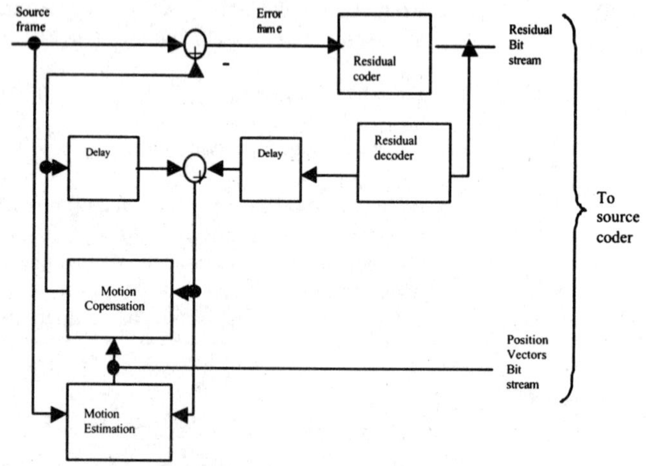

Figure 9.12 A typical video compression scheme

There are several sub-systems in this diagram:

1. *Motion Estimator:* This block uses successive frames of video to determine which pixels from successive frames are most like one another.

2. *Motion Compensator:* This block attempts to build an image from motion estimator data to try to "predict" what the current frame of video might look like.

3. *Residual Encoder:* This block has a structure much like a still-image transform coder. See the discussion on JPEG image compression above.

4. *Residual Decoder:* This block is essentially a still-image transform decoder.

These four sub-systems work together to try to reduce the redundancy in the image sequence. The motion estimator tries to predict where specific objects or object patches have moved in successive frames of a video image sequence. This information, along with the images that are reconstructed from the residual image decoder, to compute a motion-compensated image that looks as much as possible like the current image frame. Subtracting these two images pixel by pixel, we get a residual image frame of the components of the image that cannot be predicted by motion alone. This residual frame is then encoded with a residual encoder. The bits that are actually saved store

- The estimated directions of motion in the image, also known as the *motion vectors,* and

- The compressed version of the encoded residual image.

42

An example illustrates how one portion of the video coder—the motion estimator—works.

Example: Motion Estimation—the Basics

The goal of motion estimation is to try to figure out where the pixels that make up objects have moved from one image to the next. Consider the two images shown in Figure 9.13(a) and (b). In these images, we see similar objects, but their spatial positions are different. In some cases, the orientations of the objects have moved. Also shown in the original figure are the boundaries of the square pixels, which we have chosen to be overly-large to illustrate the concept of motion compensation

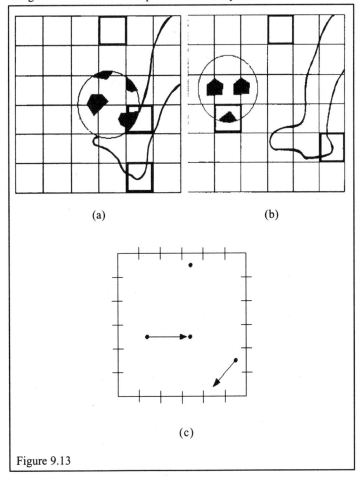

(a)   (b)

(c)

Figure 9.13

If we lay one of the images over the other, we can see how the two images are related. If we consider the centers of the pixels of the first image as points of origin, we can draw arrows from these points to the positions in the new image to which the points correspond. These arrows are shown for several pixels in the first image. Note that in some cases, the object has rotated slightly, or it has gotten farther away. These types of changes are much harder to determine and track.

The motion vectors are the pairs that define the offsets of each pixel in the image pair. For example, for the pixel at location

$$x = 5, y = 1$$

the patch of information in the original image has moved to the location

$$x' = 6, y' = 2$$

Thus, the motion vector for the original pixel is the number pair

$$m_x = x' - x = 1$$
$$m_y = y' - y = 1$$

How does a motion compensator find these points of correspondence between two image frames? Most image compression standards (such as the various MPEG standards) do not specify how this point matching should be done, leaving the design of the system up to the clever engineers who design video compression systems. A fairly straightforward but computationally-costly method would be to take each pixel in the original image and try to find its best match in all of the pixels of the second image. This full search method requires too many comparisons of pixels, however. For example, if we have a video sequence made up of 640x480 = 307200 pixels, comparing each pixel in the original image with all 307200 pixels in the new image would require 307200x307200 = 94 trillion calculations to get the best match. Fortunately, there are easier ways to do this search, so that compression of a video sequence need not be so difficult.

Decoding Motion-Compensated Video

Figure 9.14 shows a block diagram of the video decoder for a motion-compensated video signal. This system consists of a residual image decoder followed by a motion compensator. The residual image decoder produces the residual image frames from the compressed version of these frames. These images are then fed into the motion compensator, which reconstructs the large-feature portions of the image sequence. Adding these two images together gives the reconstructed image.

The compression capabilities of motion-compensated video compression are significant. A short example illustrates the achievable data rates for one popular motion-compensated video coding standard.

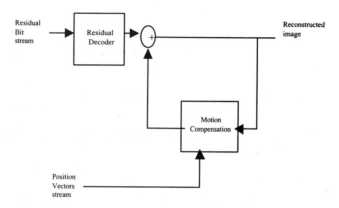

Figure 9.14 Video decoder for a motion compensated video signal

MPEG-1 Video Coding

The MPEG-1 Video Coding standard was developed to allow full-motion video sequences to be stored on CDs. MPEG-1 Video supports a special form of forward- and backward- motion compensation to allow for a high compression ratio. Figure 9.15 shows the sequence of frames for an MPEG-1 coded video. There are three types of frames:

- The *intra-coded* (I) image frames are compressed versions of original video frames, in which no motion compensation or residual image is computed. The coding used here is much like still image compression in the JPEG image compression standard.

- The *predicted* (P) image frames are compressed image frames that are computed using a system similar to the motion-compensated video compression scheme in Figure 9.12.

- The *bi-directional predicted* (B) image frames are compressed in a way similar to that of the P frames, except that information from the P frames before *and* after the B frames provide motion information. This information is used to reconstruct the B frames to a greater accuracy with a more limited number of bits per residual frame.

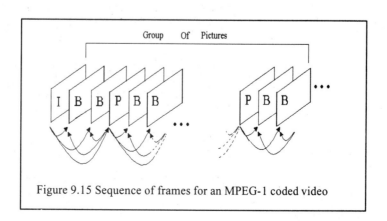

Figure 9.15 Sequence of frames for an MPEG-1 coded video

Shown in the figure are arrows that indicate the motion vector dependencies of the I, P, and B frames. As can be seen, there are effectively two different prediction paths: the sequence of I-P frames, and the finer sequence of I-B-P frames.

The data rate supported by the MPEG-1 Video Coding standard ranges from 1 to 2 megabits/second. Since standard CD-quality audio has a data rate of 1.41 megabits/second, MPEG-1 Video meets the targeted data rate. Recall that a "raw" version of a 30-frame-per-second video used

$$N_{Bytes / Frame} = 2.458 \ MB$$

per frame to store a video image sequence. Since there are 30 frames of video per second, the data rate of this "raw" video source is

$$R = N_{Bytes / Frame} \times \frac{8 \ bits}{1 \ byte} \times \frac{30 \ frames}{1 \ sec}$$

$$R = 590 \ megabits \ / \sec$$

In other words, MPEG-1 achieves a compression ratio between 250 and 500 to 1—a significant savings!

Question: Downloading a DVD-Quality Movie Using a 56k Modem

At the time of this book's writing, the most popular way to access the Internet from home is through a 56kilobit/second modem over a standard telephone line. A friend of yours has an idea for a movie-downloading service over the Internet, using standard 56k modems as the main connection method. You, however, have doubts about how good your friends' business plan is, because you think it could take a long time to download a DVD-quality movie using a 56k modem. How long would it take to download a DVD-quality movie in this way?

Answer: To figure out this answer, we need to know what the typical data rate of a DVD-quality movie is. DVD Video is compressed using an enhanced version of the MPEG-1 video coding standard called, appropriately enough, MPEG-2. MPEG-2 video supports a variable data rate between 2 and 10 megabits/second. For a two-hour movie, a storage of 4.7 GB is reasonable. Then, the amount of time it would take to download a two-hour DVD movie is

$$T_{Download} = \frac{4.7 \times 10^{9} \ bytes}{56 \times 10^{3} \ bits \ / \sec} \times \frac{8 \ bits}{1 byte} = 6.71 \times 10^{5} \ sec$$

Converting this number to days, we find that

$$T_{Download} = 6.71 \times 10^{5} \ sec \times \frac{1 \ hour}{3600 \ sec} \times \frac{1 \ day}{24 \ hours} = 7.77 \ days$$

Your friend's business plan has one problem: it will take *over a week* to download one single movie! Most of us would rather go to a local video rental store than wait that long—and tying up a telephone line in the process!

Section 9.5: Additional Topics

9.5.1 Transform and Subband Compression Methods

The audio perception example described previously points out an attribute of human hearing that can be used in compression tasks. Masking makes certain sounds more important to us than others. We can use this fact to build a more-efficient coding scheme. This method goes under the name of *transform compression* or *subband compression* depending on certain features of how it is implemented.

Subband Processing for Audio Compression

Many portable stereo systems these days have a graphical display that shows the approximate frequency content of the signal that is being played. Figure 9.16 shows the layout of a typical display. This display consists of several columns of lights arranged from left to right, with frequency values below and amplitude values on the right of the display. When music is played through the amplifier in these portable stereos, the lights "bounce" to the music. Playing sounds with lots of energy at low frequencies or bass cause the bars on the left to rise, and playing sounds with lots of energy at high frequencies or treble cause the right-most light bars to rise.

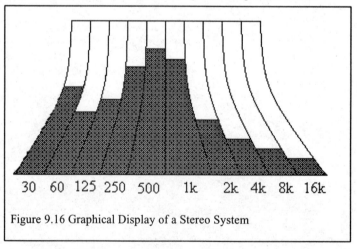

Figure 9.16 Graphical Display of a Stereo System

A circuit inside of the portable stereo processes the electrical version of the sound being played to drive the light display. A block diagram of this

48

circuit's functionality is shown in Figure 9.17. The parts of the system are described below:

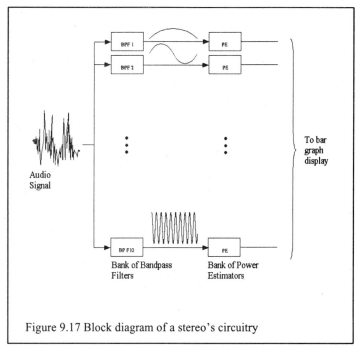

Figure 9.17 Block diagram of a stereo's circuitry

Filter pass specific frequencies from the input to the output.

- *Bank of Bandpass Filters:* This system consists of several finely-tuned frequency-selective filters that pass through certain components of the signal near the center frequencies at the bottom of the light bar display. The output of each of these filters is only large when the played sound has energy in the bandpass portion of the particular filter. Collectively, these filters are called the *analysis filter bank.*

- *Power Estimators:* Each of these systems measures the current power of the signal being sent to it over a short period of time.

The outputs of the power estimators are sent to a circuit that drives the light display, so that large powers correspond to many lit lights in a column and small powers correspond to few lit lights in a column.

Suppose now instead of 5 to 10 columns in a light bar display, there were 32 or more such columns of lights. This system could separate and track the powers of sounds at more closely-space frequencies. Subband compression uses exactly this type of system to separate sound into several parallel paths.

A block diagram of a subband coding system is shown in Figure 9.18. The input signal to be compressed is first passed through a bank of 32 bandpass filters, and the powers of each of the 32 filter output signals are estimated. These power values are then used to *quantize* the bandpassed signals, so that each signal is represented using fewer bits. These bits are then collected into one *bit frame* and then stored or transmitted.

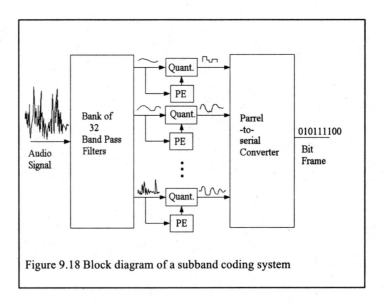

Figure 9.18 Block diagram of a subband coding system

The savings in bit storage comes through the quantization step of the subband coder, which employs information about the human auditory system to do its job. For example, the masking effect shown in Figure 9.8 is used to lower the quantization accuracy of soft bandpassed signals that are adjacent in frequency to louder bandpassed signals. This quantization process is therefore constantly changing due to the power levels sensed on each of the bandpassed signals. Of course, the information about how the power levels change must be stored in the bit frame as well, so that it can be used to decode the bits later.

The subband decoding process works essentially in a reverse fashion as compared to the encoding process. Figure 9.19 shows how a subband decoder works. The bits frames are first divided into quantized signal values and side information. The side information is sent to a decoding circuit that determines how to reconstruct the individual bandpass filtered signals from the quantized versions. This circuit sends its commands to a reconstruction device that produces the individual bandpass filtered signals from the quantized versions. Then, all of the reconstructed bandpass filtered signals are sent to a set of filters that reconstruct the original signal from the many bandpass signals. This system is collectively called the *synthesis filter bank*. The output of this system is the desired sound signal.

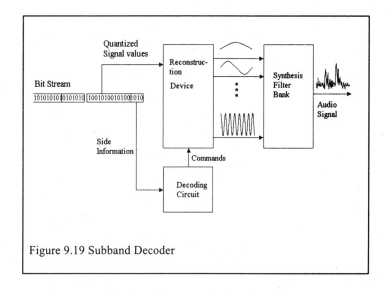

Figure 9.19 Subband Decoder

We can point out several desirable features of this subband coding method:

- The decoding process is much simpler than the encoding process, because there is no need for careful power estimation or other adaptive procedures. This feature is perfect for distributing data, because it is extremely likely that a particular signal will be decoded many more times than it is encoded. A simpler decoding process means that the decoder hardware can be made more inexpensively and with fewer components.

- The coding method is modular, allowing each particular component to be redesigned as newer hardware technologies become available.

- The accuracy of the method can be tuned to allow a wide variety of compression ratios depending on the application.

9.5.2 Transform Coding vs. Subband Coding

Transform coding differs from subband coding in the way the filtered outputs are calculated. In transform coding, a special bank of filters are used that correspond to an actual Fourier or other discrete-time transform of a block of signal samples. The encoding process uses information about the amplitudes of the transform coefficients over each block to find the best quantization strategy for the Fourier coefficients. These quantized coefficient values, along with side information about the quantization step, are what is stored.

51

In many situations, a transform other than a Fourier transform is used for this coding process. One such transform is the discrete cosine transform (DCT). The DCT is a real-valued transform; that is, the transformed data values have no imaginary component.

An example shows how transform coding is used in still image compression.

Transform Coding Example: JPEG Image Compression

The most popular compression method for still images and pictures is the Joint Pictures Experts Group (JPEG) image compression standard. JPEG compression is a repetitive process that is applied to square blocks of digital images. Because there is redundancy in local regions within an image, we can exploit this redundancy to reduce the number of bits needed to store the entire image.

The JPEG encoding process for a grayscale image is described by the following steps:

1. The digital image is divided up into 8-pixel-by-8-pixel image blocks. Each of these blocks contains 64 numbers, where each number represents the brightness of the associated pixel.

2. For each 8x8 pixel block, we divide up the rows of the block into 8 different rows and apply an eight-point Discrete Cosine Transform (DCT) to the resulting row.   This process is shown in Figure 9.20(a)

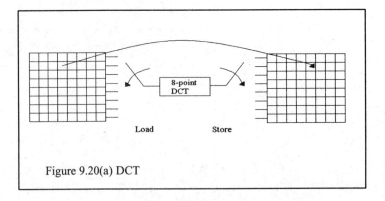

Figure 9.20(a) DCT

3. Next, we apply an identical eight-point DCT to each of the columns of the 3 row-transformed image patches. Figure 9.20(b) shows this process. The combination of the row- and column - processing produces the two-dimensional DCT for each of the 8x8 image patches.

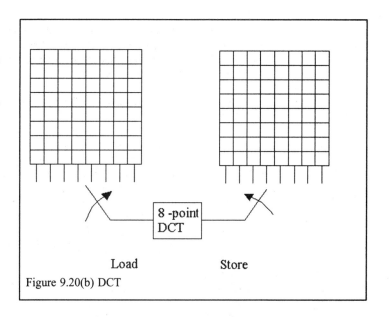

Load   Store

Figure 9.20(b) DCT

4. We take the numerical values from each transformed pixel except the uppermost left pixel and extend them out into one long 63-element row using a zig-zag pattern through the pixel positions. Figure 9.20 (c) shows the zig-zag pattern taken in this process.

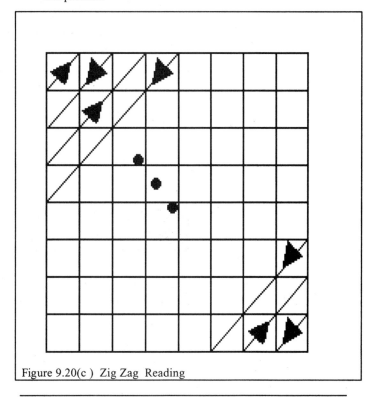

Figure 9.20(c) Zig Zag Reading

5. The values of each transformed pixel are quantized to a small number of amplitude levels using a different quantizer resolution for each pixel value. These amplitudes are then turned into a string of bit values.

6. The strings of bit values tend to have many zeros in them. We can use run-length coding to compress these values further. Moreover, Huffman coding is used to compress the number pairs obtained from the run-length coding procedure, which reduces the number of bits stored even further. More details about run-length and Huffman coding can be found in the "Lossless Compression" section that follows.

7. The uppermost left pixels are collected over several blocks and coded using another compression method called differential pulse-code modulation (DPCM).

Once compressed, the JPEG image takes up much less storage space than its original version while still having excellent visual quality. For example, it is possible to store an image to a high quality using less than one bit per image pixel! This result might seem surprising, because a one-bit-per-pixel image would appear only as a checkerboard-style image with either all-black or all-white pixels. The JPEG compression method is able to obtain this level of compression because of the redundancy that exists locally in almost all natural images.

Section 9.6:
Glossary

ASCII
an acronym for "American Standard Code for Information Interchange;" an 8-bit code used to represent alphanumeric information in almost all digital devices.

average codeword length
the average number of coded symbols it takes to store an original symbol sequence using a particular code

code
the rules the define the meanings of a set of bits; for example, the ASCII code that assigns 8-bit binary numbers to letters, spaces, punctuation, and other computer characters

codebook
the table that describes the translation of important symbols to their binary representation

codebook size
the number of entries in a particular codebook

codeword
an encoded version of an important symbol; represented by a bit string

codeword length
the number of bits in a particular codeword

compression
an encoding method that tries to reduce the number of bits needed to store important information

compression ratio
the ratio of the number of bits of the compressed version of the signal to the number of bits in the original representation of the signal

decoding
the act of recovering meaningful symbols through the translation of sets of bits

dynamic range
in human hearing, the ratio between the loudest and softest sounds that humans can hear

encoding
the act of translating meaningful symbols into sets of bits

encryption

an encoding method that tries to hide information from others

entropy

in compression, the fundamental limit to the amount that a particular signal can be compressed without any loss

error correction

a joint encoding/decoding procedure that tries to fix unknown errors in the encoded information

JPEG

an acronym for "Joint Pictures Experts Group;" an international standards committee for still image compression; also, a particular lossy image compression method

lossless compression

a class of compression techniques in which nothing about the original set of numbers or symbols is lost in the compression process

lossy compression

a class of compression techniques that throw away information about the sets of numbers or symbols being compressed

masking

a psychoacoustic effect of the human hearing system whereby soft sounds are made inaudible by louder sounds that are close in frequency

Morse code

an encoding method that uses a four-symbol language of "dot", "dash", "space" and "break/end of sentence" to represent English text

MP3

an acronym for "Audio Layer 3 of the MPEG1 Multimedia Coding Standard;" a popular internet audio coding standard

MPEG

an acronym for "Motion Picture Experts Group;" an international standards committee for multimedia

relative frequency

a measure of the number of observations of a particular event over the total number of observations of all events

run length encoding

a lossless compression method that is efficient for binary signals that contain long strings of contiguous ones or zeros

threshold of feeling

the upper limit of the human hearing system; describes the loudest sounds that humans can possibly hear

threshold of quiet

the fundamental lower limit of the human hearing system; describes the softest sounds that humans can hear

The INFINITY Project

Chapter 10: Correcting Digital Errors

How do we encode information so that we can detect and correct errors? What are the design choices behind good error correcting systems?

Approximate Length: 2 Weeks

Section 10.1: Introduction to Coding for Error Correction

Approximate Length: 1 Lecture

10.1.1 Message In A Bottle : Getting the Message Across

Most everyone is familiar with the "message in a bottle" story: A shipwrecked sailor, unable to leave his island prison, sends out a help message in a glass bottle that floats aimlessly across the sea. The sailor hopes that someone will find the bottle, open it, and read the message inside. A lot of things have to go right, however, for the sailor to be rescued:

1. The bottle has to float to "civilization," that is, to a person or people who have the means to make it to the island and rescue the sailor.

2. The civilization that gets the message has to be familiar with the language that the sailor used to write his message. Otherwise, the message can't be read and understood.

3. The message has to be reasonably undamaged so that it still can be read. If water has seeped in and washed away some of the message, or if sunlight has caused the message to fade, then the sailor's cry for help might go unanswered.

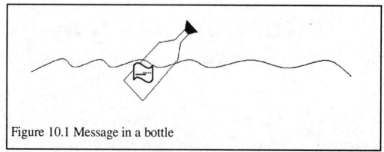

Figure 10.1 Message in a bottle

Most of us think about the first issue—the chances that the bottle actually gets into a rescuer's hands. Fewer people are likely to think about the second issue—the chances that the sailor's language is understandable to the would-be rescuer. Chance are, however, that very few of us would think about the third most practical issue—the message could be damaged on receipt. Yet, it is often the situation in digital devices such as CD players, cell phones, and Internet-based communications terminals that the latter is often the most important. It might be surprising to you to hear that in most digital devices, errors in the signals happen regularly, and sometimes the errors number in the hundreds or thousands every second! How can these systems work with this high an error rate?

Fortunately, it is possible to use clever methods to code information in bit form so that the original information can be recovered even when there are errors in the bits. Such coding methods add additional bits to a message so that any introduced errors in the message can be detected and even corrected. Since these additional bits do not add anything to the meaning of the message, these bits are called the *redundant* part of the coded message. Moreover, the procedure by which the original message is recovered from its corrupted encoded version is called *error correction*. In this chapter, we'll discuss simple forms of error correction and where error correction is used.

There are two aspects to error correction:

1. *The error must be detected.* When there are errors in the message, we usually have to recognize that an error exists. For example, the two phrases "The wood pile" and "The wood file" differ by only one letter, but both make sense to us. It is only in the context of a larger sentence, such as "The wood pile was used to dull a sharp edge," that the error becomes apparent. When dealing with bit streams, it is generally too much work to try to decode the bits to figure out whether an error exists. Instead, we develop simple ways to figure out if an error exists using some redundant information that is added to the overall message.

2. *The error must be fixed.* After detecting an error, we have to use the available information to fix the error. By cleverly encoding the information, we can actually correct for errors when they occur, usually up to some number of errors in the message.

Reading a Corrupted Message

Try out this exercise:

1. Have a friend write down a sentence without showing you the sentence. Have her or him write down numbers below each letter and space in the sentence. Ask her or him to tell you the number of characters (letters or spaces) in the sentence. Call this number *T*.

2. Pick *N* numbers between one and *T*. Choose *N* to be around 1/5th to 1/3rd the value of *T*. For example, if the sentence has 19 characters in it, choose a value of *N* between 4 and 7. Have your friend put a dash "-" through the characters associated with these numbers. Then, on a separate sheet of paper, have your friend write down the entire message with dashes in place of the marked characters. This is the corrupted

version of the sentence. The dashes indicate where the errors (in this case, omissions) have occurred.

3. Now, take a look at the corrupted version of the unknown sentence. Can you read the sentence? Do you know exactly what it means? Chances are that you can quite easily, even with $N=T/3$, because you understand the meaning of the language. In fact, in the context of complete sentences, you only need about half the letters in order to guess the correct sentence.

In this example, you are using your understanding of a human written language to correct for the omitted letters. This form of error correction is very sophisticated, and digital devices usually use much simpler ways to fix errors.

Reading a Corrupted Message—Repetition

Try out the "Reading a Corrupted Message" exercise above, with a slight twist: Instead of using the original message, have your friend repeat the message to *2T characters*. Of course, choose *2N* places for character deletion, so that the ratio of deleted and remaining characters stays the same. Is it easier to read the corrupted message now? How has repetition helped you figure out what the message means?

10.1.2 Where Do Digital Errors Come From?

Errors in digital signals are quite common. Here are some typical scenarios where errors in digital signals occur:

1. Suppose a digital device is reading information from a physical storage medium, such as from the shiny surface of a CD or DVD. Physical defects or debris such as dust, scratches, or oily fingerprints make reading this information difficult or impossible. In this case, errors are created in the bit stream from the laser reading assembly inside of the CD or DVD player or drive.

2. When two digital devices talk to one another, the connection between the two can become corrupted. A good example is in wireless cellular telephones. Interference from other electronic devices, power lines, and atmospheric phenomena disturb the digital signals emitted from the antennae of the telephones. When decoding these signals, an antenna basestation perched high on a receiving tower must try to remove as many errors as possible from the received signals.

3. When storing information for a long time, environmental effects can cause errors. A good example can be found in magnetic floppy disks. These disks cannot hold their information forever. Usually, some stray magnetic field (from another electronic device, for example) can introduce some errors. Or, physical wear and tear can cause errors to be introduced.

Wherever digital errors can occur, it is useful to consider error correction to help safeguard the data and information in use.

Exercise 10.1

1. Take a pair of sentences from any book and delete all of the vowels from the sentences. Write down the two phrases with the removed vowels, and show them to a friend. Can s/he read the sentences?

2. Many wallet-sized plastic cards—credit cards, frequent shopper cards, and ATM cards—hold information using magnetism. Describe the situations in which these cards can become corrupted.

Section 10.2:
Error Detection
and Correction

*Approximate Length: 3
Lectures*

10.2.1 Error Detection: Knowing That Something's Wrong

Many modern digital devices have ways to detect errors in digital signals. The simplest way to detect errors is to try to interpret the data and see if it makes sense, but this type of scheme doesn't always work. Adding another zero to a bank account balance still makes that balance "look like" a number, for example.

In this section, we introduce the simplest form of error detection that uses the concept of *parity*.   Parity is a noun that means "equivalence" or "of equivalent classes or types." If two things have parity, they are similar to one another. In digital signals, we can use a similar concept to classify types of signals.

A Simple Example of Parity

Let's take a simple example of parity. Suppose you have a string of 8 bits such as

$$S_1 = 00011010$$

This 8-bit string has 3 ones and 5 zeros. Ignoring for the moment the number of zeros, we see that the bit string has an odd number of ones in it. So, let us label this string as an Odd bit string. Now, if we had the bit string

$$S_2 = 00110101$$

we would count 4 ones and 4 zeros. So, this bit string is an Even bit string. Suppose we had the bit string

$$S_3 = 11111111$$

What would we label this string as? If you said Even, you're right. How about the bit string

$$S_4 = 00000000$$

For reasons that will be clear shortly, we'll label this string as an Even string (and for good reason: 0 is divisible by 2, right?)

We've used the number of ones in a bit string to label it as an Even or Odd string. Suppose now that, due to reasons beyond our control, an error in one of the bits in bit string S_1 is introduced, creating the new bit string

$$S_1 error = 00010010$$

In this case, an error can occur either by (a) a one getting changed to a zero or (b) a zero getting changed to a one. If we count up the number of ones in this new bit string, we see that there are 2 ones, and thus we recognize that bit string S_1error is an Even bit string. However, since the original bit string S_1 was an Odd bit string, then we know something is wrong in the new bit string! The concept of parity is that simple.

How do we know that the bit string S_1 was Odd to begin with? Well, we can append or add a bit to the string to indicate that it was an Odd string. For example, let a 1 denote an Odd string and a 0 denote an Even string when these bits appear at the end of the string. Thus, we would create

$$S_1 p = 000110101$$
$$S_2 p = 001101010$$
$$S_3 p = 111111110$$
$$S_4 p = 000000000$$

Don't worry that the added "1" to S_1p makes the number of ones even; remember, the last bit in each new nine-bit string is created by us. Now, if there is a single bit error (either a one "flipped" to a zero or a zero "flipped" to a one), we can detect the problem. Try it with any of the above new strings. Of course, we have to be careful, because we cannot allow our special *parity bit* at the end of our strings to be altered. Otherwise, the technique doesn't work.

The parity bit example above illustrates some general concepts about coding for error detection (and also for correction):

1. We have to change the signal or information in a special way in order to recognize that there are errors in it later.

2. Generally, the change in the information increases the number of storage locations that we use to store the information.

3. There is no new information contained in the new version; we've increased the length of the message without enhancing its content.

The above concepts work for any type of information and any added "clues" that we append to that information, not just to bit strings. In fact, we can even use parity information to provide a check for errors in collections of parity bits or parity information. We'll use this concept next to show how to both detect and correct errors in messages.

Parity Bits and ASCII Text

Try out this exercise:

1. Take the first five letters of your name and write down the ASCII 8-bit representations of each of the letters and/or spaces (if your name is shorter than 5 letters, use the "space" characters at the end).

2. Count the number of ones in each 8-bit string, and add the correct parity bit at the end of each 8-bit string (an 0 for an Even string and a 1 for an Odd string).

3. Give the five 9-bit strings to a classmate. Have her or him write the strings down again, but have her or him change one of the first 8 bits in *only one* of the five 9-bit strings. Your classmate is putting an error in one letter of your message.

4. Next, have your teacher take all of the corrupted messages and put them into a hat. Pick one of the messages out of the hat at random.

5. For the message that you've picked, first check the parity bits of each of the five bit strings. Which of the bit strings is in error? Mark this bit string with a star "*".

6. Now, using the ASCII Code, decode all of the other correct bit strings. Whose name did you pick?

In this example, you are using the parity bits to mark the letter(s) that you know have errors in them. This process makes the job of reading the name much easier than if you just decoded the message without the parity bit information (and made an error in decoding one of the letters).

10.2.2 Error Correction: Fixing the Problems

As you might have guessed, detecting the presence of errors in digital signals is not good enough for most applications. Imagine a CD player that, when playing a scratched disk, simply stopped when it came across an error and informed you through its display that "This disk has an error." A much better solution is to fix the error when it happens. How do we achieve this goal?

We can extend the parity bit check ideas presented in the last section to help fix errors as well. To see this extension, consider the following 15-bit stream:

$$S = 100110011101111$$

This bit stream has 10 ones, and as such, it is an Even bit string. Suppose we arrange these 15 bits in a table with five rows and three columns, like this:

1	0	0
1	1	0
0	1	1
1	0	1
1	1	1

Now, consider each row and column as its own bit string. We can compute parity bits for each row and column, and we'll place these bits at the bottom and at the right of the table in their own row or column, like this:

1	0	0	**1**
1	1	0	**0**
0	1	1	**0**
1	0	1	**0**
1	1	1	**1**
0	**1**	**1**	

Finally, let us compute a "parity bit" for the parity bits, by taking the last row or column as its own bit string. This action fills in the last entry with a zero, like this:

9

1	0	0	**1**
1	1	0	**0**
0	1	1	**0**
1	0	1	**0**
1	1	1	**1**
0	**1**	**1**	0

The fact that this lower-rightmost entry is a zero is no surprise; remember, the original bit string was an Even bit string. This table now contains 6x4=24 bits, so we've increased the length of the message by nine bits. We can read out the entries of this table in row order as

$$Sp = 100111000110101011110110$$

This bit stream is the one that is saved or transmitted.

Now, suppose we read the bit stream as

$$Sp = 100111000110001011110110$$

We can arrange the bits in a 6x4 table as

1	0	0	**1**
1	1	0	**0**
0	1	1	**0**
0	0	1	**0**
1	1	1	**1**
0	**1**	**1**	0

Now, we look along each column and row of the table and check to see whether the parity bits match up with the 5x3 entries of data. We immediately see that the first column doesn't check out (3 ones, but the parity bit is an 0), and that the fourth row doesn't check out either (1 one, but the parity bit is an 0). So, we immediately know that the entry in the fourth row and first column is in error! We can fix this bit easily by flipping it, because then the parity bits match the corresponding columns and rows. So, the corrected table is

1	0	0	**1**
1	1	0	**0**
0	1	1	**0**
1	0	1	**0**
1	1	1	**1**
0	**1**	**1**	0

Removing the last row and column, we obtain the original bit message again.

This form of redundancy works for numbers, too. Suppose we have a sequence of digits, such as

N=375854912

We can arrange these 9 numbers in a table as

3	7	5
8	5	4
9	1	2

Now, instead of checking whether sums of rows are even or odd, we simply add up the rows and columns and through away everything but the last digit. This type of calculation gives the bottom row and right column as

3	7	5	**5**
8	5	4	**7**
9	1	2	**2**
0	**3**	**1**	**4**

The new list of number is

$$Np = 3755854791220314$$

Now, suppose we "receive" the number list

$$Npe = 3757854791220314$$

We construct the 4x4 table as

3	7	5	**7**
8	5	4	**7**
9	1	2	**2**
0	**3**	**1**	**4**

Now, we see that the first row and the fourth column doesn't check out; 3+7+5=15, which doesn't end in 7, and 7+7+2=16, which doesn't end in 4. So, we know that the uppermost-right entry of 7 is wrong. We can also see that the correct value for the first row/fourth column entry can be found either by

1. Adding the first three entries of Row #1 and taking the ones digit, or

2. Subtracting the second and third entries in Column #4 from the parity bit in the last entry of Column #4, and then adding enough multiples of 10 to get the result over zero.

Using either method, we come up with the correct value of 5; thus, we have

3	7	5	**5**
8	5	4	**7**
9	1	2	**2**
0	**3**	**1**	**4**

Reading out only the "data" numbers, we get the original message.

The last example illustrates that there are multiple ways to reconstruct a single error in this tabular representation of information with parity values. The above example of an error correcting code is quite similar to an error correction coding method called the *Hamming Code*, first developed by Richard Hamming at Bell Telephone Laboratories (now part of Lucent Technologies). The simplest form of this coding method employs three parity bits for every four bits, and it can a single error in the four bits. Hamming codes are used in electronic switching systems and certain computing equipment.

Coding Overhead and Coding Strength

The quality of any error correction coding method can be measured by two quantities:

Coding Overhead—the amount of redundancy added to a signal to make it possible to detect and correct for errors, as measured by a percentage of the original signal length in bits or symbols.

Coding Strength—the ability provided by the redundancy to correct for errors in the signal, as measured by a percentage of the original signal length in bits or symbols.

Let us consider some exa mples:

Example #1: Remember the 15-bit signal that we considered above, given by $S = 100110011101111$? We added 9 bits to this signal for encoding; therefore, the coding overhead for this code is

$$Coding \quad Overhead = \frac{9\ bits}{15\ bits} = 0.6 \quad or \quad 60\%$$

For this overhead, we can correct any 3 of the 15 bits in the original string. Therefore, the coding strength is

$$Coding \quad Strength = \frac{3 bits}{15\ bits} = 0.2 \quad or \quad 20\%$$

In other words, we have increased the signal length by 60%, and doing so allows us to protect the signal up to 20% error.

Example #2: We now consider the 9-digit number example. Here, we added 7 numbers in the coded version, so

$$Coding \quad Overhead \quad = \frac{7 \, numbers}{9 \, numbers} = 0.778 \quad or \quad 77.8\%$$

Since we can correct up to 3 errors in the signal, we have

$$Coding \quad Strength \quad = \frac{3 \, numbers}{9 \, numbers} = 0.333 \quad or \quad 33.3\%$$

Compared to our previous example, the second coding method is more resilient to errors (33.3% coding strength vs. 20% coding strength), but we've paid by price of adding more to the message (77.8% coding overhead vs. only 60% coding overhead).

Example #3: The simplest Hamming Code adds three parity bits to every four message bits, and it can correct a single error in the four message bits. Thus, the coding overhead is

$$Coding \quad Overhead \quad = \frac{3 \, bits}{4 \, bits} = 0.75 \quad or \quad 75\%$$

The coding strength of this Hamming Code is

$$Coding \quad Strength \quad = \frac{1 \, bit}{4 \, bits} = 0.25 \quad or \quad 25\%$$

In many applications, we need not be so aggressive about protecting errors as in the above examples. For example, in wireless communications, typical error rates for communicating bits over a very demanding link are around 1%. Hence, we can design a code that protects up to a 5% error in the bits (since the 1% error rate is only an average number), and we are very certain that all of our bits will get through the communications link.

Example #4: Information on a CD-Audio disc is encoded using what is called a **Reed-Solomon code**. Reed-Soloman codes are powerful encoding procedures that are specifically designed to target *burst errors*, which are large numbers of bit errors that occur one after another in a sequence of bits. The Reed-Solomon code used in CDs decreases the amount of storage space on a CD, and in return, the CD can withstand scratches, dust, and other physical impairments that might otherwise cause it to lose data.

The information bit rate (i.e. the bit rate associated with the 16-bit stereo 44.1kHz digital recording) on a CD is 1.41 Mbits/second. The encoded bit rate using the Reed-Solomon code along with some synchronization and subcode information is 2.034 Mbits/second. Hence, the coding overhead is

$$Coding \quad Overhead = \frac{2.034 - 1.41 Mbits / \sec}{1.41 Mbits / \sec} = 0.442 \quad or \quad 44.2\%$$

With this overhead, about 2700 data bits in an error burst can be corrected, which corresponds to about a 2.5mm track length. Hence, one could block up to 2.5mm of the bit trail on the disc and still not hear it! Moreover, CDs use *interleaving* (described next) to enable more errors to be recovered from simple interpolation.

10.2.3 Error Reduction: Error Bursts and Interleaving

Good coding methods make up only part of a good error correcting procedure for important information. Another equally-important feature of coding methods is good matching of the coding method to the types of errors that are likely to occur. In this section, we describe one important method to improve the error correcting capabilities of codes.

Error Bursts and Interleaving

What makes a good error correcting code? Obviously, a good code will correct the types of errors that are likely to occur in our data. So, designing a good code means that we need to know what types of errors are likely to show up in a given application. Here, we consider one type of error that occurs quite frequently: the *error burst*. An error burst occurs when a signal gets corrupted for several successive bits or symbols. Suppose we have a text message such as:

"The meteorologist said that it will probably rain today, but it is going to be sunny tomorrow."

This message has 94 characters in it, including all spaces and punctuation. Let's label the characters using the numbers 0 through 9, as shown:

```
The-meteorologist-said-that-it-will-probably-rain-today,-
1234567890123456789012345678901234567890123456789012 34567
```

```
but-it-is-going-to-be-sunny-tomorrow.
8901234567890123456789012345678901234
```

Here, we've put dashes in place of spaces so that the spacing between the numbers and letters above them are right.

An error burst occurs when a bunch of errors are created in a row, one after another. For example, suppose that we have 12 errors from Character #41 to Character #52, so that these letters are completely lost. Then, we would have the message

```
The-meteorologist-said-that-it-will-prob************day,-
1234567890123456789012345678901234567890123456789012 34567
```

14

```
but-it-is-going-to-be-sunny-tomorrow.
8901234567890123456789012345678901234
```

Today's weather, and even the day of the meteorologist's prediction, are no longer accurately described in the message because of the error burst.

Discussion: What are some typical applications where error bursts occur?

CD Players: As described in Chapter 17, CDs store digital information in the form of an outwardly-spiraling pit trail on the disc's shiny surface. These pits encode the digital information that make up the music that you hear when listening to the CD. If a CD gets severely scratched, then the optical mechanism used to read the pits off of the CD can fail to read portions of the pit trail. An error burst occurs when a scratch or other obstruction is created in the same direction as the pit trail, along a broad arc of the CD surface. Figure 10.2 shows the type of scratches on a CD that cause such burst errors. For this reason, it is good idea to clean your CD with a soft cloth by moving along any radius (e.g. in and out from the center hole to the edge). That way, any scratched that are created, block only a few pits at a time.

TIP!
To reduce burst errors in CDs, you should clean them with a soft by moving along any radius

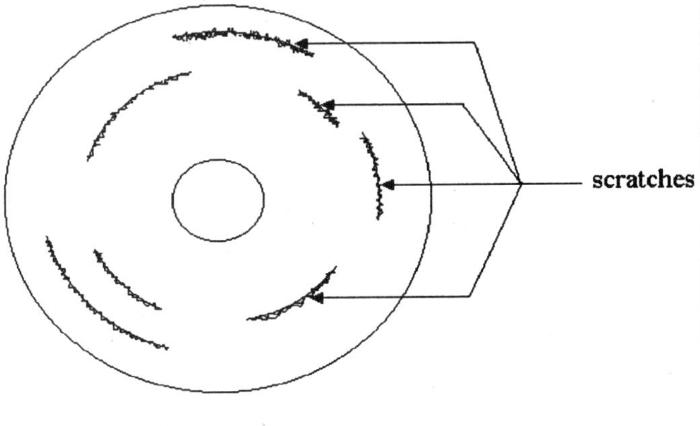

Figure 10.2 Scratches on CD

Cellular Telephones: The orientation of the antenna can greatly enhance or hinder the receiving and transmitting capabilities of a cellular telephone. When moving from one position to another, a cellular telephone user can move the antenna past interfering objects such as walls or other electrical appliances. Since the "bad" antenna position occurs only for a short time, the errors show up in a burst at the receiving end of the telephone connection.

To help address the error burst phenomenon, engineers have come up with a clever encoding scheme that effectively *distributes* the errors across the message. This method of encoding is called *interleaving*. To illustrate the concept of interleaving, consider our weather prediction message from before. However, instead of writing down the message as is, split the signal into N messages, where each message contains every Nth letter of the message. As an example, let's take N=10, so that we split the 94-character message into 10 different messages. These 10 messages can be best constructed from a 10-by-10 table, where we put the text in the table along each row, from one row to the next. This table is

T	h	e	-	m	e	T	e	o	r
o	l	o	g	i	s	T	-	s	a
i	d	-	t	h	a	T	-	i	t
-	w	i	l	l	-	P	r	o	b
a	b	l	y	-	r	A	i	n	-
t	o	d	a	y	,	-	b	u	t
-	i	t	-	i	s	-	g	o	i
n	g	-	t	o	-	B	e	-	s
u	n	n	y	-	t	O	m	o	r
r	o	w	.	-	-	-	-	-	-

Now, instead of reading out the message in row order, we read it out by the columns, which essentially takes every 10th letter and puts it in a 10 letter message, one message after another. This jumbled version is

```
Toi-at-nurhldwboignoeo-ildt-nw-gtlya-ty.mihl-yio--esa-r,s
1234567890123456789012345678901234567890123456789012 34567

-t-tttpa--bo-e--ribgem-osionuo-o-ratb-tisr-
8901234567890123456789012345678901234567890
```

We've lengthened the message to 100 characters using dashes so that we can read out the entire message from the table.

Key Idea: Let's choose this ordering for storing the message. In effect, we have put certain letters before others, which is a form of data "weaving" that is termed *interleaving*.

Suppose now that Characters #41 to #52 get corrupted as before, except that the corrupting letters have changed in our interleaved message. Our message would then look like

```
Toi-at-nurhldwboignoeo-ildt-nw-gtlya-ty.************a-r,s
1234567890123456789012345678901234567890123456789012 34567

-t-tttpa--bo-e--ribgem-osionuo-o-ratb-tisr-
8901234567890123456789012345678901234567890
```

If we put these letters back into a table row-by-row, we have

T	o	i	-	a	t	-	n	u	r
h	l	d	w	b	o	i	g	n	o
e	o	-	i	l	d	t	-	n	w
-	g	t	l	y	a	t	-	t	y

*	*	*	*	*	*	*	*	*	*
*	*	a	-	r	,	s	-	t	-
t	t	t	p	a	-	-	b	o	-
e	-	-	r	i	b	g	e	m	-
o	s	i	o	n	u	o	-	o	-
r	a	t	b	-	t	i	s	r	-

Now, we read out the message in column order, which gives us

```
The-**teorolog**t-said-t*at-it-wil*-probably*rain-toda*,-
12345678901234567890123456789012345678901234567
```

```
but-it-i*-going-t*-be-sunny*tomorrow.*-----
8901234567890123456789012345678901234567890
```

Looking at the message, it now reads

"The **terolog**t said t*at it wil* probably*rain toda*, but it i* going t* be sunny*tomorrow.*

The 12 errors have been "spread out" in the 94-character reconstructed message. Because of the natural redundancy of the English language, it is quite easy to read the sentence and get all the intended meaning out of it.

The key concepts of interleaving are:

1) **Data Shuffling:** Two or more information streams are regularly shuffled together like a perfectly-shuffled deck of playing cards. If there is only one signal, we can shuffle later portions of the signal with earlier portions, as in the example above.

2) **Use of Error Correction:** Interleaving only works with signals that have been encoded for error correction, so that single errors that happen over short data blocks can be corrected. If two data streams with no error correction encoding are interleaved, then interleaving doesn't help fix the errors later. Therefore, we should use interleaving with schemes like the parity bit check and correction schemes that we described earlier.

In the text example above, we saw how 12 errors could be spread out over a longer sequence so that only a few errors showed up in succession. If we had chosen a 12-signal interleaving, then the reconstructed signal would have only one error in every 8 letters for every 96-symbol block. Moreover, if the signal were a sequence of ones and zeros, we could use a single-bit correction scheme (like the Hamming Code) to correct for these isolated single errors. In this way, the error correction code could correct for burst errors in the signal through interleaving.

It might be surprising to learn that the CD Audio format uses interleaving to improve the system's response to scratched discs. In effect, different parts of the same sound sample are stored on different locations of the rotating disk. So, if one part of the signal is lost, powerful error correction encoding schemes can help recover the lost part, with no loss in sound quality to your ears.

What are the drawbacks to interleaving? Perhaps the largest drawback is the complexity of the interleaving and de-interleaving process. We need to store parts of the signal in order to create and undo interleaving, which takes memory. Also, if we interleave parts of the same signal, then we can only reconstruct the entire signal after a delay, because we need to process the signal in blocks. Sometimes, we cannot handle a large delay in the signal processing path. One such situation is in telecommunications, where a delay is noticeable to both telephone callers (although some delay is tolerable, more than 20 milliseconds of delay makes conversation difficult). In such applications, only a small amount of interleaving is allowed.

10.2.4 The Universal Product Code (UPC) System

All modern-day consumer products have a Universal Product Code (UPC) symbol located somewhere on their package. The UPC symbol is part of a carefully-designed standard for product labeling that was developed in 1973 by a collection of international regulatory agencies. The UPC System has revolutionized modern shopping; can you think of a time that you purchased a commercial (non-produce) product when this symbol was not scanned?

A Typical UPC Symbol

A typical UPC symbol is shown in Figure 10.3. The symbol has a number of parts to it:

1) There is a set of vertical black-and-white bars that make up most of the symbol.

2) There are a series of numbers on the bottom of the symbol, and two numbers on the left-hand-side and right-hand-side of the symbol, respectively.

18

Figure 10.3 A UPC Symbol

It turns out that the black-and-white bars encode the ten numbers on the symbol, as well as a *check number* that is sometimes listed on the right of the symbol. What do these numbers mean?

The first number on the left indicates what kind of product the item is. A zero indicates that this symbol belongs to a grocery item.

The first five numbers at the bottom of the symbol are unique for each manufacturer. In this case, the sequence 14100 belongs to Pepperidge Farm, Inc., of Norwalk, CT. Thus, there can only be 100,000 different manufacturers (because the UPC symbol only allows the numbers 00001 to 99999).

The last five numbers at the bottom denote the actual product. In this case, the sequence 07453 refers to a 6-ounce package of Pizza-flavored Goldfish snack crackers. Thus, there can only be 100,000 different products from each manufacturer (because the UPC symbol only allows the numbers 00001 to 99999). These choices of numbers are limitations of the system and will have to be revised when the number of unique manufacturers of products in any given product category exceeds 100,000.

What do the Bars Mean?

As stated earlier, the bars encode the numbers listed on the symbol plus a check number that helps catch errors when the symbol is read. The design of the bar sequence is important, because it is what allows efficient reading by a laser scanner. The laser scanner system in use in almost every modern-day grocery store employs a helium-neon (red-colored) laser reading system to detect and read this sequence of black-and-white bars. A computer inventory system is then employed to cross-reference this list of numbers with the product list in current inventory and remove one item from the inventory. In this way, automatic calculations of current inventories are performed, and from this information, accurate predictions of potential inventory shortages can be made.

We now dissect the individual meanings of the bars. There are some long start-and-stop bars on either side of the symbol. The start and stop bars are each 3 bars wide and are of the form "black-white-black". On the left, the next set of long bars encodes the first digit of the symbol. The shorter bars above the two sets of 5 numbers encode these 10 digits, and the last set of long bars before the "black-white-black" end bars encode the parity check number.

Each number in the symbol is encoded using seven segments of equal width. Figure 10.4 shows the 10 different sequences of bars corresponding to the 10 digits from 0 to 9 for the first 5 numbers of the UPC symbol. Below each bar representation, we have written a binary representation of the number, where black corresponds to a 1 and white corresponds to a 0.

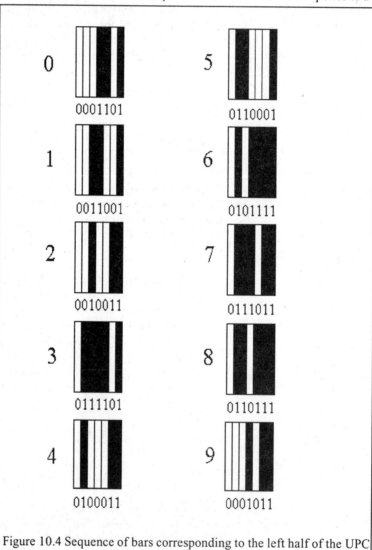

Recall: To store M different objects, we need $\log_2(M)$ bits.

0	0001101	5	0110001
1	0011001	6	0101111
2	0010011	7	0111011
3	0111101	8	0110111
4	0100011	9	0001011

Figure 10.4 Sequence of bars corresponding to the left half of the UPC symbol

As for the last 5 numbers at the bottom of the UPC symbol, they are encoded using the photographic negative of this table (that is, 1100110 corresponds to a 1, 1101100 corresponds to a 2, and so on). The reason for using different encodings for the first and last numbers is to give the symbol an *orientation* that helps the laser scanner figure out which side is the left or front of the symbol. Thus, the scanning system can work with the symbol in just about any orientation.

How is the parity check number calculated? The following formula is used for this number N, where n_0 to n_{10} denote the first 11 numbers of the symbol:

$$N = 210 - 3(n_0 + n_2 + n_4 + n_6 + n_8 + n_{10}) - (n_1 + n_3 + n_5 + n_7 + n_9)$$

From the value of N, the ones digit is stored as the check number n_{11}. The use of alternating groups of numbers in the formula above helps to avoid errors due to misreads of pairs of numbers (e.g. so that an error in n_2 and an error in n_3 have a different effect on the check number n_{11}). In our UPC symbol for Pizza Goldfish, the check number is the ones digit of

$$N = 210 - 3(0 + 4 + 0 + 0 + 4 + 3) - (1 + 1 + 0 + 7 + 5) = 163 ,$$

or 3 in this case.

The UPC symbol uses 7 bits to store each number. As we saw in our study of compression, we only need 4 bits to store the numbers between 0 and 9. Why have three extra bits been used here? The answer, not surprisingly, is closely tied to the error correction ideas that we've already studied in this chapter. Here are the reasons behind the design choices of the UPC symbol:

1) Looking at Figure 10.4, one notices that, although there are 7 bar-widths to each binary representation, each number is represented by only four alternating bars (white-black-white-black). This choice helps the scanner figure out where numbers begin and end (because all numbers begin with white and end with black). The use of only four alternating bars also helps reduce errors due to misreads, as a three-black-bar or a one-black-bar number gets automatically rejected. An alternating white-black-white-black bar symbol is also easier to print on packages.

2) Looking at all of the numbers, we find odd number of ones in each 7-bit representation. Thus, all sets of bars have Odd parity. Since the right-most 5 numbers of the UPC symbol have Even parity, the laser scanner can figure out the orientation of the UPC symbol.

From these choices, the number of different bar symbols for representing numbers is reduced from 128 (2 to the power of 7) to 10. The resulting symbol is remarkably robust to errors in reading, and reading can even be performed if there is a small amount of condensation (e.g. on a bottle of milk) or frost (e.g. on a carton of ice cream) over the symbol.

Practical Advantages of the UPC System

What are the practical advantages provided by the UPC System?

1) Clearly, the speed at the checkout counter is improved. Studies have shown that checkout proceeds at roughly double the rate with UPC scanning as opposed to manual punching of product prices.

2) Since the product prices are maintained at a central computer, individual pricing errors at each checkout stand are reduced. Short-term sales can also easily be managed, and taxable items are accurately identified.

3) Automated inventory tallies help the store manager schedule deliveries.

4) Record keeping can be automated and made more efficient.

The main disadvantage of the system is the cost and maintenance of the scanning devices, which can cost thousands of dollars per checkout stand. There are also social factors to be considered—the impact on job creation in a particular area of the city, for example, as well as the ways scanning changes the store clerk/shopper interaction. At one time, electronic scanning devices that told you the price of each and every item were commonplace—they also told you to "have a nice day" in a dry electronic monotone. Shoppers obviously didn't like being given the "automated computer treatment," although social values could always change in the future.

Exercise 10.2

1. In the ASCII code, each alphanumeric character is represented by an 8-bit string. Using the ASCII code table in Chapter 9, label the codeword for each capital letter from A to Z as an even or an odd string. Do you see a pattern associated with your labeling?

2. Suppose you have an N-bit string that you want to encode for error correction using the method in Section 10.2.2. Furthermore, suppose that the value of N is a squared number (such as 4, 9, 16, and so on).

    a. Derive an equation that describes the length of the encoded string as a function of N.

    b. Derive an equation that calculates the coding overhead from the value of N.

    c. Derive an equation that calculates the coding strength from the value of N.

    d. As N gets large, what happens to the coding overhead? What happens to the coding strength? Do you think that large values of N are useful?

3. There are many places where scanning bar codes such as the UPC symbol are used nowadays. Try to find and if possible collect as many examples of such bar codes as you can. What advantages do bar codes provide in these applications? What are some of the disadvantages?

Section 10.3: Glossary

coding Overhead
The amount of redundancy added to a signal to make it possible to detect and correct for errors, as measured by a percentage of the original signal length in bits or symbols

coding Strength
The ability provided by the redundancy to correct for errors in the signal, as measured by a percentage of the original signal length in bits or symbols

error Burst
Errors that occur one after another in a message or string

error Correction
The procedure by which the original message is recovered from its corrupted and encoded version

hamming code
An error correction code that uses parity bits in a certain way to detect and correct for errors

interleaving
A clever encoding scheme that effectively distributes bursts of errors across the message

parity
A noun that means "equivalence" or "of equivalent classes or types"

parity Bits
The redundant bits that are added to a bit string to protect it against errors

redundant
An adjective that refers to the added part of a coded string for error detection and/or correction

UPC
An acronym for "universal product code," a bar code that appears on most every modern saleable item

The INFINITY Project

Chapter 11: Keeping Data Private

How do we encode information so that it cannot be accessed by others?

Approximate Length: 1.5 Weeks

Section 11.1: Introduction to Coding for Secrecy

Approximate Length: 1 Lecture

11.1.1 "Your Secret is Safe With Me"

Privacy is an important part of many aspects of our modern world. Bank accounts, medical records, and the contents of telephone calls are all types of digital information that many of us want to be secure and private. How is this information actually protected from the prying eyes of others? In this chapter, we shall give a brief description of the basic concepts behind information security. From our discussion, we'll see how information security is both an important and challenging task.

The Internet and Information Security

The rapid rise of the Internet has make information security one of the most important challenges facing modern society. As the Internet has the ability to affect most every type of human social and commercial interaction, maintaining secure communications across the Internet is a critical issue. Software companies who design the programs and infrastructure to access the Internet are aware of the desires of many people. They are writing programs that have the ability to protect a web surfer's data. Many people, however, are not aware of these features or of Internet security issues in general. The recent rise of virus attacks by malicious computer hackers shows us how difficult information security can be in a wired world.

11.1.2 Codes, Keys, and Cryptography

We've already seen the power that coding of information provides. We can code digital information so that it takes less space to store it. We can also code information to protect it from errors. A third use of coding is to make information difficult to read or be understood by others. This form of coding is known as *cryptography.* The act of protecting information is known as *encryption. Decryption* describes the decoding process.

Cryptography has strong connections to physical security. When you lock a car door, a locker, or your front door, you are trying to prevent others from accessing the inside of the car, the locker, or the house or apartment where you live. Similarly, when you encrypt data or information, you are trying to prevent others from accessing the data or information. Unfortunately, the word "prevent" is critical in both contexts. It is almost always the case that a locked door or locker can be broken into by a malicious person with enough time, money, or other resources. Similarly, an encrypted signal can, with enough time, money, or computing resources, be decrypted and read by a law-breaking computer hacker. For these reasons, you should always view such data security methods as a form of prevention, not a form of protection.

Cryptography also shares another element with physical security: the concept of a *key.* A physical metallic key is a device that helps you gain access to something you own. Similarly, a data key is a sequence of numbers that helps you gain access to information that you encrypted. A data key generally doesn't store any information in and of itself, just like a physical metallic key doesn't hold your physical possessions inside of it.

The main difference between a physical key and a data key is that a data key can be easily copied. This difference is what makes data security so hard to maintain—imagine if anyone could copy an apartment or car key just by looking at it and remembering it!

Passwords, Access Codes, and PINs

A *password* is the name given to an alphanumeric key used to access certain computer and network systems. Most online Internet account services are accessed by a password that the user chooses when she or he sets up the account.

Access codes, also called *personal identification numbers* (PINs), are numerical sequences that many financial institutions and telephone companies rely on to secure the account access of users of their services. These numerical sequences are digital keys that are usually four to 20-numbers long. If you have a bank account, you likely have chosen a PIN number to access your savings at an automated teller machine (ATM).

The same security issues that surround data security also apply to passwords, access codes, and PINs. Since these numbers can be easily communicated, they only prevent others from accessing your computer account or monetary savings. That is why you should never choose a simple or easily-identifiable password or PIN for your account access, because a good guesser could cause trouble for you. In fact, you should get into the habit of changing your password or PIN every so often to prevent others from getting access to your digital "life."

Exercise 11.1

1. Write down all of the ways you could use cryptography to secure information in your life.

2. When picking access codes for their ATM bank cards, many people use a well-known personal number, such as their birthday. How secure is this choice?

Section 11.2: Simple Encryption Methods

Approximate Length: 2 lectures

11.2.1 Rotation Encoding

Perhaps the simplest text -encrypting method is *rotation encoding*. Rotation encoding is used in certain puzzles in the entertainment section of a daily newspaper, so you might have seen this type of encoding method.

To introduce the concept, consider the following sentence, written in all capital letters to avoid capitalization issues:

SALLY USUALLY WEARS SNEAKERS WHEN SHE IS RUNNING OUTSIDE.

A rotation encoder chooses a substitution letter for every letter appearing in the sentence. Thus, every letter "A" will be replaced by a letter different from "A", every letter "B" will be replaced by a letter different from "B", and so on. We have to be careful, however. We cannot assign the same replacement letters to different original letters, or else we won't be able to get back the original words (if this were the case, how would we decide which letter should be substituted?). Also, we need a simple way to describe the substitution of letters so that we can define a useful key to help us decode the message.

To make the assignment unique and easy-to-describe, we'll use the following rule: whatever letter we choose to replace A, we'll use the next letter to replace B, and the next letter after that to replace C, and so on. If we get to the end of the alphabet, we'll just "wrap around" the assignment starting with "A" again. For example, if we assign an "F" for every "A",
    ll assign "G" for every "B", "H" for every "C", and so on up to assigning "Z" for every "U". Then, we'll assign "A" for every "V", "B" for every "W", and so on. Figure 11.1 shows this type of assignment, where we have used a pair of concentric letter-wheels to show the letter assignments. The our ring of letters correspond to the letters in the original sentence, and the inner ring of letters correspond to the encrypted sentence.

The difference between these two letter-wheels is a rotation, which is how the encoder gets its name.

With this substitution in place, our sentence becomes

XFQQD ZXZFQQD BJFWX XSJFPJWX BMJS XMJ NX WZSSNSL TZYXNIJ.

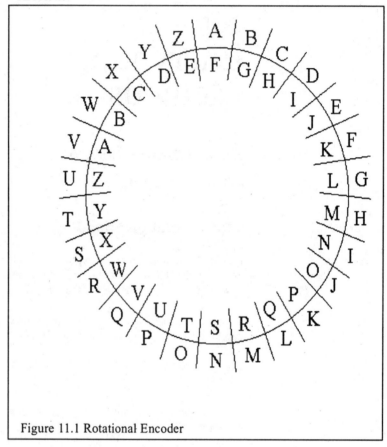

Figure 11.1 Rotational Encoder

At first glance, this sequence of letters doesn't make any sense. So, we've more or less achieved our objective of preventing casual readers from understanding the sentence.

What is the key for this encoder? We can use the number of counter-clockwise shifts of letters between the outer and inner letter-wheels of Figure 11.1 to define the key. In this case, since the wheels are offset by 5 positions counter-clockwise, the key for this encoder is $K=5$. Knowing the key and the encoding method immediately allows decoding of the message as we'll see next.

To decode the sequence, we simply rotate the inner wheel by K positions clockwise from the so-called *identity* coding, in which the inner and outer letter-wheels match. This decoder setting is shown in Figure 11.2. Now, we use the letters in the encrypted sentence with the outer-most letter-wheel to find the corresponding decrypted letters on the inner-most letter-wheel. We see that "X" maps to "S", "F" maps to "A", and so on.

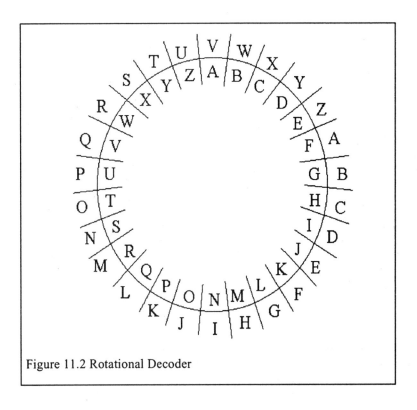

Figure 11.2 Rotational Decoder

The rotation encoder illustrates the important concepts of an encryption method:

1. Symbols in a message are mapped back to the same symbols set in the message; only the assignments of the symbols change.

2. There exists a well-defined procedure to both encrypt the original message and decrypt the encrypted message. The encryption key is used within these procedures.

3. There exists a correct key that decrypts the encrypted message properly. An erroneous key produces a useless result.

Breaking a Rotation-Encoded Sentence

The rotation encoder is a simple encryption method—so simple, in fact, that it doesn't provide a lot of security. Consider the following facts:

1. The encryption algorithm only allows one of 25 different keys. It is a simple task to try all possible rotation decodings to see which one produces a reasonable sentence at the output.

2. The English language has a structure that can be used to figure out the assignments of letters. For example, it is well-known that certain letters appear more often than others. So, one way to try to break the code is to guess rotation decodings that assign frequently-appearing letters to the encrypted letters that appear most often in the sentence. For example, in the encrypted phrase, we find that the letter "X"

appears 8 times, and the letter "J" appears 6 times. We could try substituting "E" for "X" (which assumes a key of $K=19$) or substituting "E" for "J" (which assumes a key of $K=-5$), as "E" is the most-common letter in the English language. This type of procedure reduces the number of trial rotation decodings considerably.

It is a useful exercise to see how easily an encryption method can be decoded. Code-breaking is not necessarily an evil exercise. On the contrary, it helps us determine the level of security that a particular encryption method provides and can even suggest new coding methods that are harder to break. Such efforts are only problematic and potentially illegal when other peoples' private data are involved.

11.2.2 Permutation or Substitution Encoding

The rotation encoder is a special case of a *permutation* or *substitution encoder,* a slightly more-secure encoding method. To illustrate the substitution encoder, we first describe the rotation encoder in another form.

All substitution codes map a set of symbols back to the same set of symbols. Such a mapping can be given in the form of a table. Figure 11.3 shows the table that describes a rotation encoder with a key of $K=5$. On the left are the original symbols from A to Z, and on the bottom are the encoded symbols from A to Z. To express the mapping, we put a single one (1) along every row or column. Filling in the table in this way defines the code.

```
A  0 0 0 0 1 0 0 0 0 0 0 0 0 0 0 0 0 0 0 0 0 0 0 0 0 0
B  0 0 0 0 0 1 0 0 0 0 0 0 0 0 0 0 0 0 0 0 0 0 0 0 0 0
C  0 0 0 0 0 0 1 0 0 0 0 0 0 0 0 0 0 0 0 0 0 0 0 0 0 0
D  0 0 0 0 0 0 0 1 0 0 0 0 0 0 0 0 0 0 0 0 0 0 0 0 0 0
E  0 0 0 0 0 0 0 0 1 0 0 0 0 0 0 0 0 0 0 0 0 0 0 0 0 0
F  0 0 0 0 0 0 0 0 0 1 0 0 0 0 0 0 0 0 0 0 0 0 0 0 0 0
G  0 0 0 0 0 0 0 0 0 0 1 0 0 0 0 0 0 0 0 0 0 0 0 0 0 0
H  0 0 0 0 0 0 0 0 0 0 0 1 0 0 0 0 0 0 0 0 0 0 0 0 0 0
I  0 0 0 0 0 0 0 0 0 0 0 0 1 0 0 0 0 0 0 0 0 0 0 0 0 0
J  0 0 0 0 0 0 0 0 0 0 0 0 0 1 0 0 0 0 0 0 0 0 0 0 0 0
K  0 0 0 0 0 0 0 0 0 0 0 0 0 0 1 0 0 0 0 0 0 0 0 0 0 0
L  0 0 0 0 0 0 0 0 0 0 0 0 0 0 0 1 0 0 0 0 0 0 0 0 0 0
M  0 0 0 0 0 0 0 0 0 0 0 0 0 0 0 0 1 0 0 0 0 0 0 0 0 0
N  0 0 0 0 0 0 0 0 0 0 0 0 0 0 0 0 0 1 0 0 0 0 0 0 0 0
O  0 0 0 0 0 0 0 0 0 0 0 0 0 0 0 0 0 0 1 0 0 0 0 0 0 0
P  0 0 0 0 0 0 0 0 0 0 0 0 0 0 0 0 0 0 0 1 0 0 0 0 0 0
Q  0 0 0 0 0 0 0 0 0 0 0 0 0 0 0 0 0 0 0 0 1 0 0 0 0 0
R  0 0 0 0 0 0 0 0 0 0 0 0 0 0 0 0 0 0 0 0 0 1 0 0 0 0
S  0 0 0 0 0 0 0 0 0 0 0 0 0 0 0 0 0 0 0 0 0 0 1 0 0 0
T  0 0 0 0 0 0 0 0 0 0 0 0 0 0 0 0 0 0 0 0 0 0 0 1 0 0
U  0 0 0 0 0 0 0 0 0 0 0 0 0 0 0 0 0 0 0 0 0 0 0 0 0 1
V  1 0 0 0 0 0 0 0 0 0 0 0 0 0 0 0 0 0 0 0 0 0 0 0 0 0
W  0 1 0 0 0 0 0 0 0 0 0 0 0 0 0 0 0 0 0 0 0 0 0 0 0 0
X  0 0 1 0 0 0 0 0 0 0 0 0 0 0 0 0 0 0 0 0 0 0 0 0 0 0
Y  0 0 0 1 0 0 0 0 0 0 0 0 0 0 0 0 0 0 0 0 0 0 0 0 0 0
Z  0 0 0 0 1 0 0 0 0 0 0 0 0 0 0 0 0 0 0 0 0 0 0 0 0 0

   A B C D E F G H I J K L M N O P Q R S T U V W X Y Z
```

Figure 11.3 Rotation Encoder with K=5

To see what each original letter corresponds to in the encrypted message, we find the letter to be encrypted on the left, and we follow along the row of that letter to the right until we come to a 1. Then, we move down that column to get to the encrypted letter.

The rotation encoder has a particular structure in this table format. As you can see, the pattern of ones in the 26x26 element table is quite regular and moves diagonally from left to right until the end of the table is reached. At that point, the diagonal pattern of ones continues on the left again. The shift to the right from a totally-diagonal set of ones that map each letter to the same letter defines the key.

We can come up with a more-general encryption method. The more-general method would make arbitrary assignments from original letter to encrypted letter, with the provision that *each original letter gets mapped to only one encrypted letter*. We can guarantee such a rule pretty easily by making sure that only one 1 appears along any row and column of the table. A table with only one 1 along any row and column is called a *permutation table*, because all it does is shuffle the positions of the letters. For this reason, this generic encoder is called a *permutation encoder*. Figure 11.4 shows an example of a permutation encoder table.

9

```
A  0 0 0 0 0 1 0 0 0 0 0 0 0 0 0 0 0 0 0 0 0 0 0 0 0 0
B  0 0 0 0 0 0 0 0 1 0 0 0 0 0 0 0 0 0 0 0 0 0 0 0 0 0
C  0 0 0 0 0 0 0 0 0 0 0 0 0 0 0 0 1 0 0 0 0 0 0 0 0 0
D  1 0 0 0 0 0 0 0 0 0 0 0 0 0 0 0 0 0 0 0 0 0 0 0 0 0
E  0 0 0 0 0 0 0 0 0 0 0 0 0 0 0 0 0 0 0 0 0 1 0 0 0 0
F  0 0 0 0 0 0 0 0 0 1 0 0 0 0 0 0 0 0 0 0 0 0 0 0 0 0
G  0 0 0 0 0 0 0 0 0 0 0 0 0 0 0 0 0 0 0 0 0 0 0 1 0 0
H  0 0 0 0 0 1 0 0 0 0 0 0 0 0 0 0 0 0 0 0 0 0 0 0 0 0
I  0 0 0 0 0 0 0 0 0 0 0 0 0 0 0 0 0 0 0 0 1 0 0 0 0 0
J  0 0 0 0 0 0 0 1 0 0 0 0 0 0 0 0 0 0 0 0 0 0 0 0 0 0
K  0 0 0 1 0 0 0 0 0 0 0 0 0 0 0 0 0 0 0 0 0 0 0 0 0 0
L  0 0 0 0 0 0 0 0 0 0 0 0 0 0 0 0 0 0 0 0 0 0 0 0 0 1
M  0 0 0 0 0 0 0 0 0 0 0 0 1 0 0 0 0 0 0 0 0 0 0 0 0 0
N  0 0 1 0 0 0 0 0 0 0 0 0 0 0 0 0 0 0 0 0 0 0 0 0 0 0
O  0 0 0 0 0 0 0 0 0 0 0 0 0 0 0 0 0 1 0 0 0 0 0 0 0 0
P  0 0 0 0 0 0 0 0 0 0 0 0 0 0 1 0 0 0 0 0 0 0 0 0 0 0
Q  0 1 0 0 0 0 0 0 0 0 0 0 0 0 0 0 0 0 0 0 0 0 0 0 0 0
R  0 0 0 0 0 0 0 0 0 0 0 0 0 0 0 0 0 0 1 0 0 0 0 0 0 0
S  0 0 0 0 0 0 0 0 0 0 1 0 0 0 0 0 0 0 0 0 0 0 0 0 0 0
T  0 0 0 0 0 0 0 0 0 0 0 0 0 0 0 0 0 0 0 1 0 0 0 0 0 0
U  0 0 0 0 0 0 0 0 0 0 0 0 0 0 0 1 0 0 0 0 0 0 0 0 0 0
V  0 0 0 0 0 0 0 0 0 0 0 0 0 0 0 0 0 0 0 0 0 1 0 0 0 0
W  0 0 0 1 0 0 0 0 0 0 0 0 0 0 0 0 0 0 0 0 0 0 0 0 0 0
X  0 0 0 0 0 0 0 0 1 0 0 0 0 0 0 0 0 0 0 0 0 0 0 0 0 0
Y  0 0 0 0 0 0 0 0 0 0 0 0 0 0 0 0 0 1 0 0 0 0 0 0 0 0
Z  0 0 0 0 0 0 0 0 0 0 0 0 0 0 1 0 0 0 0 0 0 0 0 0 0 0

   A B C D E F G H I J K L M N O P Q R S T U V W X Y Z
```

Figure 11.4 Permutation Encoder Table

In this table we've assigned five letter-letter pairs such that the phrase "Howdy" becomes "Great" in the encrypted message. In fact, we have a lot of freedom to choose the permutation encoder table.

Using this table, we can encode our original sentence as

LFZZT PLPFZZT IXFSL LCXFDXSL IGXC LGX WL SPCCWNY RPULWAX

An immediate question to be asked is: What is the key for a permutation encoder? Since the encoding procedure employs a shuffling, we need to save the shuffling information within the key. Thus, the key must contain the positions of all ones inside of the table (this is the smallest amount of information that we need to save in order to easily decode the encrypted message). This can be done using standard matrix notation, by simply listing the locations of all the ones. For the above example, the key is given by the sequence

K = [(1,6), (2,10), (3,17), (4,1), (5,24), (6,11), (7,25), (8,7), (9,23),(10,8), (11,4), (12,26), (13,13), (14,3), (15,18), (16,14), (17,2), (18,19), (19,12), (20,21), (21,16), (22,22), (23,5),(24,9), (25,20), (26,15)].

This key is much longer than the key for the permutation encoder; this is the price paid for a more-secure encryption format. In fact, the number of different keys is quite large. We can calculate the number of different keys as follows: In the first row we have 26 possible locations. In the second row, we have 25 possible locations (recall that we can't place a 1 in the same column or else we couldn't uniquely decode all encrypted messages). Continuing through all the rows, we arrive at

$N = 26(25)(24)(23)(22)...(2)(1) = 400000000000000000000000000$ (approx.)

Even so, this encoding method is not that secure. Remember, breaking this type of encryption scheme is an exercise that is played out almost every day on the entertainment pages of newspapers across the country. Like rotation codes, permutation codes can be broken using the statistical properties of the English language and some good word-guessing. So, we have to develop more powerful methods to encrypt our data.

Exercise 11.2

1. Consider a slightly-modified rotation encoder that reverses the order of the letters in the inner wheel of Figure 11.1. That is, the outer ring of letters goes around clockwise, but the inner ring of letters goes around counter-clockwise. Draw this rotation encoder for shifts of 5, -5, and 12 characters. Then, encode a simple message using this rotation encoder, and compare the encoded version with that of the original rotation encoder that doesn't reverse the letter order. Which one of the two encoding methods provides more protection of the message?

2. It is possible to develop permutation encoders that are easier to describe as compared to the completely-random encoding table in Figure 11.4. Consider the key given by

    K = [(1,4), (2,8), (3,12), (4,16), (5,20), (6,24), (7,2), (8,6), (9,10),(10,14), (11,18), (12,22), (13,26), (14,3), (15,7), (16,11), (17,15), (18,19), (19,23), (20,1), (21,5), (22,9), (23,13),(24,17), (25,21), (26,25)].

    Draw the encoding table for this permutation encoder. Do you notice any structure to the encoding method?

3. Make a rotational encoder using cut-out paper and a pencil (punch the pencil through the center of the paper to hold the two discs together) or some other center fastener. Then, encode a message. Hand it to a friend (foe?) and see if s/he can decode the message.

4. Make a permutation-modified rotational encoder by first designing a permutation table such as in Figure 11.4 and assigning the letters to the edge of the decoder portion. Then, rotate the wheels and write out the permutation table. How has the table changed? Would this code be easier or harder to break than the simple rotational encoder?

Section 11.3: Public Key Cryptography

*Approximate Length: 3
Lectures*

11.3.1 Sharing Keys With Everyone

The concept of a key usually implies privacy—only a few people have a copy. Most people would think that you were crazy if you started copying an important key and giving it to everyone. In the world of digital data, however, one has to assume that a key made up of bits can be easily copied over and over. Because of this fact, clever methods have to be developed to make data secure even when keys are available. Amazingly, methods to secure data in exactly this type of situation have been developed; they are called *public key encryption* methods. Such methods are often used to secure financial transactions on the Internet, for example.

In this section, we describe how public key cryptography works for binary numbers or data.   Before developing the method, we describe a less-secure way to communicate information using keys.

11.3.2 The Exclusive-OR Operation and Cryptography

The exclusive-OR operation is a way to combine two sequences of bits into one. Suppose you have an important bit sequence that you want to keep private. As an example, we'll take the sequence

$$S=1011000101001000001010010$$

To keep this sequence private, you generate another random bit sequence that is the same length as the original sequence for the key K:

$$K=0010011010011101000110101$$

The exclusive-OR operation combines these two sequences bit-by-bit by comparing the values one-by-one. If each bit in the pair is the same, a zero (0) is generated. If the bits are different, a one (1) is produced. The table below shows the exclusive-OR operation.

First Bit	Second Bit	Exclusive-OR
0	0	0
0	1	1
1	0	1
1	1	0

The exclusive-OR operation is essentially the parity bit for each pair of bits—the Exclusive-OR bit indicates whether the number of ones is Even or Odd.

Using this operation, we have

```
S=10110001010010000101010010
K=0010011010011101100110101
E=10010111110101010101100111
```

where E is the encrypted bit sequence. Notice how E looks nothing like S and K, implying that the signal is encrypted. We can therefore make E known to everyone, but we keep the key K private.

To decrypt the encoded signal E, we need to use the key K again. It turns out that the Exclusive-OR operation can be used to get back the original message from the encrypted one by combining E and K. Employing the exclusive-OR, we have

```
E=10010111110101010101100111
K=0010011010011101100110101
S=10110001010010000101010010
```

So, the original secret message S is found.

This form of encryption works, but it has some disadvantages:

3. The key K must be exactly the same length as the signal S to be encrypted. This requirement puts demands on storage and transmission capabilities, not to mention the fact that we have to generate long sequences of random-looking bits for the key.

4. Security is lost if the key K becomes available.

In the next two sub-sections, we see how both of these difficulties are overcome by clever engineers.

11.3.3 Pseudo-Random Number Generators

The cryptographic method discussed above uses long sequences of random-looking bits for the key. Generating truly-random bit sequences is actually a hard task. If you were to just bang on the keyboard and try to generate bits, you might easily fall into a pattern after a while. Fortunately, we only need random-*looking* bits; we'll call these bits *pseudo-random* (because they only appear to be random to us). The method that we now describe

---

Modulo-N: An operation that takes the remainder of a number when divided through by N. For example the modulo-12 value of 77 is 5 (77 divided by 12 is 6 with a remainder of 5).

13

produces long sequences of pseudo-random numbers. Hence, it is called a *pseudo-random number generator.*

The pseudo-random number generator employs four different numbers A, B, N, and a starting value X(0) called the *seed* of the generator. The generator then produces N random numbers given these four numbers. So, it effectively generates an N-digit number using only four numbers. The numbers are generated using the following equation:

$$X(n+1) = [A\ X(n) + B] \bmod (N),$$

where mod (N) denotes the modulo-N operation. The modulo-N operation takes the remainder of the quantity in the brackets after dividing this quantity by N. Thus the pseudo-random values stored in X(n) will be integers between 0 and N-1.

The pseudo-random number generator above only produces integer numbers in the range from 0 to (N-1), because of the modulo-N operation. To generate binary sequences of numbers, it is useful to pick N as a power of 2, as then we can take the binary representation of the resulting values that are produced.

Let's see how the pseudo-random number generator generates numbers.

Example:

For our generator, let's choose the values of A=533, B=227, N=64, and X(0)=125. Then, the first four numbers from the pseudo-random number generator are.

$$X(1) = [533\ (125) + 227] \bmod (64) = [66852] \bmod (64) = 36$$

$$X(2) = [533\ (36) + 227] \bmod (64) = [19415] \bmod (64) = 23$$

$$X(3) = [533\ (23) + 227] \bmod (64) = [12486] \bmod (64) = 6$$

$$X(4) = [533\ (6) + 227] \bmod (64) = [3425] \bmod (64) = 33$$

If we wanted to produce bit values from this sequence, we can use a 6-bit representation, because (2x2x2x2x2x2 = 64). Therefore, we have

Decimal	Binary
36	100100
23	010111
6	000110
33	100001

Putting the bits next to each other, we generate the 24-bit key

K=11011001011100011010001

We could easily generate more numbers in the sequence using this method and thus make longer and larger keys K.

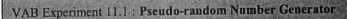

VAB Experiment 11.1 : **Pseudo-random Number Generator**

Use VAB to generate a sequence of Pseudo-random Numbers.

Why is the length of random bits L ?

$\log_2(N)$ is the number of bits we need to represent a pseudo-random number and since we have N different pseudo-random numbers , we have in total , the length of

bits = $N \log_2(N)$

An important feature of the pseudo-random number generator is the number of pseudo-random numbers it can produce. It turns out that, because of the modulo-N operation, only N random numbers are produced before the sequence *begins to repeat*. Therefore, the choice of N is critical. We need to choose N large enough so that we produce a long-enough key. The length of bits that are generated from a pseudo-random number generator is

$$L = N \log_2(N)$$

Therefore, in our previous example where N=64, we can generate a bit sequence that is L=64(6)=384 bits long before the sequence repeats.

The sequence of numbers produced by a pseudo-random number generator depend on the four values A, B, N, and X(0). In practice, the values of A, B, and N are usually made available to everyone, and only the seed value X(0) is kept secret. Thus, X(0) represents our key K. Here again, we can see the connection between N and the randomness of the sequence. As it turns out, the number of different "random" sequences that a pseudo-random number generator can produce is equal to N (because a seed value outside the range [0,N-1] is equivalent to some seed value in this range through the modulo-N operation). So, good data security is only achieved when large N values are chosen. Most commercial encryption methods today use 56-bit keys or seeds, corresponding to an N value of

$$N = 2^{56} \approx 10^{17}$$

10^{17} is big!
Remember there are less than 10^{10} humans on the earth. This means that there are 10^7 or 10 million keys for every human using 56 bit keys!

Searching through all possible 56-bit seeds is a lengthy process even for a very fast computer. So, if one doesn't know the seed, one has a difficult time decoding a bit sequence encoded using the pseudo-random number generated from the seed.

A Simple Cryptography System

A simple cryptography system would use the pseudo-random number generator in the following way:

1. To communicate a message, both the sender and receiver of the information need to know A, B, N, and the seed X(0).

2. The sender generates the pseudo-random sequence of numbers using the values of A B, N, and X(0), and converts this sequence into bits.

15

3. The sender then performs the Exclusive-OR operation on successive pairs of bits from the message to be sent and the pseudo-random number. This is the encrypted version of the message.

4. The sender transmits the encoded message to the receiver.

5. The receiver generates her or his own version of the pseudo-random number using the values of A B, N, and X(0), and converts this number into bits.

6. The receiver then performs the Exclusive-OR operation on successive pairs of bits from the encrypted message and the pseudo-random number, recovering the original message.

This entire process is shown in Figure 11.5.

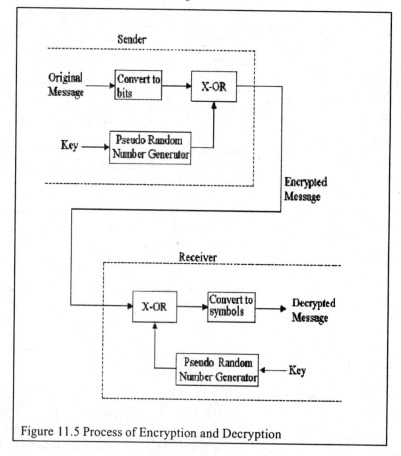

Figure 11.5 Process of Encryption and Decryption

11.3.4 Public Key Cryptography

The coding system described above still requires a secret seed corresponding to the seed X(0) of the number generator. If this seed is known to others, they can also generate the pseudo-random number and decode your encrypted message with the Exclusive-OR operation. So, another method must be used to generate a secret seed.

There is an extremely-useful way to generate a seed that involves both the sender and receiver of the information. The method generates a seed that is only easily decodable by both sender and receiver. Since the encoded message can only be decoded with use of the private seed, the security of the information is maintained.

Here's how the system works. Both receiver and sender choose an integer value C between 0 and N-1. Generally, this value is kept secret, although it doesn't have to be. Next, the sender chooses a large integer I that is known only to her or himself, and the receiver chooses a large integer J that is known only to her or himself. Neither I nor J are communicated publicly or transmitted. Then, the sender computes the number

$$P = [C^I] \bmod (N)$$

and the receiver computes the number

$$Q = [C^J] \bmod (N)$$

Both sender and receiver send the numbers P and Q to one another. Both of these numbers can also be publicly available. Now, since the sender received Q from the receiver, she or he computes the value

$$[Q^I] \bmod (N)$$

and the receiver, having received P, computes the value

$$[P^J] \bmod (N)$$

It can be shown that

$$[Q^I] \bmod (N) = [P^J] \bmod (N)$$

In other words, the two numbers just computed by the sender and receiver are identical. Thus, the sender can use this number as the seed X(0) to generate a pseudo-random number to encode the message with the Exclusive-OR operation. The recipient can then use this number to generate the same pseudo-random number and decode the message with the Exclusive-OR operation.

Some examples illustrate how this works:

Example: Suppose N=11, I=3, and J=4 are chosen. The value of C chosen is 9. Then,

$$P = [C^I] \bmod(N) = [9^3] \bmod(11)$$
$$P = [729] \bmod(11) = 3$$

and

$$Q = [C^J] \bmod(N) = [9^4] \bmod(11)$$
$$Q = [6561] \bmod(11) = 5$$

So, the numbers 3 and 5 are passed publicly. Then, processing these values, we get

$$[Q^I] \bmod(N) = [5^3] \bmod(11) = 4$$

$$[P^J] \bmod(N) = [3^4] \bmod(11) = 4$$

The key value of 4 is the same in both cases.

Example: Suppose N=23, I=5, and J=7 are chosen. The value of C chosen is 15. Then,

$$P = [15^5] \bmod(23) = [759375] \bmod(23) = 7$$
$$Q = [15^7] \bmod(23) = [170859375] \bmod(23) = 11$$

So, the numbers 7 and 11 are passed publicly. Processing these values,

$$[P^J] \bmod(N) = [7^7] \bmod(23) = 5$$

$$[Q^I] \bmod(N) = [11^5] \bmod(23) = 5$$

The key value of 5 is the same in both cases.

The important feature of this seed-generating message is that the values of I and J *are never communicated*. Therefore, it is very difficult for anyone else to generate the seed X(0), because it depends on both of the unknown values I and J. If N is large enough, the number of possibilities of any seed is so large that a supercomputer is needed to search through all of them

18

What is the main disadvantage of this public key cryptographic system? The two people must send the values of P or Q to each other before a private connection is made. This might be inconvenient in some situations, such as in broadcast communications where the desired information need only flow "one way". Where two-way communication of information is possible, however, public-key encryption is a very useful method. In fact, it is widely-used on the Internet to make online financial transactions, such as credit card purchases from a Web site, more secure.

VAB Experiment 11.2 : Secret Decoder Ring
Use VAB to implement a simple encoder, that is, one that maps one character to another.

Exercise 11.3

1. Generate the first 4N bits of the bit sequences that correspond to the following pseudo-random number generators: (a) A=13, B=7, X(0)=2, N=16, (b) A=25, B=14, X(0)=152, N=128, (c) A=151, B=39, X(0)=305, N=256.

2. Generate a bit sequence using a pseudo-random number generator above, and use this bit sequence to encode a secret message to your friend, using ASCII for the letter encodings and the system in Figure 11.5. Then, have your friend decode your message.

Section 11.4:
Glossary

cryptography
A form of coding whose goal is to make information difficult to read or be understood by others.

decryption
In cryptography, a generic name for the decoding process

encryption
In cryptography, a generic name for the encoding process

permutation or substitution encoding
A particular type of encryption method for encoding letters in English text; not very secure

pseudo-random number generator
A mathematical device for generating long strings of random-looking bits

rotation encoding
A particular type of encryption method for encoding letters in English text; not very secure

seed
The starting value in a pseudo-random number generator

The INFINITY Project

Chapter 12: Digital Processing of Signals

What is frequency response? What is a filter? How does a computer filter a signal?

Approximate Length: 2 Weeks

Section 12.1:
Introduction

We saw in Chapter 2 that every signal can be thought of as a sum of sinusoids with various frequencies, amplitudes, and phases. This chapter explores the modification of frequency content, or **filtering**. Although filtering can be applied to images, we will focus on the filtering of 1-D waveforms, which is most common. Why might we wish to modify the frequency content of a signal? Certainly there are reasons to do so in the processing of musical signals, to make new sounds, for example. There are much broader uses of filtering, however, that are seen in numerous applications. We shall discuss three of these uses in some detail: frequency selective filtering, spectral shaping, and echo cancellation. All three of these uses might arise in the design of a digital hearing aid, to assist persons with hearing losses. Throughout this chapter we will frequently return to the hearing aid as a relevant application for the filtering techniques that we shall develop. However, the applications of filtering are so widespread and so varied that it would be misleading to focus on just one example. Therefore, we will comment on numerous other applications of filtering as well. In the following subsections, we introduce the three main uses of filtering that we will consider.

12.1.1 Frequency Selective Filtering

There are many important applications where it is necessary to extract a desired signal from background noise or interference. If the frequency content of the desired signal and noise are different, then the desired signal can be extracted through frequency selective filtering. Call the desired signal $s_1(t)$ and refer to the contaminating noise or interference as $n(t)$. If $s(t)$ is the noisy signal that has been received or collected, then

$$s(t) = s_1(t) + n(t)$$

and our goal is to extract $s_1(t)$ from $s(t)$. As an example, suppose $s(t)$ is the signal from a television cable or radio antenna, $s_1(t)$ is the signal from the station that you wish to tune in, and $n(t)$ is the sum of signals from all other stations. The channel selection button on your television, or the tuning knob on your radio, control a frequency selective filter that captures the frequency band where your station of interest lies, while rejecting the signals from all other stations. Without such a filter, you would not be able to watch or listen to just one station at a time. Instead, the stations would interfere with one another, making the video on your television and the audio on your radio unintelligible.

Acoustic signals that travel through the air are also contaminated with interference that extends across a range of frequencies. The interference might come from wind or machinery noise, from multiple persons talking, or from a variety of other sources in the environment. Persons with normal hearing are adept at filtering out the interference and "listening" to only the signal (e.g., speaker) of interest. This is accomplished through some combination of physiological signal processing in the auditory system along with higher-level processing in the brain. Persons with imperfect hearing often struggle with interference. In cases where some of the frequencies of the interference $n(t)$ are different from the frequencies of the sound signal of interest $s_1(t)$, a hearing aid that implements frequency selective filtering can remove some of the unwanted noise and greatly improve intelligibility.

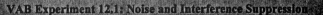

VAB Experiment 12.1: Noise and Interference Suppression

- Use simple filters to remove interference from speech and music signals.

There are some applications where it is desired to simultaneously extract multiple components of a signal, corresponding to several different frequency bands. This is the case in high-fidelity speaker systems. You likely know that most speaker cabinets contain speakers of an assortment of sizes, often within the same enclosure. (Just take the front grill off a large speaker enclosure to see the speakers of various sizes.) Small speakers most accurately reproduce high frequencies and large speakers are best at reproducing low frequencies. Thus, it is the job of the smallest speakers, called tweeters, to reproduce the highest frequencies whereas the largest speakers, called woofers, produce the low frequencies. The speaker enclosure may contain in-between sized speakers to produce mid-range frequencies.

Speaker systems require that the audio signal $s(t)$ be separated into two or more signals that feed (serve as input to) the various speakers. This form of filtering, sometimes called a **cross-over network**, is depicted in Figure 12.1. The term cross-over derives from the fact that adjacent output signals in Figure 12.1 come from adjacent frequency bands. Thus, in moving from one output signal to the next, you cross over from one frequency band to the next. Figure 12.2 depicts the frequency content of the output signals from the cross-over filter. This scenario is similar to tuning of television and radio stations, except that now we wish to capture all stations simultaneously.

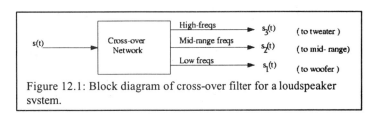

Figure 12.1: Block diagram of cross-over filter for a loudspeaker system.

3

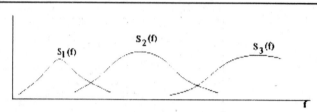

Figure 12.2: Frequency content of signals s $_1$(t), s$_2$(t), and s$_3$(t) from Figure 12.1.

There are many other applications where it is desired to extract signals from noise and interference. Telecommunication systems suffer from thermal noise generated in the receiver and from interference caused by other users. Hospital electronics are susceptible to 60 Hz interference generated from florescent lights. It is important that such signal contamination be filtered out. Frequency selective filtering is a prime tool for accomplishing this.

12.1.2 Spectral Shaping

Most audio systems have bass and treble controls. As you are aware, the bass control permits you to change the amount of low-frequency content in the audio signal. Likewise, the treble control lets you decide how much high-frequency content you wish to have. The loudness button on your stereo allows you to boost both bass and treble simultaneously by a fixed amount. Some audio systems have an equalizer that provides much finer control over the frequency content of the audio signal. Equalizers are especially common on sound boards used in theaters and at rock concerts. Equalizers typically have many slider knobs that allow you to precisely adjust the frequency content within many adjacent frequency bands covering the audio spectrum. The term **equalization** is used because an equalizer can compensate for non-ideal room acoustics by flattening the room frequency response. As suggested in Figure 12.3, the positions of the slider knobs on an equalizer can be thought of as a plot of the frequency response or frequency transfer function that is applied to the audio signal

4

before it is amplified and sent to the loudspeaker. Since the frequency content of a signal is referred to as its spectrum, this type of filtering is an example of **spectral shaping**.

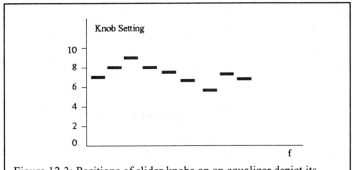

Figure 12.3: Positions of slider knobs on an equalizer depict its frequency response.

Spectral shaping arises in many contexts outside high-fidelity music. For example, a person who is hard of hearing typically has a frequency dependent hearing loss. A modern hearing aid will amplify each frequency by just the right amount to restore the hearing response to a nearly flat profile across the audio frequency band. This is illustrated in Figure 12.4, where the *product* of the two curves is nearly constant. This type of filtering is referred to as equalization, because the effect of the filter is to equalize or make nearly constant the overall frequency response.

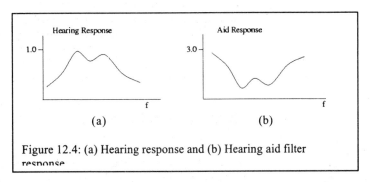

Figure 12.4: (a) Hearing response and (b) Hearing aid filter response.

Spectral shaping filtering is not restricted to audio. For example, communication systems must compensate for imperfect (i.e., nonflat) frequency responses of communication channels. That is, signals sent across communication channels have their frequency content changed, which in turn distorts the time-domain waveform. Filtering is necessary to restore the original signal. Interestingly, an important application of this idea is in the reading of data off the hard drive in your computer. The operation of reading data off a hard drive can be modeled as transmitting a signal through a communication channel. This channel has an imperfect (nonflat) frequency response. The introduction of sophisticated equalization filters has helped to dramatically increase the storage density of hard drives and to correspondingly reduce their physical size and price.

12.1.3 Echo Cancellation

In addition to frequency selectivity and spectral shaping, filters can be used to insert or remove echoes. Chapter 2 discussed how echoes or reverb can be intentionally added to musical signals as special effects or to help simulate the acoustic environment of a particular performance location (e.g., concert hall, stadium, etc.). Sometimes echoes are a large problem in the audio world. Modern-day teleconferencing systems use sophisticated filters to cancel echoes that arise from acoustic reflections within the conference room. Such filters are called **echo cancellers**. Persons with hearing losses often find echoes to be particularly bothersome. Echo cancellation for hearing aids is an active research area.

In communication systems, echo is an undesired effect that is introduced when part of a received signal is leaked into the transmitter and then sent back, with a delay, to the original transmitter. In such cases, the echoes are removed via filtering. Although echo cancellers modify the frequency content of their input signal, echo cancellers are most easily understood in terms of their time-domain effect.

Exercise 12.1

1. What is *filtering*?

2. Name three types of filtering that are considered in this chapter.

3. What is the meaning of *frequency selective filtering*?

4. Give five example scenarios of a desired signal and contaminating interference in the frequency selective filtering problem.

5. What is a *cross-over network*?

6. What is *spectral shaping*?

7. Give four example scenarios where spectral shaping would be useful.

8. What is *equalization*?

9. Give four example scenarios where an echo canceller would be useful.

Section 12.2:
Frequency Response
of a Filter

The frequency response of a filter describes how the filter modifies the frequency content of its input signal. Filtering can be used to separate one signal from another, or for spectral shaping as in the hearing aid equalization example pictured earlier in Figure 12.4(b), or for echo cancellation. In each of these cases, it is possible to plot the frequency response of the filter. Before proceeding further, it will be important to precisely define the notion of frequency response.

12.2.1 Definition of Frequency Response

Consider the block diagram of a filtering operation in Figure 12.5.

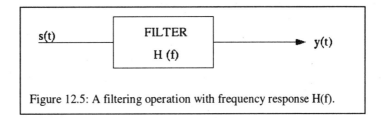

Figure 12.5: A filtering operation with frequency response H(f).

We will soon describe the inside of the filter and explain how the filter is implemented using a computer or DSP chip. For now, however, we shall focus on the relationship between the output signal y(t) and the input signal s(t), and the definition of H(f).

Most filters, including all filters discussed in this chapter, share a remarkable property. Namely, if the input signal is a sinusoid

$$s(t) = \cos(2\pi ft)$$

then the output signal from the filter will also be a sinusoid and will have the form

$$y(t) = H\cos(2\pi ft + \phi)$$

where both H and ϕ depend on the input frequency f. That is, for a sinusoidal input, the filter output is a sinusoid with:

7

1. Exactly the same frequency as that of the input

2. Amplitude given by H(f), which is independent of t but which depends on the frequency of the input, i.e., different frequencies are amplified or attenuated by different amounts.

3. Phase $\phi(f)$, which depends on the frequency of the input. Different frequencies are shifted in phase by different amounts.

The shapes of H(f) and $\phi(f)$, as a function of f, depend on the filter. Our only concern will be the function H(f) since it describes how the filter modifies (amplifies or attenuates) the frequency content of the input signal. We define H(f) to be the **frequency response** of the filter. In Figure 12.5, if S(f) is the spectrum (frequency content) of the input signal s(t), then it can be shown that the spectrum of the output signal y(t) will be

$$Y(f) = H(f)\ S(f).$$

That is, *the spectrum of the output signal is the product of the frequency response of the filter and the spectrum of the input signal*. This is a fundamental signal processing principle!

We usually study filtering in the frequency domain, i.e. we look at how the filter weighs different frequencies of the input signal by different amounts. It is important, however, to realize that this spectral weighting causes the output signal y(t) to be a modified version of the input signal s(t). The modification may be quite complex and difficult to describe if one compares the input and output signals in the time domain. However, by analyzing the filtering process in the frequency domain, the filtering operation is surprisingly simple.

We next gain some experience with the concept of frequency response by considering several examples.

Example 12.1

Recall that the easiest signal to analyze from the frequency point of view is a sinusoid. This signal is made of only one frequency. The spectrum of a sinusoid with frequency 1000 Hz and amplitude 2 is shown in Figure 12.6.

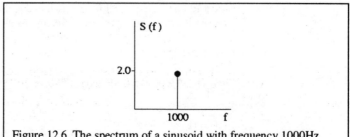

Figure 12.6. The spectrum of a sinusoid with frequency 1000Hz and amplitude 2.

8

Now, the frequency response of a filter is a function of frequency f and it tells us how the filter will scale an input sinusoid having that frequency. If we consider the hypothetical frequency response H(f) given below in Figure 12.7, we see that at 1000 Hz, the filter will scale a sinusoid by a factor of 3. Therefore, since Y(f) = H(f) S(f), we immediately know that the spectrum of the output signal y(t) will look just like the spectrum of s(t), but scaled by 3. The output signal will be a sinusoid with frequency 1000 and amplitude 6.

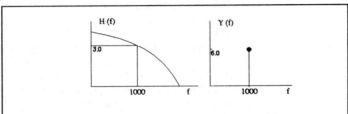

Figure 12.7: A filter frequency response and the resulting output spectrum.

This example illustrates a fundamental property of most filters. As we stated earlier, the frequency content of the output y(t) of a filter is the frequency content of the input signal s(t) multiplied by the frequency response of the filter.

So, all we need to characterize a filter is its frequency response H(f)! Once we have this function (or plot), it is easy to determine how input signals are modified. To establish this point, let's study some more examples.

Example 12.2

Let s(t) again be a sinusoid of frequency 1000 and amplitude 2, but suppose the filter has the notch-type frequency response shown in Figure 12.8. Then since H(f) = 0 when f = 1000, the output signal y(t) will be identically zero. If the input were an audio signal, we would not hear any output from this filter for a 1000 Hz sinusoidal input.

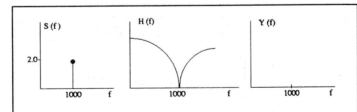

Figure 12.8: Spectrum of a sinusoidal s(t), a filter frequency response, and the resulting spectrum of y(t).

9

Example 12.3

Now, suppose s(t) is a periodic signal with fundamental frequency 200 and
the harmonics shown in Figure 12.9. Then for the given filter frequency
response, the spectrum of the output y(t) will be as shown. (Remember that
Y(f) = H(f) S(f).) Thus, y(t) will also be periodic with the same
fundamental frequency (why?), but the amplitudes of its harmonics will be
different from those in s(t) and, therefore, y(t) will look different from s(t).

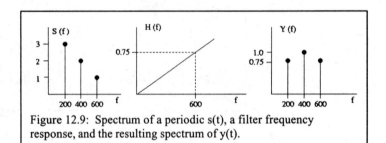

Figure 12.9: Spectrum of a periodic s(t), a filter frequency
response, and the resulting spectrum of y(t).

12.2.2 Ideal Frequency Selective Filters

In this section we consider several examples that help us define the concept
of ideal frequency selective filter. We consider low-pass, high-pass, band-
pass, and notch filters. As the name implies, low-pass filters pass
frequencies that are below some threshold, and zero out higher frequencies.
High-pass filters pass frequencies that are above a threshold and zero out
lower frequencies. Band-pass filters pass frequencies within some
intermediate band and zero out both lower and higher frequencies. Notch
filters zero out a single frequency within some intermediate band and pass
frequencies that are either higher or lower.

Example 12.4

Suppose that s(t) is not periodic and that it has the spectrum shown in
Figure 12.10. The given frequency response, H(f), is called an **ideal low-
pass filter**, since it passes all frequencies up to a given cut-off frequency,
500 Hz, with unit amplitude and completely attenuates all higher
frequencies. The resulting spectrum for y(t) is shown.

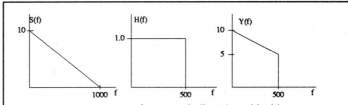

Figure 12.10: Spectrum of a nonperiodic s(t), an ideal low-pass filter frequency response, and the resulting spectrum for y(t).

Example 12.5

Let s(t) be the same as in the previous example, but now filtered by the **ideal high-pass filter** frequency response in Figure 12.11. Here, H(f) completely attenuates all frequencies below 300 Hz and passes all higher frequencies with unit amplitude. The resulting spectrum for y(t) is shown.

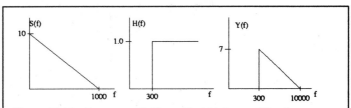

Figure 12.11: Spectrum of s(t), an ideal high-pass filter frequency response, and the resulting spectrum for y(t).

Example 12.6

Again, let s(t) be the same, but now filtered by the **ideal band-pass filter** frequency response shown in Figure 12.12. Here, H(f) completely attenuates all frequencies below 400 Hz and above 700 Hz. Frequencies within the passband (400, 700) are passed with unit amplitude. The resulting spectrum for y(t) is shown.

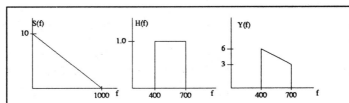

Figure 12.12: Spectrum of s(t), an ideal band-pass filter frequency response, and the resulting spectrum for y(t).

In the last three examples we have encountered three types of ideal, frequency-selective frequency responses. We summarize these, plus an ideal notch filter, in Figure 12.13 for convenient reference. f_c, f_{c1}, and f_{c2} are called **cutoff frequencies**. f_o is called a **notch frequency**. For the ideal notch filter, the notch should be infinitesimal (arbitrarily small), but we

11

have shown it as having finite width for purposes of illustration. We shall see later in this chapter that we can only approximate the ideal responses of Figure 12.13 in practice.

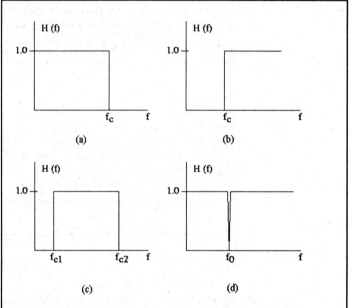

Figure 12.13: Ideal frequency-selective frequency responses: (a) Low-pass, (b) High-pass, (c) Band-pass, (d), Notch.

Exercise 12.2

1. What is the definition of *frequency response*?

2. Sketch the frequency responses of ideal low-pass, high-pass, band-pass, and notch filters.

3. What is a cutoff frequency? What is a notch frequecny?

4. Assuming an input signal $s(t) = \cos(2\pi(5000)t)$, specify the frequencies and amplitudes of the outputs of the two filters having the frequency

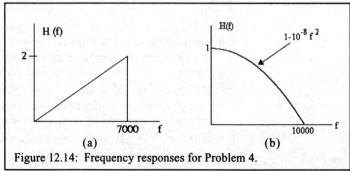

Figure 12.14: Frequency responses for Problem 4.

responses shown in Figure 12.14 Hint: The frequency of s(t) is 5,000 Hz.

5. Repeat Problem 4, except with s(t) = cos(2π(8000)t).

6. Assuming an input with spectrum given in Figure 12.15, plot the output spectrum Y(f) for the four filters with frequency responses in Figure 12.16.

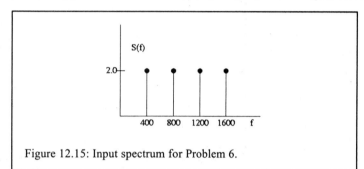

Figure 12.15: Input spectrum for Problem 6.

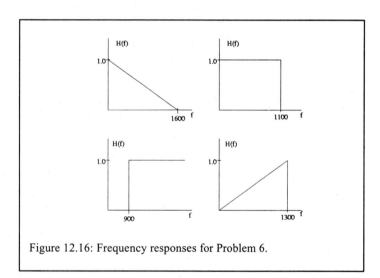

Figure 12.16: Frequency responses for Problem 6.

12.2.3 Cascaded Filters

It is not unusual to have two filters or systems connected in cascade, where the output of the first system forms the input to the second system. The top half of Figure 12.17 shows two systems in cascade, with frequency responses $H_1(f)$ and $H_2(f)$.

13

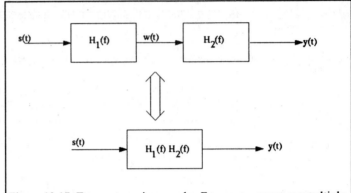

Figure 12.17: Two systems in cascade. Frequency responses multiply.

To assist in finding the frequency response of the overall system we have labeled the output of the first system (and input to the second system) as w(t). Now, we know that the spectrum of w(t) is given by

$W(f) = H_1(f) \, S(f).$

Likewise, for the second system we have

$Y(f) = H_2(f) \, W(f).$

Putting these two equations together yields

$Y(f) = H_1(f) \, H_2(f) \, S(f)$

which has the form

$Y(f) = H(f) \, S(f)$

with

$H(f) = H_1(f) \, H_2(f).$

That is, the cascade combination is equivalent to a single system having frequency response $H(f) = H_1(f) \, H_2(f)$. This equivalence is illustrated in the bottom of Figure 12.17. Remember that *when systems are in cascade, frequency responses multiply.* If there are more than two systems in cascade, then the frequency responses still multiply. Furthermore, interchanging the order of the systems in the cascade has no effect on the overall frequency response.

Example 12.7

Suppose two cascaded systems have the frequency responses $H_1(f)$ and $H_2(f)$ given in Figure 12.18. Then the overall system has the frequency response $H(f) = H_1(f) \, H_2(f)$ as shown. $H_2(f)$ is an ideal low-pass filter and it cuts off the frequency response at 400 Hz. $H_1(f)$ creates the shape of the frequency response within the band of frequencies that are passed.

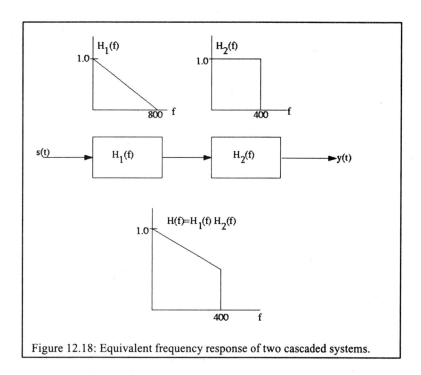

Figure 12.18: Equivalent frequency response of two cascaded systems.

Example 12.8

Suppose three cascaded systems have the frequency responses $H_1(f)$, $H_2(f)$, and $H_3(f)$ given in Figure 12.19. Then the overall system has the frequency response $H(f) = H_1(f) H_2(f) H_3(f)$ as shown. The combination of the low-pass $H_1(f)$ and high-pass $H_2(f)$ act as a band-pass filter, passing frequencies within the band (100, 300). Within this band, $H_3(f)$ determines the shape of the frequency response.

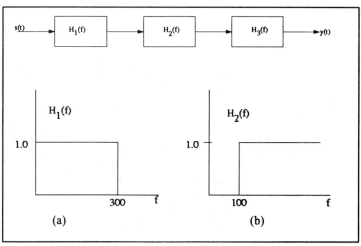

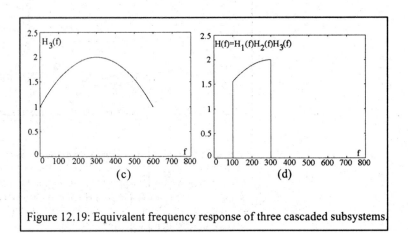

(c)   (d)

Figure 12.19: Equivalent frequency response of three cascaded subsystems.

Exercise 12.3

1. What is the definition of *cascaded* systems? How is the frequency response of the overall system related to the frequency responses of the individual systems?

2. Suppose that four systems with frequency responses $H_1(f)$, $H_2(f)$, $H_3(f)$, and $H_4(f)$ are cascaded together. What is the frequency response of the overall cascaded system?

3. Suppose that an ideal low-pass filter with cutoff frequency 4,000 Hz is cascaded onto an ideal high-pass filter having cutoff frequency 2,000 Hz. Sketch the frequency response of the overall cascaded system.

4. Suppose that an ideal low-pas filter with cutoff frequency 500 Hz is cascaded onto an ideal band-pass filter having cutoff frequencies 200 Hz and 800 Hz. Sketch the frequency response of the overall cascaded system.

5. Plot the overall frequency response $H(f)$ of three cascaded systems, where $H_1(f)$ is the response of a notch filter with notch frequency $f_0 = 400$ Hz, and $H_2(f)$ and $H_3(f)$ are shown in Figure 12.20

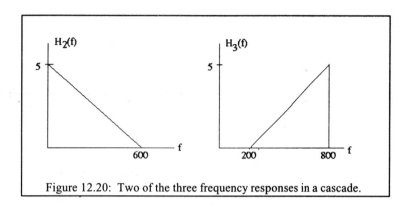

Figure 12.20: Two of the three frequency responses in a cascade.

6. Feedback systems, where part of the output is filtered and then fed back to the input, are commonly employed in control systems, such as those that control the speed of machinery or the angle of the ailerons on an aircraft wing. Find the equivalent frequency response $H(f)$ for the feedback system in Figure 12.21. Hint: Write $Y(f)$ in the form $Y(f) = H(f) S(f)$ where $H(f)$ is written in terms of $H_1(f)$ and $H_2(f)$.

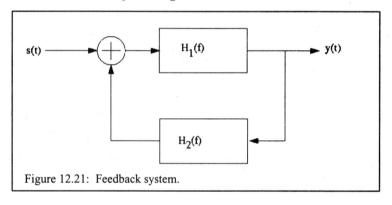

Figure 12.21: Feedback system.

Section 12.3: Digital Filtering

We have discussed filtering and frequency response at length, but we have yet to describe how a filter is actually built. In this chapter, we will focus on filters implemented within a computer or DSP chip, which are called **digital filters**. Digital processing usually takes place within the larger context of the analog world where both s(t) and y(t) are analog or continuous-time signals. How can a computer or DSP chip process an analog signal? As we have discussed in earlier chapters, digital processing requires both analog-to-digital (A/D) and digital-to-analog (D/A) conversion. The overall digital filtering scheme is shown in Figure 12.22.

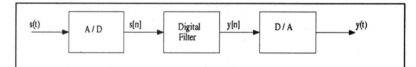

Figure 12.22: Filtering scheme employing a digital filter sandwiched between an A/D and a D/A.

Recall that the A/D samples and quantizes the signal s(t) to produce the sequence of numbers s[n]. We assume that the input s(t) is bandlimited and that it is sampled above the Nyquist rate. The D/A converts the digital signal y[n] into a smooth, bandlimited analog signal y(t) that passes through the computed set of samples y[n]. The digital filter computes the output sequence y[n] from the input sequence s[n]. That is, the digital filter is nothing more than a computer algorithm that takes the sequence of numbers s[n] and produces a new sequence of numbers y[n]. Digital filtering algorithms are designed so that in Figure 12.22 the analog signal y(t) will be an improved or modified version of s(t.

Why filter an analog signal with a computer (digital filter)? One reason is that we can do so very precisely. A second reason is that digital filters can be "reprogrammed" to act in a variety of different ways on signals. Analog filters, constructed out of standard electronic hardware, must be rewired if we wish to change their operation (frequency response). So, today most sophisticated processing of signals by filters is performed in the digital domain by computers. Let us next look at how a digital filter works.

12.3.1 Digital Filters

We will focus primarily on the type of digital filter shown in Figure 12.23.

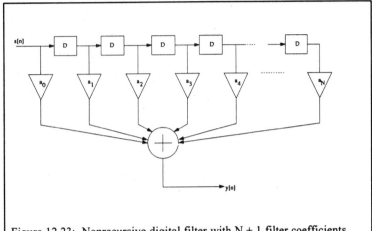

Figure 12.23: Nonrecursive digital filter with N + 1 filter coefficients.

Each square box in Figure 12.23 represents a memory location that can be thought of as a delay register. Proceeding left to right, the numbers stored in the delay registers are delayed samples s[n-1], s[n-2], s[n-3], ..., s[n-N]. The triangles represent multiplications by the coefficients $\{a_n\}$. The plus sign in the circle indicates addition. Overall, the filter implements

$$y[n] = a_0 s[n] + a_1 s[n-1] + a_2 s[n-2] + \ldots + a_N s[n-N]$$

which in the mathematics world is called a **difference equation**. This computation is simply a series of multiplications and additions that are performed in a computer or DSP chip. By iterating through various values of n we see that

y[0] is computed from s[0], s[-1], s[-2], ... s[-N]

y[1] is computed from s[1], s[0], s[-1], ..., s[-N+1]

y[2] is computed from s[2], s[1], s[0], ..., s[-N+2]

y[3] is computed from s[3], s[2], s[1], ..., s[-N+3]

etc. Thus, it is as if a sliding window of length N + 1 is moved across the input sequence s[n] and each value within the window is weighted by the appropriate multiplier coefficient a_n prior to summation of the results. In Figure 12.23 this is conceptually accomplished by shifting the s[n] sequence through the series of delay registers from left to right, one position each time the value of n is incremented. This filter is called **nonrecursive** because the current output sample y[n] depends only on the present and past inputs s[n], not on values of past outputs (i.e. not on y[n-1], y[n-2], etc.). In Section 12.5 we will see an example of a **recursive** filter used to cancel echoes.

Example 12.9.

Suppose that the digital filter in Figure 12.23 has only two coefficients, given by $a_0 = 3$ and $a_1 = -5$. Write the difference equation that describes the filtering operation. Answer:

$y[n] = 3s[n] - 5\ s[n-1]$

Example 12.10.

Suppose that the digital filter in Figure 12.23 has four coefficients, given by $a_0 = 1.3$, $a_1 = -2.4$, $a_2 = 1.9$, and $a_3 = -1.6$. Write the difference equation that describes the filtering operation. Answer:

$y[n] = 1.3s[n] - 2.4s[n-1] + 1.9\ s[n-2] - 1.6\ s[n-3]$

Example 12.11

In Example 12.9, if $s[n] = 1$ for $n = 0, 1, 2, 3$, and $s[n] = 0$ for all other values of n, find the output sequence $y[n]$. Answer:

$y[0] = 3$, $y[1] = -2$, $y[2] = -2$, $y[3] = -2$, $y[4] = -5$, $y[n] = 0$ for all other n

Example 12.12

In Example 12.9, if $s[n] = (-1)^n$ for all values of n, i.e. $s[n]$ alternates between $+1$ and -1, find $y[n]$. Answer:

$y[n] = 8$ for n even and $y[n] = -8$ for n odd

12.3.2 Frequency Response of a Digital Filter

If $s(t)$ in Figure 12.22 is a sinusoid, then the output $y(t)$ is a sinusoid of the same frequency, with an amplitude that depends on the frequency response of the overall system. *The frequency response of the system in Figure 12.22 depends in a direct way on the digital filter coefficients a_n.* By choosing the digital filter coefficients in the right way, we can design the frequency response of the system in Figure 12.22 to have a desired shape. It can be shown that the frequency response of the system in Figure 12.22 is

$$H(f) = \begin{cases} \sqrt{R^2(f) + I^2(f)}, & f \leq f_s/2 \\ 0, & f > f_s/2 \end{cases}$$

where

$$R(f) = a_0 + a_1\cos(2\pi T_s f) + a_2\cos(4\pi T_s f) + a_3\cos(6\pi T_s f)$$

$$+ \ldots + a_N\cos(2N\pi T_s f)$$

and

$$I(f) = a_1\sin(2\pi T_s f) + a_2\sin(4\pi T_s f) + a_3\sin(6\pi T_s f)$$

$$+ \ldots + a_N\sin(2N\pi T_s f).$$

Given a set of digital filter coefficients $\{a_n\}$, the above equations can be used to calculate the frequency response of the digital filtering system. In fact, we will use these equations in the next section to calculate frequency responses of some simple filters. Furthermore, engineers have determined how the filter coefficients should be chosen to make H(f) have a desired shape. This subject is called **digital filter design**. In applications such as communications, radar, or consumer electronics, we are often interested in *designing* filters to have specified frequency responses.

Before considering filter design, we first examine three very simple digital filters, the averager, the differencing filter, and a notch filter, which can be useful for modifying the frequency content of waveforms. In each case we show a diagram of the filter, write the corresponding input-output relation (difference equation) and plot the frequency response of the system in Figure 12.22.

Exercise 12.4

1. What is a digital filter?

2. How can a digital filter, which processes numbers, be used to process an analog signal?

3. Suppose the digital filter in Figure 12.23 has 2 filter coefficients, $a_0 = 1$, $a_1 = -1$. Write the difference equation that describes the filtering operation. If $s[n] = 4$ for $n = 0, 1, 2, 3, 4, 5$, and $s[n] = 0$ for all other values of n, find the output sequence $y[n]$.

4. Repeat Problem 3, except for $a_0 = 2$, $a_1 = 3$, and $s[n] = n^2 + 1$ for all values of n.

5. Suppose the digital filter in Figure 12.23 has 4 filter coefficients, $a_0 = 1$, $a_1 = -1$, $a_2 = 1$, $a_3 = -1$. Write the difference equation that describes the

filtering operation. If s[n] = 2 for n = 0, 1, 2, 3, 4, 5, 6, 7, 8, and s[n] = 0 for all other values of n, find the output sequence y[n].

6. Same as Problem 5, except s[n] = n for all values of n.

7. What is digital filter design?

Section 12.4: Simple Digital Filters

In this section, we consider three simple digital filters, the averager, the differencing filter, and a notch filter. In each case, we write the corresponding input-output relation (difference equation) and plot its frequency response.

12.4.1 Simple Low-pass Filter (Averager)

Perhaps the simplest useful digital filter is the averager, which acts to reduce the high frequency content of a signal. The averager computes the output sequence y[n] from the input sequence s[n] according to the following simple equation:

$$y[n] = \frac{1}{2} s[n] + \frac{1}{2} s[n-1].$$

> **Low-pass filters** keep only the low frequencies in the input signal.

Each element of the output sequence is the average of two adjacent elements in the input sequence. Thus, the output signal will be a smoother version of the input signal and, as we shall see below, the higher frequencies in the signal s(t) will be attenuated (or made smaller) by the system in Figure 12.22. This is similar to turning down the treble on your stereo. As we will soon see, the frequency response for this filter crudely approximates the ideal low-pass shape. Thus, this filter could help remove high-frequency noise.

To fully characterize this filter, we compute its frequency response. More precisely, we calculate the frequency response of the digital filtering system in Figure 12.22 for the case where the digital filter is an averager. We expect the response to be low-pass in nature since the digital filter output is smoother than its input (i.e. high frequency wiggles are filtered out). For the averager, we have the filter coefficients:

$$a_0 = \frac{1}{2}, \quad a_1 = \frac{1}{2}.$$

Substituting these into the expressions for R and I in the previous section, we have

$$R(f) = \frac{1}{2} + \frac{1}{2} \cos(2\pi T_s f)$$

$$I(f) = \frac{1}{2} \sin(2\pi T_s f).$$

So,

$$H(f) = \begin{cases} \sqrt{R^2(f) + I^2(f)}, & f \le f_s/2 \\ 0, & f > f_s/2 \end{cases}$$

$$= \sqrt{\frac{1}{2} + \frac{1}{2} \cos(2\pi T_s f)}, \qquad f \le f_s/2$$

where we have used the trigonometric identity $\cos^2 + \sin^2 = 1$. . Recall that T_s is the sampling period of the digital system. Figure 12.24 shows a plot of this frequency response, which has a crude low-pass shape.

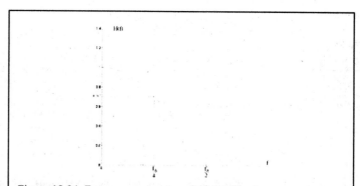

Figure 12.24: Frequency response of digital filtering system using an averager.

Sinusoids with low frequencies are passed with nearly unit amplitude, whereas input frequencies near half the sampling frequency are greatly attenuated. We can use this plot to determine the amplitude of the output for any sinusoidal input. More generally, the spectrum S(f) of any input signal s(t) will be shaped according to Y(f) = H(f) S(f) where H(f) is the frequency response characteristic in Figure 12.24. The fact that low frequencies are passed with larger amplitudes than are high frequencies suggests that the output y(t) will be smoother than the input s(t) in Figure 6.22. This agrees with our earlier observation that the digital filter output y[n] is smoother than the filter input s[n].

24

Exercise 12.5

1. Suppose two averagers are cascaded, so that the output of the first averager is the input to the second averager. Show that this overall system produces the same output as a digital filter having three multiplier coefficients: $a_0 = 1/4$, $a_1 = 1/2$, $a_2 = 1/4$. Hint: Solve this problem in the sequence domain by working with the difference equations.

2. We have seen that the frequency response of cascaded systems is the *product* of the individual frequency responses. Thus, the frequency response for the system in Problem 1 is the square of H(f) in Figure 12.24. Make this plot and comment on how the frequency response of the cascaded system is different from that of a single averager. Is the response closer to that of an ideal low-pass filter?

3. Repeat Problem 2, except for a cascade of three averagers.

4. Suppose that the sampling frequency in Figure 12.24 is $f_s = 1000$ samples per second. For sinusoidal inputs s(t) having amplitudes = 1 and frequencies f = 100, f = 250, and f = 400, find the amplitudes of the outputs y(t) assuming that the digital filter is (a) an averager, (b) a cascade of two averagers, (c) a cascade of three averagers.

VAB Experiment 12.4: Averaging Digital Filter
• Listen to various signals filtered by the simple averager.

12.4.2 Simple High-Pass Filter (Differencer)

A very slight change to the averaging filter gives us a filter having completely different characteristics. The digital filtering algorithm used in a simple differencing filter is

$$y[n] = \frac{1}{2} s[n] - \frac{1}{2} s[n-1].$$

> **High-pass filters** keep only the high frequencies in the input signal.

The only difference between this equation and the equation describing the averager is that the second filter coefficient has been changed from +1/2 to −1/2. For the differencer, each element of the output sequence is proportional to the difference between two adjacent elements in the input sequence. Thus, whenever the input signal stays relatively constant from one sample to the next, the output signal will be approximately zero. Conversely, if the input signal changes greatly from sample to sample, then the output signal will be large (either positive or negative.) We will next see that the frequency response for this filter crudely approximates the ideal high-pass shape. Thus, this filter could help remove low-frequency noise.

To fully characterize the differencing filter, we find its frequency response. As for the case with the averaging filter, we actually compute the frequency response of the complete system in Figure 12.22. We expect the response to be high pass in nature since we decided that the digital filter would preserve high-frequency wiggles in s[n] and remove smoother sections of the signal. For the differencer, the filter coefficients are

$$a_0 = \frac{1}{2}, \quad a_1 = -\frac{1}{2}.$$

Substituting these values into the expressions for R and I gives

$$R(f) = \frac{1}{2} - \frac{1}{2}\cos(2\pi T_s f)$$

$$I(f) = -\frac{1}{2}\sin(2\pi T_s f).$$

So,

$$H(f) = \begin{cases} \sqrt{R^2(f) + I^2(f)}, & f \le f_s/2 \\ 0, & f > f_s/2 \end{cases}$$

$$= \sqrt{\frac{1}{2} - \frac{1}{2}\cos(2\pi T_s f)}, \quad f \le f_s/2$$

where we again have used the trigonometric identity $\cos^2 + \sin^2 = 1$. Figure 12.25 shows a plot of this frequency response, which has a crude high-pass shape.

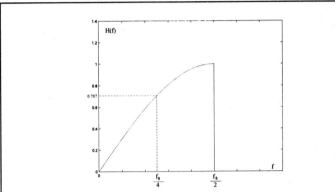

Figure 12.25: Frequency response of a digital filtering system using a differencer.

Sinusoids with high frequencies (near $f_s/2$) are passed with nearly unit amplitude, whereas sinusoids having low frequencies are greatly attenuated. Since high frequencies are passed with larger amplitudes than are low frequencies, the output y(t) in Figure 12.22 will be more "jumpy" than the input s(t). This agrees with our earlier observation that the digital filter output y[n] would be nearly zero for n where s[n] is almost constant and would be much larger for n where s[n] varies greatly from one sample to the next.

Exercise 12.6

1. Suppose two differencers are cascaded. Show that this overall system produces the same output as a digital filter having three multipliers: $a_0 = 1/4$, $a_1 = -1/2$, $a_2 = 1/4$. Hint: Solve this problem in the sequence domain by working with the difference equations.

2. We know that the frequency response for the cascaded system in Problem 1 is the square of H(f) in Figure 12.25 for a single differencer. Make a plot of $H^2(f)$ and comment on how its appearance differs from the response H(f) in Figure 12.25. Is the response closer to that of an ideal high-pass filter?

3. Repeat Problem 2, except for a cascade of three differencers.

4. Suppose the sampling frequency in Figure 12.22 is $f_s = 50,000$ samples per second. For sinusoidal inputs having amplitudes = 1 and frequencies f = 5,000, f = 12,500, and f = 20,000, find the amplitudes of the outputs y(t) assuming that the digital filter is (a) a differencer, (b) a cascade of two differencers, (c) a cascade of three differencers.

12.4.3 Simple Notch Filter

Notch filters eliminate a narrow range of frequencies in the input signal.

The third type of simple filter we will examine is the notch filter. This filter removes frequencies from signals in some small region around the notch frequency f_o. This filter is the type to use if the signal you have received is corrupted by an interference signal with a very limited frequency range (such as the buzz, caused by a motor, in a radio signal). Unlike the averager and differencer, the notch filter has 3 coefficients. The digital filtering algorithm for the notch filter is given by

$$y[n] = \frac{1}{2} s[n] - \left(\cos(2\pi T_s f_o)\right)s[n-1] + \frac{1}{2} s[n-2].$$

where the frequency f_o is set equal to the frequency of the interference to be eliminated. This is an example of a filter whose capabilities are difficult to assess without studying its frequency response.

To see the effect of the notch filter, we derive its frequency response. The coefficients of the notch filter are

$$a_0 = \frac{1}{2}, \quad a_1 = -\cos(2\pi T_s f_o), \quad a_2 = \frac{1}{2}.$$

Substituting these into the expressions for R and I gives

$$R(f) = \frac{1}{2} - \cos(2\pi T_s f_o)\cos(2\pi T_s f) + \frac{1}{2}\cos(4\pi T_s f)$$

$$I(f) = -\cos(2\pi T_s f_o)\sin(2\pi T_s f) + \frac{1}{2}\sin(4\pi T_s f).$$

Substituting R and I into the expression for H(f), and performing some algebra yields

$$H(f) = \left|\cos(2\pi T_s f) - \cos(2\pi T_s f_o)\right|, \quad f \le f_s/2$$

where the vertical bars indicate absolute value or magnitude. Figure 12.26 shows a plot of this frequency response for the specific case with the notch frequency chosen in the middle of the frequency band, i.e. $f_o = f_s/4$. We can clearly see that frequencies around f_0 will be completely or nearly eliminated, while other frequencies in the input signal s[n] will be passed through the filter. That is why this type of filter is called a *notch filter*.

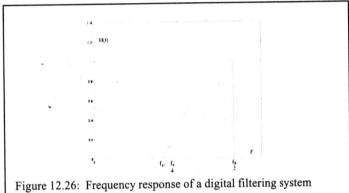

Figure 12.26: Frequency response of a digital filtering system using a notch filter with $f_0 = f_s/4$.

This filter can be used to remove unwanted sinusoids, but the frequency response in Figure 12.26 is not a particularly good one. The problem is that all frequencies, except $f = 0$ and $f = f_s/2$ have their amplitudes attenuated at least somewhat, and by an amount that depends on f. This spectral shaping will distort the signal that is to be preserved (although the offending sinusoid will be removed). Designs are available for better notch filters whose frequency responses come very close to the ideal characteristic in Figure 12.13. These digital filters require many more multiplier coefficients than the above simple filter.

Section 12.5: Better Filters – The Complexity/Quality Tradeoff

We studied the filters in Section 12.4 primarily to help us better understand the connection between digital filtering and frequency response. In practice, few signal processing systems use such simple filters. It is more common to encounter filters that use tens, hundreds, or even thousands of memo ry locations and multiplier coefficients. More filter coefficients allow for a closer approximation of the desired frequency response.

A general nonrecursive filter uses N memory locations and N+1 multipliers as shown earlier in Figure 12.23. The equation describing the input-output relationship of this filter was

$$y[n] = a_0 s[n] + a_1 s[n-1] + a_2 s[n-2] + \ldots + a_N s[n-N] \; .$$

Obviously, increasing N, the "length" of the filter, increases the cost of the implementation. For each value of the output index, n, computation of $y[n]$ requires $N + 1$ multiplications and N additions. If the sampling rate is f_s, then the number of multiplications or additions per second will be roughly $N f_s$. Suppose that N is in the hundreds and that f_s is high (it could be in the tens of millions of samples per second in certain communication systems or in radar, where the signal bandwidths are wide – remember that the Nyquist rate must be satisfied). In such cases, the digital filter requires a huge amount of computation, which drives up the cost and power requirement of the processor that implements the filter. In return for the extra complexity associated with a long filter, the filter design engineer can produce a filter design having a frequency response that is closer to the desired response. Thus, there is a **tradeoff** between complexity (cost) and quality. This is a tradeoff that engineers work with every day! The more complicated (and expensive) we make the filter, the closer the filter can be made to ideal.

There are many ways to design (specify) the coefficients of digital filters. An assortment of design methods have been developed because when trying to approximate a desired response, there are many ways to mathematically define measures of closeness between the frequency responses of the desired and the designed filter. Different measures of closeness yield different filter design methods, which in turn produce slightly different filter coefficients a_n. In this section, we have no intent on becoming expert filter

30

designers. Instead, we simply wish to illustrate the complexity/quality tradeoff. We will illustrate the tradeoff by examining the design of low-pass filters. We will present a formula for the filter coefficients derived using a standard design method from the digital signal processing literature, called the "window design technique" with a Hamming window.

12.5.1 Approximation of Ideal Low-Pass Filters

The formula for the multiplier coefficients of a length $N + 1$ low-pass filter having cutoff frequency f_c is

$$a_n = 2f_c T_s w[n] \sin c\left[2\pi f_c T_s \left(n - \frac{N}{2} \right) \right] \qquad 0 \le n \le N$$

where

$$w[n] = 0.54 - 0.46 \cos\left(\frac{2\pi n}{N} \right)$$

is the Hamming window sequence and

$$\sin c(x) = \frac{\sin(x)}{x} \,.$$

The sinc function comes up very often in communications and signal processing and is pronounced "sink." It is defined to have the value 0 at $x = 0$ and is symmetric about $x = 0$. It looks like a decaying sinusoid as x becomes larger in either the positive or negative direction.

Do not be disturbed by the complexity of the formula for the a_n! The important point is that there *is* a formula that can give us values for the a_n that will produce a digital filter with a very high-quality frequency response. The values for a_0, a_1, a_2, and so forth are literally calculated by substituting $n = 0, 1, 2$, etc. into the above formula. These are the numerical values for the multiplier coefficients in the filter in Figure 12.23.

Let us now look at the coefficient sets and frequency responses for low-pass filters for two different filter lengths N. As in Section 12.4, we will be plotting the frequency response for the overall filtering system in Figure 12.22, which includes the A/D and D/A. Figure 12.27 shows the filter coefficients and frequency response for a low-pass filter with $N = 20$ and cutoff frequency $f_c = f_s/6$.

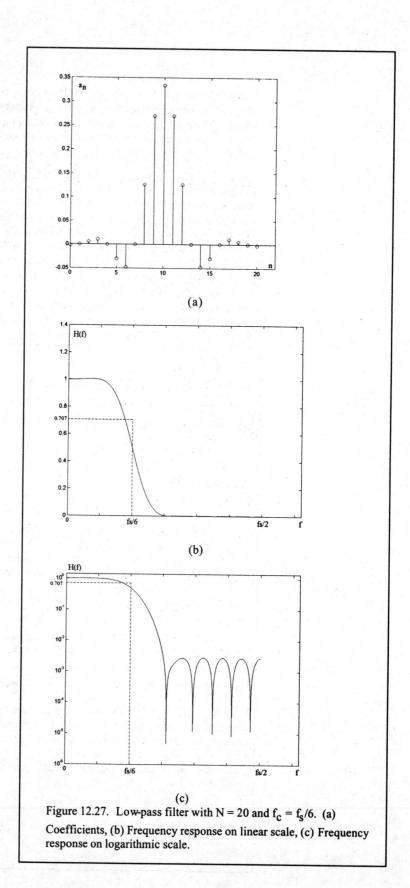

(a)

(b)

(c)

Figure 12.27. Low-pass filter with N = 20 and $f_c = f_s/6$. (a)
Coefficients, (b) Frequency response on linear scale, (c) Frequency
response on logarithmic scale.

We have plotted the frequency response on a logarithmic vertical scale in Figure 12.27(c) to make the smaller values in the plot visible. The range of frequencies below f_c, where H(f) is nearly one, is called the **passband** of the filter, because sinusoids are passed with significant amplitude in this band. The range of frequencies above f_c, where H(f) is small, is called the **stopband**, because sinusoids in this band are almost completely attenuated. Although the frequency response in Figure 12.27 is starting to approach the ideal low-pass shape, at the expense of a more complex filter we can improve the response by increasing N. Figure 12.28 shows the coefficients and frequency response for a low-pass filter of length N = 100, with the same cut-off frequency $f_c = f_s/6$. Notice that the responses in both the passband and stopband are improved, with a sharper transition between the two bands. This improvement has been bought at the expense of a five-fold increase in filter complexity. In practice, the job of the filter designer is to produce the least complex filter having sufficient quality (in terms of frequency response) for the application at hand.

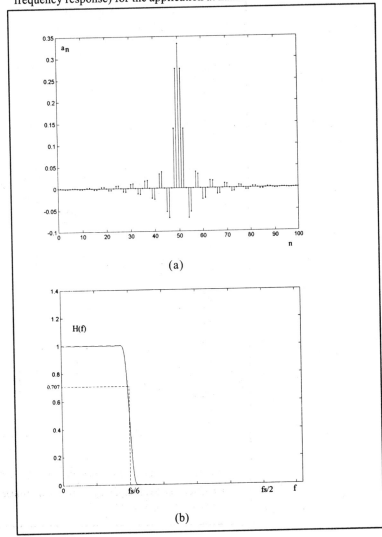

(a)

(b)

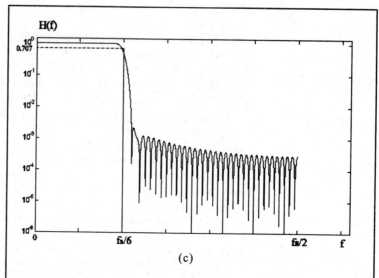

Figure 12.28: Low-pass filter with N = 100 and $f_c = f_s/6$. (a) Coefficients, (b) Frequency response on linear scale, (c) Frequency response on logarithmic scale.

12.5.2 Approximation of Ideal High & Band-Pass Filters

For completeness, and for possible use in student projects, we present the design formulae for the high-pass and band-pass cases. The formula for the coefficients of a length-N high-pass filter having cutoff frequency f_c is

$$a_n = 2f_c T_s w[n](-1)^n \operatorname{sinc}\left[2\pi\left(f_c T_s - \frac{1}{2}\right)\left(n - \frac{N}{2}\right)\right] \qquad 0 \le n \le N.$$

The Hamming sequence w[n] and the sinc function are as defined in Section 12.5.1. The most important difference between this formula and the one for the low-pass case is the additional factor $(-1)^n$. This is simply a sequence that alternates between +1 and −1.

The formula for the coefficients of a band-pass filter is

34

$$a_n = 2 f_b T_s \, w[n] \cos\left(2\pi f_o T_s \left(n - \frac{N}{2} \right) \right) \sin c \left[2\pi f_b T_s \left(n - \frac{N}{2} \right) \right]$$

$$0 \leq n \leq N$$

where f_o is the center of the passband (f_{c1}, f_{c2}), i.e.

$$f_o = \frac{f_{c1} + f_{c2}}{2}$$

and f_b is the half bandwidth

$$f_b = \frac{f_{c2} - f_{c1}}{2} .$$

The difference between this formula and the one for the low-pass case is the additional cosine term. It is this term that determines where the passband is centered. If we select $f_o = 0$, $f_{c1} = 0$, and $f_{c2} = f_c$, then the above formula reduces to that for the low-pass case.

VAB Experiment : Design of Frequency Selective Filters
• Students "design" various filters and apply them to speech and music.   Inputs: Upper and lower cutoff frequencies, filter length N   Outputs: Plot of frequency response, sound

Exercise 12.7

1. Explain the tradeoff in designing digital filters.

2. Suppose a digital filter has 50 multiplier coefficients and the sampling rate is $f_s = 40{,}000$ samples per second. How many multiplications per second must be computed by the filter? How many multiplications per second would be required for the same filter length, but a sampling rate of 20,000,000 samples per second?

3. What are the definitions of *passband* and *stopband*?

- Students design hearing aid filter as described below and apply it to distorted musical waveform.
    Inputs: Filter length N, desired response via Sketchwave
    Outputs: Plot of frequency response, distorted and equalized music

Mr. Smith attends a choral concert in a very difficult acoustic environment. He has a significant frequency-dependent hearing loss and he is in a very live (reverberant) concert hall. In addition, there is an annoying low-frequency rumble from the air-conditioning compressor, below 150 Hz, and significant high-frequency noise from the blower, above 3 KHz. Fortunately, Mr. Smith is wearing a new digital hearing aid. Although reverb cancellation is not possible, the frequency response is adjustable. He figures that with the appropriate filtering, he will be hearing the concert better than anyone else! Mr. Smith's hearing response is shown in Figure 12.29. Obviously, some serious equalization is in order.

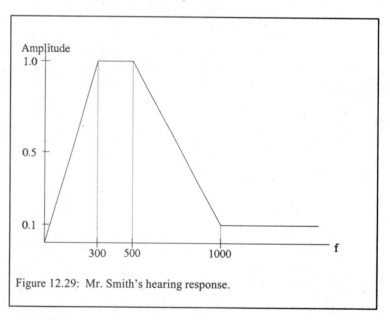

Figure 12.29: Mr. Smith's hearing response.

Your job is to use VAB to design Mr. Smith's digital hearing aid to provide the best sounding music possible. You are encouraged to subdivide the filter design problem as follows.

Use sketchwave in VAB to sketch a desired frequency response satisfying:

a) The response should be zero below 150 Hz to remove the low frequency rumble

36

b) The response should be zero above 3 KHz to remove the high-frequency blower noise.

c) For frequencies between 150 Hz and 3 KHz, the response should be the reciprocal of Mr. Smith's hearing response to compensate for his hearing loss.

After using Sketchwave to sketch the necessary frequency response, specify a filter length (start with 11) and use VAB to apply the filter to some distorted music to witness the improvement offered by the digital hearing aid.

Section 12.6: Echo Cancellation

Echoes are delayed signals. In the audio world, echoes arise when a signal reflects off a non-absorbing surface and the listener hears both the direct-path signal, which arrives first, and the reflected signal, which arrives (usually) a fraction of a second later. Concert halls are full of echoes but the amount of echo is controlled to a desired level through the placement of sound absorbing material. Some echo is good because it adds "warmth" to sounds within a room.

Since an echo involves signal delay, it is not hard to determine how an echo can be produced digitally. A digital filter employs a series of memory locations, so a signal can be delayed by simply holding it in memory for a short while before reading it out to the D/A. Figure 12.30 shows a block diagram of a digital echo generator that produces an output y[n] that is equal to the input s[n] plus a delayed and scaled (echoed) version of the input. The equation describing this system is

$$y[n] = s[n] + \alpha s[n\text{-}N].$$

The delay introduced in the computer is N sampling times, T_s, so that the overall delay is NT_s. The value of α is typically between 0 and 1, with the echo generally having a smaller amplitude than the direct-path signal.

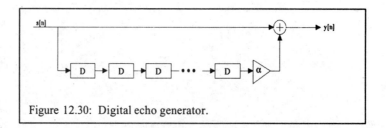

Figure 12.30: Digital echo generator.

Certainly echoes are not always desired. Echoes can make speech hard to understand and can turn a choral group into a "wordless choir." Highly reverberant rooms pose a special problem for those persons with hearing loss, who are already at a disadvantage in trying to understand conversations.

Echoes can have serious effects in telecommunication systems. Suppose Person 1 and Person 2 are communicating by phone and that Person 1 is speaking. Due to the way that the phone network is constructed, it is common for some of the signal that reaches Person 2's receiver to reflect back through the electrical network to Person 1. Suppose that the communication channel has a delay, as it would if the communication were taking place over a satellite channel (satellites are far above the earth and it takes a while for the communicated signal to travel up and come back down). Then Person 1 will hear himself speaking, plus a delayed version of himself. This is very disconcerting! The story is worse, however. The delayed signal can be reflected a second time so that Person 2 also hears Person 1 plus a delayed version of Person 1. In this case, the speech can be totally unintelligible to the person trying to listen.

This exact scenario used to be common on overseas phone calls, which often use satellite channels. The problem was solved, however, by the introduction of digital echo cancellers. That is, it is possible to estimate the echo and to then remove it! We will not go into the theory of echo cancellation, but by examining the echo equation, we can begin to understand how a digital filter can remove echo. We had the echo modeled as

$$y[n] = s[n] + \alpha s[n-N]$$

where y[n] is the signal containing the echo. Rearrange this equation to read

$$s[n] = -\alpha s[n-N] + y[n].$$

This equation says that we can recover s[n] from the echo signal y[n] via the filtering operation shown in Figure 12.31. Indeed, this equation describes the input-output relationship of the filter, where the echo signal y[n] is the filter input and the clean signal s[n] is the output.

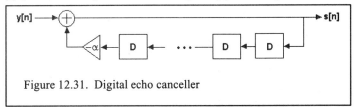

Figure 12.31. Digital echo canceller

The filter in Figure 12.31 is called *recursive* because it employs feedback so that the output at time n depends on a past output, as well as on the current input. In practice, the amount of delay N and the scaling factor would have to be estimated from the echo signal y[n]. In addition there may be multiple echoes with different delays and scaling factors, so that the filter structure actually must be more complicated. When all factors are considered, it turns out that nonrecursive filters such as in Figure 12.23 are generally preferred for echo cancellation, because there are stable algorithms for computing the coefficients a_n needed to remove echo. Furthermore, as the structure of the echoes changes, the coefficients can be easily updated. Thus, we are suggesting that the filter can be *adapted* to the input signal. Such filters, where the coefficients are changed with time as a function of the input, are called **adaptive filters**. The ability to adapt is a powerful advantage of a digital filter over classical analog approaches where the filter consists of "hard-wired" resistors, capacitors, and transistors. Adaptive digital filters are used throughout the communications network, in audio teleconferencing systems, in the signal processing circuitry that reads bits from computer hard drives, and in many other applications.

VAB Experiment: Echo Cancellation

- Students listen to speech and music with an echo (use FIR filter in Figure 12.30). Delay and amplitude of the echo are adjustable by the students.
- Students then listen to result of applying the echo canceller in Figure 12.31. Delay and amplitude are adjustable by the students.

Exercise 12.8

1. Name four situations where it is desired to remove echoes from signals.

2. How does an adaptive digital filter differ from a regular digital filter?

3. Name three situations where adaptive filters are used.

Section 12.7:
Glossary

adaptive filter
a digital filter whose coefficients are changed with time according
to the filter input

averager
a crude low-pass filter that computes the output as the average of adjacent
elements in the input sequence

band-pass filter
a filter that emphasizes mid-range frequencies and attenuates both lower
and higher frequencies

cascaded filters
two or more filters in tandem, where the output of one filter is the input to
the next

cross-over network
an analog filter in a speaker system that separates the electrical signal into
various frequency bands, often bass, midrange, and treble

cutoff frequency
the frequency above which or below which a filter is designed to either pass
or block signal energy

difference equation
equation that describes the operation of a digital filter

differencer
a crude high-pass filter that computes the output as the difference between
adjacent elements in the input sequence

digital filter
filtering algorithm implemented on a computer

digital filter design
selection of the digital filter multiplier coefficients a_n

41

echo canceller
digital filter designed to cancel echoes

equalization
spectral shaping applied to flatten the frequency response of a communication channel or other system

filter
a device used to modify a signal's frequency content

frequency response (H(f))
the frequency characteristic of a filter; the output spectrum Y(f) is the *product* of H(f) with the input spectrum S(f)

frequency selective filtering
a type of filtering where a desired signal lying within part of the frequency band is extracted from noise or interference

high-pass filter
a filter that emphasizes higher frequencies and attenuates lower frequencies

low-pass filter
a filter that emphasizes lower frequencies and attenuates higher frequencies

non-recursive filter
a digital filter whose output y[n] depends only on the input s[n] and past inputs, and not on past outputs

notch filter
a filter that passes all frequencies except those near a designated notch frequency f_0 where $H(f_0) = 0$

notch frequency
the frequency at which a notch filter is designed to block energy

passband
the range of frequencies where a filter lets through signal energy

recursive filter
a digital filter whose output y[n] depends on past outputs

spectral shaping
modification of a signal's frequency content via filtering

stopband
the range of frequencies where a filter blocks signal energy

Authors: John Treichler
and Sally Wood
Version: 4
Date: 5 May 2001

The INFINITY Project

Chapter 13: Communication Channels

Approximate Length: 1 Week

Section 13.1:
The Transmission
Channel

In Chapter 7 we designed several methods for communicating digital information. We assumed that this data might have originated from a multimedia source and might be carrying any mixture of text, images, music, speech, or even computer-generated bit streams. All of the methods we looked at sent the digital information using sound waves, which passed through the air from the transmitter to the receiver. We used this example because of its close parallel with human vocal communication. Most practical data communications systems do not use acoustical transmission, however. In fact, to communicate with greater speed or over greater distances, most communications systems use some other type of medium to carry their signal. The medium over which a communications signal is carried is called a **channel**. In this chapter we will explore a number of different kinds of channels, and we will examine what level of performance we might expect from them.

Section 13.2:
How is Digital
Data Conveyed?

The transmission of digital data might seem to be a modern technology, but, in fact, some versions of it have been around as long as humans have been able to communicate. As a result, a large variety of digital communication methods exist, and a representative list appears in Table 13.1. Some are still in daily use while others are only of historical interest. In this chapter we focus not so much on how these transmitters and receivers work, but rather on what medium is used to carry their signal from the transmitter to the intended receiver and why different media are used for communications systems. To do this we will now examine a few specific systems, and determine what they have in common. As we do this we will find that they fall into three fundamental classes —those that carry signals optically, those that carry signals electrically, and those that use radio waves to convey their messages.

The acoustical systems we examined in Chapter 7 to introduce the fundamental concepts of communications systems are rarely used in practical data transmission systems. However, both the basic structure of a communications system and the basic data representation alternatives discussed in these chapters do apply directly to communications systems using other media. Some examples of optical, electrical, and radio communications systems will be described in the next sections.

13.2.1 Optical Transmission Systems

13.2.1.1 Wigwag Flags

The Boy Scout shown in Figure 13.1 is sending a textual message using a single red and white flag. He translates each character in the message into Morse code and he waves the flag to his left to indicate a "dot" in the Morse code and to the right to indicate a "dash." He lowers the flag to indicate the break between letters and the space between words. His counterpart watches the flag, detects the transmission of dots and dashes, converts the flag's movements into characters, and passes on the received message to the proper person. On a clear day with an unobstructed view and a good telescope, it is possible to send a message for a distance of up to a mile or more. A Scout skilled with the flag can send at a rate of about one character per second, although the typical rate is slower.

Figure 13.1: A Boy Scout using a single "wigwag" flag to transmit character-oriented messages using a binary version of the Morse Code

It is common for the receiving Scout to send acknowledgements as words are received, and to ask for retransmissions when they are not. This procedure for obtaining error-free transmission is a simple example of a **data transmission protocol** of the type that we discussed in Chapter 8. Both of the Scouts can either send messages or receive messages, but they cannot send and receive at the same time. This mode of operation in which we can transmit only in one direction at a given time is called **half-duplex** transmission. Most of the systems listed in Table 13.1 are half-duplex. In contrast, when you are having a conversation on the telephone you could simultaneously talk and listen in what is called **full-duplex** communication, although excessive use of that capability would be considered rude.

Digital Transmission System	The Channel	Purpose
The human nervous system	Ionic conduction in nerves	sensing and control
Wigwag flags	visual/optical	character-oriented messages
Semaphore flags	visual/optical	character-oriented message
Flag hoists	visual/optical	character-oriented message
Blinker or "flashing light"	visual/optical	character-oriented messages
TV remote control	infrared/optical	control
Lasers operating through an optical fiber	Infrared/optical	general binary data transfer
Electrical pulses over coaxial cable	electrical	digitized telephony signals
A letter written on paper	human conveyance	character-oriented messages
The telegraph over wire	electrical	character-oriented messages
Braille text	tactile	character-oriented messages for visually impaired
Smoke signals	visual/optical	sensing and control
A deep space probe out beyond Pluto	radio/ electromagnetic	sensing and control
"One if by land, two if by sea"	Visual/optical	Signaling

Table 13.1: A variety of communications systems, highlighting the transmission channel over which the signals pass.

4

13.2.1.2 Semaphore Flags

The Boy Scout shown in Figure 13.2 uses a pair of flags to transmit a text message. He sends out the message character-by-character and uses the positions of the two flags to indicate to his receiving counterpart which letter is being sent. There are 26 different positions that indicate the letters and a few additional ones used to control the transmission and ask for repetitions. As with the wigwag flag scheme, the receiving signalman writes down the message and passes it on to the intended recipient. The time required to send a message is much less than when using the wigwag method because the Scout using the semaphore flags sends a complete character every time he positions the flags, while the wigwag signaler must move the flag up to six times to send a character. Clearly the semaphore method can be expected to send messages substantially faster.

In the context of Chapter 7, the wigwag method can be seen to send one bit per flag movement while the semaphore method sends a whole multi-bit character or symbol each time the flags move.

Figure 13.2: A Boy Scout using two flags and the "semaphore code" to transmit character-oriented messages

13.2.1.3 Blinker Lights

The Navy signalman shown in Figure 13.3 is using a blinker light to communicate text messages to another ship. He uses the Morse code and opens the light's shutter for a short time to indicate a "dot" and for a longer time to indicate a "dash." A good signalman can send two or three Morse characters per second, and on a clear, dark night the signal can be seen for more than five miles.

Figure 13.3: A sailor using a blinker light to transmit character-oriented messages between two ships using the International Morse Code

13.2.1.4 Lasers Transmitting Through an Optical Fiber

Optical fiber transmission systems, shown in simplified form in Figure 13.4, can be considered to be a modern version of the Navy's flashing light system with a few small differences. The signal to be transmitted on a fiber optic system is typically in purely binary form, rather than grouped into Morse-coded characters. The incoming bits are used to turn a laser or light-emitting diode (LED) on and off, instead of a big blinker light. They are guided to the receiver via a thin piece of highly transparent glass called an "optical wave guide," or, more commonly, an "optical fiber," rather than being allowed to pass through the atmosphere. The light stays within the fiber because of a special coating that makes all light reflect at the fiber's surface. The receiver determines when light is present and when it is not, sending out a 1 in the former case, and a zero in the latter.

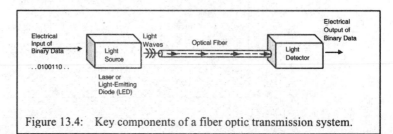

Figure 13.4:   Key components of a fiber optic transmission system.

Figure 13.5 shows two examples of optical transmission systems. On top is a cable carrying five plastic tubes, each of which holds twelve fibers. Below is a ribbon of twelve optical fibers, each about the diameter of a human hair. Some designs for fiber optic cable contain a number of these ribbons. These cables may carry from a few to as many as 432 fibers, each capable of carrying billions of bits of data per second.

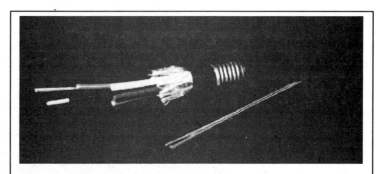

Figure 13.5: Optical fibers used to carry digital multimedia signals over long distance telecommunications systems.

Even though the details are different, the underlying concept for digital fiber optic transmission is exactly the same as the flashing light system. The transmitter encodes information into flashes of light, which are conveyed to the receiver, where they are detected with a photodetector. The photodetector converts the responses to the light flashes back into the original information, which is then relayed to the recipient at the destination point. Although the basic communication approach is the same, these two methods are different in several important operational ways. A modern fiber optic transmission system can reliably send 40 billion bits per second for a distance of more than a hundred miles in all weather conditions. The blinker light can handle the equivalent of ten bits/s for five miles under nearly perfect optimal weather conditions.

13.2.2 Electrical Transmission Systems

The first commercially successful system for communicating data over long distances was the electrical telegraph, invented in the United States by Samuel F. B. Morse in the mid-1830s. It was originally called the magnetic telegraph to distinguish it from proposed optical telegraph methods using semaphore-style long distance communication [1]. The magnetic telegraph used open-wire transmission systems of the sort shown in Figure 13.6 to carry their signals. The same open-wire scheme was later used to carry telephone signals.

Figure 13.6: An open-wire transmission system, first used for telegraphy circuits and later for telephony

The circuit diagram of a simple version of the telegraph is shown in Figure 13.7. It consists of a battery, shown symbolically as 6 vertical lines, a manually operated switch called a **key**, and a **sounder** connected together in one loop of wire. The sounder is a device based on an electromagnet that makes a clicking sound whenever the key is closed or opened. A photograph of a key and a sounder of the type used in the early telegraph networks is shown in Figure 13.8. The circuit in Figure 13.7 is exactly the same as the circuit used for a flashlight except that the flashlight has a lightbulb or light emitting diode (LED) in place of the sounder. When the flashlight switch is closed current flows in the loop causing the light to be on. Similarly, when the switch key of a telegraph circuit closes to complete the loop, the electromagnet causes the sounder to move and make a click. It should be noted that light bulbs had not yet been invented when the telegraph was first used in the 1840s.

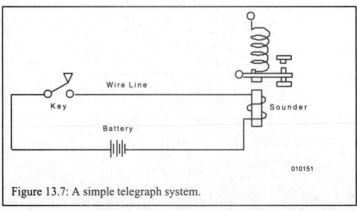

Figure 13.7: A simple telegraph system.

The simple communications systems that we examined in Chapter 7 converted each textual character into a set of one or more tone bursts and sent them over the air to the receiver. It is possible, and in fact far more common, to send them electrically or optically, rather than acoustically.

The loop connecting the battery, key, and sounder could be as long as 200 miles and still carry the electrical signal controlled by the key well enough to operate the sounder at the distant end of the circuit. The circuit, as shown in Figure 13.7, can carry information in only one direction. By adding a key with the sounder at the receiving site and a sounder with the key at the transmitting site, information can be transmitted in either direction as long as the key which is not sending a message is closed. This is another example of half-duplex operation.

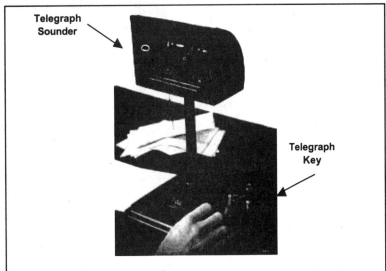

Telegraph Sounder

Telegraph Key

Figure 13.8: A telegraph key and sounder of the type used in the early American telegraph networks.

13.2.3 Transmission Systems Using Electromagnetic Waves

Figure 13.9 shows a picture of the Voyager 1 spacecraft. It was launched in the late 1970s to explore Jupiter, Saturn, Uranus, and Neptune. Along the way it made scientific measurements, took photographs of the planets and their moons, and sent all of this information back to earth using a radio transmission system. After passing those planets it continued past the edge of the solar system. It currently takes more than 21 hours for radio signals from Voyager 1 to reach the earth.

Why was a radio system employed to send the data home? As a practical matter there were no alternatives. A system based on wire or fiber is clearly out of the question, and laser-based optical transmission was not and still is not feasible. The vacuum of space prevents the use of acoustic methods. Radio transmission was the only choice. When Voyager 1 was near the earth it could send data back at a rate of up to 115 kilobits/s, but now, past the edge of the solar system, its signals can be reliably received at rate no faster than 160 bits/s, even when all the antennas worldwide in NASA's Deep Space Network are used to collect it.

Figure 13.9: An artist's conception of the Voyager deep space probe as it passed Pluto

13.2.4 A Basis for Comparison

For all communications methods, analysis begins with a common set of basic question as shown in Table 13.2 A very important attribute of a communications system is the rate at which it can transmit data. Some of the transmission rates achievable by various communications systems are listed in Table 13.3, along with the factor that limits the speed of each. We observe that for older systems the limit on the achievable transmission rate was often due to the equipment used at each end, not the transmission medium itself. With more modern electronically implemented "terminal equipment" the limits are now more often imposed by the transmission medium or the noise and interference that are unavoidably received along with the desired signal.

In general we find that some methods are fast and some are slow. Some are inherently binary in their transmission mode and some are "multi-level," that is, they are capable of carrying symbols with more than two values. Some use waves that are guided with a wire or fiber, while others propagate their signal through the atmosphere or through space. Whenever a new transmission system is encountered, it is informative to characterize it in the ways listed in the table.

Characterization of a transmission system

Is it based on binary transmission?

Does it use electrical transmission?

Does it operate optically?

How fast can the method transmit information?

What are the factors limiting transmission rate?

How hard is it to receive data?

Can a machine do it, or does it require human operator?

What are the principal causes of transmission errors?

What error detection and error recovery methods are possible?

For what kind of service is the system best suited?

Table 13.2: Fundamental questions about the properties of various transmission systems.

Communications Method	Typical Rate	Reason for Limit
Wigwag	2 b/s	Human dexterity
Telegraph over wire	10 b/s	Mechanical instruments and human dexterity
Voyager at the edge of the solar system	160 b/s	Noise
Coaxial cable carrying digital signals	135 Mb/s	Dispersion and noise
Single-laser fiber optic system	40 Gb/s	Dispersion in the fiber, speed of the electronics
Multi-laser fiber optic system	1600 Gb/s (as of April 2000)	Ability to multiplex many lasers onto the same fiber

Table 13.3: A comparison of the transmission rates of various communications schemes.

Exercise 13.1

1. From Table 13.1, select a digital communication method that was not discussed in this chapter and evaluate it based on the questions in Table 13.2.

2. Find three additional digital communication methods not discussed in the chapter or listed in Table 13.1. Be careful to distinguish digital communications from analog methods. This distinction may require some careful investigation since some methods, such as communication by telephone, began as analog methods but later were converted to digital methods.

3. Draw a circuit for a half-duplex telegraph system as described in Section 13.2.2 with both a key and a sounder at each end. Keep in mind that current can only flow when there is an uninterrupted loop, and that to make a clicking sound at the other end of a circuit a key must be able to turn the current on and off.

4. A telegraph circuit could operate over a length of 200 miles. Suppose that three sites within a 200-mile range wished to communicate. Could the circuit from Exercise 13.1.3 be modified to allow three key and sounder pairs at three different locations if everyone agreed to keep their key closed when not is use? What would happen if the operator at one location accidentally left his key open?

Section 13.3:
The Art of
Signal Detection

Signal detection is defined as using the received signal to determine exactly which of the possible symbols was actually sent by the transmitter. The ability to perform signal detection is usually quantified in two ways: one is the rate at which data can be sent, and the other is the rate at which errors are made in the process. The former determines the maximum **transmission rate** of the system. The latter, usually called the **bit error rate** (**BER**), is the number of errors made at the receiver as a fraction of the total number of bits sent. Obviously the best BER ever achievable is zero and it would not be a surprise to know that few transmission systems achieve this. Why do receivers make errors and what factors play into the rate at which they do?

Consider again the comparison of the wigwag and semaphore flag methods. Although they are far from state-of-the-art, modern, high-speed communications methods, they conveniently demonstrate some important aspects of transmission rates and bit error rates. From our earlier investigation we know that transmission with wigwag flags is slower than when the semaphore method is used. This is because the semaphore method encodes a whole letter or character into each transmission whereas the wigwag flag requires several flag movements to send each character. But which is the more reliable at a distance or in the fog? In fact, the wigwag method is more reliable, since it requires the receiver to determine only that the flag has moved to the left or right, and not exactly where the flags are positioned. Thus, when obscured, the method using the fewest possible choices works more reliably, albeit more slowly, than the method providing more choices.

Two important observations that apply to all communications systems can be made. The first is that no matter how cleanly or clearly the transmitted signal is sent, there will always be degradations that confuse the decision process at the receiver. These degradations may "muddy" the signal before it gets to the receiver so that the signal does not look like any possible transmitted signal should look. Degradations may also confuse the receiver because the degraded signal may appear too similar to another possible signal. The specific nature of these degradations depends on the actual transmission system but they do exist for every practical system. Because of these degradations, some errors will be made and the average rate at which these errors occur is an important performance measure.

The second observation is that the transmission speed and error rate compete with each other. Sending symbols faster and placing more

13

information on each one (for example, the semaphore) leads to a faster transmission rate, but it also leads to more confusion at the receiver and therefore a higher probability of making errors. Using simpler modulations types (for example FSK, the "serial binary" method) and slower symbol rates makes the detection process more reliable and therefore lowers the bit error rate. This phenomenon is well known on a subjective level to those learning a foreign language. Information is most reliably transmitted when the speaker talks slowly, clearly, and with a limited vocabulary. This applies to data transmission systems as well.

Exercise 13.2

1. Consider a system with a bit error rate of 0.001, or one error per 1000 bits transmitted. If each character is sent using 8 bits, what is the character error rate (the ratio of the number of incorrect characters to the total number of characters sent) for this system? Assume that two errors never occur in the same character.

2. Consider the system described in Exercise 13.2.1 with a bit error rate of 0.001. If each character is sent using 8 bits and each word is composed of five characters, what is the word error rate (the ratio of the number of incorrect words to the total number of words sent) for this system? Assume again that two errors never occur in the same word.

Section 13.4: Multiplexing – Sharing the Transmission Medium to Increase Data Rates

When the capacity of a transmission system must be increased so that more information can be communicated, one alternative is to build more systems communicating between the same two points. Depending on the type of system, there will be a cost associated with adding more electrical wire or more fiber optic cable or using more bandwidth. However, if the existing system can be used to send more messages using the existing physical resources, there would be a great cost savings. Sending more than one message at the same time using the same transmission media is called **multiplexing**. This situation is analogous to a road linking two cities that is used so much that it cannot handle the traffic between the two cities. The obvious, but expensive, solution to problem is to build a new road between the cities to handle the additional traffic. However, if the existing road is wide enough, it can be subdivided into several lanes of traffic in each direction. This allows more traffic to use an existing resource instead of building a new road.

Multiplexing is an important component of most communications systems. By definition it is the sharing of a transmission link in a way that permits it to carry the messages, or "traffic," for more than one user without any interference between the messages. It draws its name from the Greek roots "multi" and "plex," and means, literally, putting many things into one place. For example, a *multiplex* theater allows many little movie theaters to share one big building, parking lot, and concession stand. In the case of a telecommunications system, the "things" are conversations, messages, or data streams, and the "place" is the transmission system. The primary motivation for multiplexing is the economical use of resources. Transmission systems require resources that cost time and money. If more users can be served by a system, then, at least in principle, each user should have to pay less for the service than if fewer users divided the same costs.

How might one perform this multiplexing? We will examine two fundamental methods. The first is shown in schematic form in Figure 13.10. Suppose that we have built a transmission system capable of transmitting 2 million bits/sec and that we have a number of customers with binary files to send. A natural strategy for sharing the link is to let each of them use the link when their files are ready to send. If we suppose that the total average transmission rates of the bits to be sent by all of the users is less than 2 Mb/s, then the link can handle them all in a timely manner. The bottom portion of Figure 13.10 shows how the blocks of data might look as they pass down the link from the left side to the right.

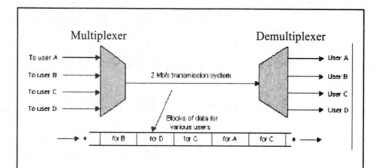

Figure 13.10: Using time-division-multiplexing (TDM) to share a digital transmission system among many users.

This method of sharing the time on the link is called **time-division-multiplexing** or **TDM**, and it can be subdivided into two classes. The type discussed above is often called **statistical multiplexing** because the users might show up randomly and ask to use the link. This is quite efficient, and works well, as long as the users are prepared to wait if they are told that the link is already occupied. The other type is called **synchronous TDM**. It allocates the link into fixed sized portions for each user and makes these portions available on a regular rotating basis. Telephone systems use this method because it is simple and all paying users are guaranteed to get their portion at exactly the time they want it.

Experiment 13.2: Time Division Multiplexing (TDM)

Time division multiplexing permits two or more users to share the same transmission link by letting each user send its data exclusively for a while and then wait for the others to send theirs before sending more.

Another multiplexing method appropriate to optical fiber systems is shown in Figure 13.11. It is called **wavelength-division-multiplexing** or **WDM**, and it is based on the idea that a good optical fiber can carry light in many colors at the same time without mixing them up. To exploit this we can replicate the scheme shown in Figure 13.4 several times, once for each color (that is, frequency or wavelength) that we would like to add. In Figure 13.11 three input streams of digital data enter the transmitter. Each is used to turn a laser or light-emitting diode (LED) on and off. These lasers or LEDs are assumed to produce light at frequencies (wavelengths) sufficiently different from each other that they can be cleanly separated from each other when they arrive at the receiver. In the figure they are shown as red, green, and blue in the visible range, although in practical fiber optical systems they are typically different shades of infrared. The light sources are combined into a single fiber and sent toward the receiver. At the receiver a prism or similar device is used to separate the colors, and each goes to its own optical detector.

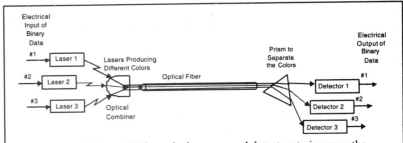

Figure 13.11: Using multiple optical sources and detectors to increase the transmission capacity of an optical fiber system.

This approach can be used in wire and radio systems as well. In that context the method is usually called **frequency division multiplexing (FDM)** since the available frequency spectrum is divided up among many users. An example of this appears on the AM radio dial where the 550 kHz to 1650 kHz band is split up into channels for more than 500 radio stations. All stations can broadcast simultaneously without interfering with each other as long as each stays within its allocated frequency range. When a user selects a specific radio station, a bandpass filter in the receiver is tuned to select the frequency range of the desired station while rejecting all other frequency ranges. The receiver behaves very much like the prism in Figure 13.11 in that it separates the incoming signal into the part that is in the desired frequency range and the part that is not in that range. Then it moves the desired part to a frequency range where the user can hear it as a normal audio frequency signal.

Experiment 13.3: Frequency Division Multiplexing (FDM)

Frequency division multiplexing (FDM) permits two or more users to share the same transmission link by giving each user exclusive use of a portion of the link's bandwidth that does not overlap any others.

Section 13.5: Summary

In the previous chapters we demonstrated that communications systems could transmit data from one location to another and that there were a variety of methods for generating signals to carry the message information. In this chapter we explored a variety of ways to convey or transport the information signal.

The key to the reliable operation of a communications system is the ability to generate signals that can be distinguished from each other at the receiver. A failure to detect the presence of a signal, or to distinguish which one it is, leads to errors in the information passed on to the intended recipient.

However, even if the transmitter produces signals that could be distinguished at the receiver if received perfectly, errors often occur anyway since the signals are degraded by their transmission channel in some way or because noise or interference enters the receiver along with the signal of interest, thus confusing the detector inside the receiver. We also discovered that techniques used to increase the data rate tend to make the problem of detecting the data reliably more difficult. Slowing down often results in a lower bit error rate.

Economic realities will encourage the owner of a digital telecommunications system to share the system among as many users as possible. The techniques that permit this are called multiplexing and are widely applied. Time-division multiplexing splits the time on the transmission medium into nonoverlapping time segments and allocates the segments among the users. Frequency-division multiplexing splits the total bandwidth of the medium into separate nonoverlapping frequency bands that are each allocated to different users. The same concept when applied to optical systems is called "wavelength division multiplexing (WDM)" and is directly responsible for the explosion in the availability of "bandwidth" for advanced Internet and telephony systems.

With all of this background we turn in the next chapter to study in detail a very important class of transmission systems, those that use radio waves to convey signals from the transmitter to the receiver without wires or optical fibers. These systems are generally called wireless systems.

Section 13.6:
Glossary

bit error rate (13.3)
the ratio of the incorrect bits to total bits in a received data signal

channel (13.1)
the medium over which a communications signal is carried

data transmission protocol (13.2)
a prearranged procedure for transmitting data including how to start and end
a message and what to do if an error is detected

frequency division multiplexing (FDM) (13.4)
sending two or more signals at the same time by assigning nonoverlapping
frequency ranges to each one; the receiver can then separate them by using
a filter to select only at the frequency range of the signal of interest

electromagnet
a device that becomes a magnet when current flows though its coil and
ceases to be a magnet when the current stops

full-duplex (13.2.1.1)
an operational communications mode in which the transmitters at each end
of the communications link can operate simultaneously; this allows data
transmission in both directions at the same time so that at either end the user
can both "talk" and "listen"

half-duplex (13.2.1.1)
an operational communications mode in data tranmission in both directions
is possible, but only one transmitter can operate at a time; this allows data
transmission in one direction at a time so that one user can "talk" while the
other "listens," and then they can switch roles

key (13.2.2)
an electrical switch that can be opened or closed manually; it was first used
to send telegraph signals

multiplexing (13.4)
using the same transmission medium to send two or more different signals
at the same time; it literally means "many things in the same place"

sounder (13.2.2)
a device used at the receiving end of a telegraph system to make sounds
corresponding to the dots and dashes sent by the transmitting telegrapher
with his or her key; the receiving telegrapher records the message by

listening to the sounds; the movement of the sounder is controlled by an electromagnet

time-division multiplexing (TDM) (13.4)
the sharing of a transmission medium among multiple users by permitting each user to have time intervals of exclusive usage

transmission rate (13.3)
the rate at which information can be communicated over a transmission link; it is measured in communication units per time interval such as words per minute, characters per second, or bits per second

wavelength-division multiplexing (WDM) (13.4)
sending two or more message signals at the same time along a light fiber by using a different color for each message stream

Section 13.7: References

[1] Lebow and Irwin, *Information Highways and Byways*, IEEE Press, Piscataway, NJ, pp 9, 10, 1995.

Figure References:

[13.1] <u>Boy Scout Handbook,</u> courtesy of the Boy Scouts of America, Inc.

[13.2] <u>Boy Scout Handbook,</u> courtesy of the Boy Scouts of America, Inc.

[13.3] Courtesy of the United States Navy

[13.5] Courtesy of Rich Taylor, Applied Signal Technology, Inc.

[13.6] Courtesy of Sally L. Wood

[13.8] Courtesy of <u>The World Book Encyclopedia,</u> © 1955

[13.9] Courtesy of the National Autonautics and Space Administration (http://vraptor.jpl.nasa.gov/voyager/l_voyager.html)

The INFINITY Project

Chapter 14: The Wireless World

Approximate Length: 1 Week

Section 14.1: Introduction

Telecommunications transmission systems fall into two broad classes. The first class uses some physical connection between the transmitter and receiver to carry or guide the user's signal. Telephone wire and optical fiber are two examples of this class. The second class uses electromagnetic waves to span the distance from the transmitter to the receiver wirelessly. We call the systems that fall into this class radio communications systems. This chapter will examine how such radio systems are built and identify the key issues associated with the design and operation of radio communications systems.

Section 14.2:
The Structure of a Radio Communications System

The block diagram of a simple radio communications system is shown in Figure 14.1. The input is assumed to be a stream of bits and might represent any type of multimedia data. The bit stream is the input to the **transmitter**. The transmitter uses this bit stream to produce a signal that can be efficiently transmitted by the **antenna**, usually through air or space. This transmitting antenna accepts the radio frequency electrical signal from the transmitter and converts it into a "radio wave," an electromagnetic wave that travels to the receiving antenna. The receiving antenna converts the electromagnetic energy impinging on it into an electrical signal, that is then fed to the **receiver**. As described in Chapter 7, the receiver's function is to recover the transmitted bit stream from the received signal. These recovered bits are then sent by the receiver to the intended user at the destination. The major difference between radio communications systems and the acoustic communications systems of Chapter 7 is that the information is carried by high frequency electromagnetic waves rather than low frequency sound waves. The transmitter and receiver will have components that allow them to send and receive information using these high frequencies instead of the audio frequencies in the frequency range that humans can hear them.

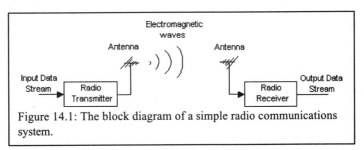

Figure 14.1: The block diagram of a simple radio communications system.

14.2.1 Components of a Radio Communications System

The Transmitter

Figure 14.2 illustrates a radio communications system in more detail. The transmitter typically consists of three key elements. The first is the **oscillator**, a circuit that generates a sinusoidal signal at the **radio frequency (RF)** which will be transmitted. This signal is often referred to as the **carrier wave**, since it is the signal that "carries" the originator's information. In commercial AM and FM broadcast radio, the frequency of the carrier wave is used to identify the station. Typical radio frequencies are in the thousands, millions, and billions of Hertz, or cycles per second.

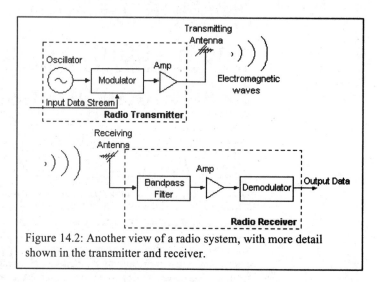

Figure 14.2: Another view of a radio system, with more detail shown in the transmitter and receiver.

The sinusoidal signal from the oscillator carries no information about the bit stream to be transmitted. This is added by the **modulator**, which influences the amplitude, frequency, or phase of the sinusoid from the oscillator. This modulation, if done properly, modifies the transmitted signal in a unique way for each possible set of transmitted bits. This makes it possible for the receiver to accurately recover the bits and pass them on to the user.

The third element of the transmitter is the **power amplifier**, which takes the signal from the modulator and boosts its power to the level needed to reach the intended receiver. The output of this amplifier is conveyed to the transmitting antenna where is it converted into electromagnetic waves.

The Receiver

Figure 14.2 also shows the receiver in more detail. Like the transmitter, the receiver consists of three key elements. The electrical signal from the receiving antenna is first **filtered** so that the desired radio signal is not changed, but signals at other radio frequencies are removed. Obviously the frequency range of the receiver bandpass filter must match the frequency range of the selected transmitted radio signal for this signal separation to be effective.

After being filtered, the signal is amplified to make it strong enough for the third stage, which is the **demodulator**. The demodulator examines the incoming signal and extracts the information from the carrier wave that was put there by the transmitter's modulator. Once extracted, these bits are sent out to the user.

It is important to note that the transmitter and receiver must be designed and deployed as a pair. As noted above, the frequency of the transmitter's oscillator, and therefore the desired signal, must match the frequencies that the bandpass filter accepts at the receiver. In addition, the type of modulation applied at the transmitter must be known at the receiver so that it can be properly removed and the data can be recovered. This is another form of the **prearrangement** discussed in Chapter 7. For example, if the transmitter were to send information as frequency variations, a receiver looking for amplitude variations to extract the information would be unable to recover the transmitted data.

14.2.2 Example of a Radio Communications System

Figure 14.3 shows a well known example of a radio communications system, a modern cellular telephone and an associated cellular base station. Cellular telephones contain both a receiver and a transmitter which each communicate with a corresponding transmitter and receiver at a base station. The cellular telephone transmitter and receiver operate at the same time but at different frequencies so they will not interfere with each other. This bidirectional communication allows the phone's user to talk and listen at the same time, a mode of operation usually called **full duplex communication.**

Figure 14.3: (a) A cellular telephone handset and (b) a typical cellular base station.

The transmitter of this cellular phone uses carrier waves operating at about 850 MHz and it communicates with a receiver at the closest base station. It adds the speech information to the carrier by changing the frequency of that carrier wave slightly in response to the user's speech signal using a modulation technique called **frequency modulation** or FM. The cellular telephone transmits with a maximum power level of about 300 milliwatts.

The cellular phone's receiver operates at a slightly different carrier frequency and communicates with a transmitter at the base station. The receiver extracts the speech from the incoming amplified, filtered signal, and sends it to the earphone of the cellular handset. The antenna of the cellular telephone, which is shared by both the transmitter and the receiver, is designed to operate simultaneously at both the transmitting and receiving frequencies.

While cellular phones in common use convey voice signals, we might suspect that they could be used to carry data signals as well, since voice, images, and data can all be represented by bit streams. For example, we could use the tone-based schemes from Chapter 7 to encode text messages as acoustic signals, send them from one cellular phone to another, and then extract them at the receiving end from the cellular telephone's audio output.

A closer examination shows this to be a case of double modulation and demodulation. The text is first converted into one or more tones, which is the first modulation step. Then these tones are frequency-modulated onto a radio-frequency carrier wave just as if they were speech for the second modulation step. In principle it should be possible to directly alter the carrier wave to carry the text information or its binary equivalent. Most practical systems, including more modern digital cellular phones, use this simpler one-step process.

Exercise 14.1

1. Why must the transmitter and the receiver of a cellular telephone operate at slightly different frequencies? What would happen if they operated at the same frequency?

2. Identify for each of the following communications systems the transmitter, the receiver, and the distance over which the system usually operates:

   - Broadcast television

   - Cellular telephone

   - TV remote control

   - Garage door opener

   - Pager

   - Wireless personal digital assistant (PDA)

   - Global Positioning System (GPS)

   - Cordless telephone

   - AM broadcast radio

   - FM broadcast radio

   - Electronic door lock for an automobile

Section 14.3:
The Wavelength of an Electromagnetic Wave

Radio communications systems employ a band of operating frequencies that is only a small part of the complete frequency range over which electromagnetic waves can be generated, **propagated**, and received. Table 14.1 lists many of these and their associated operating frequencies, so that radio communications can be put into perspective. Note that infrared, visually perceived light, and even cosmic rays are electromagnetic waves, as well as radio signals. All of these forms of electromagnetic waves propagate at the same speed (at least in a vacuum) of about 300,000,000 meters/second. The letter **c** is commonly used to represent the speed of light, and hence all of these forms of electromagnetic radiation.

An important consideration in the design of any radio system is the type of antenna to be used. In order to design an antenna we need to know more about electromagnetic waves as they travel, since it is the antenna's goal to produce and receive such waves.

	Type of signal	Frequency (Hz)	Period (secs)	Wavelength (cm)
	Extremely long wave	$< 30 \times 10^3$	$> 3.33 \times 10^{-5}$	$> 10^7$
↑	**Long wave**	$30 \times 10^3 - 300 \times 10^3$	$3.33 \times 10^{-6} - 3.33 \times 10^{-5}$	$10^7 - 10^5$
Radio	**Medium wave**	$300 \times 10^3 - 3 \times 10^6$	$3.33 \times 10^{-6} - 3.33 \times 10^{-7}$	$10^5 - 10^4$
	Short wave	$3 \times 10^6 - 30 \times 10^9$	$3.33 \times 10^{-7} - 3.33 \times 10^{-8}$	$10^4 - 10^3$
↓	**Microwave**	$3 \times 10^9 - 3 \times 10^{12}$	$3.33 \times 10^{-10} - 3.33 \times 10^{-13}$	$10 - 0.01$
	Infrared	$3 \times 10^{12} - 4.3 \times 10^{14}$	$3.33 \times 10^{-13} - 2.33 \times 10^{-15}$	$0.01 - 7 \times 10^{-5}$
	Visible	$4.3 \times 10^{14} - 7.5 \times 10^{14}$	$2.33 \times 10^{-15} - 1.33 \times 10^{-15}$	$7 \times 10^{-5} - 4 \times 10^{-5}$
	Ultraviolet	$7.5 \times 10^{14} - 3 \times 10^{17}$	$1.33 \times 10^{-15} - 3.33 \times 10^{-18}$	$4 \times 10^{-5} - 10^{-7}$
	Xray	$3 \times 10^{17} - 3 \times 10^{19}$	$3.33 \times 10^{-18} - 3.33 \times 10^{-20}$	$10^{-7} - 10^{-9}$

Table 14.1: Relationship of wavelength, period, and frequency for electromagnetic waves.

14.3.1 Wavelength and Frequency

Figure 14.4 shows the "traveling" electric fields from radio waves produced by a transmitters operating at three different frequencies. Notice that the cosines shown in this figure are functions of distance, not just time as in earlier chapters. All of the waves travel at the same speed, that is, the speed of light, **c**. Because they are operating at different carrier frequencies, each completes a different number of cycles from positive to negative and back to positive again during one second. The signal with the highest frequency oscillates the fastest, while the signal with the lowest frequency oscillates the slowest.

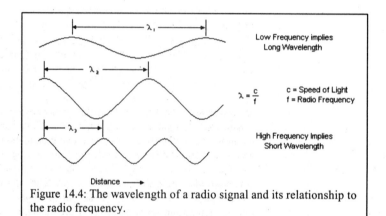

Figure 14.4: The wavelength of a radio signal and its relationship to the radio frequency.

In order to understand these variations for a wave that travels, we must explore the relationship between the wavelength and the temporal period of a propagating wave. The **wavelength** of a propagating signal is defined as the **distance** between adjacent amplitude peaks in a traveling electromagnetic wave when viewed at the same instant in time. This differs from the temporal **period** of a signal, which is the **time** between adjacent amplitude peaks in the wave when viewed at the same physical position. When the frequency of a signal and its speed of propagation are known, the wavelength can be computed. This is an important characteristic of a propagating wave because this measure of distance is related directly to the physical size of an antenna that will best respond to electromagnetic waves at the radio frequency of the signal.

The period T is simply the reciprocal of the carrier frequency f, or $T=1/f$, since the period measures seconds per cycle and frequency measures cycles per second at one location. For waves travelling or propagating at the same speed, **c**, the distance between the adjacent peaks in meters per cycle can be computed easily as the propagation speed in meters per second multiplied by the period in seconds per cycle. The Greek letter λ is used to represent the wavelength as shown in Equation 14.1.

$$\lambda = c*T = c/f \qquad\qquad 14.1$$

The effect of the definition in Equation 14.1 is that signals with low frequencies have long wavelengths while those with high frequencies have shorter wavelengths. This inverse relationship is shown in Figure 14.4. A

8

signal with high frequency completes each cycle in a brief time, and therefore has not traveled very far, while signals with low frequencies take a relatively long time for each cycle, and therefore travel further between the peaks. Thus they have longer wavelengths. Table 14.1 shows the relationship of frequency, period, and wavelength for several classes of electromagnetic waves.

14.3.2. A Mathematical Description of Traveling Waves

When engineers and scientists study radio communications systems they find it useful to have a mathematical description for the signal that will be carried on traveling electromagnetic waves. In re-examining Equation C.1 we find it inadequate since it only describes a signal in one location, that is, s(t) is a function of only time, and not time and distance. We will now expand our signal definition so that it includes distance as well.

$$s(t) = a * \cos(2\pi f t + \phi)$$ C.1

As described in Chapter 7, the general equation for a sinusoidal signal defines the instantaneous amplitude of the signal s(t) as a function of one variable, time, which uses the symbol t, and three parameters which are the maximum amplitude a, the frequency f, and the phase angle ϕ. A propagating or "traveling" wave will be a function of two variables, time and position, which uses the symbol x, and can be written as s(x, t). It will also be a function of a fourth parameter, the speed of propagation c. To represent a propagating sinusoidal wave mathematically, we must have a cosine that is a function of both time and distance. In addition, the value must be unchanged by an advancement in time of exactly one period, T, or by a movement in space of exactly one wavelength, λ. This relationship is expressed in Equation 14.2:

$$s(x,t) = s(x,t+T) = s(x+\lambda,t)$$ 14.2

When (t) in Equation C.1 is replace by (t - x/c), as shown in equation 14.3, this relationship will be valid since cos(θ +2π) = cos(θ). From the definition of the period, it is clear that $2\pi f T = 2\pi$, and from Equation 14.1 we see that $2\pi f(\lambda/c) = 2\pi$.

$$s(x,t) = a * \cos\left(2\pi f \left(t - \frac{x}{c} \right) + \phi \right)$$ 14.3

A more intuitive way to write this equation is shown in Equation 14.4 where f has been replaced by 1/T and moved inside the inner parentheses. Here (t/T) represents the number of periods or cycles in time interval t, and (x/λ) represents the number of wavelengths or cycles over the distance x. Additional observations about these equations will be found in the exercises at the end of this section.

$$s(x,t) = a * \cos\left(2\pi \left(\frac{t}{T} - \frac{x}{\lambda} \right) + \phi \right)$$ 14.4

9

14.3.3 The Effects of Wave Propagation on the Best Frequency for an Application

While, theoretically, electromagnetic waves can have very low frequencies and very high frequencies, most radio waves of practical interest fall in the range between about 75 Hz and 40 GHz. The upper limit is almost a billion times higher than the lower. Why would we be interested in such a wide range of frequencies? Why won't just one nominal set of frequencies satisfy all of our communications needs? The answer to that hinges on the fact that radio waves of different frequencies, and hence wavelengths, interact with the earth and its atmosphere in different ways. Specific applications may benefit from the behavior of one frequency range compared to another. These behaviors are illustrated in Figure 14.5.

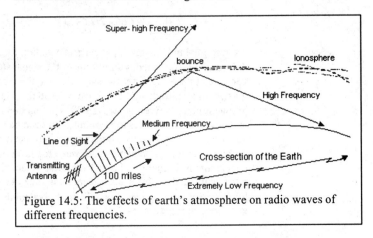

Figure 14.5: The effects of earth's atmosphere on radio waves of different frequencies.

Radio waves of very low frequencies (about 100 Hz) tend to travel right through the earth, and through the seawater that covers 70% of the earth. The wavelengths of such radio waves are on the order of 3000 Km, which is about half the radius of the earth. In contrast, radio waves with frequencies of about 1 MHz tend to follow the earth's curvature for distances of about one hundred miles. Radio waves with frequencies from about 3 to 30 MHz, often called short waves, will "bounce" off the ionosphere and return to the earth's surface, while signals with frequencies above 50 MHz or so are often termed "line-of-sight" signals since they travel in straight lines as light would and tend not be reflected, ducted, or refracted by the earth or its atmosphere.

How might an engineer take advantage of these differences? Table 14.2 shows a key application within each of these frequency ranges. The ability to communicate through the earth and seawater makes the extremely low frequency (ELF) band well suited to communicating with submerged submarines. The ability of high frequency (HF) signals to bounce off of the ionosphere and return to earth thousands of miles away makes them well suited to short wave communication to distant points in the world. The ability of superhigh frequency waves to pass right through the ionosphere without reflection and keep on going into space makes them well suited to communication with satellites. This microwave frequency range has wavelengths that are less than 10 cm, very short compared to the other ranges that we just discussed.

	Frequency Range	Typical Applications
Extremely low frequency	30-300 Hz (ELF)	Through-the-earth communications with submarines
Medium frequency (MF)	300-3000 kHz Ground wave	AM broadcast radio
High Frequencies (HF)	3-30 MHz "short waves"	Inexpensive long distance communications
Superhigh Frequencies (SHF)	3-30 GHz "microwaves"	Satellite communications, radar

Table 14.2: Taking advantage of the propagation properties of radio waves of different frequencies.

In general, the radio frequency chosen by the designer for a particular communications application will be strongly influenced by its propagation behavior, which is the effect that the earth and its atmosphere will have on the propagating signal. The second major effect that the choice of radio frequency has on a radio communcations system is on the physical size of the antennas used in the system. To understand why this would be true, we need to understand how antennas work.

Exercise 14.2

1. How long would it take an electromagnetic wave travelling at the speed of light to reach a receiver 200 km away from the transmitter?

2. A cellular telephone uses transmitting and receiving frequencies near 850 MHz. Compute the period and the wavelength for this frequency if the propagation speed is the speed of light.

3. Newer digital telephones use frequencies of about 1900 MHz. Repeat Exercise 14.2.2 for this frequency.

4. In equation 14.3, show that (t) and (x/c) are expressed in the same units of measurement.

5. In equation 14.4, find the units of measurement of (t/T) and (x/ λ).

6. Show that the relationships of equation 14.2 are true for equation 14.4. (Hint: Let θ be the whole argument of the cosine in equation 14.4 so that $s(x,t) = a* \cos(\theta)$. Then see how θ changes when $t+T$ is used in place of t.)

7. A radio wave is generated by a transmitter that has a carrier frequency of 10 MHz. What is the period T of this carrier wave? What is the wavelength of the electromagnetic wave once it leaves the antenna?

8. The radio waves produced by a particular type of radar has a wavelength of 3 cm. What frequency does this radar operate at? Using the information in Table 14.2, identify the name of the band of frequencies that this radar operates in.

9. When visible light travels through a vacuum or the atmosphere, its speed is very close to 300,000,000 meters/second. When it travels through glass, however, it slows to about two-thirds of that speed. What is the wavelength of the light when traveling though the glass, compared to its wavelength through air or a vacuum?

Section 14.4:
The Design of Antennas

14.4.1 Dipole Antennas

Antennas for radio communications systems come in virtually all sizes and shapes. However, there are some basic principles that guide how they are designed. Virtually all antennas have as their root design the simple **dipole** shown in Figure 14.6. The dipole consists of two metal rods, both of the same length. They are positioned in line with each other and almost touching. The transmitter sends an electrical signal to the antenna on a pair of wires called the **transmission line**. These wires connect to the dipole at the point where the rods almost touch. One of the wires is connected to one rod and the other line to the other rod.

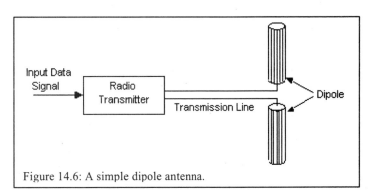

Figure 14.6: A simple dipole antenna.

When the total length of the dipole has a special relationship to the radio frequency or signal wavelength used by the transmitter, then the antenna is most efficient at creating a strong signal. Under this circumstance virtually 100% of the energy reaching the transmitting antenna via the transmission line is converted from electrical signals into propagating electromagnetic waves. Since almost all of the transmitter's energy is converted into radio waves, this is the most efficient operation possible. The same relationship applies to the receiving antenna. A propagating radio wave impinging on a receiving antenna of the proper length will be completely converted into an electrical signal that can be conveyed over a short pair of wires to the receiver.

The laws of physics can be used to explain why this relationship between wavelength and antenna size exists and to determine what the relationship must be for the most efficient operation of a dipole antenna. However, it is also possible to understand this relationship by imagining the physical dimensions of the propagating waveform relative to the physical dimensions of the dipole as shown in Figure 14.7. The answer is that the dipole's total length must equal exactly one-half of the wavelength, λ. This can be related to the operating frequency of the transmitter and receiver pair since, from Equation 14.1, $\lambda = c/f$. Thus we see that the end-to-end length of the perfect dipole is given by

$$L_{dipole} = \lambda/2 = c/2f. \qquad\qquad 14.5$$

Figure 14.7 shows an electric field portion of a radio wave with a wavelength of exactly twice the dipole's length, which is the relationship needed for the most efficient operation. Note that in this case the far ends of the dipole correspond in location to the most negative peak and the mo st positive peak of the electric field. If one thinks of the antenna as sampling the radio wave as it passes by, the biggest signal the antenna can sense occurs when the most positive value is on one monopole and the most negative value is on the other. This puts the largest possible voltage difference across the antenna. No other length of the dipole can sample a greater difference in the electric field. Hence, this is the optimal dipole length for converting a signal on a transmission line into a propagating signal and also for the reciprocal conversion of a received wave into an electrical signal.

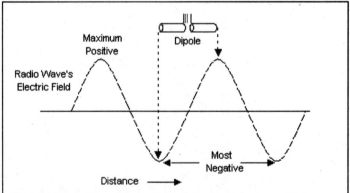

Figure 14.7: An illustration of how a dipole of the most efficient length corresponds to the radio wave it is intended to transmit or receive.

14.4.2 Antennas Based on Dipoles

Few radio systems use dipoles directly for their antennas, but virtually all antenna designs start with the dipole as the bas is for the design. A few of these designs are worth noting here because we see them in our daily experience. An example of each is shown in Figure 14.8.

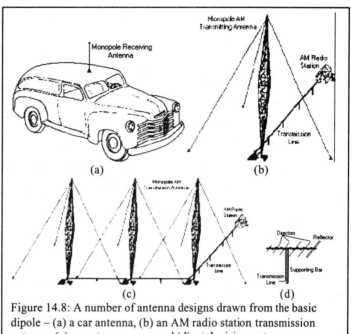

Figure 14.8: A number of antenna designs drawn from the basic dipole – (a) a car antenna, (b) an AM radio station transmission antenna, (c) an antenna array, and (d) a television antenna.

Figure 14.8(a) shows a receiving antenna on a car. This is often called a **monopole**, since it can be recognized as the top half of a dipole. The wires to the receiver in the car are connected to the base of the monopole antenna. Monopoles are almost always positioned over a large piece of metal, such as a car, or over the moist earth itself. In these cases, the earth or the metal tends to act like the missing monopole which would have been the other half of the dipole. Since the monopole length is half of the effective dipole length, its most efficient length can be computed by equation 14.6.

$$L_{monopole} = 0.5 * L_{dipole} = \lambda/4 = c/4f. \qquad 14.6$$

Conversely, if one sees a monopole antenna, one can estimate its most efficient operating frequency using the formula of equation 14.7 where f is the frequency in Hz and $L_{monopole}$ is the length of the monopole in meters.

$$f = c/ (4\ L_{monopole}). \qquad 14.7$$

Figure 14.8(b) shows another monopole. This one is the transmitter for an AM radio station. Suppose that we measured the height of the tower to be 100 meters. From Equation 14.7 we would deduce that the probable operating frequency of the radio station is about

$$(300000\ km/s) / (4 \times 100\ m) = 750\ kHz,$$

which is a frequency in the lower end of the AM radio band.

Figure 14.8(c) shows another variation, an **array** of antennas. In this case three monopoles are lined up near each other and the electrical energy from the transmitter is shared between them. The best operating frequency is still determined by the length of each monopole. Why then would a designer build such an array rather than use a single monopole? A single **monopole** or dipole propagates its energy uniformly in all directions. This is called

omnidirectional propagation. However, by carefully distributing the energy to individual monopoles arranged in an array with different relative time delays or phases, it is possible to focus or shape the distribution of the electromagnetic waves leaving or entering the array. The array shown in Figure 14.8(c) belongs to an AM radio station that uses the focusing effect at night to send its signal much further in one direction, presumably toward customers, at the expense of sending less signal in other directions, such as the ocean where presumably there are fewer listeners.

The last common variation of the dipole is shown in Figure 14.8(d). It is a receiving antenna for a television set receiving broadcast transmission rather than cable. A careful examination of the picture shows it to be a large number of dipoles, all of different lengths and all mounted on the same bar. Why is this antenna designed this way? The answer comes from the fact that television signals are sent out on a wide range of radio frequencies from as low as 50 MHz (channel 2) to as high as 700 MHz (channel 60). Like radio stations, each TV channel has a specific carrier frequency assigned to it, although television channels are commonly identified by their channel numbers rather than their carrier frequencies. A single dipole of only one length cannot receive all TV channels with equal and high efficiency because such a wide range of frequencies requires an equally wide range of dipole lengths. A design solution for this case is to provide many different dipole sizes, corresponding to the range of frequencies to be expected from TV transmitters.

14.4.3 Antennas for Very High Frequencies

When radio transmitters operate at high enough frequencies, it is common to use concepts and approaches from optics for the design of the antennas. An important example is the **parabolic** antenna, often referred to simply as "the dish." A drawing of one is shown on the left side of Figure 14.9, while the right side shows a photograph of a very large parabolic reflector used for radio astronomy to receive signals from naturally radiating celestial objects. Instead of using a properly phased set of dipoles to focus the transmitter's energy in a specific direction, the metallic dish reflects the energy spreading away from a single dipole in such a way as to form a directional beam. The concept of the parabolic reflector is applied to other types of electromagnetic radiation as well. Flashlights and automobile headlights use the same principle to focus optical radiation into tight beams. The degree of focusing by the parabolic reflector is proportional to the size of the dish and inversely proportional to the wavelength of the signal being generated by the transmitter.

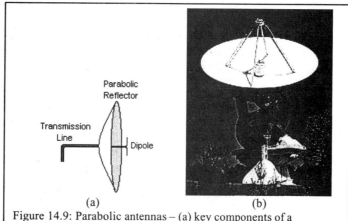

Figure 14.9: Parabolic antennas – (a) key components of a parabolic dish antenna and (b) a picture of a parabolic antenna used for radio astronomy.

The parabolic antenna is effective at directing the omnidirectional output of the dipole because of the mathematical definition of its reflecting surface. Consider a simple two-dimensional parabolic curve in which $y = ax^2$. It can be shown (see Exercises 14.3, 4-6) that a parabolic curve represents the set of all points $Q = (x_q, y_q)$ such that the sum of the distances from a horizontal line $y = C$ to point Q and from point Q on to the "focal point" $P = (x_p, y_p)$ is constant. A curve defined by this relationship has the property that all lines emanating from the focal point P and reflected from the parabolic surface will travel in a direction parallel to the vertical line passing through the focus point P.

These parabolic reflection properties are illustrated in Figure 14.10. The parabolic curve represents all points Q as defined in the previous paragraph. Two specific points are shown in Figure 14.10a. Here two rays travel in *different* directions from the focal point P to the parabolic surface. At the surface they are reflected with the angle of incidence equaling the angle of reflection. Then the two reflected rays both travel in the *same* direction parallel to the vertical axis. On the right side of the figure, ten different rays are shown reflecting from the surface at ten different Q points and then travelling away in the same direction.

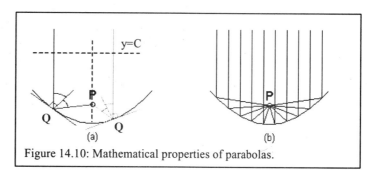

Figure 14.10: Mathematical properties of parabolas.

There are two important properties of the parabolic reflecting surface. One is that rays may leave the focal point P traveling in any direction, but after reflection they all travel in the same direction so the energy can be focused in a specific direction. This is clearly illustrated in Figure 14.10. The second is that the path lengths to a distant point will be the same length, so the signal focused in this direction will arrive coherently. The implication of these two properties is that a dipole transmitting antenna positioned at the focal point of a parabolic surface will have most of its energy directed along the axis of the parabola. Similarly, all of the incoming signals from a distant source parallel to the axis of a parabolic surface will be reflected to converge on a receiving dipole antenna at the focal point.

14.4.4 General Observations About Antennas

Three general observations about the design and use of antennas should be noted. The proper physical size of an antenna depends directly on the wavelength of the signal to be transmitted or received. This is obvious in the case of dipoles and monopoles, but it is also true for arrays and parabolic dishes as well. Assuming that performance and efficiency are held constant, the size of the antenna is inversely proportional to the operating frequency, f.

However, not all antennas are operated at their most efficient length. For example, an AM radio antenna for a car cannot be the 150 meters in length needed for optimal performance. The antenna is made smaller, so that it fits on the car. The fact that it operates inefficiently but the complete system still works is due to compensating factors such as more powerful transmission from the source of the radio signal.

It should be noted that virtually all antennas are "reciprocal" in that they can transmit as well as they receive, at a given frequency, and in a given direction. It is the design of the rest of the system around the antenna that determines which of the two functions it will perform.

Exercise 14.3

1. What percentage of the maximum possible signal amplitude difference would appear across the antenna if the dipole were exactly one wavelength long? How would this change if the length were increased to 1.5 λ?

2. Compute the most efficient length for a dipole antenna for a radio communications system using a frequency of 1200 kHz.

3. How long should a monopole antenna be for a cellular telephone as described in Exercise 14.2.2? How does this length compare with your personal experience with cellular telephones?

4. A parabolic surface will be defined with the focal point P at the origin. This means that for all points Q = (x_q, y_q) on the parabola, the sum of the distance from point Q to the origin and the distance from point Q to the line y = C is a constant value. Find the mathematical relationship between x_q and y_q. (Hint: If the distances are equal, then the squares of the distances are equal.)

5. A parabola is often defined mathematically as the set of all points that are equidistant from a focal point P and a specific line. For the parabolas shown in Figure 14.10, this line would be below the parabolic surface. Show that this definition is equivalent to the definition in Section 14.4.3 of a parabola as the set of points where the sum of the distances from the point to the focal point and from the point to a line is a constant. For the parabolas shown in Figure 14.10, this line would be above the parabola's focal point, such as the line connecting the end points of the rays shown.

6. The tangent line to a parabola y = ax^2 at point Q = $(x_q, a*x_q^2)$ makes an angle α with the x-axis such that $\tan(\alpha) = 2a*x_q$. Show that this is also the angle of reflection at that point on the parabolic surface. (Hint: You do not need to explicitly identify the focal point to do this.)

7. Find the frequency bands set aside for AM broadcast, FM broadcast, UHF television, and "personal communications services (PCS)" using Table 14.3.

8. Suppose that an antenna is mounted on a tower with height h above the earth's surface and it is used as part of a radio system that communicates using line-of-sight frequencies. Find the antenna's effective range to the earth's horizon by computing the length of a line from the top of the antenna to the line's tangent point to a spherical earth. Assume that the earth has a radius r which is much larger than h so that terms in h^2 can be ignored relative to terms in $r*h$.

19

9. Assume that power from an antenna is radiated uniformly in all directions so that the power is reduced by a factor of r^2 where r is the distance from the antenna. Assume further that a transmitter operates with a power level of 10 watts, and that at a distance of 1 mile an antenna receives 1 milliwatt.

   (a) If the same type of antenna is used, what power is delivered to a radio receiver 10 miles away? 100 miles away?

   (b) A large clear channel station may be broadcasting with a power of 50 kW. What power would it deliver to receivers at those same distances used in part (a)?

10. A submarine is trailing a long antenna wire to receive a 75-Hz signal. Compute the appropriate length for a monopole antenna at this frequency and comment on the feasibility of using a trailing antenna of this length.

11. Is a parabola the only shape that can be used to focus radio waves? Hint: Check out a DirecTV receiving antenna and determine whether or not it can be parabola.

12. Navy ships use vertical monopole antennas that are 10 meters tall. At what frequency will these antennas most efficiently transmit and receive?

Section 14.5:
Considerations in the
Design of Radio Systems

The design of radio systems depends on a broad number of considerations, both economic and technical. While generalizations are difficult to make, there are some that are worth noting. They are also useful when trying to understand why an existing communications system was designed the way it was.

Older radio communications systems used analog modulation to put information onto the carrier wave, while more modern systems tend to do this digitally. There are many schemes for both approaches. It is interesting to note that the first commercial use of radio transmission was for wireless transmission of a digital signal, Morse code, to ships at sea.

Some radio systems are used for broadcasting to users in all directions, while others are used for "point-to-point" applications. The antennas needed for these systems are usually chosen with the application in mind. For example, omnidirectional antennas are often used for broadcasting systems while parabolic dishes are used almost exclusively for point-to-point systems. In radioastronomy the antenna dish is pointed at a particular celestial object while an automobile monopole radio antenna receives signals from all nearby broadcasting stations regardless of their location relative to the position and orientation of the vehicle.

Radio signals spread out from the transmitting antenna and become progressively weaker with the square of the distance from the antenna. In order to receive the signal properly, enough power must be transmitted to overwhelm the noise and interference present at the receiver's input. The power level at which a transmitter operates must therefore grow with the distance the signal must traverse on its way to the receiver. Some operating frequencies are better than others for a particular type of application, owing to considerations of the required antenna size and the resulting signal's propagation through the atmosphere and the earth.

The ability of a carrier wave to carry data is roughly proportional to its frequency. The rule of thumb among engineers is that the number of bits per second carried on a radio signal should not exceed one-hundredth (1%) of the carrier frequency. This is not a strict limit, but the design of the radio's transmitter and receiver becomes much harder when it is violated. As a practical matter this rule means that building a communications system to handle higher data rates encourages the engineer to use higher radio frequencies. It was this consideration that first pushed radio systems into the microwave range.

14.5.1 Utilization of the Radio Frequency Spectrum

In this chapter we have shown that there are three important technical specifications that determine the best radio frequency for a communications system for a particular application. The propagation behavior of the radio waves will determine what frequency range is most suitable. The physical size of the antenna determines the operating frequency that will be most efficient. The required "bandwidth" of the signal to be carried will limit the choice of operating frequencies. However, there is another important consideration in choosing frequencies, and that is the law. Specifically, it is legally permissible to use only certain bands or ranges of frequencies for certain types of communications services, and transmitters often must be licensed.

An international organization that is a part of the United Nations meets once every four years and agrees to which parts of the radio spectrum shall be used for various applications. Each country then interprets these rules for its own citizens There are a number of reasons for this type of regulation , but the most important one is easy to understand. If anyone, in any country, were allowed to transmit in any direction, at any power level, then every receiver's input would be filled with interference. Radio communications would be impossible for virtually everyone. By imposing rules and regulations, globally and nationally, it is possible to minimize interference to the point that radio communications system can be operated reliably.

The level of regulation done at the international level is "the big picture" level. Sections of the radio spectrum are set aside for the whole world for communications systems of various types of general use. Examples include AM broadcast, as well as non-communications uses of radio waves such as radar and navigation systems like the Global Positioning System (GPS). Within these limitations, organizations in individual countries subdivide the allocated spectrum bands into smaller ones for different kinds of roughly similar services. In the United States the organization that does this interpretation is the Federal Communications Commission (FCC), which is in turn a part of the Department of Commerce. An internationally allocated band for mobile communications, for example, might be divided into four by the FCC, one for police radios, one for taxi radios, one for cellular phones, and one for the expensive phones on airplanes.

A second type of regulation is often imposed. Not only must a particular type of service be offered in only certain portions of the spectrum but in many cases the transmitter must be licensed. The owner of a transmitter and usually its operator must have a license to operate it. In many cases the license will further specify the frequency and power at which the transmitter can legally operate. Every transmitter then falls into one of the following three classes:

- Licensed transmitter operating at a specific frequency – examples: AM, FM, and television broadcasters (1 to 250 kilowatts)

- Licensed transmitters operating within a specific band of frequencies – such as cellular phones (0.3 to 3 watts) and amateur radio operators ("hams")

- Unlicensed transmitters operating within a specific band of frequencies – such as the wireless phones you might have at home or radio controlled toy cars. These very low power transmitters are limited to less than 100 milliwatts.

Systems with very powerful transmitters are tightly regulated, systems with moderate to low power transmitters are moderately regulated, and very low power systems are essentially unregulated. Powerful transmitters create more interference for others and at greater distances, and therefore mu st be regulated more carefully. This is particularly true when operating in radio frequency bands where the signal easily propagates over much of the world. Weak transmitters, and ones operating at frequencies where the signal is quickly absorbed by the earth or the atmosphere, are much less likely to interfere with others and are therefore regulated much less rigorously.

A portion of the FCC's spectral allocations chart for the United States is shown in Table 14.3. One can find it all at

http://www.fcc.g ov/oet/faqs/freqchart.html

Frequency Band	Application
300 – 325 kHz	Aeronautical navigation, maritime radio beacons
325 – 405 kHz	Aeronautical mobile systems
415 – 495 kHz	Maritime mobile systems
495 – 505 kHz	Mobile distrss frequency
535 – 1605kHz	Broadcasting AM radio
1615 – 1800 kHz	Radio location
1800 – 1900 kHz	Amateur radio
2495 – 2505 kHz	Standard frequency and time signal

Table 14.3: A Portion of the Federal Communications Commission's Spectrum Allocation Chart for the United States.

14.5.2 Extension to Optical and Infrared Communications Systems

Lightwaves are electromagnetic waves just as surely as radio waves are. From Table 14.1 we can see that they are at a higher frequency range than radio waves. Thus infrared systems (such as the TV remote control) and optical fiber systems might be considered to be "radio" systems as well. Theoretically they are, although in practice they are not. As a practical matter, radio's upper limit of carrier frequency is about 300 GHz. Above this frequency the wavelengths are shorter than a millimeter, making it hard to build the electronics and the antennas needed for radio systems. Above this frequency, and at wavelengths shorter than this, systems tend to be built using optical concepts and are considered optical systems.

23

Similarly, the radio spectrum extends all the way down to zero Hertz in theory. As a practical matter though, very few radio systems are built to operate at frequencies much lower than 500 kHz. Below this frequency the required antenna sizes are too large for practical use and higher power levels are required for high-quality reception. An important counter example is the submarine communications system discussed earlier. Its operating frequency, about 75 Hz, was determined completely by its need to propagate through the whole earth to reach the intended receivers.

Exercise 14.4

1. Experiment with the signal from a remote control unit for a television or CD player.

   a.) Aim the controller directly at the TV or CD player and determine how much you can rotate it to the left and right and still control the device.

   b.) Try pointing the controller away from the device so that the control signal can reflect off of a wall or window to get to the device. How does this work? Try reflecting the signal off of draperies or a sofa. Does that work?

2. Find an inexpensive AM radio receiver and position it near a photocopying machine while it is copying. What do you hear? Why is the photocopier not licensed as a transmitter?

3. Using the FCC's spectrum allocation chart, find the transmitter frequencies for channels 2, 3, and 4 for broadcast television. Where would you expect to find the frequency for channel 1 if it existed? What is the designated use of that frequency range? *Extra credit*: Why is there no channel 1 for television?

Section 14.6: Summary

Radio transmission systems are an important part of telecommunications, and therefore of multimedia engineering. In this chapter we have explored the design of radio communications systems and the underlying physical principles that determine design decisions.

Radio communications systems place data on a radio frequency carrier wave and apply it to an antenna. If the antenna has been properly designed, most of the transmitter's output will be converted into electromagnetic waves. These waves will propagate in directions determined by the antenna's design. When these waves reach the receiving antenna, they are converted back into an electrical signal from which the user's data is extracted with a demodulator.

The quality of the data extracted by the demodulator is a function of the strength of the received signal, the amount of noise and interference also arriving at the receiver, and the way that the data was modulated onto the carrier wave.

There are several important considerations in the choice of the operating frequency for a radio communications system. The signal must be capable of propagating to the intended receiver. The frequency must be high enough to carry the required data rate, and the required antennas cannot be either prohibitively large or too small to be useful. A very important consideration, and one enforced by substantial regulation, is that there must be low enough interference at the operating frequency for the signal to be accurately received.

Antennas are an integral part of radio communications systems. Their optimum physical size scales directly with the signal wavelength, which is inversely proportional to the operating frequency of the transmitter and receiver. An antenna does not need to be the perfect theoretical size to receive or transmit a radio wave but it operates less efficiently if it is not. The simplest antenna is the dipole. Most antennas which operate below about 1 GHz are based directly on the dipole, while ones which operate above that often combine the dipole with "optical" devices such as lenses and parabolic reflectors.

One of the main reasons for making antennas more elaborate than a single dipole is that the dipole sends its electromagnetic energy equally spread in all directions. By creating an array of dipoles or by placing a parabolic reflector next to a dipole, it is possible to direct the energy from the antenna nonuniformly. This ability to "focus" the transmitted or received signal is highly desirable in many applications. It permits transmitters to operate with less power, and receivers to reject interference that comes from directions other than that of the desired signal.

We have established that it is possible to communicate data over wires and over radio communications links. Designers of such systems are concerned with how to create the best systems possible. Designers want to know how to compensate for noise, interference, and degradations incurred on the way to the receiver and how to efficiently pack the most data in the least amount of bandwidth. The following chapter will address these issues.

Figure 14.11: The main communications tower in London, England – It carries antennas for many types of communications systems and services.

Section 14.7: Glossary

antenna (14.2)
a device that converts electrical energy into electromagnetic waves or electromagnetic wave into electrical signals

carrier frequency (14.2.1)
operating frequency of the oscillator in a radio transmitter; when a sinusoidal signal at this frequency is modified to contain the information to be broadcast, the frequencies in the broadcast signal are close to the carrier frequency

demodulator (14.2.1)
a circuit that separates the information-bearing part of the received signal from the carrier wave and then converts the information-bearing part into a form that can be understood by the intended recipient

dipole (14.4.1)
a particular kind of antenna consisting of two equal-length metal rods positioned end to end with a small gap separating them

frequency modulation (14.2.2)
a modulation method that adds information to the carrier wave by changing its frequency in response to changes in the amplitude of the information to be transmitted; FSK (16.2.3) is a special case in which the information is binary and can have only two values, so the transmitted frequency modulated signal can have only two frequencies

modulator (14.2.1)
a circuit that modifies the sinusoidal carrier wave so that it will contain the information to be transmitted

monopole (14.4.1)
an antenna consisting of one half of a dipole; it is usually placed over the earth's surface or a large metal surface called a "ground plane" so that it will function as a dipole and attain the efficiency of a dipole

omnidirectional (14.4.2)
when used to describe an antenna, omnidirectional means that it transmits or receives with equal performance in all directions; that is, that there is no preferred direction of transmission or reception

oscillator (14.2.1)
a circuit that generates a sinusoidal electrical signal

parabolic antenna (14.4.3)

an antenna that uses a parabolic reflector to focus energy of a single dipole antenna in a direction determined by the orientation of the reflector; it makes an omnidirectional antenna into a highly directional antenna

period (14.3.1)

the time between two peaks of a sinusoidal signal observed at the same physical location

radio frequency (14.2.1)

a frequency in the range used for communication by transmission of electromagnetic waves, typically between 300 kHz and 300 GHz; this includes medium wave, short wave, and microwave frequencies

wavelength (14.3.1)

the distance between two peaks of a propagating sinusoidal signal observed at the same time

Section 14.8: Figure Credits

[14.3a] Courtesy of Nokia, Inc.

[14.3b] Courtesy of John R. Treichler

[14.9b] Courtesy of the National Radio Astronomy Observatory (www.gb.nrao.edu)

[14.11] Courtesy of Jennifer Treichler

The INFINITY Project

Chapter 15:
Bandwidth

Approximate Length: 2 Weeks

Section 15.1:
Introduction

In this chapter we explore how well a transmission system can be made to perform. We will find that performance depends on a number of factors— how clever we are as designers, the amount of noise and interference that intrudes on the signal we desire to receive, and the amount of **bandwidth** available for sending our signals. We begin by defining the concept of bandwidth, explore its various interpretations, and use it to examine approaches to obtaining the best performance that we can from a transmission system.

Section 15.2:
Bandwidth

15.2.1 Definitions of Bandwidth

Sometimes people speak colloquially of a person's "bandwidth." The term is generally recognized to mean a measure of that person's ability to absorb and understand information, and, in particular, new ideas. The word **bandwidth** actually originated with electrical and communications engineers. It has specific meanings in that context relating to the ability of a communications system to convey information, and, therefore, new ideas, which is the same as familiar conversational usage.

Figure 15.1 illustrates the first and simplest definition of the bandwidth of a signal. As we learned in earlier chapters, a signal can be described in spectral terms. Specifically a given analog signal (music or voice, for example) can be thought of as the combination of a number of sinusoidal components, each with different frequencies and amplitudes. In those chapters we also learned that engineers and scientists often represent a signal by its **spectrum**, a plot of the amplitude or power of each sinusoidal component versus its frequency. Figure 15.1 shows two such plots. With these plots we can construct our first definition.

> **Definition 1:** The **bandwidth** of a signal is the difference, in Hertz, between the frequency of the highest frequency component and that of the lowest frequency component in the signal's spectrum.

The high and low marks for the signals in Figure 15.1 are indicated on the graph. Figure 15.1(a) illustrates the case in which the signal is composed of many distinctly separate spectral components, while Figure 15.1(b) shows the case where there is only one wider component. In both cases the bandwidth is the difference between the highest and lowest frequencies comprising the signal.

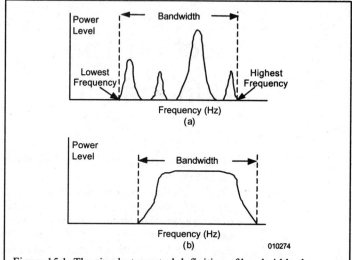

Figure 15.1: The simplest spectral definition of bandwidth: the difference, in Hertz, between the highest and lowest frequencies comprising a signal – (a) signal with multiple well defined spectral components and (b) signal with a single dispersed spectral component.

The spectrum of a real signal is shown in Figure 15.2. The signal in this case is from a V.32 modem, once commonly used in personal computers to call an Internet service provider (ISP). An examination of the signal's spectrum shows that it uses frequencies as low as 350 Hz and as high as 3300 Hz. Using the first definition, we conclude that the bandwidth of this modem signal is 3300 - 350 = 2950 Hz.

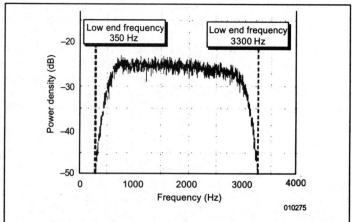

Figure 15.2: The spectrum of a modem signal of the type used in a personal computer, with marks to indicate the highest and lowest frequencies needed to convey the signal.

4

The second definition of bandwidth is applied to the transmission channel through which the signal will travel rather than to the signal itself.

> **Definition 2:** The **bandwidth** of a transmission channel specifies a frequency range over which signals passing through the channel will not be significantly changed. Thus, any signal sent through the channel will not be impaired if all of its frequency components are completely contained within the bandwidth of the transmission channel.

This second definition is best explained using Figure 15.3. The dashed curve shows the degree to which a particular transmission channel weakens or attenuates sinusoids as a function of their frequency. The curve is often called the **transfer function** of the channel since it indicates the ability of the channel to "transfer" sinusoids of various frequencies to its output. In the example shown in Figure 15.3, signals with sinusoidal components with frequencies higher than F1, about 500 Hz, and lower than F2, about 3500 Hz, will pass through without reduction in amplitude. As the frequency of a tone increases above F2, the channel causes its amplitude to steadily decrease until at 4000 Hz it is reduced to 20% of its input value. As the frequency component of a tone decreases below F1, its amplitude decreases much more sharply than the amplitude for the frequencies above F2. Using Definition 2, the bandwidth of this channel equals F2 – F1.

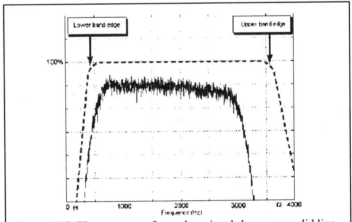

Figure 15.3: The spectrum of a modem signal shown as a solid line within the transfer characteristics of the channel provided by a typical telephone circuit indicated by a dashed line.

Clearly the two definitions are closely related. We know from previous chapters that a receiver can accurately demodulate a transmitted signal only when the received signal is strong compared with any noise or interference also arriving at the receiver. In order to retain the needed signal strength and avoid significant changes in the signal, two requirements must be met. First, the signal's bandwidth must be less than or equal to that of the transmission channel. In addition, the signal's spectrum must be frequency-aligned aligned with the "passband" of the channel These two conditions require that the signal spectrum must fit within the channel bandwidth. The signal must be properly positioned spectrally and its bandwidth must not be too great. If either of these conditions is not satisfied, then the signal is attenuated by the channel, and proper demodulation is jeopardized.

Satisfaction of these conditions is illustrated in Figure 15.3. The transmission channel in this case is a telephone circuit of the type used for home service, and the dashed line shows its transfer characteristics. The V.32 modem's spectrum, taken from Figure 15.2, is plotted on the same frequency scale and can be seen to fit within the telephone line's passband. For this reason one might expect that the modem could operate well over the telephone line designed to transmit acoustic signals such as speech. The modem does in fact work well on the telephone line.

Now we can consider how these two definitions relate to each other. Definition 1 can be seen to define the "spectral occupancy" of a signal, while Definition 2 defines the ability of a channel to accommodate the reliable transfer of signals. This ability must exceed each signal's need.

15.2.2 Bandwidths of Common Communications Signals and Transmission Channels

Table 15.1 lists the nominal bandwidths of some common communications signals and channels. We observe that AM radio uses less bandwidth per station that FM, and much less than a television channel. The television bandwidth must be large compared to radio because the television signal must carry both an audio signal, or several audio signals, and the video signal. A video signal alone has a much higher bandwidth than an audio signal, even a high-fidelity one.

Signal or Channel Type	Typical Bandwidth
Human voice	4,000 Hz
AM radio broadcast	10,000 Hz = 10 kHz
FM radio broadcast	150,000 Hz = 150 kHz
Television broadcast	6,000,000 Hz = 6 MHz
Telephone transmission channel	3,100 Hz
Fifty miles of twisted pair wire	100,000 Hz = 100 kHz
Five miles of coaxial cable for cable TV	1,000,000,000 Hz = 1 GHz
Optical fiber	100,000,000,000,000 Hz = 100 THz

Table 15.1: A number of signals and channels, and typical bandwidths for each.

Note also the bandwidth progression in the transmission systems. The attractiveness of modern optical fiber transmission systems becomes quickly obvious. The **twisted pair** of wires is, as the name implies, a set of two wires running parallel to each other except that one is twisted around the other. The twisting decreases the amount of interference received by the pair from other nearby pairs. Even with this advantage, twisted pair can not carry even a single television signal. Coaxial cable has enough bandwidth improvement over twisted pair to carry more than a hundred television signals. A coaxial cable consists of a long central wire surrounded by a long metallic tube. The tube acts as the second wire in the pair. This "coaxial" structure greatly interference from nearby wires or coaxial cables. In contrast, a single optical fiber has the transmission bandwidth to handle more than 10 million television signals.

15.2.3 What happens when the channel and the signal do not have the same spectral bandwidth?

In general the natural bandwidth of a signal will be different from the natural bandwidth of a transmission channel, and, a practical matter, one is bigger than the other. If the signal's bandwidth is greater than the channel's bandwidth, we know from our previous discussion of the definitions of bandwidth that the communications system will not work well. For proper demodulation the channel must be able to "transfer" all of the spectral components of the transmitted signal. When this is not true, reliable reception and demodulation cannot occur. In the opposite case the transmission channel's bandwidth would be larger than that of the signal. If the signal's center frequency places the signal within the channel passband, then the receiver's demodulator should be able to accurately recover the transmitted signal.

In many cases the channel's bandwidth is much higher than the signal. Consider, for example, the amount of unused bandwidth on a twisted pair carrying one human voice signal using the values in Table 15.1. In these cases, it is common to subdivide the bandwidth of the transmission channel so that many signals can use the transmission channel simultaneously. This is the multiplexing we discussed in Chapter 13.

An example of this is shown in Figure 15.4, which illustrates the transmission characteristics of a pair of long distance telephone wires. This pair, part of the first practical long distance telephone transmission system, was capable of carrying sinusoidal signals with frequencies from about 4000 Hz to about 70 kHz for about 50 miles. This transmission bandwidth of roughly 60 kHz is obviously more than enough to carry a 4000-Hz voice signal. In the telephone system, this transmission bandwidth was actually divided into twelve voice channels of 4 kHz bandwidth each, as shown in Figure 15.4. Twelve individual voice signals were moved up in frequency to a different voice channel using some electronic equipment. Then the group of 12 voice channels was transmitted across the span of telephone wire to a destination site. At that site each individual voice signal was moved back down to its normal frequency range, where it would audible and meaningful to a listener, and placed on its own individual wire. In this way twelve voice signals could be carried at the same time over a long distance on a single wire. None interfered with the others, as long as each stayed within its own channel, which was its own allocated frequency range.

This technique, called **frequency division multiplexing**, was discussed in Chapter 13. It is this technique that first made long distance telephony economically feasible. Using one pair of wires for each long distance caller was just too expensive, and the ability to share the pair among twelve users simultaneously allowed the cost to be spread across all of them.

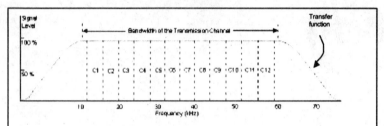

Figure 15.4: The transmission characteristics of fifty miles of telephone wires and the twelve "Voice Channels" of 4-kHz bandwidth each constructed from the transmission system.

This scheme of dividing up a wide transmission channel into a large number of artificially smaller ones is very common in telecommunications systems. The argument for doing it is the same as in that early telephone system. If a large number of users can share a single transmission medium, then more people are served and the cost for each is lower. Usually the subdivision is done by a regulatory organization like the FCC, particularly when the medium being subdivided is a public one like the airwaves. To the designer of the transmission equipment, it means that each signal must be filtered to make sure that it uses only its allocated share of the bandwidth and that it must be moved in frequency up to the "channel" to which it has been directed. Other examples of frequency division multiplexing include AM and FM radio broadcast and both broadcast and cable television.

Experiment 15.2: The Impact of Inadequate Channel Bandwidth

Multimedia communications systems fail to operate properly when the bandwidth of the transmission channel is smaller than the bandwidth of the signal. We find that it is also necessary for center frequency of the signal to be aligned with the "passband" of the channel.

Exercise 15.1

1. A conventional analog television broadcast actually consists of three different signals — one that carries the brightness of the image, another that carries the color of the image, and a third that carries the sound. The bandwidths of these separate components are about 3 MHz, 2 MHz, and 0.5 MHz, respectively.

    (a) Assuming that the frequency spectra of the three components do not overlap, determine the minimum bandwidth of the complete broadcast television signal.

    (b) Given the result from (a), what is the minimum bandwidth that must be allocated for the transmission of a television signal?

2. The manually operated telegraph system described in Chapter 8 generates a signal whose spectrum extends from zero to about 100 Hz.

    (a) What is the bandwidth of this signal?

    (b) Is this value less than or greater than the bandwidth of the voice channel shown in Figure 15.3?

    (c) Should the voice channel shown in Figure 15.3 be expected to carry the telegraph signal? Why or why not?

3. Can a telephone voice channel, again shown in Figure 15.3, be expected to carry a television signal? Why or why not?

4. Using the information provided in Table 15.1:

    (a) How many voice channels could be carried within the bandwidth available on a coaxial cable of the sort used for cable television?

    (b) How many 6 MHz-wide video channels could be carried within the bandwidth available on an optical fiber?

5. The adjacent figure shows the spectrum of a V.22 modem, the type used a decade ago in personal computers to produce two-way data communications at rates of up to 2400 b/s.

   (a) Estimate the bandwidth of the signal from the spectral plot.

   (b) Can this signal be carried within a telephone voice channel?

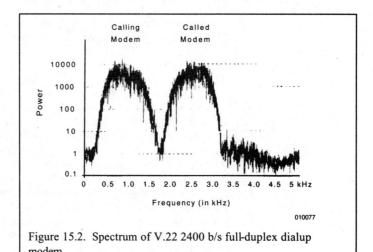

Figure 15.2. Spectrum of V.22 2400 b/s full-duplex dialup modem.

Section 15.3: Making Efficient Use of Bandwidth

15.3.1 A Measure of Bandwidth Efficiency

Using the relationship between the first two definitions of bandwidth for the signal and the channel, we can create a measure of the efficiency of the bandwidth usage. Such a measure could be used to indicate that the channel has unused capacity or it could be used as a performance measure to compare competing designs. Using Definition 1, a communications signal has some measurable bandwidth, and, if a communications signal is to be received and demodulated properly, its bandwidth must be less than that of the transmission channel according to Definition 2. If the channel is subdivided for multiple users, then the bandwidth of the signal must be less than the portion of the channel bandwidth allocated to each user by a regulatory organization. However, these definitions do not address how efficiently the channel's bandwidth is being used, and efficient usage is important if we are endeavoring to build communications systems in an economical way. We will define the term **spectral efficiency**, E, of a transmission system using Equation 15.1. Here E is the efficiency,

$$E = R/B \qquad\qquad 15.1$$

measured in bits per second per Hertz of bandwidth, R is the bit rate at which data is transferred over the channel in bits/second, and B is the bandwidth of the channel, or our allocated portion of the channel, in Hz.

For example, this definition can be applied to the one-tone-per-character audio transmission scheme we designed and experimented with in Chapter 7. To find the transmission rate R, we need to know how many pulses we send per second and how many bits of information each pulse represents. If we send ten pulses per second and each of them carries 5 bits, then the air modem is transferring 50 bits/second to the receiver. Using Definition 1, the bandwidth is 3100 Hz, which is the difference between the frequency of the highest tone and that of the lowest tone. The resulting spectral efficiency R is therefore (50 bits/second)/(3100 Hz), or roughly 0.016 bit per second per Hertz.

The efficiency computed in the previous example for the air modem used in the laboratories is well below normal operational standards. For comparison, the V.32 modem, whose spectrum is shown in Figure 15.2, sends 19.2 kb/s using a bandwidth of about 3000 Hz. The V.32 modem's spectral efficiency is about 6 ½ bits per second per Hz, which is roughly 400 times better than the air modem.

11

In general, efficiency is an important measure of performance for any system. The inefficient use of a commodity usually implies that money is being wasted that could be spent or invested in other ways or that a system is over-designed for its application. How important the efficiency measure is compared to other performance measures depends on how much the commodity costs and how much it would cost to modify a system to make more efficient use of the commodity. We will examine this issue more in the next section.

15.3.2 Bandwidth as a Scarce Commodity

Historically bandwidth has been considered a scarce commodity, in the sense that there was more demand for it than there was a supply of it. When a commodity is scarce, its price rises and usually efforts are made to use it carefully. To explore this further we need to consider why bandwidth is scarce, whether or not it is always scarce, and how we should react as designers and users when it is scarce. There are two important situations which create a scarcity of bandwidth. We'll examine both, and then see what measures might be taken in response.

Interference in radio systems: When a radio signal is transmitted, energy is radiated in all directions, even when directional antennas are used to focus the energy in a particular direction. Similarly, the receiving antenna picks up signals from any transmitters operating on the same frequency. The receiver must recover the desired data from the sum of all of those signals arriving at its antenna. We term all of the signals other than the one we want to be "interferers." The traditional method of solving this problem is to permit only one transmitter to operate on a given frequency at a time in a given geographical area. Since the strength of a signal at the receiver will decrease as the distance to the transmitter increases, this traditional approach of restricting transmitters will insure that one strong signal will dominate the many much weaker signals at the receiver. However, if more users want to transmit than there are "radio channels" in which they can operate, then those users consider the available radio bandwidth to be scarce. This is how commercial broadcasting of television and radio is actually handled.

Limited bandwidth in wire-based transmission systems: Transmission systems that use copper wire all have some practical limit to the highest frequency that can be reliably sent through them. Over the past hundred years, electrical engineers have developed amplifiers and other devices that permit the bandwidth to be extended from roughly 3000 Hz to more than 100 MHz. But although the upper limit is higher, the bandwidth is still limited and thus the number of customers who can simultaneously use the transmission system is also limited.

There are fundamentally three approaches for dealing with bandwidth scarcity. One approach is to introduce regulation, for example by the Federal Communications Commission, to make sure that whatever bandwidth there is used fairly. A second approach, which may be used along with the first, is to design our communications signals so that they use what bandwidth there is as efficiently as possible. Specific ways of doing this are discussed in Section 15.5.1. A third approach is to develop alternative transmission systems that have higher bandwidths or evade interference in a more clever way. Table 15.2 provides a few examples of each of these remedies.

General Approach	Specific Examples
Introduce regulation	AM radio stations: Frequency allocations are created, e.g., every 20 kHz between 530 kHz and 1650 kHz. The FCC assigns frequencies to some of the license applicants. The FCC limits the power that can be transmitted and, in some cases, the directions in which it can be transmitted.
Improve bandwidth efficiency	More efficient modulation (e.g., V.34 and V.90 modems) Data compression , e.g., PKZIP for file transfer, Digital cellular systems
Creatively avoid the problem	Reuse radio frequencies aggressively, e.g., cellular telephones Develop systems with enormous bandwidth, e.g., optical fiber transmission

Table 15.2: Several approaches to the problem of scarce bandwidth.

15.3.3 The Limit to a Channel's Capacity — the Shannon Limit

Before we leave the topic of bandwidth and its implications to communications systems, we might reasonably ask a simple question. Is there a limit to the number of bits that can be reliably transmitted and received by a communications system? And if there is, on what does that limit depend? It turns out that there is a straightforward answer. Such a limit does exist. However, the path to that answer is fairly complicated and uses some sophisticated mathematics.

Finding a clear mathematical answer requires that we ask a clear mathematical question. Here's the one we'll ask. Suppose that we're given a transmission channel with bandwidth B, that there is noise present at the receiver's input with power of N watts, and that the maximum power we'll allow the transmitter to have will cause P watts to arrive at the receiver. Figure 15.5 shows these assumption in the form of a block diagram. Suppose further that we're allowed to be just a clever as we can to design the signal, the transmitter, and receiver. We're free to do the best we can. What would this best be?

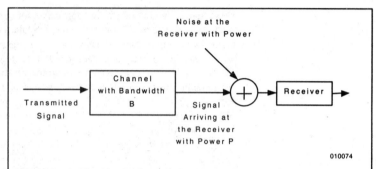

Figure 15.5: The Simple Model of a Communications System assumed by Claude Shannon for his famous theorem.

From the previous chapters we know that a very important measure of the performance of a system is its ability to recover the symbols that have been transmitted, and to distinguish each one from all others. The mathematical question then becomes "How closely can I pack the symbols to each other without the receiver making too many errors?" A famous mathematician from the Bell Telephone Laboratories, Claude Shannon, showed in the 1940s that this question could be phrased in a way that leads to the picture shown in Figure 15.6(b). The figure shows a large sphere whose radius is determined by the maximum signal power P. The big sphere is filled with smaller spheres. These are all of equal size and each has a radius determined by the noise power level N. Figure 15.6(a) shows a two-dimensional view of the same idea where the spheres have become circles. Shannon reasoned that the noise tends to confuse the receiver within a volume equal to that of the small spheres, while the signal could be the size of the large sphere. The number of different symbols that can be reliably distinguished at the receiver is then the ratio of the volume of the big sphere to that of the little ones. He then argued that the spheres should not just be the two dimensional ones shown on the left of Figure 15.6 or the three-dimensional ones on the right. They should be allowed to be n-dimensional, where n is as large as you might imagine. Using this approach, he developed what is known to this day as the Shannon Limit, given by

$$C = B \log_2 (1 + P/N). \hspace{3cm} 15.2$$

The variables B, P, and N are as already defined. The variable C is called the transmission channel's **capacity**, measured in bits/second, and is the theoretical upper bound on the number of bits that can be sent through the channel no matter how clever the designer of the system. The bandwidth efficiency, defined earlier this chapter, for a channel "operating at capacity" is given by

$$E = C/B = \log_2 (1 + P/N). \hspace{2.5cm} 15.3$$

We observe that these formulas suggest results that match our intuition. The capacity grows as the channel's bandwidth B grows. The capacity, and with it the bandwidth efficiency, grow as the noise becomes weaker with respect to the power of the desired signal at the receiver. This **signal-to-noise ratio**, P/N, can be seen to be a very important parameter, since it has a crucial influence on the channel's ability to convey data to the receiver.

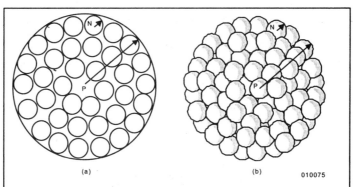

Figure 15.6: The "sphere packing" model that Claude Shannon used to determine the maximum transmission capacity of a channel – (a) packing 2-dimensional *spheres* (circles) with radii determined by signal power P and (b) extension to three dimensions. The radii of the smaller spheres is determined by the noise power N while radius of the larger sphere is determined by the signal power P.

When Shannon first derived this result, it was generally believed to be of interest only to theoreticians. No communications signal at the time (mid-1940s) even came close to "achieving capacity," that is, reliably transmitting data through a channel of bandwidth B at a rate even nearing close to C. For example, the most common method of sending messages at the time, the electromechanical teleprinter, used a telephone channel with a bandwidth of 3100 Hz and was able to carry 1200 bits/sec. Telephone channels at that time supported signal-to-noise ratios of about 1000:1, that is, the power ratio between the strongest allowable signal and the noise present in the channel was about 1000 to 1. If we compute the telephone channel's capacity, we find it to be

$$C = 3100 \log_2 (1 + 1000) \cong 31000 \text{ bits/second.}$$

The channel was capable of sending 31 kb/s, at least theoretically, while the transmission system only achieved 1200 bits per second.

This situation is no longer true. Modern digital signal processing algorithms implemented with modern semiconductor technology have made it possible to develop better modulators and more robust receivers, allowing designers to attain reliable transmission at rates quite close to the Shannon limit. For example, a modern V.34 modem of the type used in a home computer can send data at a peak rate of 33,600 bits/sec. This would not even be possible on the telephone lines of the 1940s. Telephone channels are better now, supporting signal-to-noise ratios of 4000:1 or so. This leads to a theoretical capacity of about 37000 bits/s, only slightly more than can actually be achieved with the V.34 modem. As a result, the Shannon limit is often the first thing that a communications engineer computes when confronted with a new system to design. It serves as an excellent guide to what can be achieved given today's high performance signal processing technology.

15.3.4 Intermediate Summary

The word **bandwidth** occurs frequently in the discussion of telecommunications systems. We've learned here that one must be careful in its use. When applied to a signal, it means the signal's spectral extent, which is the difference between its highest frequency component and its lowest. When applied to a transmission channel, it means the spectral width of the band in which a signal can be conveyed without significant degradation in amplitude.

In addition we made the following observations.

- Different signals have different bandwidths, and so do different types of transmission channels.

- A signal's spectrum must "fit" within the "passband" of a channel for reliable reception to occur.

In many cases the transmission channel is substantially wider than the bandwidth of the signals desiring to pass through it. It is common in these cases to subdivide the transmission channel into a number of non-overlapping spectral bands. Unfortunately, these are often termed "channels" as well. Then different users are allocated to different channels.

Bandwidth is often considered to be a scarce commodity, and therefore something that needs to be conserved and carefully husbanded. This is true when operating in interference-prone radio environments or when a transmission system hasn't the bandwidth needed to handle all of the customers who desire to use it. In contrast, there are some transmission systems, ones based on optical fiber, for example, that have so much available bandwidth that it is considered to be essentially free, and is therefore often wasted.

Transmission channels do have an intrinsic limit to their ability to reliably transfer data. Under a simple set of assumptions Claude Shannon proved that this limit depends on the bandwidth of the channel and the signal-to-noise ratio of the signal of interest when it arrives at the receiver's input. Shannon's limit was once only of academic interest, but advances in theory and implementation have made it possible to almost reach that limit in many practical communications systems.

Exercise 15.2

1. A data modem called the V.29 was built for use over telephone lines and was capable of a maximum transmission rate of 9600 b/s. Its signal occupied 3000 Hz of bandwidth. Compute the spectral efficiency E for this modem signal.

2. Suppose that we use the "serial binary" method of transmission described in Chapter 7 to send a data stream. (It is more commonly termed "frequency shift keying (FSK)"). Assume that the frequency 1180 Hz is used to represent a binary one while the frequency 980 Hz is used to represent the binary value zero.

    (a) Long ago a man named Carson found that the bandwidth of an FSK signal was approximately the sum of the frequency separation between the tones representing 1 and 0 and the bit rate at which the FSK modulated was being keyed. Write a formula for the spectral efficiency of the signal produced by this modem.

    (b) Compute the efficiency when keyed at 100, 200, 400, and 1000 b/s.

    (c) What is the highest possible efficiency for an FSK signal? The lowest? Describe the conditions that produce the two limits.

3. Use Shannon's formula to compute the capacity of a channel that is 2400 Hz wide when the signal-to-noise ratio at the receiver is equal to 1023.

4. Use Shannon's formula to determine what signal-to-noise ratio at the receiver is needed to theoretically achieve a spectral efficiency E of 4.

5. Redraw Figure 15.6 for the one-dimensional case.

6. Many cities have a television broadcast station operating in Channel 2, the spectral band between 52 and 58 MHz. Why don't these signals interfere with each other?

Section 15.4:
Maximizing the
Data Rate

Communications design engineers strive to build systems that get close to the performance predicted by Shannon's capacity limit. The simple systems that we constructed in Chapter 7 fall far short. In this section we examine how we might improve these systems and, in the process, get closer to our objective of efficient, cost-effective communications for multimedia data.

15.4.1 Approaches

Looking back at the various schemes that we developed in Chapter 7, we see that we can represent the data transmission rate that each of them achieved using the following simple formula:

$$R = r\,d, \qquad\qquad 15.4$$

where R is the transmission rate in bits/second, r is the pulse rate, in pulses per second, and d is the number of bits carried on each pulse. For example, using the one-tone-per-character method, with a choice of thirty-two characters, and a pulse rate of 10 pulses per second, we find that the transmission rate is

$$R = 10 \times 5 = 50 \text{ bits/s,}$$

since a choice of 32 characters means that 5 bits are conveyed with each one of them.

If the signal-to-noise ratio is about 1000:1 (which is typical for telephone channels in many countries), then Shannon showed that theoretically

$$C = 3200 \times \log_2 (1 + 1000) \cong 32000 \text{ bits/sec}$$

could be reliably carried in the same channel. Clearly we have a long way to go before we even begin to get close to Shannon's limit.

Claude Shannon developed a formula in the 1940s which determines the maximum rate at which information can be reliably sent and received given the bandwidth of the channel, the signal-to-noise ratio at the receiver's input, and the ingenuity of the designer. In this experiment we determine how well several of the air modem schemes when received with noise, and how they compare with the Shannon Limit.

How might we modify our basic schemes in order to improve the transmission rate? We will find that improvements fall into four major categories. We also will find that the "real world" often limits our ability to use the four techniques to their maximum potential.

- Increasing the pulse rate r

- Increasing the number of bits carried on each pulse, d

- Reducing the number of bits to be sent in the first place, by "compressing" the input data

- Increasing the resistance to errors using "error correction" techniques

Before we begin these investigations, we should make our technical objective clear. We assume that we have been given a channel with a bandwidth of B Hz, which can tolerate signals with a maximum power of P watts. We'll further assume that the signal is received along with some disruptive noise, which has a power of N watts. Our objective is to obtain the maximum possible transmission rate of the user's input data, while allowing no more than a specified number of errors. This may seem overly technical, but a clear description of the engineer's objective is necessary if clear and useful answers are to be obtained.

15.4.2 Faster Pulses

Our first approach is to increase the rate at which we send pulses. A re-examination of Figure 15.7 provides a simple intuitive result. The first step is to squeeze the pulses together so that the duration of each pulse D is exactly the same as the time T between the beginning of each adjacent pair of pulses. This implies that there is no "dead space" between the pulses. This is shown in Figure 15.8. The second step is to decrease the pulse duration D.

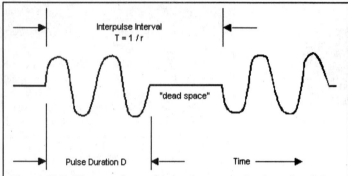

Figure 15.7: The waveform of a simple communications signal that encodes binary data on the phase or frequency of pulses of energy.

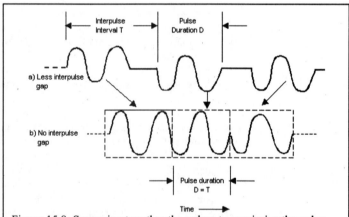

Figure 15.8: Squeezing together the pulses to maximize the pulse rate and hence the bit rate of a transmission system.

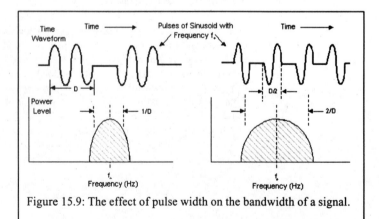

Figure 15.9: The effect of pulse width on the bandwidth of a signal.

So, how much can we reduce D? It would seem that we could make D as small as we like, and therefore send data as fast as we'd like. In fact, this isn't true. To see why this is, we now consider Figure 15.9. The top left portion of the figure shows a pulsed signal, much like that shown in Figure 15.8. The lower left portion shows its spectrum. This can be computed using the ideas presented in previous chapters. We will not repeat that calculation here, although the reader should do this. But the result is that a signal centered on the sinusoidal frequency f_0, which is pulsed on and off with a pulse duration of D seconds, has a bandwidth of about 1/D Hz centered around f_0, as shown in the bottom left of Figure 15.9. As the right side of Figure 15.9 shows, the effect of halving the pulse duration is to double the bandwidth of the pulses and therefore the signal.

What then is our limit on the rate at which we can send pulses? There are in fact two limits:

- Our ability to squeeze the pulses next to each other: It appears at first that it should be possible to do this completely, thus removing the "dead-time" altogether. However, some transmission degradations, specifically effects called "dispersion" and "multipath propagation," prevent us from doing this without causing a lot of errors at the receiver.

- The bandwidth of the transmission channel: We agreed earlier that the signal had to be confined to a bandwidth of B Hz so that it would "fit" into its natural or regulatory transmission channel. This limit means that the minimum duration for the pulse, D, is 1/B seconds, and therefore the maximum pulse rate, assuming that the pulses can be completely squeezed together, is 1/D = B pulses/second.

Thus, unless we have to space the pulses apart to deal with some transmission problems, we can send roughly B pulses per second through a channel of bandwidth B.

> **Experiment 15.5: The Effect of Pulse Duration on Signal Bandwidth Demo**
>
> In the quest to get the highest transmission rate through a channel, it seems clear that we should try to send as many pulses as possible per second. In this experiment we show that the signal bandwidth depends on the narrowness of the pulses, not on their rate.

15.4.3 More Choices

The second degree of freedom we have in the design of the transmission system is how many bits to attempt to send on each pulse. Reviewing again the methods we developed in Chapter 7, we see that several of them sent about 5 bits on each pulse, which was enough to completely specify the capital letters in the Roman alphabet. One of the methods, the "serial binary" scheme, only sent one bit per pulse.

Clearly then we have a choice. The first choice is to attempt to send only one bit per pulse, that is, d =1. Since the maximum pulse rate achievable within the channel bandwidth of B Hz is B pulses/second, this limits the maximum bit rate R that we can achieve to B bits/sec. The alternative is to send more bits on each pulse, thus raising the achieved bit rate proportionally. We'll discuss two issues here. How, in principle, are multiple bits sent on each pulse? Second, what is the limit to our ability to do this?

We already suspect the answer to the first question. If each pulse can have only two different values, and the receiver's demodulator can only distinguish those two, then the transmitter can only send one bit on each pulse — conveying the bit 1 with one of the two values and the bit 0 with the other. If the pulse can have four different values, and the receiver can reliably distinguish among them, then the transmitter can place two bits of information on each pulse. This trend continues, as shown in Table 15.3. More distinguishable values make it possible to carry more bits on each pulse. Mathematically, this is simply stated

$$d = \log_2 \{\text{\# of values or choices}\} \text{ and}$$

$$\text{\# of values or choices} = 2^d.$$

Number of Values	Bits Conveyed
2	1
4	2
8	3
16	4
32	5
64	6
128	7
2^d	d

Table 15.3: The number of distinguishable values that a transmitted pulse must have to convey a specified number of bits with each pulse.

To this point we have been deliberately vague. Choices of what? Amplitude? Frequency? Phase? Color? A combination of these? The answer is that it doesn't matter. If a transmitter can be built and the receiver's demodulator can be correspondingly built to distinguish reliably the different values that the transmitter sends, then an arbitrary combination of these can be used. We can examine some of the examples we already understand and then extend them a bit.

One-tone-per-pulse Method

The "one-tone-per-character" method discussed in Chapter 7 sent 27 different values, each encoded as a different frequency. This was enough to specify any of the capital letters and the space character. We later extended the number of tones to 32, permitting the system to carry any of 32 characters, or, according to Table 15.3, any arbitrary set of 5 bits with each pulse.

What if we desired to send only four bits? A quick calculation shows that only 16 tones are needed. To do this we would use half of the tone frequencies listed in Table 7.1. Which half? Every other one? The top half? The bottom half? An arbitrary set? It doesn't matter as long as there are 16 and they are all distinguishable from each other at the receiver.

Now what if we desire to send 6 bits on each pulse? Consulting Table 15.3, we see that we need 64 distinguishable values. How might we do this? The simplest extension of what we've already done would be to add 32 more tone frequencies, for a total of 64? Assuming that the receiver is similarly extended, and can distinguish all 64 from each other, we can achieve our goal.

Summarizing, this technique encodes bits into the frequencies of the pulses, and there must be 2^d possible frequency choices to carry d bits on each pulse.

Two-tone-per-pulse Method

The "two-tone-per-pulse" method described in Chapter 7 uses a different method to create signals that are distinguishable at the receiver. Each character, or, equivalently, each set of bits, were "mapped" into a combination of two tone frequencies. A pulse was then sent which was composed of the sum of two tones, one at the first frequency and the other at the second. Even though a different approach was used, the underlying concept is the same — permitting more choices allows the transmission system to carry more bits. For example, a dual-tone scheme that chooses one of its frequencies from a set of eight, and the other from a different set of eight, can send 8 x 8 = 64 distinguishable pulses. Using this arrangement the transmitter could place six bits on each pulse and, hopefully, the receiver could recover them accurately.

Extending the Single-tone Method

Now consider doing something a bit different. Return again to the "single-tone-per-character" method and assume that we are using 32 possible tone frequencies. Assume now that our transmitter has the ability to send out each selected tone burst not at just one standard amplitude, and hence one power level, as we have implicitly assumed all along. Assume that it can send it at twice the amplitude as well. What does this new freedom allow us to do? Thinking narrowly we might say, "Oh, good. Now I can send the lower-case letters as well as the capital letters. I'd use one power level for one set and the other to indicate the other!". This is true, of course, but it misses the bigger point that we've now doubled the number of choices that the transmitter has in selecting the pulse to send. Every pulse now conveys 6 bits, since there are now 64 allowable combinations of frequency (32) and amplitude (2). Since the amplitude and frequency can be chosen independently, there are 2 x 32 = 64 combinations.

Can we go further with this idea? What if we built the transmitter so that it is capable of sending pulses at any of 32 frequencies and with any eight amplitudes? What if we could build a receiver that could reliably distinguish all of these amplitude and frequency combinations? If so, we could send 8 bits on each pulse, since there are 32 x 8 = 256 possible values for the transmitted pulse.

While we have only looked at the combination of frequency and amplitude here, the principle extends. Combinations of amplitude, phase, and frequency have all been used to build communications systems. The amount of data that is conveyed with each pulse is easily determined, using Table 15.3 or the logarithm formula, from the number of choices or values that each transmitted pulse can attain.

How far can we push this? If 64 choices works, carrying 6 bits, why not 1024 (for 10 bits/pulse)? And if that, why not use 1048576 choices and carry 20 bits pulse? What is the limit?

There is a limit and it is imposed by the demodulator's ability to reliably distinguish the different symbol values that the transmitter might employ. When we add more amplitude choices within the fixed amplitude range that the transmitter has available to it, or we add more frequency choices within the limited bandwidth B, we make it harder for the receiver to determine which of the possible symbol values has been sent. When the demodulator becomes confused, then reception errors are often made.

How does the demodulator become confused? Assuming that it is operating properly and not broken, there are only two reasons for confusion.

• The transmitted symbols were not constructed properly by the designer.

• Noise, or other degradations acting like noise, unintentionally received along with the signal confuse the demodulator.

We'll assume that we're smart enough to avoid the first problem and that the second is the real issue here.

15.4.4 The Effects of Received Noise on the Detection of Multilevel Signals

The effects of noise can be illustrated with a simple example. Suppose that we send pulses at only one frequency, but that these pulses can have any of eight amplitudes. From our previous calculations and from Table 15.3 we know that this system could carry three bits per pulse. The demodulator for such a signal builds the electronic equivalent of a ruler, and measures the height, technically, the "amplitude," of each pulse. It then determines which of the eight expected amplitudes that the received signal is closest to by "rounding" the ruler's measurement to one of eight allowed values. A different set of three bits is associated with each of these eight values.

Figure 15.10 shows the amplitudes of a train of such pulses as they are processed by the demodulator. These pulses are assumed to be received "cleanly," that is, without distortion from the transmission channel and without any significant amount of noise. As each of the pulses is received, its amplitude is measured, the closest of eight expected amplitudes is determined, and the appropriate set of three bits sent out. The ruler on the right-hand side of the figure indicates how this electronic process is performed. The amplitude is measured accurately, and then rounded to one of the big black marks. Received amplitudes of 5.8 and 6.2 would both be rounded to 6.0, and the bits 110 would be sent on to the user.

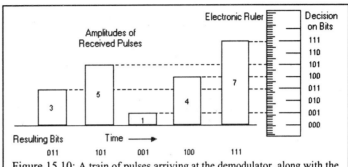

Figure 15.10: A train of pulses arriving at the demodulator, along with the electronic ruler used to measure the amplitudes of the pulses and determine which symbols they represent.

Now consider the received pulses seen in the top half of Figure 15.11. They are the same as those in Figure 15.10 except they are received in the presence of a moderate amount of noise. This noise has the effect of changing the amplitudes of the pulses. In this case, however, the noise is not strong enough to change any of them enough for the demodulator to make a mistake. Mathematically this means that the maximum value of the noise's contribution is less than half the distance between any two adjacent big black marks, or "decision levels," on the electronic ruler. When this condition is true, the demodulator can still recover the transmitted signal accurately.

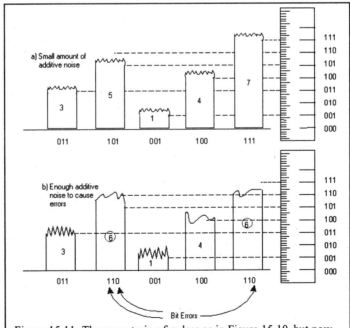

Figure 15.11: The same train of pulses as in Figure 15.10, but now received in the presence of a moderate and then an extreme amount of additive noise.

What happens when it is not true? The bottom half of Figure 15.11 again shows the same train of pulses, but now the noise's contribution is significantly higher. Note that the amplitudes of the second and fifth pulses have been changed so much by the noise that the electronic ruler now draws the incorrect conclusion about the amplitudes of the transmitted pulses. Those incorrect estimates of the underlying amplitude turn into incorrect decisions about which symbols were transmitted, and that turns into one or more bit errors for every symbol that is incorrectly assessed by the demodulator.

How does this tie back to our question about the number of choices that a modulator can safely use? The signal power P, the noise's power N, and the number of choices 2^d are related in the following way.

The signal's maximum allowable power P determines the highest mark on the decision ruler.

The number of choices, 2^d, determines the number of "big black marks" on the ruler, that is, the decisions that can be made by the demodulator.

The noise's power N determines the greatest perturbation that the noise can make on the received signal. If that maximum perturbation is significantly less than the amplitude difference between the big black marks, then the received pulses will be accurately demodulated. When this is not true, that is, when the noise amplitude exceeds that critical amount, the demodulator begins to make errors. As the noise's power grows greater, the probability of making errors grows greater as well.

From this description it should be clear that trying to send more bits on each pulse makes the decision levels on the ruler closer together and increases the probability of making a mistake for a given amount of noise. Figure 15.12 shows the electronic rulers for systems using 8 and 32 possible pulse amplitudes with 3 and 5 bits on each pulse, respectively. Note how much closer the decision levels are on the second one and therefore how much more prone that system would be to the effects of noise. The designer of the communications system must assess the maximum signal power P that he or she can legally or economically use and must assess the amount of noise N unavoidably present at the receiver's input. Then the designer must pick a modulation scheme and number of bits per pulse, d, that maximize the amount of data sent without exceeding the customer's tolerance for errors in the received data.

This type of information is often precomputed and made available to engineers in the form of charts of the type shown in Figure 15.13. The x-axis is the "signal-to-noise ratio," the ratio of P to N. The y-axis indicates the expected number of erroneously detected symbols, expressed as the "symbol error ratio," the number of symbols expected to be detected incorrectly as a fraction of the total number transmitted. The particular chart shown in Figure 15.13 shows the performance that can be expected from a communications system that sends its digital information using only the amplitudes of pulses. For example, by following the dashed line from the y-axis to the first curve, and then down to the x-axis, an engineer can determine that an SNR of about ** is required to make one error in a billion when transmitting one bit per pulse.

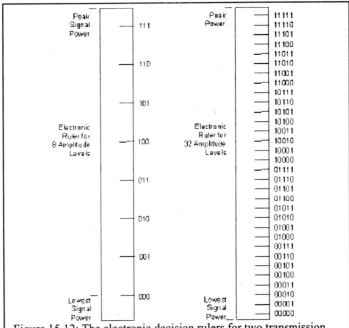

Figure 15.12: The electronic decision rulers for two transmission systems, one that carries three bits on any of eight possible pulse amplitudes and another that carries five bits on the pulses with any of 32 amplitudes.

From this graph it is easy to verify the observations made earlier.

• As the signal's power grows with respect to the noise, that is, as P/N gets bigger, the error rate goes down.

• For a given signal-to-noise power ratio P/N, systems that try to carry more bits on each pulse make more errors.

Figure 15.13 shows the expected performance of a system using only pulse amplitudes. As we've learned earlier, it is possible to impress digital information on a signal by modifying its amplitude, phase, or frequency, and, of course, even a combination of them. The performance of every one of those schemes can be characterized using a graph of the type shown in Figure 15.13. As might be expected, the exact shapes of the curves are different depending on the modulation scheme being used, but they all have the same basic characteristics.

- The error rate declines as the ratio of signal power to noise power increases, and

- Trying to carry more information per pulse increases the probability of errors being made.

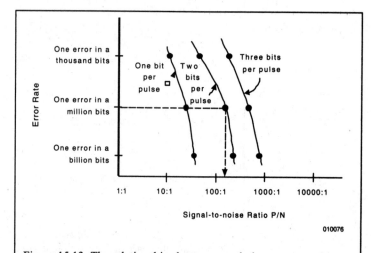

Figure 15.13: The relationships between symbol error rate and input signal-to-noise ratio for pulse amplitude modulation systems with various numbers of bits carried on each symbol.

Exercise 15.3

1. Compute the spectral efficiency E using Equation 15.1 for a V.27 modem. This modem, still used in facsimile machines, transmits 3 bits on each symbol, sends symbols at a rate of 1600 symbol/second, and occupies 2400 Hz of bandwidth.

2. The cable and broadcast systems developed for analog television signals allocate 6 MHz of bandwidth for each TV channel, of which the center 5 MHz can be used without distorting the signal. Suppose that we now desire to send binary data over the same channel so that we can introduce high-definition television (HDTV) and other advanced multimedia services.

   (a) What is the maximum symbol rate that can be used over one of these channels?

   (b) Suppose that our goal is to send binary data at a rate of 30 Mb/s. How many bits must each symbol or pulse carry in order to achieve this goal?

   (c) In order to combat errors made at the receiver because of noise and interference, the designer makes the decision to add extra bits at the transmitter so that "forward error correction" can be performed at the receiver. His calculation indicates that in order to include these extra bits, the total bit rate to be transmitted must increase by 15%. Assuming that the symbol rate is already as fast as it can be, how many bits must now be carried on each symbol?

3. A common problem that occurs in data transmission is "multipath propagation," which means that the transmitted signal takes more than one path to the receiving antenna. Using the air modem as an example, the receiving microphone can often hear both the signal coming straight from the transmitter and a version of the signal which bounces off of a wall. The "echo," the reflected signal, reaches the receiver some time after the directly arriving signal and is usually weaker in amplitude. The two add together at the receiving microphone. In this problem we examine the effect that this multipath propagation will have on the signal seen at the receiver.

   (a) Assume that the air modem transmits a single sinusoidal burst whose frequency is 500 Hz and whose duration is 50 msec. Assume further that the burst is followed by a "dead space" of at least 50 msec. Use your calculator to graph this burst and the dead space following it.

   (b) Assume now that the reflected burst arrives 20 milliseconds after the directly received burst and that its amplitude is only half of the directly received burst. Graph the sum of these two.

   (c) How does the multipath propagation affect the apparent length of the received signal?

(d) Now regraph the sum of the two assuming that the delay is now 20.5 msec. Repeat again with delays of 21 and 22 msec. Explain why the received signal changes shape.

(e) Sound travels at approximately 330 meters/sec at ground level. What is the additional distance that the reflected signal must be traveling to cause it to arrive 20 milliseconds after the directly arriving signal?

4. In a simple communications system using sinusoidal bursts, it is necessary to wait until all echoes have been received for a given burst before sending the next.

(a) If a sinusoidal burst has a duration of D seconds and echoes might arrive up to Δ seconds later, and the time between burst transmissions is T seconds, write an inequality that defines the minimum value that T can have.

(b) Assume now that the acoustical delay for an air modem setup is 10 milliseconds. What is the very fastest rate at which the air modem would be allowed to send pulses?

5. The V.29 modem sends pulses that can have any one of 16 different choices of amplitude and phase. It sends those pulses at 2400 pulses or symbols per second and occupies about 3000 Hz of bandwidth.

(a) At what bit rate does the V.29 modem operate?

(b) What is its spectral efficiency E?

(c) What is the Shannon capacity for the same signal bandwidth with the assumption that the signal-to-noise ratio at the receiver is 511?

(d) Compare the V.29 modem's achieved rate with Shannon's capacity for the channel.

6. Figure 15.13 allows us to figure out the signal-to-noise ratio (SNR) required for obtain a particular bit error rate.

(a) Assume that our target bit error rate is one in a million. Use Figure 15.13 to estimate what SNRs are needed to achieve this goal when using one, two, and three bits/pulse.

(b) From this trend, estimate the SNR required for four and five bits/pulse.

7. Again referring to Figure 15.13:

(a) If SNR is good enough to receive 5 bits/symbol at the target error rate, is the same SNR good enough to received 4 bits per symbol?

(b) Conversely, if the SNR is not good enough to achieve 4 bits/symbol at the target error rate, can it support 5 bits/symbol?

8. Suppose that the SNR at the receiver might dip at low as 100. Using Figure 15.13, what is the maximum number of bits/symbol that can be used and still guarantee an error rate of one in a million or less?

Section 15.5: Maximizing the Transmission Rate

We can now summarize what we've learned to this point about maximizing the transmission rate that can be achieved through a communications system. In general we will do the following.

- Send pulses as fast as we can, and

- Carry as many bits on each pulse as we can.

What are the limits to our ability to arbitrarily make these values as big as we'd like, thus making the transmission rate as high as we'd like? The limitations we found are the following.

- The spreading or smearing of the transmitted pulses by the transmission channel itself can limit the desirability of squeezing the pulses completely together.

- The maximum pulse rate cannot exceed the channel bandwidth B or the portion of it that we've been allocated.

- The number of choices of amplitude, frequency, or phase for each pulse is limited by the amount of noise present at the receiver's input and the number of errors that the user of the communications link is willing to tolerate.

Two examples can illustrate these points.

- HF radiotelegraphy – The first reliable systems for communicating data over high-frequency (HF) short-wave radios sent 50 pulses per second and used two-level "frequency-shift-keying (FSK)" to carry one bit on each pulse. Thus the transmission rate was 50 bits/second. The pulse rate was limited by the effects of the ionosphere which tends to smear the pulses out in time, and the limited number of symbol choices (specifically, two) was imposed by the noisiness of the HF radio environment.

- A V.34 voiceband modem – The V.34 modem is the most common dialup modem used in personal computers. This strange designator was chosen by the International Telecommunications Union, which has the responsibility to set standards for telecommunication equipment around the world. This modem is capable of sending up to 33,600 bits/second through a telephone line. It does that by sending up to 3200 pulses per second, essentially filling the 3300-Hz bandwidth of the typical telephone channel, and by sending up to 11 bits on every pulse, encoded as 2048 combinations of amplitude and phase angle. This large number of bits per pulse is achievable because the noise power N on modern telephone lines is quite low, and because a scheme not yet discussed called "forward error correction (FEC)," is used to correct some errors when they do occur.

As we have seen, the number of pulses per second, as limited by the channel's bandwidth, and the number of bits per pulse, as limited by the noise and level of errors the user can tolerate, determine the fundamental limit of the rate at which data can be transmitted through a channel. Two additional techniques are commonly used, however, to improve the system's throughput. We'll discuss both of them shortly. We'll find, not surprisingly, that we've already studied both of them when we examined storage media and how it could be most efficiently used.

15.5.1 Data Compression

The transmission schemes we've talked about so far are capable of carrying arbitrary streams of random data. In many practical circumstances, however, the input to the transmission system is not random, and the fact that it is not permits us to find a way to improve the effective transmission rate of the system.

A simple example will suffice. Suppose the input to the transmission system is a message composed of Roman capital letters, just as we explored in Chapter 7. Suppose further that a typical message is given by:

... A B C D E A B C R F T H A B C W I A B C N M Y I A

B C G A B C P A B C ..

At first glance this message appears to a random list of letters, but a second look reveals that the three letters **A B C** appear in sequence again and again. Suppose we adopt the following strategy. Every time the transmitter sees the "trigraph" **ABC,** it substitutes the character # instead. If we did this, the transmitter would actually send the following:

... # D E # R F T H # W I # N M Y I # G # P # ...

A quick count shows that the "compressed" string of characters is only half as long as the original string, and can therefore be sent in half the time required for the original message. Assuming that the receiver knows how the transmitter compressed the message, it can "decompress" the message, returning it to its original form before delivering it to the destination's user.

Why does this scheme work? It relies on the fact that the input is not random, that the lack of randomness can be recognized at the transmitter, and that some method can be devised to send any patterns that have been recognized down to the receiver using fewer bits than the pattern itself.

Experiment 15.6: Using Data Compression to Increase the Apparent Transmission Rate

The apparent transmission rate of a multimedia communications system can be improved if it is possible to "compress" the outgoing data before it is transmitted. As seen in earlier chapters, compression is often possible because the outgoing data is not perfectly random. When the data is not random, a clever algorithms can be used to send only the "new" information. The receiver will use this new information to update the "older" information that it has previously stored in order to reproduce the original input data.

There are many practical methods for compressing signals, and several will be listed. However, there are two points that need to be emphasized first.

- Efficient **data compression** schemes tend to take advantage of special characteristics of the type of signal they were developed for. As a result, a good compression method for voice will probably work very poorly on text.

- There must some method for coordinating how the transmitter compresses and the receiver decompresses the signal. In some cases the necessary information is built into the equipment before its use. In other cases the general method is agreed to, but the transmitter must send the remaining information as a part of the user's message.

There are many common examples of communications systems that use compression to increase the system's data rate, or, equivalently, to reduce its transmission time or its required transmission bandwidth.

- A facsimile machine uses compression algorithms to exploit the fact that most of a sheet of paper is white, and that each raster-scanned line often looks very much like the one above and below it. This compression permits each page to be sent in about 20 seconds rather than the two or three minutes that would otherwise be required.

- A digital cellular telephone uses compression to reduce the number of bits transmitted to carry a human voice from 64 kb/s to about 8 kb/s. This reduction permits the bandwidth needed for each caller to be reduced. With this reduction, more cellular calls can be made simultaneously within the band of frequencies allocated to the cellular telephone service provider by the Federal Communications Commission.

- The V.42bis modem is not a modem at all, in spite of its official name. It is an algorithm that operates inside of most dialup modems such as the V.34 discussed previously. It watches the data coming into the modem for patterns. When it sees a pattern, it sends information to its counterpart at the other end of the link identifying the pattern and what character it plans to substitute for it. It is very sophisticated version of the example that we examined at the beginning of this section. For typical textual traffic it is capable of increasing the effective transmission rate of a modem by a factor of two or more.

15.5.2 Error Detection and Correction

Another very common enhancement to a basic communications system is the ability to detect the fact that errors have been made, and, in some cases, to correct them. While there are many ways to do this, virtually all of them work in the same general fashion. The input to the transmitter is applied to some mathematical function. The result of that computation is sent along with the transmitted data. The receiver, after recovering both, recomputes the mathematical function. If its result matches the information sent to it by the transmitter, then the "payload" is judged to be received properly. If not, then the receiver infers that an error has been made somewhere. Historically two approaches have been used when an error is detected –

- Ask the transmitter to retransmit the data thought to be in error – which assumes, of course, that the communications link is a two-way one and that you have time to wait for the corrected information.

- Attempt to use the "extra information" to not only detect the presence of errors but to infer where they are and then correct them.

Experiment 15.7: Using Error Correction to Improve Reception

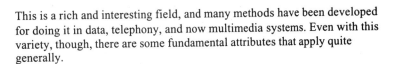

Error correction techniques are used to improve the quality of the data delivered to the ultimate user. Even though they require the transmission of extra bits to the receiver, their use surprisingly can improve the system's transmission rates since it compensates for the use of faster, but more error-prone, signal designs..

This is a rich and interesting field, and many methods have been developed for doing it in data, telephony, and now multimedia systems. Even with this variety, though, there are some fundamental attributes that apply quite generally.

- **Error detection** and **error correction** techniques reduce, rather than increase, the transmission rate of the user's data since they require the transmission of extra information to the receiver.

- More information must be sent to the receiver to correct errors than merely to detect them.

Error correction systems permit more bits per symbol to be used at a given noise level while maintaining the customer's tolerable error rate. This and the first point interact to create the very reason for using error correction coding — even though extra bits must be sent to perform this correction, it can be so effective in counteracting the noise's effect on the demodulator that the net transmission rate can actually be increased!

15.5.3 The Extended View of a Communications System

Figure 15.14 shows the block diagram of a more complete communications system, and the starting point for the designers of those systems. The incoming data is first compressed, using some technique appropriate to the type of multimedia data at the input. The output of the compressor is usually processed to determine the information that should be attached to permit error detection or correction at the receiver. The resulting bit stream is then broken into symbols of d bits each. Each symbol is used to modulate the phase, amplitude, frequency, or a combination of a pulse that is transmitted via wire, radio, light, or even sound waves, to the receiver.

The receiver must reverse all of the actions performed by the transmitter. The value of each received symbol must be determined as well as it can be in the presence of whatever noise and interference might be present at the receiver's input. Any error correction information that has been sent is used at that point to reduce the impact of the noise. Once the best possible estimates of the transmitted symbols have been obtained, they are fed to the decompression algorithms which recover the original input to the transmitter. The output of the decompressor is sent on to the user.

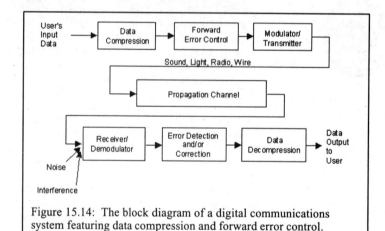

Figure 15.14: The block diagram of a digital communications system featuring data compression and forward error control.

Exercise 15.4

1. Suppose that we desire to send the message **THE RAIN IN SPAIN FALLS MAINLY ON THE PLAIN** and we would like to use the compression scheme discussed in this section.

    (a) Write the message after substituting # for all occurrences of **AIN**.

    (b) Is there a second candidate for substitution? If so, write the compressed message again using # for **AIN** and @ for the repeated sequence that you have found.

2. A very simple way to increase the reliability of a data transmission system is to repeat the data to be sent some number of times. Suppose that we do use this scheme and repeat each transmitted pulse three times.

    (a) Assuming that the total pulse or symbol rate is limited by the channel bandwidth to B pulses per second, what effect does this repetition have on the amount of actual data being sent?

    (b) Assume that each pulse has a one-in-a-hundred chance of being received incorrectly. What is the chance of all of them being received correctly?

    (c) A common strategy when using this repetition scheme is to do a "majority" vote at the receiver to determine the most likely transmitted system. In this case that would mean that if any two received symbols are the same, then their values would "win." Again assuming a one-in-a-hundred error rate for each received pulse, how often would this majority rule scheme produce the right answer? Is it better than the performance for each pulse by itself?

Section 15.6:
Summary

- From all of our study in this chapter it is clear that the communications system designer engineer has to make many, many choices in the pursuit of an approach that meets the needs of the users. To determine what is "best," the designer must balance these needs with the laws of physics, the mathematics of communications theory, and the hard realities of what the customer will pay.

- We have found that the **bandwidth** of the channel and the **signal-to-noise ratio** P/N are the two main factors in determining how efficiently we can transmit data through a channel.

- We've also observed that **forward error coding** can be used to further combat the noise and that it is possible to use "redundancy" in the input data to compress it, making the transmission link more efficient yet.

Virtually all modern systems for communicating multimedia data, from high definition television (HDTV) to Internet-ready personal digital assistants (PDAs) use all of these techniques.

Section 15.6:
Glossary

bandwidth of a signal(15.2.1)
is the difference, in Hertz, between the frequency of the highest component in the signal's spectrum and that of the lowest

bandwidth of transmission channel(15.2.1)
the range of frequencies over which a signal passing through it might be spread without impairing its quality; also the number of bits that can be accurately carried through the channel per unit of time

capacity (15.3.3)
the theoretical upper limit of the amount of information that can be transmitted with no errors through a communications channel at a given signal-to-noise ratio as seen at the receiver

data compression (15.5.1)
a technique for reducing the number of bits to be transmitted by exploiting patterns or other non-random characteristics of the data stream

error correction (15.5.2)
techniques for correcting errors induced into a communications system by noise, interference, or propagation difficulties. Most work by adding extra information at the transmitter that can be used at the receiver to detect the presence of errors and then to correct at least some of them

error detection (15.5.2)
techniques for detecting errors induced into a communications system by noise, interference, or propagation difficulties. Most work by adding extra information at the transmitter that can be used at the receiver to detect the presence of errors

frequency division multiplexing (15.2.3)
a technique for allowing the use of a single transmission medium by many users by the strategy of giving each of them a spectral band to operate in that overlaps with none of the other users

signal to noise ratio(15.3.3)
the ratio of the signal power to the noise power in a channel; this ratio determines the rate at which information can be communicated using a "real
ications channel

spectral efficiency(15.3.1)
the ratio of the rate at which data can be sent through a channel to the bandwidth of the channel

transfer function(15.2.1)
the amount of the energy of a sinusoidal signal that can be transferred through a channel

The INFINITY Project

Chapter 16:
Designing Networks

Approximate Length: 2 Weeks

Section 16.1:
Introduction

In chapter 8 we learned that a communications network can be defined as a set of transmission links that are connected by switching or routing nodes to provide connectivity to and from a large number of users. A network will typically be designed to meet a specific service objective at the lowest possible cost. In this chapter we examine a number of techniques that are used to improve the performance of modern data networks.

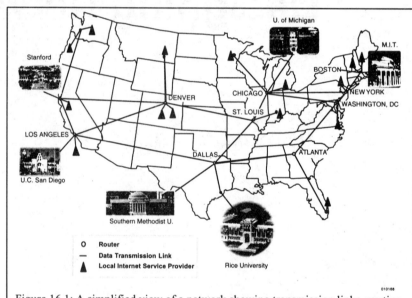

Figure 16.1: A simplified view of a network showing transmission links, routing nodes, and several users.

Section 16.2:
Performance Issues

Our objective at the beginning of chapter 8 was to use our knowledge of digital communications to build a system that would allow a large number of users to interchange data with any other. Our first attempt was a "fully-meshed network," one in which every pair of users was connected with a two-way communications link. In all matters other than cost, this approach was very attractive. Every user had a dedicated path to every other, and could therefore send data at any time with no delay other than that caused by the actual speed of transmission. However, in most cases this would result in a low utilization rate for the individual transmission links, making the cost of this arrangement very high. This issue encouraged us to search for alternatives that would lower the cost by sharing the transmission links among the users. To make sharing possible, we introduced switching or routing nodes that could interconnect the links.

The sharing of network assets, which is crucial to the lowered cost of the complete network, degrades the system's performance as seen from the perspective of the users. From a technical point of view, there are many potential problems in a network due to limitations of capacity and reliability of the individual nodes and links. For example, an individual user can wish to send more data than his access link can handle. An individual link may also cause problems by failing completely, slowing down, or creating errors. It is also possible that more traffic can arrive at a switching or routing node than it can send on, due to limitations of the node itself or its outgoing transmission links. Switching or routing nodes can also fail completely or misdirect messages through the network because of inaccurate information about how the elements of the network are connected.

The problems with individual links could, of course, occur even in the fully meshed network, but the problems with nodes appear when we introduce the routing or switching. The users usually would not know whether node or transmission link problems affected their service. The degradations they might observe include delayed reception, errors in reception, or, in some cases, no reception at all.

Many design choices are made when a network is built. The service provided by a network to its users can be improved if some simple objectives are considered when design choices are made. One simple objective is to minimize the end-to-end delay from one user to any other. A second goal might be to improve the network reliability, both in terms of minimizing the number of errors made and in terms of delivering all messages. Performance would be improved if a network were dynamically responsive to unpredicted changes such as the failure or appearance of a link, node, or user. In general, the quality of service should be controlled to meet the customers' objectives.

16.2.1 Quality of Service (QoS)

Before investigating how each of these technical objectives can be met, the issue of the customer's objectives should be discussed further. When customers buy communications services from a network, they have in mind a level of service or a quality of service (QoS) that they expect to receive for their money. In general these service levels fall into two categories.

- **best effort service:** The network makes an honest attempt to deliver the user's data in a timely fashion and with no errors, but with no guarantees.

- **guaranteed service:** The network delivers the user's traffic with a very high probability of success and within strict performance limitations.

Common sense tells us that best effort service is easier to provide and therefore probably costs much less than service with guarantees on delivery and performance. This is, in fact, true. Further, the type of service a network is to provide leads to fundamental differences in how the network is designed and operated. One of the principal challenges facing network designers of today is how to build a network that can do both, that is, deliver best effort service very cheaply, while providing more stringently controlled service for those who need it and are willing to pay extra for it.

16.2.2 Packets and Cells

The input to modern data networks can be as short as a few bytes, or as long as gigabit rate streams which go on for a month or more. In all of our descriptions so far, we have spoken of sending one user's message and then another message from another user, and then yet another message from still another user. This is reasonably practical when all of the users are generating short messages, but it becomes impractical, for a number of reasons, when the "messages" become long files, or, even worse, continuous streams. To accommodate this variety in "message sizes," a strategy was developed in the mid-1960s for splitting messages into **packets** if they exceeded a certain length. Each of the packets is then treated as an individual message and transported through the network as if it were. The user to whom the packets are addressed would hopefully receive them all, put them into the right order if they were not received that way, and then "reassemble" the message or file.

4

Figure 16.2 shows this process for four users, all of whom are sharing or multiplexing the same transmission link. User #1's message is broken into three packets while User #2's file is broken into several hundred. User #3's continuous stream is broken into an unending number of packets of roughly equal size. User #4 is sending a message so short that only one packet is needed to encapsulate it.

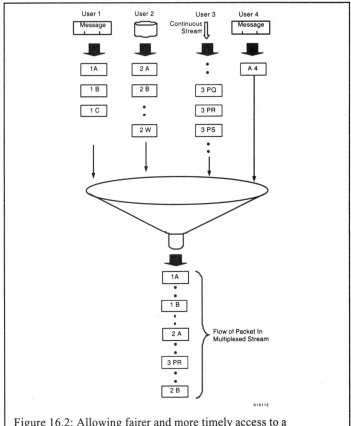

Figure 16.2: Allowing fairer and more timely access to a transmission link by breaking all messages, files and streams into shorter *packets*.

Note that each of the packets must be numbered or marked in some way so that the receiving user can reassemble the message, file, or stream properly. Thus, every packet needs to include the address of the destination, just as the original message did, so that the routers or switches can send it toward its destination, and a "sequence number" to permit reassembly at the destination. The sequencing in Figure 16.2 is indicated by giving each packet in each stream a different letter.

The primary reason for packetizing the traffic is to permit each of the users who are sharing the link to get quicker delivery of their traffic since they do get hung up waiting in line behind another user who has a large, or even infinite, amount of data to send. Even though this is the principal reason, there are two additional excellent reasons, either one of which is important enough to encourage the use of packetization even if the main reason didn't exist. For example, if only one user has exclusive access to the link, the primary reason for packetizing traffic would not be relevant.

One benefit of packetizing is that error detection and/or correction is simpler and less costly to do for shorter packets than for longer and infinitely long streams of data. This is particularly important if the error correction technique is based on asking the transmitter to re-send information when a receiver detects an error.

A second benefit is that breaking the data into packets reduces the amount of storage needed at each switch or router since they are required to store only one or more packets rather than the whole file or stream.

The use of packetization is virtually universal in any modern data communications system that carries messages or files with lengths greater than 200 bytes or so. The terms **packet** and **frame** are typically used when the data is broken up into units that don't necessarily have the same length. The term **cell** is typically used when all of the units have exactly the same length.

Exercise 16.1

1. The average size of a packet built using the Internet Protocol (IP) is about 800 bytes. How many packets would be required, on average, to transfer a 1 megabyte file?

2. A protocol called asynchronous transfer mode (ATM) breaks messages and streams into **cells** carrying 48 bytes each. Suppose that we use ATM to carry the simplest type of multimedia call, a 64 kb/s stream carrying a voice telephone call. If the telephone call lasts for three minutes, how many ATM cells must be generated and transmitted?

3. Packets built using the Internet Protocol carry a **header** that includes the source and destination address for the packet, plus a variety of other information. The length of the header is typically 40 bytes or so. Assuming this value is accurate, and that the average length of the whole packet is 800 bytes, what percentage of the packet is used to carry the user's data?

4. ATM cells carry 5 bytes of address in addition to their 48 bytes of "payload." What percentage of each cell is the user's data?

Section 16.3:
Improving the Probability of Accurate Delivery

Having introduced packetization to minimize the average user delay on shared transmission links, we now turn our attention to the problem of making delivery more reliable and accurate. To address this problem, we must first identify the problem's root causes. We find that they fall into a small number of categories.

- Errors made by the transmission links themselves, owing to noise, interference, or degradations introduced by the propagation channel,

- Duplicated packets, which can happen when error control systems send extra copies of a packet to ensure the delivery of at least one,

- And missing packets, which can happen when so many transmission errors are made that the transmission system gives up on a packet, or when the router's memory "overflows." Memory overflow happens when more packets arrive at a router than it can process or than its output transmission links can handle.

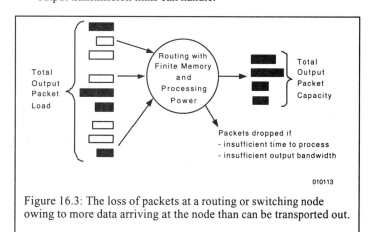

Figure 16.3: The loss of packets at a routing or switching node owing to more data arriving at the node than can be transported out.

Addressing this last problem we consider Figure 16.3. A router requires a certain amount of processing to examine the address of each incoming packet and to determine on which transmission link it should be sent out. In certain cases packets arrive at the router faster than this processing can occur. The packets waiting to be processed are stored. If they require more storage than the router has, the router overflows and some number of packets are dropped, never to be seen again and certainly not to be delivered. The second reason for overflow is also shown in Figure 16.3. Suppose that many of the inputs to the router are carrying traffic that can

7

only be delivered via one output path, and that the sum of the input loads exceeds the capacity of the one output link. The router will respond by buffering as many packets as it can to level the load, but when the amount of traffic to be stored exceeds the available memory, then, once again, the memory overflows and packets are dropped on the floor.

There are five strategies that are commonly applied in modern multimedia networks to solve, or, at least, reduce the effects, of these problems. We will address each in the following sections.

16.3.1 Adding Error Control to Individual Transmission Links

In Chapter 15 we learned that the error rate on a transmission link could often be reduced by adding extra bits to the transmitted data which could be used by the receiver to detect the fact that errors had occurred, and, in some cases, to correct them. These procedures are used very frequently in data and telephony communications systems. When the transmission rate is very fast, it is most common to attempt to correct any errors at the receiver. When the transmission rate is slower, or the user's tolerance to delay is higher, it is common to use the scheme shown in Figure 16.4.

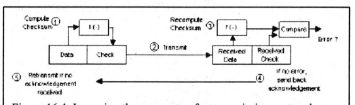

Figure 16.4: Lowering the error rate of a transmission system by retransmitting packets found to be in error by the receiver.

HF: High frequency radio operates at frequencies between 3 MHz and 30 MHz; it also referred to a "short wave" radio.

Before a packet is sent over the transmission link to the receiver on the right of Figure 16.4, a set of bits, often called the "check sum," are attached to the end of the packet. This set of bits is computed from all of the data in the packet. At the receiver the same function is used to recompute those bits. Once computed, they are compared with the ones actually received. If they match, the data in the packet is judged to be valid and is sent on to ward the ultimate user. In addition, a message is sent back to the transmitter (over the other half of the two-way link) to indicate that the packet was accurately received.

If the packet was not received, or if the check sum does not match, no acknowledgement message is sent. After waiting for this acknowledgement for a specific period of time, and not receiving it, the transmitter re-sends the packet. This procedure continues until all of the packets are accurately sent and acknowledged.

In older times this method was called **automatic repeat request (ARQ)** and was widely used in HF radio data transmission systems to bring the error rate down to an acceptable level. In more modern networks it is usually termed **positive acknowledgement of receipt (PAR)**.

16.3.2 Adding Alternative Routings to Handle Transmission Link Overflows

The phenomenon of routers dropping packets due to their memory overflowing is the principal cause of errors in the Internet. Other than the obvious tactics of making the routing processor faster and adding more memory, the solutions to this problem have to be introduced at the system level. The first approach to this is simple to understand but hard to apply. If the data rates of the sources of data and destinations of the data can be accurately estimated, then sufficient link capacity and processor capacity can be installed. The problem of not having enough output bandwidth can be resolved by adding additional output links that can be used as alternatives to the overloaded ones.

The problem with this "engineering" approach to the problem is that it requires knowledge about how the users behave, and the willingness to invest in more equipment and transmission capacity than are needed on a normal day (or microsecond). It is the approach used, however, for telephone networks and for data networks which must be highly reliable.

16.3.3 Adding Admission Control to the Network's Control Logic

An alternative strategy to handling the router overload problem is to make sure that it doesn't happen. This can be done by simply not letting traffic onto the network unless there is capacity to handle it. This technique is called **admission control**. In all of the discussion so far, when a user wanted to send traffic, and the link from it to the router or switch that serves it is available, then it sends its data and assumes that the network will handle it properly. What happens if the third router in the chain to the destination overflows and drops some or all of the originator's traffic? The originator may never know, or it may be told sometime later to retransmit its traffic. In any case, its use of the routers and transmission links has been wasted.

In a system with admission control, the originating user does not start sending its traffic until the network has agreed that it has the capacity to haul the traffic expeditiously and with low probability of loss. How would admission control work? In a restaurant, as illustrated in Figure 16.5, admission to the dining room is controlled by the maitre'd.

> **Interesting fact**: Admission control is used on some busy freeways where stop lights control the entry to on-ramps.

Figure 16.5: The maitre'd at an expensive restaurant as an example of **admission control**.

In data networks, arranging for admission is a little more complicated. In such a system the originating user sends a special message, often termed **signaling** or a **connection set-up message**, to the switching node that serves it. The signaling message is then passed along over a path to the destination. When all of the switches or routers along the way agree that they have the capacity needed to handle the user's traffic, the originator is sent a message to that effect and the transfer begins. The originator is not "admitted" to the network until this happens.

The telephone network works this way, as do modern data networks that have strict quality-of-service requirements. An important case is those networks carrying interactive multimedia traffic in which fast delivery and quick response are absolutely essential.

16.3.4 Adding Transport Control

While admission control and overbuilding are common ways to improve the QoS of a network, they are expensive, more complicated, or require a level of user knowledge or system-level coordination that is just not available. An alternative to those methods is to use **transport control**, a layer of software algorithms that supervises the transfer of information from the originator to the destination user in order to assure reliable delivery.

In order to provide fair allocation of a link's bandwidth to its various users we already break the "file" of data to be sent into packets. As discussed in the previous section, these packets are sequentially numbered so that the receiver can rearrange any packets that arrive out of order. **Transport control** uses this numbering more aggressively to insure the integrity of the received data. Specifically, a message is sent from the originating terminal to the destination terminal announcing the desire to provide reliable transport. Knowing that the outgoing terminal is prepared to offer this service, the receiving terminal watches the arrivals of the numbered packets and then asks the originator to retransmit any packets it believes to be missing. This scheme is shown in Figure 16.6. Suppose that the outgoing message was broken into six packets, that they were numbered sequentially, and then they were sent. Suppose further that Figure 16.6 shows what arrives – six packets, but not exactly as sent. From the numbering we see that packets 1, 5, and 6 are received where we would expect them to be. Packet #3 is received before either of two copies of packet #2. Packet #4 is not received at all.

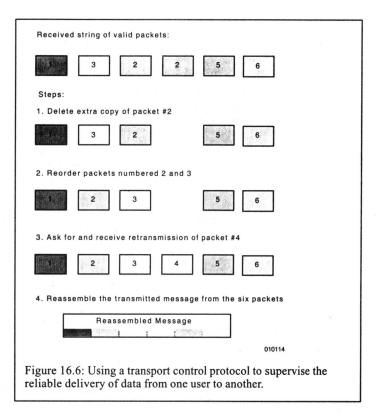

Figure 16.6: Using a transport control protocol to supervise the reliable delivery of data from one user to another.

The transport control algorithm will take the following actions:

- Delete the extra copy of packet #2.

- Reorder packet #3 and the remaining copy of #2.

- Request from the originator a retransmission of packet #4.

When packet #4 is subsequently received, the complete message can be "reassembled." When this is complete, the transport control protocol sends the complete and accurate message on to the destination's user.

16.3.5 Improving the Network's Response to Changes in Topology and Load

Fundamental to the concept of using relay points to reduce the cost of a network is the idea that each of them must be intelligent enough to send each message or packet on toward its ultimate destination. Imparting this intelligence proves to be difficult for complex networks, and is even more difficult when the network is changing with time. To see why this is, and to understand how modern networks deal with this, we must first drop back and examine, first, a simple case.

Network topology: the physical interconnection configuration of the terminals in the network.

11

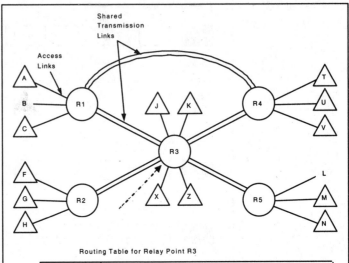

Routing Table for Relay Point R3

Destination	Next Relay	Expected Hop Count	Alternative Route
A	R1	2	R4
B	R1	2	R4
C	R1	2	R4
F	R2	2	–
G	R2	2	–
H	R2	2	–
J	–	1	–
K	–	1	–
L	R5	2	–
M	R5	2	–
N	R5	2	–
T	R4	2	R1
U	R4	2	R1
V	R4	2	R1
X	–	1	–
Z	–	1	–

Figure 16.7: The topology of a simple network, illustrating the "routing table" used by relay point #3 to determine how messages and packets should be forwarded toward their destinations.

Consider the simple network shown in Figure 16.7. It consists of five relay points and five inter-relay communications links, and it serves sixteen users. Intuitively it is clear how this network should work. Suppose that User B wants to send a message or a packet to User L. Through its "access link," User B sends the packet to relay point #1, its only choice. We would expect relay point #1 to send the message on to relay point #3, and for relay point #3 to send it on to relay point #5. Relay point #5 would then be expected to deliver it to User L through its access link. How would this sequence of actions be implemented? The standard method for doing this relies on the concept of a **routing table,** which is a list kept by each relay point that suggests the next transmission step to be taken for every possible destination. Each relay point will have such a table and in general they are all different from each other. A reasonable table for relay point #3 is shown in Figure 16.7. Its relationship to the connectivity of the network is clear. For example, any message or packet reaching relay point #3 and destined to Users L, M, or N should be sent toward relay point #5. The table must contain directions for every possible destination. In more sophisticated systems the table might also contain a list of alternatives to be used if the first, and best, choice isn't available. For example, messages from A to T

12

might be routed toward relay point #3 if the best and most direct path, toward relay point #4, isn't available or is too congested.

How should these routing tables be chosen? In cases when the network is very simple, the choice of the best table may be obvious and it is usually manually inserted into the router or switch at the relay point. For more complicated networks, the best choice for the table might be determined by performing a set of mathematical optimization calculations using a computer.

The methods just discussed are termed "static" since the tables are determined once, plugged in for use, and then not changed. Experience has shown, however, that using static tables doesn't perform well in practical networks such as the Internet. The reason is that the environment changes. Specifically, in practical networks we see the following phenomena:

- The load of messages and packets carried by the network can change from that expected when the tables were designed,

- Transmission links can "go down" and return to operation, as can the relay points,

- New relay points and transmission links can appear in the network as it grows.

This kind of dynamic behavior means that a fixed set of routing instructions will not perform as well as ones that are chosen appropriately to the current status of the network. This fact has led to the development of a number of different approaches to letting the routing tables change in response to the network's changes.

Four principal techniques for **dynamic routing** have evolved.

1) Automatic centralized control: Each of the relay points reports regularly to a central processor with information about how heavily it is loaded and the status of its transmission links. The central processor performs a new optimization calculation and sends new routing tables to each of the relay points. This whole activity is done periodically, with that period being short enough to track the changes in the network.

2) Isolated table determination: The intent of this class of methods is to build and revise the routing table without having to explicitly share information with the other relay points. The most common approach toward this goal is to exploit information already present in the packets or messages as they pass through the relay point. Recall, for example, that we attach to each packet not just the destination address but also the source address. Suppose that we keep track of the source addresses for all packets as they pass through, and the transmission link on which they appear. Let's assume that the best direction in which to send a packet destined for a particular address is that path on which traffic from that address most commonly arrives. If that assumption is reasonable, then we can build the routing table for that node by noting the most common arrival path for each source, and then using that path for any outgoing packets to that address.

3) Distributed table determination: All of the methods examined so far are attractive in some regard but also have important disadvantages. Static routing is simple to execute but unresponsive to changes. The centralized method offers near optimal performance, but at the cost of a central processor and using up bandwidth to communicate status information about the relay points and their new tables. The isolated routing approaches use no transmission bandwidth to update its tables but generally don't perform as well as the others. The failure of any of these three to meet the combination of our objectives has led to the development of a fourth class of schemes. These are termed distributed methods. Their goal is to perform almost as well as the centralized method, but without a central processor. The approach is to permit the relay points to interchange information with each other. Based on this information, each relay point then revises its own routing table.

An example of how this might work is shown in Figure 16.8. It illustrates the same network that we saw in Figure 16.7. It also shows the routing table for each of the relay points. Each table has been augmented to contain the number of hops, or transmission links, that it believes lie between it and each listed destination. Relay point #3's best route to destination L, for example, can be seen to be toward relay point #4 and it expects destination L to be 2 hops away. We will assume that periodically all relay points send their own routing table to each of their neighbors, and vice versa. Thus, as shown in Figure 16.8, relays 1, 2, 4, and 5 will send messages containing their tables to #3 and it will send its table to them. Relays #1 and #4 also exchange their tables in this network. Once received, each of the relay points uses the information contained in their neighbors' tables to update and, hopefully, improve their o wn. By passing this information around, each of the relays can become aware of changes in the network and revise their own tables appropriately. The process of sharing the tables, updating, and then re-sharing continues forever in order to continuously track changes in the network. In practice it is common for this updating to be done every two minutes or so.

There are other ways to perform distributed routing beyond this one. Another common approach relies on every relay to send a small amount of information to all other relays, not just its neighbors, with the goal of getting the word out about network changes faster than the nearest neighbor scheme we first examined. Yet another method is based on the first one but allows the network's system administrator to enforce policy, that is, to make sure that certain paths are not used. The reason for avoiding a path is usually not a technical one, but rather a financial one or a security matter. We don't want to pay someone else to carry our traffic for us, or we want to carry theirs! Another example is when we don't want our information to travel in a given direction for fear of who might see it as it passes through.

One can argue that distributed routing is really the worst of all three previous methods since it consumes some bandwidth for inter-relay communications and it requires significant computation to update the tables. Even so, it does not require a central processor and it does accommodate changes in the network topology and network load. For these reasons distributed computation of routing tables is the basis for how modern Internet routers work. The Cisco router shown in Figure 8.10 is capable of using any of the three distributed methods that we just discussed. It can also be statically routed if the network administrator so chooses.

4) Real-time : Modern telephone networks have carried routing technology even one more step. Older networks use some combination of static and centrally updated tables to select paths through their switches. Newer networks use **real-time network routing**, which is AT&T's name for it. Each call is attempted along the path that the outgoing switch's static routing table says is best. If bandwidth on that path is available, then the call is completed to the destination subscriber. If there is not enough bandwidth on that path, however, the originating switch sends message to the destination switch asking about its loading to and from all other switches in the network. Armed with the reply, the outgoing switch compares its view of the network with that of the destination switch and finds a "via" path that they both believe to be available. The call is then connected through that two-hop path. The whole process takes less than a second and permits virtually any two-hop path in the whole network to be used in completing the call.

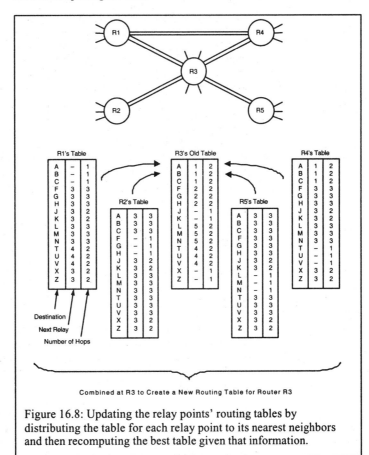

Figure 16.8: Updating the relay points' routing tables by distributing the table for each relay point to its nearest neighbors and then recomputing the best table given that information.

Figure 16.9 illustrates real-time routing for a simple network consisting of five switches. Suppose A desires to send data to user W. The switches directly connecting A and W check to see if adequate bandwidth is available to meet A's transmission needs. If so, the connection is made. If not, the three alternatives shown in Figure 16.9(c) are examined.

No tables are exchanged when using real-time routing and each call is routed based on a real-time view of the network. It makes very efficient use of the network's bandwidth, at the expense of sending a signaling message when congestion is encountered on the path of first choice.

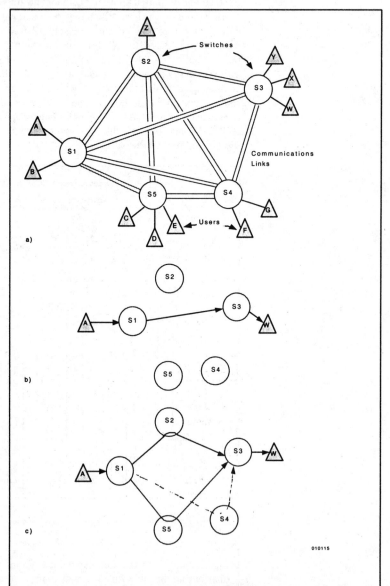

Figure 16.9: A network using **real-time network routing** to arrange for a connection to send data from one user to another. (a) a simple fully meshed network with five switches, (b) the initial attempt to make a connection between user A and user W, (c) the three alternative paths available if the direct path between A and W is not available.

No single modern network uses all of the methods just described to improve its performance. It is rare, however, that a modern network won't use several.

Using the public Internet as an example, we find that it does the following:

- Breaks messages and files into packets,

- Uses error correction and error detection techniques to control bit errors,

- Uses a transport control protocol (TCP) to manage the end-to-end quality of a data communications connection, and

- Uses distributed updating of router tables to respond effectively to changes in the network topology and load.

Conversely, the designers of modern public telephone networks have made other choices, but with the same goal of making the quality the best it can be for interactive multimedia data. They have done the following:

- Allocate each call to an individual "voice channel," reserving for it a fixed amount of bandwidth for the duration of the call,

- Use admission control to exclude calls for which is does not have enough bandwidth,

- Use alternative routing to increase the probability of finding a path to the destination,

- Use switching, rather than routing, to reduce the signal's delay from the originator to the destination, and back.

- Use "real-time network routing" to find a path to the destination even as network conditions change.

Class Activity

Here is the spot for the class to analyze, in groups or together, some real networks and determine what design choices have been made. Some examples:
  1. Telegrams
  2. A telephone party line
  3. An airline reservation system
  4. Credit card validation at the gasoline pump

Exercise 16.2

1. Suppose that the three data links entering the router in Figure 16.3 from the left are carrying a total of three megabits/second of data. Further assume that the transmission link exiting from the right can carry only 2 megabits/second, and that the router has 64 megabytes of temporary storage to hold packets until they can be transmitted.

    a) How much faster is data entering the router than can exit from it?
    b) Assuming that the buffer memory is empty to begin with, how long will it take to fill up?
    c) What happens once the buffer fills?

2. One of the simplest schemes for error detection is to add a "parity bit" to the end of a block of data. If the data has an even number of binary ones, then the parity bit is given a value of zero. Conversely, if the numbers of ones is odd, then the parity bit is given a value of one.

    a) What parity bit would be attached to the binary stream 00110101?
    b) What would the parity bit for 1111111111111111 be?
    c) Assume that you received the stream 010111010101 followed by the parity bit 0. Would you acknowledge the data stream positively?

3. A communications system is assumed to use the "positive acknowledgement of receipt" method, sending an acknowledgement whenever it determines that a valid packet is received. Suppose that it takes 10 milliseconds to send each packet, but 500 milliseconds to receive the acknowledgement. (This delay is typical of a transmission link that passes through a geosynchronous satellite!).

    a) What percentage of the time can data be transmitted if each packet must be acknowledged before the next one can be sent?
    b) How does the efficiency improve if the packet length is extended to 100 milliseconds?
    c) Can you think of another method to improve the efficiency of the transmission system?

4. List several examples of admission control, including some in communications systems (for example, the dial tone for a telephone) and some from other aspects of life (for example, the metering lights permitting entrance to a crowded freeway).

5. Assume that a router delivers packets marked with the sequence numbers 1 4 2 2 7 6 8 3 9 5. Write down the sequence of actions taken by the Transport Control Protocol (TCP) in its attempt to deliver a complete and accurate block of data to the intended recipient.

6. Using the network in Figure 16.7:
    a) Write down the routing table for Router R1.
    b) Suppose an additional link is installed between routers R4 and R5. How should the routing table for router R3 be revised?

7. Again using the network in Figure 16.7:

   a) Assume that a new user, designated W, is connected to Router R4. If the scheme for distributed routing shown in Figure 16.8 is used, how many table updating cycles must go by before router R3 knows about W?

   b) When will router R1 and R2 get the word on the new user W and how to reach it?

8. Real-time network routing counts on the availability of a fully meshed network, that is, all N switches are connected to all others. Suppose that a call attempt is made and that the direct path is not available. How many alternative two-step paths are available for consideration?

Section 16.4:
Issues in the Design of a New Network

After studying all of the concepts in this chapter it is reasonable to ask how a new network should be designed. It should be clear at this point that the answer to that question depends on what the network will be expected to do. Specifically, we ask questions like the following:

- How many users are there?
- How much data do each of the users expect to send and receive?
- How close are the users to each other and what is their geographical distribution?
- How much delay can be tolerated by the users?
- What is their tolerance to errors?
- How quickly must data transmission be initiated?
- How much are the users willing to pay?
- How secure must the system be?
- What will government regulators permit?
- Must I use equipment from a specific vendor?
- Must the new system work with older "legacy" equipment and software?
- And probably many more.

As we observed in Chapter 8, we find, of course, is that there is no single "best" design for a network. The definition of best depends on the application and the needs of the customer. The engineer's design problem is to best match the available technology with the desires and needs of that customer.

Section 16.5:
Glossary

admission control (16.3.3)

a strategy for limiting congestion in a network by admitting a new user only after verifying that the network has the capacity to handle it

automatic repeat request (16.3.1)

a method for correcting transmission errors in which the receiver asks the transmitter to repeat a message that the receiver believes to have been received in error

best effort service (16.2.1)

a description of network performance that implies no guarantees on performance measures such as data transfer rate and bit error rate, but rather indicates that the "best effort" will be made

cell (16.2.2)

a fixed-length packet, as distinguished from a "frame" which is a packet whose length might vary based on a variety of conditions. High-speed switches tend to use cells since their regular length permits hardware implementations, while routers tend to use frames

dynamic routing (16.3.5)

a strategy used in routers and switches in which the routing tables are permitted to change over time, accommodating variations in the traffic load from users and changes in the network itself

frame (16.2.2)

a data packet whose length is not uniform from packet to packet, as distinguished from "cells" which are of constant length

guaranteed service (16.2.1)

a description of network performance which indicates that measures are taken to guarantee the level of performance (for example, transmission rate, error rate, delay, and so on) to a user

header (16.2.2)

the portion of a packet (frame or cell) that carries the information needed to identify the packet stream and direct it to its destination. This information usually appears at the beginning or "head" of the packet, and is therefore termed the "header"

packets (16.2.2)

the subdivided portions of longer message or data stream; these packets, when reassembled, contain the complete message or stream

positive acknowledgement of receipt (16.3.1)

a method for correcting transmission errors in which the receiver sends the transmitter an acknowledgement for every properly received message or packet. The transmitter re-sends those messages or packets that have not been acknowledged within a selected time interval

quality of service (16.2.1)

a collective term describing the performance of a network in terms of a number of measures, including availability, bit rate, bit error rate, latency, variability in delay, and so forth

real-time network routing (16.3.5)

a strategy for selecting paths through a network for data transfer that relies on communications between the two terminal switches to find the best path at that moment

routing table (16.3.5)

a table held and maintained at each router or switch in a network that provides information about the next transmission path to be taken in order to route a packet or complete a path to its destination. The table is a list of entries that contains, at the least, every possible destination and, for each of those, the output path on which the data should travel to best get there

signaling (16.3.3)

special messages that travel between users and switches and between the switches. They are not used to send the user's data but rather to set up a path for transferring the data

transport control (16.3.4)

strategies used between the originator and the destination of a stream of data to make sure that the data is delivered completely and accurately. In the Internet this function is performed by a protocol called the Transport Control Protocol (TCP)

Postscript:
A Summary of
Communications and
Networking Concepts

This appendix provides a summary of the chapters in the communications section of the textbook.

Communications is defined as the movement of information, assumed to be digital here, from one physical location to another.

- Communications signals can be conveyed on a large number of types of transmission media—including wire, optical fiber, the vacuum of space, and the air around us.

- The transmitter's objective is to place one or more bits of information onto a signal which can be distinguished from all others that the transmitter might have sent.

- The demodulator's objective is to successfully distinguish each received symbol from the other possibilities and to send the set of one or more bits associated with that received symbol on to the user of the data.

- The transmitter must be appropriate to the type of transmission media to be used. In the case of radio signals, the frequency must be chosen properly to obtain the desired propagation characteristics.

There are limits to the amount of information that can be passed over a transmission channel. These limits depend on the signal's maximum power, the noise's power level, and the bandwidth of the channel.

- Within this limit, however, there are techniques available that can be used to improve the transmission rate, including forward error control (FEC), compression, channel equalization, and clever modulation.

There are also factors that tend to reduce the achievable data rate—for example, interference, dispersion, and multipath propagation.

- Claude Shannon provided us with a theoretical limit for the transmission bit rate. It was historically very hard to build a communications system with performance that even began to approach the "Shannon limit" but modern digital-signal-processing has made it feasible in several practical cases, for example, voiceband dialup modems.

- The bandwidth available for transmission of a signal must be the natural bandwidth of the media, or it might be a subdivision of it. These subdivisions are sometimes imposed by regulatory agencies to fairly share a scarce resource, and sometimes by design engineers, who desire to allocate no more bandwidth to a user than that which the user needs (or has paid for!).

There are various types of service that can be offered beginning with the simple transmitter-to-receiver communications link. They include the following:

- radio and television broadcast—where one transmitter sends unidirectionally to many receivers,

- point-to-point links—where two counter-propagating transmitter-receiver pairs are used to build a bidirectional ("full-duplex") communications system between two fixed or moving points,

- and networks—which use a number of point-to-points artfully constructed to connect many users to each other in a cost-effective way.

The concept of a network is based on the sharing of the transmission links among many users in order to minimize the cost of providing connectivity between them.

The INFINITY Project

Chapter 17:
Hardware for Digital
Storage :

It is all about ones and zeros

Approximate Length: 3 Weeks

Section 17.1:
Introduction

Approximate Length: 1 Lecture

17.1.1 How Do We Store All Those Bits?

As we've seen, all the fun signals in the world—music, movies, pictures, drawings, email, even your conversations with your friends on the telephone—are simply numbers when used in or created by high-tech digital systems. There's a potential problem with all these digital signals, however: how are we going to store them all? Unless we use the signals right away, we'll need ways to store them, because signals take up space. In this chapter, we'll learn about the physical systems and methods we use to store many different types of digital information. These systems include

- optical methods that use compact discs (CDs) and digital versatile discs (DVDs),

- magnetic methods used in most hard disks and floppy drives in modern PCs, and

- chip-based methods such as flash RAM and electronically erasable programmable read-only memory (EEPROM).

The name **storage** is the common term used to describe all of these different techniques for keeping bits around until we need them again.

An Example of Storage: DVD

The digital versatile disc, or DVD as it is commonly known, actually had a different name when a team of engineers at Sony and Philips developed it in the late 1980's. Originally called the ``digital video disc,'' DVD was designed to replace the prerecorded videotape as the medium of choice for watching prerecorded home videos. The name was changed when engineers realized the usefulness of DVDs for storing other types of digital information such as computer programs, music, and other important data. The DVD resembles a CD in both size and shape, but a DVD can store up to 17 gigabytes (GB) of data. That's about 20 times the amount of information that a CD can store. Imagine a DVD that holds every song that your favorite musical artist has ever recorded—along with the words, sheet music, and even a music video or two.

Ones and Zeros:
All things you want to store.
- Text
- Audio
- Image
- Video
- Numbers

2

Although the DVD name was changed for future applications, the most prevalent use of DVD technology today is in prerecorded movies. In fact, DVD-Video is the most successful consumer product launch in modern history. Over 16 million standalone DVD-Video players were sold in the U.S. alone in the three years after the format's introduction in March 1997, and millions more of DVD movie discs have been sold. DVDs are also being used to store and distribute large computer programs and database information, multi-channel surround-sound music, and high-quality video games.

While storage describes the physical form that digital data takes, **coding** is the name we give to the mathematical process of converting the numbers that we want to save into the bits that we physically store. There are several ways to encode the same information into bits. Each of these methods has advantages and disadvantages. One advantage of some coding methods is the ability to get back the correct stored information when the original storage object is damaged (through scratches, dirt, and the like). We'll talk about coding methods in the next chapter. For now, we'll assume that we have converted the information we want to store into 1's and 0's, the language of all things digital.

Exercise 17.1

1. Is copying a file from the hard disk on a computer to a floppy disk an example of storage or an example of coding?

2. CDs are examples of what storage technique? (a) Optical (b) Magnetic (c) Electronic.

Section 17.2:
A Brief History of Physical Storage

Approximate Length: 1 Lecture

17.2.1 Analog Storage

Before there were CD players, videogames, and computers, people stored information in many different ways. In fact, most digital storage methods are the end result of centuries of human invention. In many cases, these storage methods were designed to store analog signals—pictures, drawings, and other waveforms—as opposed to letters or numbers. For this reason, we shall call these methods *analog storage methods*. Most digital storage methods have strong connections with some form of analog storage. In the digital age, we have modified these analog storage methods so that they are better suited for digital storage than analog storage. It is important to study the history of analog storage methods, as it gives us a better understanding of the origins of our modern storage devices.

The Oldest Method of Storage – Cave Drawings

The oldest method of storage in human history is most certainly the drawings on cave walls made by ancient cavedwellers in prehistoric times. The pictures and symbols created by our ancestors tell stories about their experiences as well as important information about survival (such as when to plant and harvest crops, for example). Different methods were used by these peoples to create these drawings. In some cases, paint was used, whereas in others the drawings were carved into the rock as well as painted.

As might be imagined, the drawing process involved some physical effort and skill of the drawing's author. Moreover, once the drawing was made, it was difficult to change, and clearly it is not very portable! The main advantage of this form of storage is its lasting power. We can still find such records in well-preserved caves throughout the world.

Portable Storage – Paper, Canvas, and Wood

Paper is a finely-woven fabric of mainly wood strands that is then dyed to a light, usually white color. It has been used throughout ancient and modern human history to store historical, scientific, and religious writings. Paper is and continues to be the most popular storage device in the world. In fact, it is hard to find a facet of human endeavor that is not affected by its use. We dedicate entire buildings to paper documents (i.e. libraries), use it to record our daily business transactions (i.e. cash register receipts), create our currency out of it (i.e. paper money), and read millions of news stories and articles on it every day (i.e. newspapers and magazines).

What makes paper so useful? Paper has a number of features that make it ideal for storing information:

1. Paper is strong;

2. Paper is light;

3. Paper can be written on using many different types of pigments, such as ink, graphite within lead pencils, and the plastic used in photo reproductions; and

4. Paper lasts a long time when properly stored.

The main disadvantage of paper as a storage device is the difficulty in copying information written on it. Photocopiers can only approximately reproduce the information on a piece of paper, and there is always some defect introduced in the copy. Thus, paper is great for creating original documents, but it is not so good for recopying information.

Technical Term Focus: *Generations of Copies*

For copies of a document, audio recording, or video signal, a terminology has been developed to indicate how many times the information has been copied. A *first generation* copy is a reproduction of a piece of information, drawing, or work that employs the original information, drawing, or work to create it. For example, this page is a first generation copy of a book page that the book publisher has in its possession. If one were to make a copy of this page using a photocopier, the photocopied page is a *second generation* copy. Subsequent "copies from copies" would be called third generation, fourth generation, and so on.

Generally, when dealing with analog copies, there is some distortion that is introduced in the copying process. For this reason, earlier generations of copies will be closer to the original intent of the artist or author than later generations of copies. Such need not be the case with digital copies, an issue we will discuss later.

Canvas is the material of choice for painters. Canvas has been used throughout modern history to record portraits of people, landscapes, and even the imagined imagery of artists. Like paper, canvas is woven; however, its chief component is wool fabric strands somewhat similar to much of the clothing we wear. Like paper, canvas has excellent strength and is portable; however, its reproducibility is even worse than paper. For this reason, artists mainly use it today.

Wood is another form of storage that has features slightly different canvas and paper. Wood can be carved into useful shapes, and these shapes store information. Since wood holds its shape reasonably well, the carvings can record information for later use. The main drawback of wood is its size; imagine the lumber that you would need to do your homework assignments every week!

Early Audio Storage, Part I – The Phonograph

Thomas Edison, probably the greatest inventor of the Western world in the 19th and 20th Centuries, came up with an ingenious instrument that took sounds and stored them for later playback. His invention , the phonograph, is at the heart of all modern audio and video recording and playback systems, including CD and videotape players. It was a simple idea: turn sound into motion, and then store the movement into a long thin track, much like a trail through the mountains. The material that he used in his invention to store the "audio trail" was wax, which turned out to have poor long-term storage properties (For one thing, wax changes its shape too easily.) But the general concept was sound, and only the storage material needed to be changed. Today, his invention is most closely related to the modern-day *record player*, a device we now describe.

Figure 17.1 shows a diagram of a record player. It has a 12-inch diameter platter called a *turntable*, with a thin metal stalk called the *spindle* in the center. At one corner, a rotatable arm sticks out over the top of the platter. At the end of this arm is the phonograph *needle*, which is usually housed inside of a device called the cartridge. The phonograph player plays records, which are thin flat discs generally made of a black plastic with a 12-inch diameter.

Thomas Alva Edison :
Born Feb. 11, 1847, Milan, Ohio, U.S. - Died Oct. 18, 1931, West Orange, N.J. American inventor and holder of 1093 patents. Some of Edison's inventions are: the light bulb, the phonograph, the first electric power system, and the moving frames notion used in film and video display.

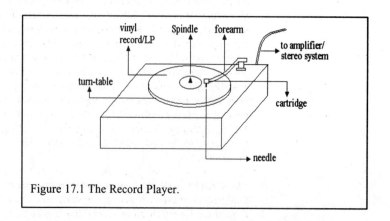

Figure 17.1 The Record Player.

To play records, one puts the record onto the turntable and turns on the turntable motor, which begins to spin the platter in a clockwise direction. Then, one positions the arm such that the needle is just at the outside of the record. There, the needle begins riding the small grooved spiral trail contained on the record as the platter spins. The small wiggles within the grooved spiral trail on the record show up as motions to the needle, and little magnets located within the cartridge help turn these motions into electrical signals that are then sent via wires to a set of speakers via an amplifier or receiver. Phonographs are in rare use today, although many hobbyists still collect, sell, and listen to records of their favorite musical artists.

The phonograph established a way of storing information that influenced the designers of many modern digital storage methods. The key features of phonograph that have been used in both optical (CD, DVD) and magnetic (computer floppy drives, hard disks) media are:

5. A nearly-circular spiral recording track, where one long path is wrapped in a circular fashion around one surface of a flat disc;

6. A playback mechanism that rotates the disc while reading the information; and

7. A reading device that moves along the radius of the disk in an inward or outward motion.

In other respects, however, the phonograph is different from optical and magnetic digital storage devices. For example, the information stored in a CD is read out using optical techniques rather than via mechanical motion; more details about the way optical storage works is in the next section. Moreover, because the physical grooves of a record contain the audio information, the record could be easily damaged through dust, dirt, and wear and tear. The fact that records can be easily damaged is one of the chief reasons that consumers have replaced records by more-durable CDs.

Early Audio Storage, Part II– The Tape Recorder

Another audio recording device that was invented around the same time as Edison's phonograph is similar to the cassette tape recorder that we use today. The first U.S. patent on a magnetic recording device was awarded to Vlademar Poulsen in 1899. Poulsen's recorder didn't use tape at all, however. Instead, it employed steel wire very similar to the wire used to hang up framed pictures in your home. These "wire recorders" stored the information magnetically by translating the sound into a magnetic field that gets impressed onto the wire as it is moved through the field. Think of how a needle can be turned into a magnet by exposing it to an actual magnet; now, imagine that the wire is simply a series of flexible needles. The technique used to read and write the signal will be discussed further in the Magnetic Storage technology below. In the 1930's, steel wire was replaced by plastic tape coated with oxidated metal, which is what cassette tapes are made of today.

The main advantage of tape recorders over record players was that one could erase and re-record over the same portion of the tape. Records were best suited for distribution of pre-recorded material, although some companies did manufacture "record recorders" in the early days of the phonograph. These recorders are similar to CD-Recordable drives of today; they could only write the material once into the disc.

Another advantage of tape recorders is the editability of the medium. It is easy to rewind and record new material at the end of old material on a tape. Moreover, it is even possible to physically "cut and paste" the tape sections together, which is another form of editing. Famous recording artists of the 20th century, most notably the singer Bing Crosby, used this technique to create taped radio shows in the 1940's and 1950's, which was unusual during those days of live radio broadcasts. In this way, only the best material was saved for the broadcast. Today, the "cut and paste" editing is done digitally, using computer hard disks and RAM as the intermediate storage media.

Videotape player/recorders are quite similar in their general function to cassette player/recorders, except that both the magnetic recording head and the tape move in a videotape player/recorder.

The main drawbacks to magnetic tape players/recorders are 1) tapes can only be searched by running through the tape itself, which takes more time than the *random access* provided by a disc-based technology, 2) tapes are prone to breakage and mechanical failures, and 3) tapes can be erased by a stray magnetic field such as a speaker magnet. Today, magnetic storage technology is much more prevalent in computers, where floppy and hard disk drives employ a refined version of the same technology.

17.2.2 Digital Storage

When the first electronic computers were developed in the 1950's, engineers needed some way to store the digital information that was to be processed. The earliest digital computers used a very primitive form of storage-- the position of a physical switch. Imagine a wall of toggle switches similar to modern-day light switches mounted on a wall. The "on" position of the switch stores a one (1), whereas the "off" position of the switch stores a zero (0). To load in a number or command to the computer, the computer operator needed to set each switch to its proper position before turning on the computer's power and letting the machine do its calculation. Today, we don't use physical switches to store information; more often than not, we use electronic switches—the transistors contained in almost every digital system.

As computers became more complex, computer operators needed a better way to store programs and numbers. Before optical methods (CD-ROMs) and magnetic methods (hard disks and floppy disks) became popular, the most popular storage medium was paper! To program a computer, a programmer would develop a binary code whose ones and zeros would get represented on paper cards a little bit longer than a 3" x 5" index card. The representation method was physical holes; that is, a one was represented as a cutout rectangle in the card, and a zero was a non-cutout portion. Figure 17.2 shows the layout of such a card system.

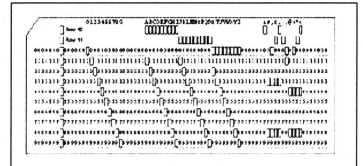

Figure 17.2: Punched card for older computers - Dimensions : 7 (3/8) inches X 3 (1/4) inches.

Today, programs are stored using optical, magnetic, or electrical storage technologies. We shall explore the forms of these storage methods in the next several sections.

Exercise 17.2

1. Name three analog storage methods, and compare their advantages and disadvantages.
2. Suppose you wrote down a message on a piece of paper that was written on a wall and then created a photocopy of the message. Is the photocopy a first generation, second generation, or third generation copy of the message?
3. Do videotape players provide random access of the information contained on videotapes? How about CDs?
4. Imagine the following situation: You took a photograph of someone famous and made a photocopy using a photocopier. Then, from that photocopy, you made another photocopy. Repeating this action, you make a 100th generation copy. Do you think you could recognize the famous person from the 100th generation copy of the photograph? What if you had a digital picture of the same person and made a 100th generation digital copy?
5. The storage material used in computer hard disks and floppy disks is the same as that used in (Choose one) (a) Phonographs, (b) CDs and DVDs, (c) Cassette tapes and videotapes.

Section 17. 3: Optical Discs

Approximate Length: 2 Lectures

17.3.1 Introduction

If you like music, videogames, and even movies, you probably know what CDs and DVDs are. These shiny round plastic discs are used to store all of the information and entertainment that we love. In this section, we'll find out how the information is stored on CDs and DVDs, how information is read out, and a little bit about the manufacturing of CDs and DVDs.

17.3.2 Physical Parameters

Both CDs and DVDs are round discs that are 13.33 cm (about 5.25 inches) in diameter and 1.2 mm (about $1/20^{th}$ of an inch) thick. The disk actually has a sandwich structure, in which two or more thin metal layers are placed between clear plastic layers. The physical makeup of this sandwich structure for CDs is shown in Figure 17.3. The corresponding structure of a DVD is similar. In each disc, there is at least one layer of metal that contains the digital bits that are stored on the disc. In the case of the DVD, there can be multiple layers of bit-carrying metal. These multiple layers suggests that DVDs can store more information than CDs; however, that is not the only reason why DVDs can hold more information than CDs, as we will soon find out.

Micron: a millionth of a meter – a length that is about $1/100^{th}$ the thickness of a single strand of human hair.

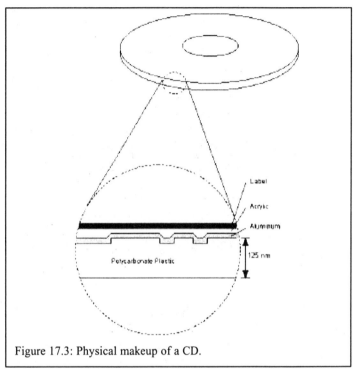

Figure 17.3: Physical makeup of a CD.

Both of these shiny discs carry information in the form of long miniature bumps and flat spots into which the metal layer(s) are shaped. These bumps are arranged as long trails in a spiral fashion around the disc, starting from the innermost edge of the disc and spiraling outward. The width of these trails is incredibly small—for a CD, each one is only 1.6 microns (millionths of a meter) wide, and for a DVD, each one is only 0.74 microns wide. The bumps in these trails are actually much thinner—roughly 1/4th the width of the trail. The lengths of the bumps vary and range from 0.822 to 3.560 microns for a CD and 0.400 to 2.054 microns for a DVD. Figure 17.4 shows the layout of these bump trails for a CD, along with their critical dimensions.

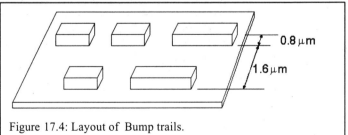

Figure 17.4: Layout of Bump trails.

Why are CDs and DVDs round? The answer is clear if you've ever looked inside a CD or DVD player that has a clear window to the disc holder tray: the disc is spun to read out its information. The speed of rotation is between 200 to 500 revolutions per minute (rpm) for a CD, and 570 to 1600 rpm for a DVD. The speed of rotation is actually changed depending on which part of the disc is currently being read. The disc is spun faster initially when the innermost part of the disc is being read, and the spin speed is reduced as the outer portions of the disc are read. Usually, only one portion of the bump track is read out at any one time.

Question: How many times is a CD rotated in a CD player to read out all of its information? In other words, how many revolutions does a CD undergo when it is played from beginning to end?

Calculation: To find an answer to this question, we draw a picture of the important features of the problem. Figure 17.5 shows a diagram of the CD, along with some important dimensions.

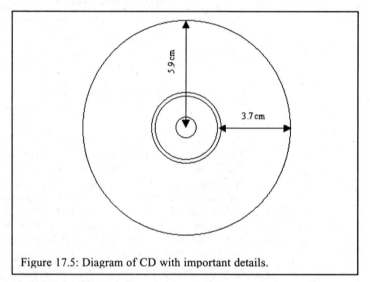

Figure 17.5: Diagram of CD with important details.

There is only one information trail that spirals outward on the CD. However, if we were to travel along this trail for exactly one complete rotation of the disc, we would end up right next to the place where we started. Now, looking along the *radius* of the CD, we can count the number of trailwidths in the usable area of the CD, that is, the area of the CD that contains the spiraling trail. It is easily seen that

$$N_{TrailWidths} \times W_{Trail} = W_{UsefulArea}$$

Solving for the number of trailwidths gives

$$N_{TrailWidths} = \frac{W_{UsableArea}}{W_{Trail}}$$

The critical quantities that we need to know to evaluate the above equation are

1.   the width of each trail, which we know is

$$W_{Trail} = 1.6 \mu m.$$

1.   the width of the usable area of the CD or the area where the bumps are stored, which from the figure is

$$W_{UsableArea} = 3.7cm.$$

From these two quantities, we can calculate the number of trailwidths that make up the width of usable area of the CD. Since there are 10 mm in 1 cm and 1000 microns in 1 mm, we find that.

$$N_{TrailWidths} = \frac{3.7cm}{1.6 \mu m} \frac{10mm}{1cm} \frac{1000 \mu m}{1mm} = 23100$$

In other words, a 74.5-minute CD spins over 20000 times when it is played from beginning to end.

Question: How long is the bump trail on a 74-minute CD? In other words, if we could unwrap the spiral trail of bumps on a CD, to what distance could you stretch it out?

Calculation: The answer to the last question can help us in figuring out the answer to this new question. First, however, we need to think about how the spiral trail is laid out onto a CD. Figure 17.6 illustrates the way the trail is wrapped around the CD.

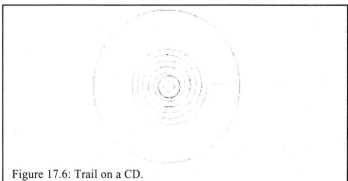

Figure 17.6: Trail on a CD.

If we travel exactly one revolution around the CD, we come back nearly to the same spot, but the arc we make is not exactly a circle. For purposes of calculation, though, we can *approximate* the length of this arc by a circle of radius equal to the average distance away from the center of the CD along this arc. Now, we know that the circumference of a circle is related to its radius by the equation

$$C = 2\pi r$$

Therefore, we can approximate the length of the entire CD bump trail by the sum of the circumferences of 23100 different circles, each with a slightly different radius.

Suppose we know the radius of each of these circles as *r(n)* for *n* =1,2,…,23100. Then, the total length of the bump trail is

$$L_{Trail} = 2\pi[r(1) + r(2) + \ldots + r(23100)] = 2\pi\sum_{n=1}^{23100} r(n)$$

All we need now is an expression for the radius of each circle in our model. The circles start at a radius of

$$r(1) = 5.9cm - 3.7cm = 2.2cm$$

The trail width is

$$W_{Trail} = 1.6\mu m \times \frac{1cm}{10000\mu m} = 0.00016cm$$

We can see that

$$r(2) = r(1) + W_{Trail}$$

$$r(3) = r(2) + W_{Trail} = r(1) + 2W_{Trail}$$

$$r(4) = r(3) + W_{Trail} = r(1) + 3W_{Trail}$$

Therefore, the general formula for *r(n)* is

$$r(n) = r(1) + (n-1)W_{Trail}$$

14

Plugging this expression into our equation for the length of the trail, we get

$$L_{Trail} = 2\pi \sum_{n=1}^{N_{TrailWidths}} [r(1) + (n-1)W_{Trail}]$$

$$L_{Trail} = 2\pi [N_{TrailWidths} r(1) + W_{Trail} \times \sum_{n=1}^{N_{TrailWidths}} (n-1)]$$

From our knowledge of sums of sequences, we know the formula

$$\sum_{n=1}^{N} (n-1) = \frac{N(N-1)}{2}$$

The total length of the CD bump trail is then

$$L_{Trail} = 2\pi [N_{TrailWidths} r(1) + W_{Trail} \frac{N_{TrailWidths}(N_{TrailWidths}-1)}{2}]$$

$$L_{Trail} = 2\pi N_{TrailWidths} [r(1) + \frac{N_{TrailWidths} W_{Trail} - W_{Trail}}{2}]$$

We can simplify this expression even further by noting that

$$N_{TrailWidths} W_{Trail} = W_{UsableArea}$$

This simplification gives the equation

The values for our problem are

$$r(1) = 2.2 cm$$
$$W_{Trail} = 0.00016 cm$$
$$W_{UsableArea} = 3.7 cm$$
$$N_{TrailWidths} = 23100$$

Plugging these values into the expression above produces

$$L_{Trail} = 2(3.1416)(23100)[2.2 + \frac{3.7 - 0.00016}{2}]cm$$

$$L_{Trail} = 5.88 \times 10^5 cm$$

We can convert this expression to miles as

$$L_{Trail} = 588000cm \times \frac{1m}{100cm} \times \frac{1mile}{1600m} = 3.68miles$$

In other words, the length of the bump trail on a CD is over 3 miles long!

Alternate Calculation: In many engineering problems, there is often one or more ways to come to a similar answer that is usable for our problem. We show one such alternate calculation here We know that for a rectangle

Length x Width = Area

So, if the CD were rectangular with the bump trails going lengthwise next to each other in parallel tracks, we could compute the total length of the bump trails as

$$L_{Trail} = \frac{A_{CD}}{W_{Trail}}$$

What happens if we simply use the actual area of the CD in the above calculation? We can show (see the example in Section 17.3.4) that

$$A_{CD} = 94.15cm^2$$

Plugging this expression into the above one, we have

$$L_{Trail} = \frac{94.15cm^2}{0.00016cm} = 588500cm$$

which is close to the value we got from our previous calculation! This method is a lot easier than before, and it gets us *close* to a good answer.

17.3.3 How Do CD and DVD Players Work?

Both CDs and DVDs employ *optical* means of storing information. By the term optical, we mean that light is used to read out the information from the disc. Obviously there is no light bulb or other source of light in the disc, so the way we get information off of the disc is by shining light onto the disc and looking at the light's reflection. The light source for this procedure is known as a *laser*. Lasers are special devices that emit only one concentrated color of light in a thin beam. They are great for use in optical storage because:

1. They can be inexpensive to manufacture, and

2. They last an incredibly long time (several years of constant illumination before burning out).

Most importantly, they provide an important optical property: they can create a distinctive light pattern from the disc through the principle of *interference*.

Figure 17.7 shows a diagram of an optical reading mechanism for CD and DVD players. There is a laser light source, a diffraction grating, a beam splitter, some optical lenses, and a photodetector array at the bottom. The laser light source produces *coherent* light of a certain color (or frequency) that passes through the diffraction grating, the beamsplitter and the optics. The diffraction grating spreads out the light as it passes through the beamsplitter. The beamsplitter is a lot like the "one-way" mirrors that you see on police dramas; it lets light through in one direction but reflects light in the other. The light then bounces off of the CD surface and goes back towards the beamsplitter, which now acts like a reflector. Then the light shines onto the photodetector array.

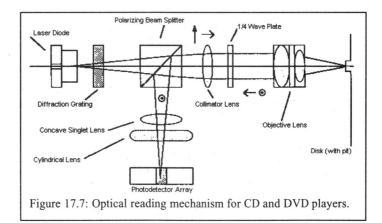

Figure 17.7: Optical reading mechanism for CD and DVD players.

Figure 17.8: A domestic DVD player.

The height of the reflective surface—in this case, each of the bumps of the CD or DVD being read---is designed to be *(1/4)* wavelengths high as measured from the reflective surface of the CD or DVD. When the light is shined on the disc surface, the optical reading mechanism is calibrated so that the flat portions of the disk look bright. As the laser beam is scanned across the disc, bumps are encountered, as shown in Figure 17.9(a). Notice how the laser beam covers a spot that includes both the bump and the flat portion of the disc. The light collected at the photodetector array thus contains light that is reflected both from the flat disc portion as well as from the top of the bump. The wave interactions in this situation are shown in Figure 17.9(b). Here, the light reflected from the flat portion of the disc travels 1/2 wavelength farther than the light reflected off of the top of the bump. When these two waves combine, they cancel each other out, such that the bump edges appear dark under the illumination of the laser beam light. In this way, each bump edge produces a dark region by destructive interference. These bump edges encode the "ones", whereas the flat portions encode the "zeros" of the CD or DVD bit stream.

To obtain several bits in succession, the photodiode is attached to a switch that stores which bit (1 or 0) it is currently observing on the disc. Then, as the disc is rotated, a new area of the CD or DVD is illuminated, and a new bit is read out. In this way, bits are read out from the disc one by one. Careful control of the rotational speed of the disc and the position of the laser reading mechanism is needed to allow the reading of the spiral trail of bits on CDs and DVDs.

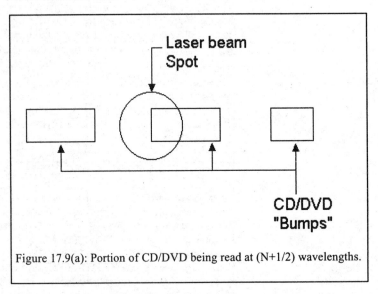

Figure 17.9(a): Portion of CD/DVD being read at (N+1/2) wavelengths.

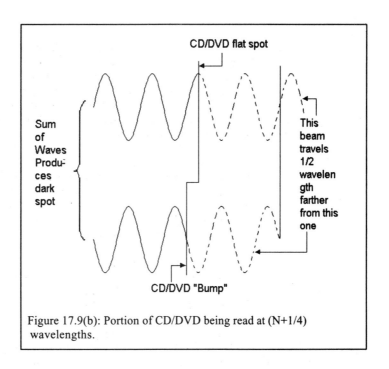

Figure 17.9(b): Portion of CD/DVD being read at (N+1/4) wavelengths.

For DVDs, there can be an additional layer of information contained on a single side of the disc. The first layer of information is contained in a half-reflective portion of the disc. Usually this half-reflective layer is made of gold; this is why a two-layer DVD is gold as opposed to silver in color. When reading out this information, the lens is moved so that the laser focuses on this first layer. When reading out the second layer of information, the lens position is changed so that the laser is focused on the second fully-reflective layer. The change in lens positions is quite small, so that the refocusing is done in a fraction of a second. Most modern DVD players are quite good at this change, so that the pause in the middle of the bit stream (e.g. the digital movie) is minimized.

17.3.4 Data Capacity and Areal Density

An important feature of any storage medium is its *storage capacity*, sometimes called its *data capacity*. The data capacity of a storage technology is the amount of data in bits that it can hold in one package, such as a disc or cartridge.

The storage capacity of a CD is approximately 650 megabytes (650 MB) of information. In the case of a DVD, the storage capacity is approximately 4.7 gigabytes (4.7 GB) for a single-layer disc and 9 GB for a dual-layer disc. Moreover, while the CD is strictly a one-sided medium, a DVD can be a two-sided medium; hence, one can store almost 18 GB of data on a dual-layer, dual-sided disc.

Question: What is the average length of the space used to store each audio-carrying bit on a CD?

Calculation: We can use the answers we've found for the length of a CD bump trail to figure out the answer to this question. Recall that we calculated the length of the CD bump trail as

$$L_{Trail} = 588000 \ cm$$

The number of audio-carrying bits stored on a CD is

$$N_{Bytes} = 788 \times 10^6 \ bytes$$

Therefore, the average length of the bump trail used to store each bit is given by

$$L_{Bit} = \frac{L_{Trail}}{N_{Bytes}}$$

$$L_{Bit} = \frac{588000 \ cm}{788 \times 10^6 \ bytes} \times \frac{1 \ byte}{8 \ bits} = 0.0000933 \ cm$$

Converting this answer to microns, we find that

$$L_{Bit} = 0.0000933 \ cm \times \frac{10000 \ \mu m}{1 \ cm} = 0.933 \ \mu m$$

This value is only an average number. The number doesn't take into account the overhead or extra information needed to keep track of where important bits are (such as the format of the disc, to be described shortly). Another important factor is the amount of extra information contained in the form of a *code* to correct for errors if some bits are lost in the reading process. Coding is the subject of the next chapter.

A more-important feature of a storage technology is its *areal density.*, also sometimes called *bit density.* The areal density of any two-dimensional storage device is defined as

$$D = \frac{(Total \quad Number \quad of \quad Bits)}{(Total \quad Storage \quad Area)} = \frac{N_{Bits}}{A}$$

In this equation, the area A refers to the active area of the disc or other medium, not the area of the disc and its package. For example, the active area of a CD or DVD is the area where bits are stored, neglecting the inner edge portion that holds no data.

Areal density is useful for comparison of different storage methods because it normalizes out the size of the storage medium. An easy way to store more on a CD is simply to make it bigger, but that doesn't make the CD any better in its ability to store information. For example, bigger CDs would also require more material to made and would be harder to carry around. They would also be harder to spin at precise speeds due to their increased mass.

Question: What is the areal density of a CD? Of a single-layer DVD?

Calculation: Since a CD stores about 788MB of data, we only need to calculate the total square area of the storage part of the CD surface. Figure 17.5 shows the dimensions of the CD and what the usable area is.

We can see that the area we want to find is in the form of a ring, and its value can be found from the difference of the areas of two circles of radius

$$R_1 = 2.2cm$$
$$R_2 = 5.9cm$$

The formula for the area of a circle is

$$A = \pi R^2$$

Therefore, the total usable area of a CD is

$$A_{CD} = A_2 - A_1$$

where

$$A_1 = \pi R_1^2$$
$$A_2 = \pi R_2^2$$

Plugging in the values, we get

$$A_1 = 3.1416(2.2cm)^2 = 15.21cm^2$$
$$A_2 = 3.1416(5.9cm)^2 = 109.36cm^2$$
$$A_{CD} = 109.36 - 15.21cm^2 = 94.15cm^2$$

We can then find out the areal density of a CD as

$$D_{CD} = \frac{N_{Bits}}{A} = \frac{788 \times 10^6 \, bytes}{94.15 \, cm^2} \times \frac{8 \, bits}{1 byte} = 67.0 \times 10^6 \, bits \, / \, cm^2$$

In other words, every square centimeter on the data-carrying portion of an audio CD holds about 67 million bits of information, or about 8.3 MB of information. Again, this is only an average value; certain portions of the

21

disc hold more than others in practice. Moreover, a CD-ROM carries less information because of additional formatting that is needed to store general-purpose data as opposed to music samples.

We could repeat the above calculation for a DVD, but there is a much easier way to solve this problem. Recall that a single-layer DVD stores about 4.7 GB of information. How many times more data does a DVD store over a CD? The ratio is

$$Density \quad Ratio = \frac{D_{DVD}}{D_{CD}} = \frac{4.7\,GB}{788\,MB} \times \frac{1000\,MB}{1\,GB} = 5.96$$

Since we know the storage density of a CD, and since the working area of a DVD is the same as that of a CD, we can immediately get that

$$D_{DVD} = D_{CD} \times Density \quad Ratio$$

$$D_{DVD} = 67.0\,MBits\,/\,cm^2 \times 5.96 = 399\,MBits\,/\,cm^2$$

In other words, a single-layer DVD stores about 400 million bits of information in every square centimeter, or about 50MB of information in every square centimeter.

17.3.5 Accessing the Data

Position Control

The mechanism used to follow the trail of bits on a CD or DVD is a miniature technological marvel. The entire laser reading assembly rides on a track with a precision motor to guide the motions of the assembly. This motor uses coarse positioning methods to correctly place the assembly along any desired portion of the disc. Once there, *position control* is then used to follow the bit-carrying path of the bump trail on the disc.

All position control methods have the same general structure. The block diagram of this system is shown in Figure 17.10. In this system, there is a sensing device that senses an error in the position of the read assembly. How this signal is sensed depends on the type of storage method. In the case of optical storage, one can measure the average *reflectivity* or brightness of the light being reflected off of the disc and collected at the photodiode. If the laser assembly begins to go off track, the light output will generally increase because of the blank spaces in-between each track. If this signal shows an error in the path of the assembly, a signal is sent to the motor that controls the position of the reading device. The motor uses this signal to determine which direction and by how much the position of the read assembly should be moved to correct the position error. As a correction is made, the sensing device senses less error in position, which subsequently decreases the correction that needs to be made. Over time, the error can be reduced to a very small amount, small enough to provide accurate following of the information contained on the disc.

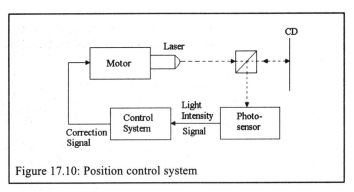

Figure 17.10: Position control system

Position control is an important component of all CD and DVD players. Position control helps to overcome small problems with the manufacturing of CD and DVD discs, such as a misplaced center hole. The wobble in position caused by even a small change in the center hole position of an optical disc can cause large changes in the position of the laser reading mechanism during reading. Of course, if the center hole is misplaced too much, the position control may not be able to track the wobble. Fortunately, such discs usually are recognized at the factory during manufacture and are rejected there as faulty.

Position control also determines another important capability of a CD or DVD reading device: its *data rate*. The data rate of a device is the speed at which bits can be read off of the disc. Players that read audio-only CDs or DVD videos need only read out the information contained on them at the bit rate of the application. For computers, however, a faster data rate means that programs and data load faster when using these devices. Position control continues to improve over time due to advances in control technology. For this reason, you will likely see faster data rates from existing disc technologies in the foreseeable future.

Format

In addition to position control, all CD and DVD players are designed to read off the information contained on their respective optical discs according to a standard layout or *format*. This formatting information includes issues such as

1. The number of successive bits that make up a particular information unit, such as a sound sample or image pixel;

2. The definition of specific access points or *tracks*, and

3. The specification of different coherent bit streams, such as the video, audio, and subtitle bit streams on a DVD video disc.

This format information depends on the application in which the disc is used. For example, the information on a CD-ROM has a different format than the information on an audio CD. The details of these format differences are important if you want to design your own CD player or computer drive mechanism.

When one inserts a CD or DVD into a player or computer, the device doesn't begin collecting the interesting information right away. The first bits read are those that describe where important data begins on the disc, such as the beginnings of the musical tracks on an audio CD or the file structure contained on a CD-ROM. These bits are used in the playing or reading device to tell the user what the disc holds. Many types of information can be stored in this way, including small programs that run when they are read. DVD video players often work in this way.

Information on a CD or DVD disc is stored in logical groups to make the job of locating and reading out the information easier. The smallest group of bits on a CD or DVD is called a *sector*. A sector contains a *header* that contains information about how the bits in the remainder of the sector should be used. The sector header is generally much shorter than the entire sector, because otherwise the amount of information that could be stored on the disc would be limited.

Did you ever put a CD-ROM disc designed for a computer into an audio-only CD player? Chances are, the player rejected the CD-ROM disc as unplayable. The fact that the CD player can figure out whether the disc it sees is audio-only or not is due to a system of *standards*. In storage terms, a standard is a specification or description of how some device or technique stores information. The standard is a reference document that all engineers and designers can refer to when building a device to read and store information. Standards are developed by a consortium of companies and technical experts to promote the use of a particular technology. Standards are important because they guarantee the sameness of a particular medium, such as a CD or DVD, so that it can be read by a player or drive that is built according to the standard. In the case of a CD-ROM disc, it is designed to a different standard than is an audio-only CD. Even so, a CD player can easily recognize the difference due to the standard that is in place.

It is important to realize that a standard does not describe the exact way to build or design a device. A standard only gives a description of how the device should work. Engineers who design CD players are free to design any system that achieves the specifications set out by the standard. That is why there is no one "right" way to build a CD player and why new CD players are always being designed and sold. In this way, a standard is similar to a map of a place that you might want to explore. A map doesn't tell you how to get there (such as what road to follow), nor does it tell you what vehicle to use (bicycle, airplane, boat, or spaceship).

CDs, CD-ROMs, and DVDs are read-only devices. That is, you cannot store your own information on them; you can only read what has been placed on them. There exist certain types of discs that are writable, so that you can store your own information on them. These discs go by several names: CD-R and CD-RW for the CD standard, and DVD-RAM, DVD-R, and DVD-RW for the DVD standard.

CDs and DVDs have spawned an information revolution that is exceeded only by the Internet in its capability to distribute information. These optical discs are cheap, light, strong, and small. Most importantly, they carry a lot of information in a small package. It is likely that CDs and DVDs will continue to be the most popular medium by which to physically distribute information for many years to come.

Exercise 17.3

1. What is the radius of a CD in inches?

2. To get an idea of the incredible scale to which information is stored on a CD, consider the following "thought experiment." Suppose a 12-inch-wide pothole in a paved circular parking lot corresponds to a 0.5μm-wide bump on a CD. What would the diameter of the parking lot have to be in miles to hold the equivalent amount of information that a CD holds? How long would the paved track have to be to read out all of this information? How fast would you have to ride a bicycle along this track to read out all of this information in 74.5 minutes?

3. How many CDs would it take to hold the same amount of information as that contained on a dual-layer DVD?

4. When reading information off of a CD or DVD, why is laser light used? What special properties does laser light have as compared to ordinary white light?

5. Why is the information on an audio CD laid out in a spiral pattern? Discuss the advantages and disadvantages of this layout.

6. Suppose that, instead of the hole in its middle, a "hole-less" CD was designed such that the entire 5.25-inch diameter disc contained information. What would the storage capacity of the disc be? Discuss how one might design a mechanical system to spin such a disc.

7. A MiniDisc is quite similar to a CD, except that (a) a MiniDisc sits permanently in a plastic carrier (much like a floppy disk), and (b) the size of the media is smaller. The diameter of the disc is 6.4 cm. What is the data capacity of a single MiniDisc, assuming that it has the same areal density and form factor (ratio of inner to outer radii) of a CD? How many minutes of CD-quality audio can a MiniDisc hold?

8. A friend of yours has an idea to build a bumpy wall that absorbs sound by destructive interference, and he's asked you to help him. He wants to design the wall so that it reduces sound at an audio frequency of 1 kHz. You learn that the wavelength of acoustic sound in air at 1kHz is about 1 ft. How high should the bumps be on your friends wall? List as many similarities and differences between optical disc and magnetic disk storage technologies as you can.

9. Calculate the areal density of a page in this book. Estimate the number of characters per page by taking any line of text in the page and counting the number of characters. Then, count the number of lines on the page and multiply the number of characters per line by the number of lines per page. Then, realize that the ASCII representation of characters holds 8 bits of information (1 byte=8 bits). Finally, calculate the area of the page using (length)x(height). You should be able to do the areal density calculation now. How does the areal density compare to that of CDs? Of DVDs?

Section 17.4: Magnetic Storage

Approximate Length: 2 Lectures

17.4.1 Introduction

Floppy discs, hard discs, and hard drives are all examples of *magnetic storage* technology. Magnetic storage is by far the most popular method for temporary storage of large amounts of information. Practically every desktop computer today has a magnetic hard drive inside of it, and many devices use some form of magnetic storage to transfer information. In this section, we'll discuss how magnetic storage works, and we'll discuss some of the capabilities and limitations of magnetic storage.

17.4.2 Physical Parameters

The Magnetic Disk(s) or Platter(s)

Like the name suggests, magnetic storage uses magnetism to store bits. If you've ever played with magnets, you know that all magnets have a North Pole, denoted as N, and a South Pole, denoted as S. The names come literally from the geographic places after which they are named, since the Earth has its own magnetic field. A *magnetic field* is an invisible force field that only acts on certain types of materials, known as *magnetic materials*, and interacting with their molecular structure. So, every magnet is like a "little Earth," with a direction of orientation to tell which end is "up." The orientation of this magnetic field is known as the magnet's *polarity*.

Imagine now that we chop up a magnet into small filings. Each filing will itself be a miniature magnet with N and S poles. Hard disks and floppy discs contain essentially flat pieces of plastic or metal that have been coated with magnetic filings or particles and then sanded down to a mirror-like smoothness. These discs have magnetic material on them, yet when they are made, there is no preferred orientation to the magnetism carried by each of the particles on the disc. This condition is that of a blank disc; no information is being stored as all the particles have random directions. The disk is mounted on a spindle that has its own rotating motor that spins the disk underneath the read/write head assembly, to be described next. Figure 17.11 shows the general layout of a single-platter magnetic disk drive.

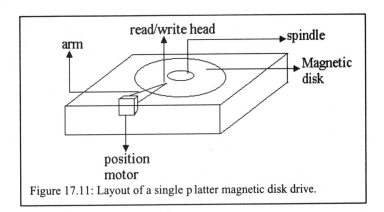

Figure 17.11: Layout of a single p latter magnetic disk drive.

The size of the particles on a magnetic disk or drive are extremely small—of the order of a micron. Although we could compare this size with the size of the bumps on a CD or DVD, such a comparison wouldn't be quite fair, because each magnetic particle doesn't store a single bit in most magnetic media. The reasons for this fact will become clear when we talk about reading and writing information onto magnetic discs.

Generally, a magnetic disk drive is used for the sole purpose of temporary storage of information that is created on the computer or other device. For this reason, the designer of a hard disk device has a lot of freedom in choosing how the system will work. For example, many hard disk drives have multiple platters, that is, there are several glass or plastic discs on the same rotating spindle, stacked on top of each other with a spacer in between to allow one or more read/write heads to access the information for each disc. Just like buildings have multiple floors to save land, magnetic hard drives have multiple "data patios" that hold useful information. Figure 17.12 shows an example of such a layout for a three-platter drive.

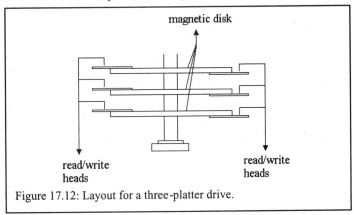

Figure 17.12: Layout for a three-platter drive.

The Amazing Flying Read/Write Head Assemblies

Along with the disks, a magnetic disk drive has one or more *read/write heads,* each of which is very much like the metal read/write head in a cassette tape player, only smaller. This read/write head is mounted on an arm which has a track that runs along one radius of the magnetic disk and is positioned by a linear motor. This read/write head assembly literally flies above the spinning disk surface, using the principles of aerodynamic flight to ride a small cushion of air above the magnetic disk. When the disk is not spinning, the read/write heads must "land" on some unused portion of the disk, so that the magnetic areas that save bits don't get damaged. This is one important difference between magnetic disks and magnetic tape technology such as videotapes and audio cassettes: the read/write head does not touch the area of the disk where the information is stored. Because of this fact, magnetic disk drives can be made very precisely and can last a long time before they wear out.

17.4.3 How Do Magnetic Disk Drives Work?

Magnetic disk drives use the properties of magnetism to store bit information. In this portion of the text, we describe how magnetic disk drives work and the physical principles that they use. Before we discuss the way the parts of a magnetic disk drive function together, we first review some important properties of magnets.

Some Useful Properties of Magnets

If you've ever played with magnets, such as kitchen refrigerator magnets, you've probably noticed that they can attract and repel each other depending on their orientation. The poles of magnets tend to attract the opposite poles of other magnets and repel similar poles of other magnets. If you put two bar-shaped magnets next to each other such that the N poles of both magnets were next to each other, they would repel. Similarly, if you put the N pole of one magnet next to the S pole of the other, the magnets would attract. Figure 17.13 shows this property in action. The reasons for this property have to do with the physics of magnetism, which is an important subject in its own right, although we won't focus on the fine details of the phenomenon here.

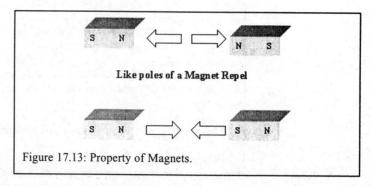

Figure 17.13: Property of Magnets.

Magnets are generally made up of metal particles that have been *magnetized,* or exposed to a magnetic field. Some types of metals, known collectively as *ferromagnetic* metals, have the ability to remember the orientation of the strongest magnetic field to which they are exposed. That is, if a large magnet of any material is brought close to a smaller piece of ferromagnetic material, the larger magnet forces a magnetic field onto the ferromagnetic material. Figure 17.14(a) shows the situation before the two objects are brought close together, and Figure 17.14(b) shows the situation after the two materials are placed next to each other. The time it takes for the larger magnet to change the polarity of the ferromagnetic material depends on the material properties of the magnets and the magnetic forces involved.

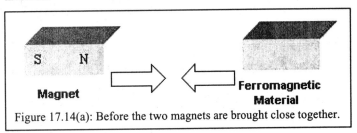

Figure 17.14(a): Before the two magnets are brought close together.

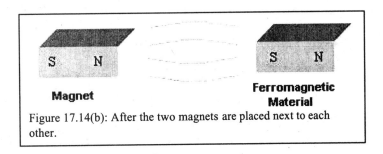

Figure 17.14(b): After the two magnets are placed next to each other.

The most common ferromagnetic metals include iron (Fe), cobalt (Cb), and nickel (Ni). These metals can be combined using chemical processes to produce magnetic materials with other useful properties. For example, by combining iron with oxygen, the magnetic compound known as iron oxide (FeO_2) is obtained. Iron oxide is ordinary metal rust. In the past, iron oxide is the most popular material for magnetic tapes such as cassettes and videotapes, because it was easy to deposit on plastic film. However, its storage capabilities are not as good as newer *thin film* magnetic materials, which can store more information per square area. The magnetic disk in a computer hard drive or floppy drive contains a thin layer of ferromagnetic particles that is protected by a thin plastic covering. The covering acts both as a lubricant and as a protective layer to prevent accidental damage of the disk surface.

How Magnetic Media Stores Binary Information

In a nutshell, magnetic disks save information by "remembering" the orientations of the magnetic fields that have been placed near them. Because magnets have a binary orientation (N-S vs S-N), they are naturally suited to storing bits. More importantly, if these magnets are made of ferromagnetic material, then the stored bits could be easily changed. This idea is best illustrated through an example.

A First Attempt at Storage Using Magnets: Suppose you had five bar magnets. Your friend asks you to store the following set of bits:

$$b_0 b_1 b_2 b_3 b_4 = 01101$$

You could save this information spatially using a sequence of magnets, as shown in Figure 17.15. In this case, a magnet with a (N-S) orientation, like the Earth, stores a 0, whereas a flipped magnet (S-N) stores a 1.

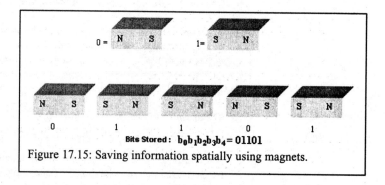

Figure 17.15: Saving information spatially using magnets.

To read the bits, you could use a device that picks up magnetic field information without imposing its own magnetic field. We'll talk about such a device shortly; for now, we assume that each magnet's orientation can be sensed without messing with its magnetic properties (such as changing its orientation).

Now, suppose that these magnets are made up of ferromagnetic material. Then, we can change the orientation of any of the magnets by bringing another larger magnet close to the magnet that we want to change. For example, if we wanted to change the third bit in our stored message from a 1 to a 0, we would bring a large magnet with the opposite orientation close to the third bar magnet—and only the third bar magnet—to cause its orientation to flip. Figure 17.16(a) shows this situation. Afterwards, the magnet's orientation will have taken the opposite orientation, producing the sequence of magnets in Figure 17.16(b). It is seen that the magnets now store the bit sequence 01001.

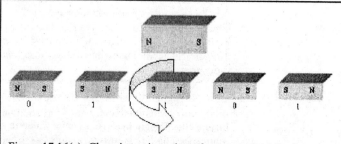

Figure 17.16(a): Changing orientation of a magnet by bringing another magnet close to it.

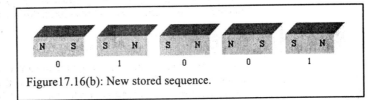

Figure17.16(b): New stored sequence.

This crude example how information could be stored using magnets. In storage, the term *write* means "store information." Storing information is exactly what you do when you write something important down on a piece of paper, so the term is fairly accurate. This example also points out two issues that arise when using magnetic material as storage:

1. The areal density depends on how closely we can pack the magnets next to each other without their affecting the others' orientations.

2. Our ability to write information is determined by how large a magnet is needed to change the orientation of one of our storage magnets—without affecting any other magnets close by.

The first issue leads us to an immediate problem with the storage technique we've just described: it is difficult to put the magnets close together without affecting the fields of other magnets. We could put a lot of space in between each of our magnets, but this would cut down on how many bits that we could store. The next example shows how we can get around this problem somewhat

A Second Attempt at Storage Using Magnets: Suppose you had six bar magnets. Your friend asks you to store the following set of bits:

$$b_0 b_1 b_2 b_3 b_4 = 01101$$

You could save this information spatially using a different sequence of magnets, as shown in Figure 17.17(a).

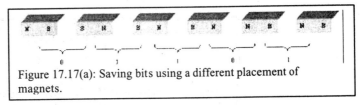

Figure 17.17(a): Saving bits using a different placement of magnets.

In this case, what matters is not the orientation of any one magnet but instead the *change* in orientation between two successive magnets. We can represent a 0 by two magnets placed next to each other whose orientations are opposite—either (N-S) followed by (S-N) as in Figure 17.17(a), or (S-N) followed by (N-S), as shown in Figure 17.17(b). We then represent a 1 by two magnets placed next to each other whose orientations are similar—either (N-S) followed by (N-S), or (S-N) followed by (S-N).

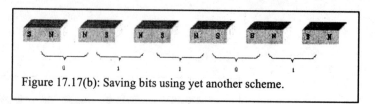

Figure 17.17(b): Saving bits using yet another scheme.

Now, suppose we want to change the third bit of the sequence from a 1 to a 0. This change means that we have to change the orientations of *all* of the magnets after and including the fourth magnet. The new sequence of magnets is shown in Figure 17.17(c).

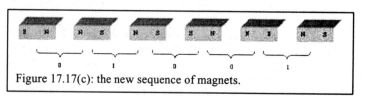

Figure 17.17(c): the new sequence of magnets.

The change amounts to flipping the orientations of all of the magnets in the sequence after the fourth magnet. Since flipping two magnets doesn't change the 1 or 0 being stored in a magnet pair, the remaining information is saved as before. If the change was in the first bit, we would have to flip the orientations of all of the magnets except the very first one, or we could just flip the orientation of the first magnet and leave the rest alone.

The second example above points out two disadvantages of the way we store the bits in this new scheme

1)   We need one additional magnet to store a message (6 magnets instead of 5).

2)   If we want to change a small number of bits in our message, we may need to change the orientations of a lot of magnets.

These two disadvantages are outweighed, however, by the advantage of being able to pack more magnetic orientations closer together on a computer hard disk or floppy disk.

Writing Information Onto A Magnetic Disk

There are some differences between this crude example and the actual way information is stored on a magnetic disk inside of a computer or floppy drive. The major differences include the following:

1)   The magnets are not discrete bar magnets on a magnetic disk or floppy drive. Instead, they are collections of ferromagnetic particles on a small patch of the disk surface. Like the CD and DVD, these particle patches are laid out in circular tracks on the disk surface. Unlike the CD and DVD, however, they are true circles; there is no one long spiral track of information.

32

2)   The device used to write information is not another fixed magnet; instead, it is the read/write head of the drive. The read/write head is almost always an *electromagnet,* a device that uses electricity to create a magnetic field.

Figure 17.18 shows the general structure of a read/write head, which is an example of an electromagnet.

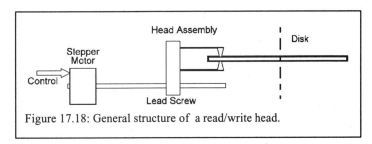

Figure 17.18: General structure of a read/write head.

The head consists of a coil of wire and a ring-shaped piece of metal which has a small sliver of metal removed to create a gap. The wire is connected to an electrical source that allows current to oscillate (move back and forth) through it. This motion of electrical charge generates a magnetic field in the metal ring, turning it into a temporary magnet. Moreover, the magnetic field at the gap doesn't travel in straight lines from one end of the magnet to the other; instead, the field bows out from the gap a slight amount. This bowing effect is called *fringing*, and the magnetic force field created by the gap is called the *fringing field.* It is important to realize that this fringing field is strong only near the gap. If you were to sense the field even a few gap distances away, the field would be so small as to be of no concern.

The fringing field can be used to impose an orientation onto the surface of a magnetic disk. In this operation, the read/write head is brought close to the disk surface with the electricity turned off. Then, by turning on the electricity, the fringing field is created, and the ferromagnetic material on the disk aligns its magnetic field in harmony with the fringing field. After the electricity is turned off, the ferromagnetic material closest to the read/write head has accepted a new orientation. The rest of the disk is not changed in this process.

We can obviously move the head over some small distance and write a new "orientation" into the disk surface without affecting what we have written. Alternatively, we can spin the disk itself and put a new "clean" magnetic patch underneath the read/write head. In this way, we can write information over the entire useful area of the magnetic disk.

Reading Information from a Hard Disk

To read information that has been written to a hard disk, we can use the same read/write head that was used to save off the bits in the first place (which is why it is called a read/write head!) The technique that the read/write head uses to save information is essentially the reverse of the electromagnetic writing process. When the read/write head gap is brought close to the moving disk surface, it senses a change in the magnetic field from the surface. This changing magnetic field creates a current in the coils of the structure, and the direction of this current around the wire loop determines the polarity of the moving magnetic field. The principle by which this system works is called Faraday's Law, in honor of the physicist Michael Faraday who discovered the property.

Although older disk drives do use this type of sensing mechanism to read bits off of the magnetic disk, newer systems employ more advanced methods. For example, there exist *magnetorestrictive* materials whose electrical properties change according to the magnetic field in which they are placed. These materials make better read heads than electromagnets-in-reverse because the resulting assembly can be made much smaller than before, and it can sense smaller magnetic fields. All of these features help to boost areal density of magnetic disks—which is the most important feature by far.

The Dimensions and Speed of Magnetic Disks and Read/Write Heads

Our previous examples give us a feel for how information can be written to a hard disk or floppy disk, but it doesn't give the dimensions of the disks or head, nor does it give an idea of the speed at which these parts move when writing information. Here are some important facts about most modern-day hard disk drives inside of computers:

1) The diameter of a computer hard disk is normally between 6.35 cm and 9.5 cm (between 2.5 inches and 3.74 inches). Its size is somewhat smaller than a CD or DVD.

2) The thickness of a computer hard disk can vary from approximately 0.8 mm to 3.2 mm (between $1/32^{nd}$ of an inch and $1/8^{th}$ of an inch). This thickness is anywhere from a fraction to about three times that of a CD or DVD.

3) The rate at which the disk spins is usually between 3600 rpm and 10000 rpm. This rate is much faster than that of a CD or DVD.

4) The distance between the read/write head and the surface of the magnetic disk is typically between 2 microns and 15 microns wide. By contrast, the thickness of a human hair is about 100 microns—in other words, the gap between the head and the disk surface is incredibly small!

5) The density of storage is much higher in modern-day hard disk drives than it is in CDs or DVDs. The following example illustrates this calculation.

Question: What is the areal density of typical modern-day PC hard drive?

Calculation: This calculation is nearly identical to one we did for the areal density of a CD in the last section. Recall that the important features we needed to know about the problem were:

1) The total working area of the storage medium.

2) The total amount of data stored on the medium.

Figure 17.19 shows the dimensions of a typical 12 GB hard disk in a PC, where 6 GB are stored on each side. The important dimensions are

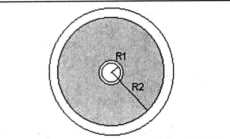

Figure 17.19: Dimensions of a typical 12 GB hard disk in a PC.

$$R_1 = 1.25cm$$
$$R_2 = 4.75cm$$

From these radii, we can calculate the usable area of the hard disk as

$$A_{HD} = A_2 - A_1$$

$$A_1 = \pi R_1^2$$
$$A_2 = \pi R_2^2$$

Plugging in the values, we get

$$A_1 = 3.1416(1.25cm)^2 = 4.91cm^2$$
$$A_2 = 3.1416(4.75cm)^2 = 70.88cm^2$$
$$A_{CD} = 70.88 - 4.91cm^2 = 65.97cm^2$$

Therefore, we have

$$D_{HD} = \frac{N_{Bytes}}{A} = \frac{6\times10^9\ bytes}{65.97cm^2} \times \frac{8bits}{1byte} = 728\times10^6\ bits/cm^2$$

35

In other words, every square centimeter on the data-carrying portion of a computer hard disk holds about 720 million bits, or about 90MB of information. Per square unit of area, a magnetic hard disk is about as good at storing information as a single-sided DVD. A magnetic hard disk is about 10 times better at storing information than an audio CD.

Magnetic Disks: Unformatted vs. Formatted

Once the magnetic disk drive is assembled, with spindle motor, read/write heads, and the magnetic film-coated platters themselves, the drive is still not ready to store information yet. There are no specific places where bits can be stored. All the magnetic particles have random orientations on the disks. This state of randomness is called *unformatted*. Think of this situation as you might an empty closet, with no shelves, no bars to hang clothes on, and no boxes to put things in. Before anything can be stored in the closet, a person should figure out a logical way to use the space and then put up the necessary structures to make use of that space. Without this structure, it is pretty hard to find things quickly and easily. In the case of a magnetic hard disk, speed is important. If it is hard to find the right information, every task a computer or other digital device does will be slowed down as well. The process of putting this structure in place on a hard disk is called *formatting*.

Figure 17.20 shows the typical structure that most hard disks get when they are formatted. The layout is different from a CD or DVD in that there are distinct rings of information, called *tracks*, and each track is divided up into several smaller arcs called *sectors*. Each sector will contain a fixed number of bits. So, the sectors nearer the center of the disk will hold fewer bits than the sectors nearer the edge of the disk. The size of each sector, as well as other values important to decoding the bits inside of each sector, are contained in the sector *header*. The sector header is a set of bits that is stored in the front of every sector. The length of the sector header is much shorter than the entire sector. If this weren't the case, we wouldn't be able to store as much useful information on the hard disk itself.

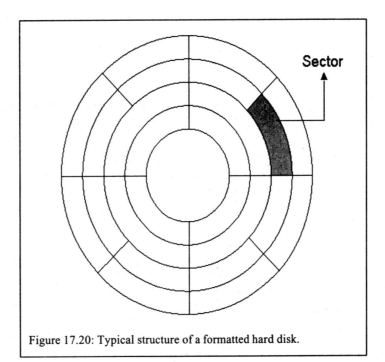

Figure 17.20: Typical structure of a formatted hard disk.

To format the disk, the read/write head is instructed to go over the entire disk and store a sequence of zeros in every sector. Then, the disk is tested by going back and reading off all of the zeros. Whenever the reading process fails for any bit inside of a sector, the entire sector is labeled as bad, and its location is stored as well. All this location information—where good and bad sectors are—is stored on the hard disk as well.

Support Systems

There are many other important systems that make up a magnetic disk drive. Some of these systems are briefly described below:

8. *Spindle motor:* This electric motor rotates the platters or disks when reading or writing information.

9. *Actuator Arm Motor:* This device positions the arm carefully over the disk when information is being written or accessed.

10. *Printed Circuit Board With Electronic Parts:* This assembly holds the control electronics and memory that are used to perform important functions, such as starting and stopping the drive, positioning the read/write head, and translating certain signals to useful signals that the computer or digital device can understand.

11. *Enclosure:* While not so obvious a part, the case that holds the entire assembly protects the parts of the drive from dust and dirt. Realize that the size of the air gap between the read/write head and surface of the disk is only a fraction of a typical dust particle! For this reason, the air nearest the disk needs to be super-clean in order for the system to work properly.

17.4.4 Speed of Rotation and Data Rate

One of the mo st important features of any storage device is its ability to read and write out information quickly. The rate at which a computer hard disk can access information depends on three factors:

1) The *rotational speed* of the magnetic disk;

2) The *linear speed* of the read/write head assembly along a track; and

3) The *diameter* of the disk itself.

The first two factors control the speed at which any sector of bits on a disk can be directly accessed. The reasons for this can be easily seen when thinking about how data is accessed on the disk surface. When accessing information, we need to position the head at the right radius at which the desired data is stored. We also need to spin the disk to the right sector.

The last factor is a limiting one; too large of a disk makes it difficult to both spin the disk and get the head assembly to the right radius along a disk to get at the information we want. This last factor acts as a magnifier of the other two; if the rotational speed of the disk and the linear speed of the head assembly are too slow, a larger disk size will make the overall access job harder.

We can find the best and worst-case scenarios for reading information rather easily:

1) Best Case Scenario: The access point is in the sector over which the read/write head is currently positioned. Then, we can potentially read and write bits at the rate at which we cover the necessary area of the disk. This rate is controlled by the rotational speed of the disk and is directly proportional to the maximum data rate that the disk can support.

2) Worst Case Scenario: The access point is in a sector that is far away from the current head position. Moreover, when we position the head at the right radius, we need to wait until the necessary area of the disk rotates to a position underneath the read/write head assembly. The time it takes to access information in this way is called the *seek time.*

An example illustrates the calculation of the seek time.

Example: Calculating worst-case seek time on a hard disk

Consider the hard disk whose physical dimensions are shown in Figure 17.21. In addition, the following parameters are known:

♦ Rotational speed: 7200 revolutions per minute (RPM)

♦ Linear Read/Write Head Speed: This is the speed at which the read/write head can travel.   A typical value is 175cm/s.

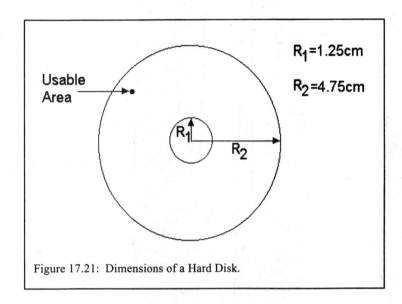

Figure 17.21: Dimensions of a Hard Disk.

The worst case access time would be the sum of the following two times:

1. The time it takes to move the linear read/write head from one edge (inner or outer) of the disk to the other edge.

2. The time it takes for the disk to spin around once.

Figure 17.22(a) and (b) show two versions of the worst-case seek time scenario.

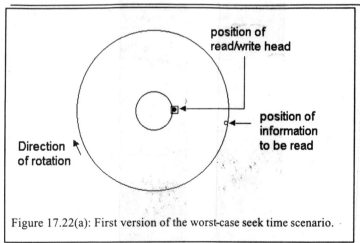

Figure 17.22(a): First version of the worst-case seek time scenario.

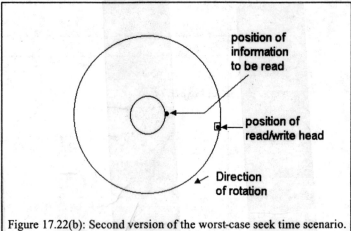

position of
information
to be read

position of
read/write head

Direction
of rotation

Figure 17.22(b): Second version of the worst-case seek time scenario.

To calculate the total time, we set up the equation as

$$T_{Seek} = T_{Linear} + T_{Rotation}$$

To calculate the first term, we note that

$$T_{Linear} = \frac{(Dis\tan ce \quad Traveled)}{(Speed \quad Traveled)}$$

$$T_{Linear} = \frac{R_2 - R_1}{V_{Linear}}$$

Putting in the known values, we have

$$T_{Linear} = \frac{4.75cm - 1.25cm}{175cm/s} = 0.02s$$

To calculate the second term, we note that

$$T_{Rotation} = 1/(72\,00 \times 60) = 0.0\,0833 \text{ s}$$

Summing the two times, we get

$$T_{Seek} = 0.020s + 0.00833s = 0.02833s = 28.33ms$$

In other words, it can take up to about 30 milliseconds to access information on a hard disk.

Exercise 17.4

1. List as many similarities and differences between optical disc and magnetic disk storage technologies as you can.

2. Compare Figures 17.1 and 17.11. How are they similar? How are they different? Do you think that the first designers of magnetic disk drives knew something about phonographs when they designed magnetic disk drives?

3. Is it possible to change the polarity of a magnet (without moving it, of course)? If so, how? If not, why not?

4. Which can store more information per square—a CD or a magnetic hard disk drive inside of a computer?

5. Every planet has a magnetic polarity (which is where the names for North and South magnetic polarities come from). How many bits could be stored using the nine planets of the solar system? If one could "hold" Jupiter next to Mercury, which planet's polarity would likely be changed?

Section 17. 5:
Memory Chips

Approximate Length: 2
Lectures

17.5.1 Introduction

In biology, memory is most often associated with the human brain. We remember past experiences, the faces of people we meet, and our favorite foods. These long-term memories are different from the short-term memory we use to remember little things like the phone number of who just called and what we had for lunch today. We can draw similar parallels with storage devices. While optical discs and magnetic disks are good for medium- to long-term storage needs, certain types of *memory chips* are ideal for short term memory needs. In this section, we'll discuss the various kinds of memory chips, the way they work, and their advantages and disadvantages.

17.5.2 Physical Parameters

Memory chips are devices that are mostly made out of silicon, the same semiconductor material that makes up the processing devices of computers and other digital devices. Unlike processor chips that can do a wide range of tasks, memory chips have only one task: to save information. For this reason, the circuitry on a memory chip is much different than that on a processing chip. Figure 17.23 shows the layout of a typical memory chip.

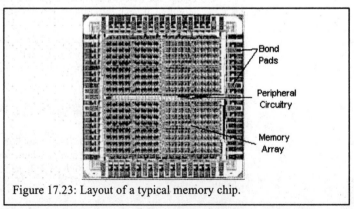

Figure 17.23: Layout of a typical memory chip.

The different parts of the memory chip are described below

1. *Memory Array:* The memory array is located in the center of the mostly-silicon chip substrate. The memory array is where the bits of information are stored in a memory chip. It consists of millions or even billions of identical cells arranged in a two-dimensional checkerboard pattern, where each cell contains identical circuitry.

2. *Peripheral Circuitry:* At the edge of the memory array, additional circuitry is used to translate the electrical signals coming to and from the memory array such that information can be stored into and read from the array. Think of the peripheral circuitry as the system of traffic lights and signs that control the flow of vehicles on city streets. Without peripheral circuitry, the memory array would be useless.

3. *Bond Pads:* Bond pads are the entry points of information onto the silicon chip substrate where the memory array and peripheral circuitry lie. Bond pads are made of metal and are much larger than the memory array cells. They are connected to the leads of the memory chip package through the bond wires.

4. *Bond Wires:* The bond wires connect the bond pads to the leads seen on the outside of the memory chip package. They are usually made of gold, because gold is one of the best conductors of electricity. (Don't go trying to collect memory chips for their gold content, however, as the amounts used are incredibly small.)

5. *Leads:* The leads are the "metal legs" of the memory chip package; in fact, most memory chip packages look like flat plastic beetles due to their protruding leads. The leads make the electrical connection from a specially-designed socket in a digital device to the bond wires, which are connected to the bond pads, which in turn are connected through the peripheral circuitry to the memory array.

6. *Package:* Made of plastic, the package holds all of the above components in a rigid, light, protective structure.

While all memory chips have this general structure, some types of memory chips have additional features, such as a small clear plastic windows in the package over the memory array, or a different package type such as a card.

17.5.3 ROM vs. RAM - What's the Difference?

Memory chips are classified according to the type of storage capability they have. The most important difference between memory chips is denoted by the acronyms ROM and RAM. What do these three-letter acronyms mean?

1. Memory chips that can only be read are called *Read-Only Memory (ROM)*. ROM chips are like CDs, CD-ROMs, and DVDs. Once the information is stored in them, that information can never be changed. Well, almost never—some types of ROMs have a limited erasing and rewriting capability. We'll cover this variation in the coming pages.

2. Memory chips that can be both read and written are called *Random Access Memory (RAM)*. That name might seem strange, because the name doesn't say anything about the writability of the memory device. Unfortunately, the acronym RAM was chosen early in the development of memory chips to emphasize a different feature than read/write capability, and the name stuck. Many technical names have a similar history; it is often just easier to keep the name that a technology is christened with than change it according to the times.

There are other differences between ROM and RAM. These differences will be illustrated when we discuss the way each stores information.

17.5.4 How Memory Chips Work

An Analogy: Tic-Tac-Toe

All memory chips hold information in a two-dimensional array of binary cells. A simple example of a two-dimensional array of binary cells is the tic-tac-toe board shown in Figure 17.24. On this board, we have shown a game that has ended in a draw, so that all the spaces have either an "X" or

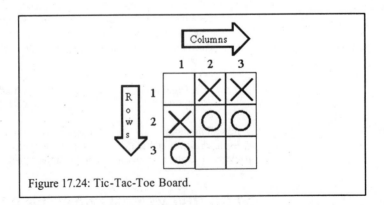

Figure 17.24: Tic-Tac-Toe Board.

We've put numbers along the left side and at the top of the board. The numbers along the left side of the board label the *rows* of the board, and the numbers at the top of the board label the *columns* of the board. These numbers can be used to point to a particular "X" or "O" that lies on the board. For example, the entry at row #1 and column #2 is an "X". The entry at row #3 and column #2 is an "O".

When playing tic-tac-toe, the board is so small that it is easy to see the entire board in one view. Remember, however, that computers are dumb and need to be told what to do step-by-step by a computer program. Fortunately, our row-column numbering system gives us a way to play tic-tac-toe on a computer with a simple three-step process:

4. Pick a row number.

5. Pick a column number.

6. Place an "X" or "O" at the intersection point of the row and column that was chosen.

Similarly, if we wanted to read out information, we could replace the last step with a step that looks at the contents of the box at the row and column number chosen. Note that the read-out process need not follow the same order at the write-in process. An example illustrates this fact.

Example: Suppose you have a tic-tac-toe board on which two other players have begun to play a game. You read out the following information, which is not in the same order that the game was played:

2. Row #1, Column #3: contains an "X"

3. Row #2, Column #1: contains an "X"

4. Row #3, Column #1: contains an "O"

5. Row #2, Column #2: contains an "O"

6. Row #1, Column #2: contains an "X"

7. Row #2, Column #3: contains an "O"

What board location (row and column number) would give the " win (that is, three in a row either vertically, horizontally, or diagonally)?

How Memory Chips Work

All memory chips uses the same procedure to read out information as the tic-tac-toe game example you just read. Moreover, RAM chips can also save new information in a similar manner. The difference between memory chips and tic-tac-toe are the way in which they are implemented and the number of cells. Specifically:

2. Memory chips use microscopic circuitry to hold the binary information. The size of each cell is worked out in an example later.

3. The number of cells in memory chips is amazingly large. While every tic-tac-toe board has only nine cells, typical memory have half a billion cells or more.

4. Instead of "X's" and "O's", memory chips use either electronic switches (not unlike a typical light toggle switch) or a stored charge (like a marble under a plastic cup) to save their information. In the latter case, the presence of a charge denotes a 1, whereas an absence of charge denotes a 0.

Figure 17.25 show a simplified version of a RAM chip layout. Each memory cell contains a *capacitor*, a device that stores charge, as well as circuitry to load and read out the stored charge. The cells are connected to both *row address lines* and *column address lines*. These lines are simply wires that run horizontally (for rows) and vertically (for columns) to and from each cell along their row and column, respectively. Also running horizontally are *data lines* that provide the source of charge for the individual cells.

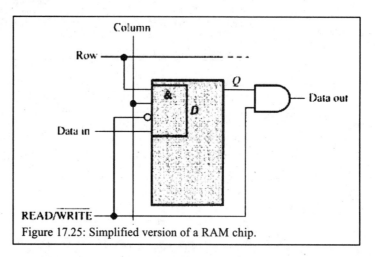

Figure 17.25: Simplified version of a RAM chip.

Notice that we've used a single cell in this example. As said before, most RAM chips store 100's of millions of bits, so this drawing is still simplified. There is one feature, however, that makes this example important and different from the tic-tac-toe board example: each dimension of the cell array is a power of two. The reason for this is so that we can label each of the rows and columns of the array with a binary number. In the figure, we have used the four binary numbers 00, 01, 10, and 11 to label the rows and columns, respectively. Using a binary number means that we can access the information using binary numbers as well. The combination of row address and column address binary numbers is called the *cell address*. These addresses are usually combined by putting the column address bits after the row address bits.

Using binary numbers to store the addresses has another advantage: we can address large arrays with a relatively short cell address. For an *NxN* cell array, we only need $log\_2(N)$ bits for each row or column, so that the cell address is only $2 \, log\_2(N)$ bits long for an *N-squared* element array.

Question: Suppose a digital device is using a 32MB RAM chip. What is the length of the cell address?

Answer: We first need to figure out how many bits are stored in the chip.

$$N_{Bits} = N_{Bytes} \times \frac{8 bits}{1 byte} = 8 \times 32 \times 10^6 bits = 256 \times 10^6 bits$$

The number of bits stored there is

The number of memory cells needed to store this many bits is found by the equation

$$N_{Cells}^2 = N_{Bits}$$

Taking square roots of both sides give us

$$N_{Cells} = \sqrt{N_{Bits}} = \sqrt{256 \times 10^6} = 8 \times 10^3$$

To find the length of the cell address, we solve the equation

$$L_{RowAddress} = \log_2(N_{Rows})$$
$$L_{RowAddress} = \log_2(8 \times 10^3) = 12.96 bits \approx 13 bits$$

The "rounding off" of the answer is actually not an approximation; in fact, the number of bits in the memory cell is only an approximation to a small number of significant digits. Since the column address bit length is the same as the row address bit length, we have that

$$L_{CellAddress} = L_{RowAddress} + L_{ColumnAddress} = 26 bits$$

In other words, we only need 26 bits to point to any one of 256 million cells in a 32MB RAM chip!

17.5.5 Erasing and Rewriting

17.5.6 Types of RAM and ROM

There are many different types of RAM and ROM, depending on how the information is stored.

Types of ROM

1. *Standard ROM:* This is the cheapest kind of ROM chip, in which the information that is placed inside of it is done at the factory where the chip is made. In this case, the circuitry is pre-wired to store a particular set of bits, and once this chip is made, the bits cannot be changed. This type of ROM is best suited to high-volume, low-cost applications, where the cost of setting up the assembly line manufacturing process are spread out over the sale of millions and millions of chips. Typical applications for this type of process include fundamental computer program information inside of a computer and some simple toys.

2. *Programmable ROM (PROM):* PROM chips are "write-once" circuits. They contain small fuses at the location of each memory cell. These fuses are like the electrical fuses in your home electrical system and either carry electricity (closed) or don't (open). To store information into a PROM chip, one uses a special writing machine that selectively "blows" the fuses inside certain cells of the chip. Once the fuses are blown, however, they cannot be replaced. PROM chips are good when building only a few systems, because the cost to manufacture a small number is much less than that of standard ROM chips. They are also

good if a designer expects to upgrade a system, because they allow
changes to be made to the memory of the device by swapping out the
old PROM with a new one.

3. *Erasable Programmable ROM (EPROM):* Although we said that
    ROM chips are "read-only," there exist certain types of ROM chips that
    can be erased re-written. These erasable devices generally cannot be
    erased by the user, however; generally, they are only changeable at the
    factory, like ROMs are. These devices have a little clear plastic
    window above the memory cell array. The cells are designed in a
    special way such that, if a certain type of light is shined onto them,
    their contents are erased. This erasure is complete over the entire cell,
    such that if one cell contains an undesirable bit, all must be erased and
    rewritten again. EPROM chips are most useful for situations where
    PROM chips are used, except EPROM chips have the advantage of
    being reusable. They are especially good for designing systems, where
    the designer can change the information contained in the EPROM chip
    as the design evolves. EPROMs generally cost more than PROMs,
    however.

4. *Electronically Erasable Programmable ROM (EEPROM):* EEPROM
    chips are like EPROM chips, except the mechanism to erase the cells is
    electronic instead of optical. These chips can be used wherever
    EPROM chips are used.   They are a little easier to reprogram because
    they use electricity instead of light in the erasing process. They have
    more circuitry on them, however, and thus tend to cost more for a given
    storage capability.

Types of RAM

1. *Static RAM (SRAM):* These chips are devices that store information in
    the form of the presence or absence charge at each cell location. The
    term "static" refers to their ability to hold information in the absence of
    any electricity. Static RAM is analogous to an array of glasses, where
    the presence or absence of water (analogous to charge) corresponds to a
    1 or 0, respectively. Like the water glasses, the cells in a static RAM
    chip hold their data when left alone. The drawback to static RAM is
    the complexity of each cell in the memory array. The circuitry needed
    to store the charge is more complicated (it usually requires between
    four and six transistors per cell). There fore, the density of the cells
    cannot be as great as other RAM types. Static RAM is most useful for
    applications where electricity or power is a concern, because they can
    "stand alone" in their ability to store information in the absence of
    either.

2. *Dynamic RAM (DRAM):* Dynamic RAM chips are the most efficient
    chip-based storage devices. They use a single capacitor to store the
    charge associated with a bit in each cell array. Because the storage cell
    is so simple, the amount of bits that can be stored per unit area—their
    areal density--is the highest of any chip-based memory device. DRAM
    chips are the most common type of memory chips used in computers

because of their high areal density. The only drawback to DRAM chips is that they lose their information over a very short period of time. Think of the array of water glasses, but suppose that the glasses each had a small hole in the bottom. Over time, one would need to pour water into the glasses that were storing water (analogous to a 1 in our DRAM). In the case of DRAMs, the individual memory cells of the chips lose charge so quickly that their memory cells need to be read out and *refreshed* with their information thousands of times each second. The refreshing process takes energy, which means that it reduces the electricity and power available to other devices in a digital system. For this reason, DRAM is most useful in non-portable devices, such as computers, although engineers are getting better at reducing the power consumption of typical DRAM chips for portable devices as well.

Exercise 17.5

1. What is the shape of most electronic storage devices? Why isn't electronic memory round like CDs and magnetic disks?

2. A new RAM chip uses 24 bits in total to access the information contained inside of it. How many bits of information can the RAM chip potentially hold?

3. Compare and contrast all of the different types of electronic memory by creating a table, where "features" are listed along the top of the table and "Types of RAM/ROM" are listed along the left side of the table.

4. A company wants to develop a 3D RAM device that has four layers of memory cells at every row and column location. Describe how the address of each bit in the memory chip would change. How many more or fewer bits would you need to access all of the information in this new device?

5. There are many types of computing devices built today that only use electronic memory (that is, they don't contain magnetic hard disks or optical disks). Name as many of these computing devices as you can. Why do you think these devices only employ electronic storage?

Section 17.6: Glossary

address lines
in a memory chip, the wires that allow direct access to the information contained in memory cells

antinode
a point in an interference pattern where two waves constructively interfere

areal density
a measure of the storage capability of a particular storage technology, as measured in bits per square centimeter

capacitor
an electronic device that stores electrical charge

CD
an acronym for "compact disc," a plastic disc that uses optical means to store information

coding
the mathematical process of converting the numbers to be saved into the bits that are physically stored

data lines
in a memory chip, the wires that carry useful information to and from the memory cells to the outside of the chip

DVD
an acronym for "digital versatile disc," a plastic disc that uses optical means to store information

electromagnet
a device that uses electricity to create a magnetic field; used in magnetic storage devices to read and write information

frst generation copy
a copy of an original document, work, or piece of information

format
the layout of bits on a storage device such as a CD, DVD, or magnetic disk

iterference
te interaction of two physical waveforms that either cancel out (destructive interference) or reinforce (constructive interference) each other

laser
a special type of optical device that emits single-color coherent light; used extensively in CD and DVD players

memory cells
in a memory chip, the portions of the chip that actually store bit information

node
a point in an interference pattern where two waves destructively interfere

polarity
the orientation of a magnet, denoted by a "North-South" designation

RAM
an acronym for "random access memory;" a shorthand for storage devices that can change the information stored within them

record player
a 20th-century device for decoding audio signals from plastic media (also known as the phonograph)

ROM
an acronym for "read-only memory;" a shorthand for storage devices that store information in a permanent format without possibility of change.

second generation copy
a copy of a first generation copy

sector
the smallest group of bits that can be stored on a rotating storage device (such as a CD, DVD, or magnetic disk)

seek time
the amount of time it takes to access a particular point on a rotating storage device

storage
the general term used to describe all methods (optical, magnetic, and electronic) to store information

The INFINITY Project

Chapter 18: Hardware for Digital Imaging

What are different image sensors? How are the film camera, the digital camera, and the human eye related to each other? How are individual pictures and movies displayed?

Approximate Length: 2 Weeks

Section 18.1: Introduction

In previous chapters we discussed how images (still and moving) are represented digitally and how they can be manipulated to better serve our needs. We also discussed how a digital image could be generated strictly within a computer as a matrix of numbers. Most frequently, however, an image or a movie is acquired via an image sensor, stored and then displayed later for our enjoyment. Figure 18.1 shows several familiar image sensing, storage and display devices.

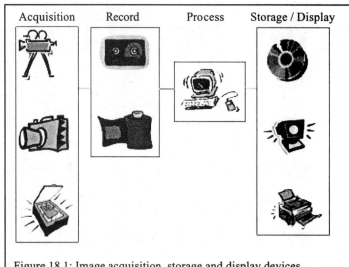

| Acquisition | Record | Process | Storage / Display |

Figure 18.1: Image acquisition, storage and display devices examples.

The most familiar image sensor (that all of us possess) is our eye. Film cameras have been commonplace for past 100 years. In the past few years, electronic digital cameras (for still pictures and for movies) are improving in quality and becoming more affordable. How are these different image sensors constructed? What are their operating principles? How are they similar and how are they different? These are the questions we will address in this chapter.

The only stored image that is of intrinsic value to us is the one that is in our own memory. An image or a movie that is stored in a computer or on a magnetic tape is of no value to us unless we can see it with our own eyes. Displaying individual or moving images is the converse operation of image sensing. Again, for well over 100 years, photographs and photographic films has been the predominant method for sensing, storing and displaying pictures and movies. Then 60 years ago the familiar television with its bulky vacuum tube and glowing phosphor burst on the scene as a way of showing movies. In the past 5 years several new technologies (such as liquid crystal displays) have come on the scene that are challenging the primacy of television as one of the most common displays. Computer printers are now available with a quality that is fast approaching the best photographic paper for displaying pictures. Display technologies for pictures and movies will be the second topic discussed in this chapter.

It should be emphasized that digital imaging systems can be designed for several different applications. Figure 18.2 shows the functional block diagram that shows three separate streams for different applications. The first one where the images captured by sensors are immediately displayed (after some simple processing) is used for real time scene monitoring in manufacturing or security applications. The second one includes an intermediate storage of the acquired images (still or movies) that are then displayed at a later time. Such scenarios are most common where the images are to be distributed to a large number of users for its entertainment or scientific value. The third stream involves substantial amount of image processing for object tracking or recognition. The output that is displayed may not be easily recognizable as an image and in some cases it could be a simple alarm. These examples are meant to illustrate the variety of applications and are not supposed to be exhaustive. They are also designed to show that regardless of the ultimate application, image acquisition and display technologies are common to all. So the performance of image acquisition and display devices often determines the performance of the overall system.

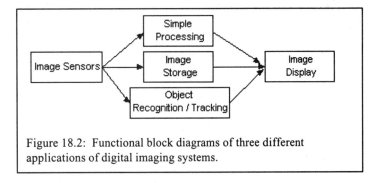

Figure 18.2: Functional block diagrams of three different applications of digital imaging systems.

Exercise 18.1

- Bring pictures to supplement the examples shown in Figure 18.1.

- Give examples from every day life of each of the digital image processing application streams listed in Figure 18.2. Identify as many details within the box as you can.

Section 18.2:
Image Sensors

Image sensing is the first step in a digital imaging system. This is the step where information about the world around us is converted into a form that is amenable to processing and storing and as such is one of the most important steps. If the sensing devices are inadequate, then certain information about the scene is lost and can never be recovered (or can only be recovered at great cost). Therefore it makes sense to perform this step "right" in order to avoid insurmountable problems later in the processing chain. In this section we will discuss three most important and most common image sensors:

- The human eye,

- The photographic film and

- The electronic sensor (still and video).

After discussing the structure and operation of each one separately, we will compare all the sensors with each other. An important category of an image input device is a scanner, which can take an image recorded on a piece of paper or photographic film and convert it into its digital representation. The section will conclude with a description of an image scanner.

18.2.1 The Human Eye

Every morning when we wake up, our first act is to open our eyes. We look around us (most of the time) in a dark room. We can make out vague outlines of the objects in our room. Relying more on our memory than what we actually see, we reach over and turn the light on. Suddenly the room is filled with light and everything in the room becomes clear and recognizable (after we put our glasses on, that is!) to us. What we just described was an operation of an imaging system that nature has engineered over millions of years to provide us with a very powerful tool to understand the world around us. The human vision is an extremely complex system that can be studied for its construction and function in a biology class. Here, we will study it from an engineering point of view.

The schematic diagram of a human eye is shown in Figure 18.3. The human eye is basically a hollow sphere that is filled with clear fluid. At one end is a transparent opening though which light is admitted. The curved bulging part of the transparent opening is called **cornea**. Inside the cornea is a colored circular region called the **iris**, which gives our eyes its characteristic color. In the middle of the iris is a circular opening, called the **pupil**. The size of the pupil responds to the amount of light available to us, opening wide to let more light in when inside a dark room and contracting to restrict the amount of light getting in the eye when outside in bright

---

Determining Brain Death:
The contraction of the pupil is an automatic reaction that does not require one to be awake or even conscious. Indeed, one of the simplest tests that can be performed to determine lack of brain activity (the so-called "brain death") is to shine bright light directly into the eye. If the pupil does not contract immediately, it is a clear indication of the lack of brain activity.

4

sunlight. Further inside the eye, is the lens, which together with the curved transparent cornea, serves to focus light on the backside of the eyeball.

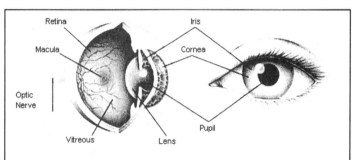

Figure 18.3: Schematic diagram of a human eye showing the basic components.

On the backside of the eyeball is a light sensitive screen, called the **retina**. The retina contains millions of light sensitive cells. The signals generated by these cells are all collected and transmitted to the brain via the optic nerve to an area of the brain dedicated to processing visual signals. There are two types of these light sensitive cells in the retina: rods and cones. Cones are much smaller in size (1 micron across versus 5 microns across for rods), they are less sensitive to light compared to rods and they can detect and discriminate between colors while the rods cannot. Cones operate mostly under bright light conditions whereas rods operate under low light levels. In total, there are approximately 7 million cones and 120 million rods distributed over an area approximately 5 cm square. It should be noted that the optic nerve that carries signals from the eye to the brain contains only about 1 million nerves (or biological "wires"). This implies that the connection between the retinal light sensitive cells and the "wires" in the optic nerve is not simple and that there is some processing that goes on in the retina itself.

The rod and cone nerve cells are not uniformly spaced along the retina but have a variable distribution. This distribution is shown in Figure 18.4 and has very important consequences.

- Exactly on the visual axis, there are maximum numbers of cones, which generates a highly detailed, color view of the scene. This spot is called the **fovea**

- 10 degrees away from the fovea is a region where the nerves from the retina exit the eyeball via the optic nerve. That region does not have either rods or cones and hence is called a **blind spot**.

- Away from the fovea, the density of cones drops very rapidly and rods dominate. This region is called **peripheral vision**.

Micron:
A micron is one millionth of a meter, i.e., 1 micron $= 10^{-6}$ meter

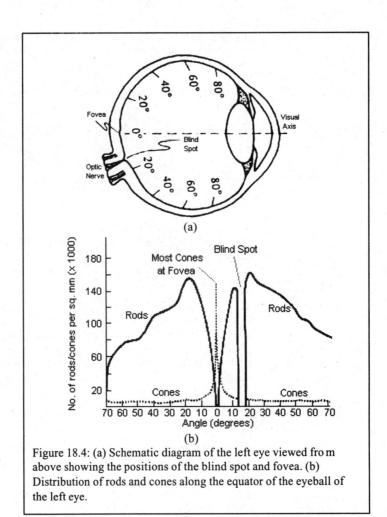

Figure 18.4: (a) Schematic diagram of the left eye viewed from above showing the positions of the blind spot and fovea. (b) Distribution of rods and cones along the equator of the eyeball of the left eye.

Our eyes adapt to different light levels by using two different mechanisms. The first mechanism is the control of the pupil – it shrinks in diameter in response to bright light and opens up wide when the light level decreases. The second mechanism is the activation of rods versus cones when the light level decreases. The first mechanism has a rapid response but the second mechanism takes several minutes. This process is called dark adaptation and is familiar to all of us. After entering a dark room, it takes more than five minutes for our eyes to get adapted before we start seeing details. Since rods do not have color discrimination, we do not see colors in a dark room. These two different mechanisms give humans the ability to see under a tremendous range of lighting conditions. The difference in light levels between a bright sunny day and a moonless starlit night is approximately a factor of 10 billion and yet we are still able to see under both lighting conditions.

The light sensitive cells in the retina have a finite reaction time. There is a small delay between the presentation of a bright flash and the light sensitive cell output, which is called **latency**. Similarly, when the bright light is turned off, the light sensitive cell output also stays high for a small length of time, which is called **persistence of response**. The persistence of response is analogous to the light exposure time in the camera. This persistence time varies with light intensity and is approximately $1/25^{th}$ of a second at low light levels to as short as $1/50^{th}$ of a second at high light levels. This persistence manifests itself in us seeing continuous motion even though the scene may be changing in jerky manner with discrete images.

Finally it should be noted that the act of "seeing" involves not just the eye, but also the brain that processes and stores the signals generated by the retina and carried along the optic nerve. The processing carried on in the brain is an integral part of the human visual system and therefore it is impossible to separate the function of the eye in sensing and the function of the brain in processing the signals generated by the eye that together determine our visual experience.

Exercise 18.2

- If the retina consisted of only cones uniformly distributed over a 5 square cm area of the retina, how many total cones will we have?

- If each nerve fiber (axon) which carries signals from the retina to the brain, is 1 micron in diameter, what is the diameter of the optic nerve?

- If the optic nerve contained one axon for each retinal cell (rod and cones), what will be the size of the optic nerve?

18.2.2 Photographic Film Camera

Until very recently, the title of this subsection would have been highly redundant – *ALL* cameras were film-based. There may be movie cameras, instant cameras, disposable cameras, and cameras costing thousands of dollars. Whatever else was different about them, there was one factor common to all of them – they used a specially treated film as the light sensitive screen to capture the image. A picture of a modern 35 mm Single Lens Reflex (SLR) camera is shown in Figure 18.5. The "35 mm" refers to the width of the film that such a camera accepts and is currently *THE* most commonly used film format around the world. The camera body contains a drive mechanism that holds and advances the roll of photosensitive film. It also contains various optical prisms and mirrors that allow us to look through the lens and see the image on a viewfinder as it will appear on the photographic film. It also houses the control electronics that measures the total amount of light that is allowed to fall on the film and controls the shutter mechanism. The mechanical shutter release button and other controls are located on the top and front part of the body. This camera body can accept a variety of lenses of different focal length. It is this flexibility of changing lenses on a camera to get different special effects in the middle of shooting a sequence that makes the SLR designs a choice of professional photographers. The design of the various controls and the design of specialized lenses are fascinating topics themselves and are often the subjects of separate textbooks. Our main goal here is to understand the *fundamental operation* of such a film-based camera.

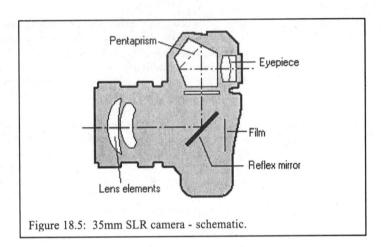

Figure 18.5: 35mm SLR camera - schematic.

We discuss briefly the common types of lenses encountered in photography. These lenses are not simple spherical lenses discussed in the appendix on lenses and imaging systems, but contain multiple lenses (concave and convex). These multiple lenses are placed at precise spacing from each other. The composite lens will have an equivalent focal length that is determined by the **focal length** and separation of the individual lenses. Thus changing the spacing between lenses can change the equivalent focal length. There are different types of lenses that are commonly available for 35 mm SLR cameras.

- **Standard Lens** : This is the most commonly used lens on a 35 mm SLR. It has fixed focal length that is typically between 35 to 50 mm. This focal length is found to be the best compromise for capturing portraits of individuals standing 2-5 meters away and for capturing natural scenes.

- **Telephoto Lens** : This lens also has a fixed focal length, but it is considerably larger than the standard lens and is typically greater than 70 mm. The longer the focal length, the greater the magnification of the image captured on the film. The telephoto lenses are primarily used to capture details of a far-away object. Very strong telephoto lenses can have focal lengths of ~ 200 mm. Since they magnify the images of far away objects, it stands to reason that on a film of fixed size, they capture a much narrower region of the scene, technically termed the **Field of View**.

- **Wide Angle Lens** : This is the opposite of a telephoto lens and has a fixed, shorter focal length than a standard lens, typically 28 mm or smaller. Such a focal length produces smaller size images of objects and hence can accommodate a much wider region of the scene on the film – hence the name!

- **Zoom Lens** : This is a versatile lens with a variable focal length. In most automatic cameras, the focal length is chosen to provide modest wide-angle performance and a modest telephoto performance (focal length range 35-80 mm) in the same lens. There are special telephoto zoom lenses that have

Focal Length:
A simple convex lens has the ability to concentrate light. Focal length of a convex lens is the distance behind the lens where parallel rays of light come together to form a point.

8

focal lengths that vary between 100-200 mm. These lenses can change the image magnification in a continuous manner to allow better framing of subjects in the picture without moving the camera position.

The focusing of the image onto the film is accomplished by moving the lens closer to the film or farther away from the film. In case of zoom lenses, changing the spacing between the component lenses changes the equivalent focal length of a multi-element lens. A high quality zoom lens will allow the user to change the focal length without losing the focus of the image.

All lenses are specified by their focal length and something called their F-number. This **F-number** is defined as a ratio of the focal length to the maximum useable diameter of the aperture in the lens (F# = f / D). As seen earlier, the diameter (or the radius) of the aperture controls the amount of light that the lens transmits on to the film. So the F# is an indication of the light gathering capacity of the lens. The light gathering capacity of a lens is related inversely to the square of the F#. Since it is desirable to collect more light, it is important to find a lens with a small F#. So a camera lens with 50 mm focal length and F#1.2 will be considerably more expensive than a lens with focal length 50 mm and F# 2.8. A smaller F# allows more light, thereby decreasing the exposure time to get the same total amount of energy on the film. So lenses with smaller F# are often termed **"fast lenses"**. Adjusting the aperture in a lens will change the F# and therefore its light gathering capacity. This in turn will change the exposure time for a picture.

Photographic Film

The photographic film has been the most significant development (prior to the advent of digital imaging) in imaging systems and it still forms the most common medium for capturing and storing images. Here, we will give a thumbnail description of its construction. A photographic film consists of a plastic (most commonly polyester) film base coated with layer(s) of photosensitive emulsion and various color sensitizing dyes. For a black and white film that does not capture the color information, there is only one emulsion layer. The layered structure of a black and white photographic film is shown in Figure 18.6.

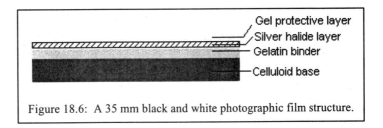

Figure 18.6: A 35 mm black and white photographic film structure.

The emulsion layer contains a polymer binder in which tiny crystals (called "grains') of silver salt (silver bromide) are embedded. These crystals are activated when exposed to even the smallest amount of light. This activation forms what is known as a "latent image" in the film. If nothing further is done, such a latent image can stay on the film for several months.

(Most of us have the experience of taking pictures and then forgetting to send the role to get developed for several months.) The film development uses chemicals that convert the activated silver salt crystals into metallic silver. Further chemical processing -"fixing" - washes away unexposed (and hence inactivated) silver salt crystals leaving behind clear film and fine metallic silver particles. These fine metallic silver particles are black and absorb light. The amount of light incident on the film determines the amount of silver salt grain that is activated and hence the amount of silver

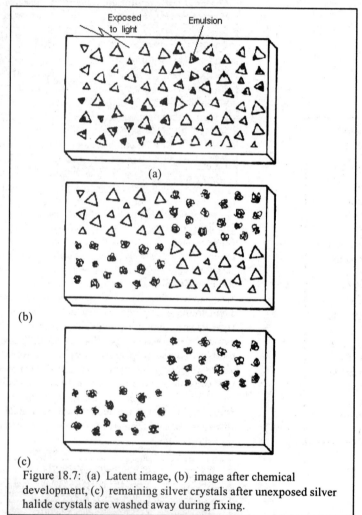

Figure 18.7: (a) Latent image, (b) image after chemical development, (c) remaining silver crystals after unexposed silver halide crystals are washed away during fixing.

particles that remain on the film after development and fixing. This whole process creates a photographic negative where brighter regions in the object come out darker. These steps are shown in Figure 18.7.

Recording color images requires a film that has multiple layers of photographic emulsion, each with a color dye designed to absorb specific color. There are many different types of color films that will produce color negatives (e.g., Kodacolor) or color positives, i.e., color slides (e.g., Kodachrome). The science and technology of color film operation is far more complex than the black-and-white film. Here we describe an example to illustrate the principles of operation. Figure 18.8 shows the layered structure of one particular type of color film.

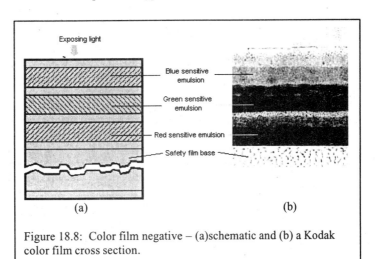

Figure 18.8: Color film negative – (a)schematic and (b) a Kodak color film cross section.

The film contains three layers, each sensitive to blue, green and red color. A latent image is formed after being exposed to the light pattern shown in Figure 18.7(a). After development, the black silver layer is replaced by a complementary color of that corresponding layer. For example, complementary color for blue layer is what one gets when blue is taken away from a white light (yellow). Complementary color for green and red are magenta and cyan, respectively. When exposed to white light, each layer takes away the respective colors from it according to the exposure pattern. So in the region that was black, all three layers are absorbing taking away yellow, magenta and cyan, respectively, from white, which leaves no light (black) In parts originally exposed to red light, the yellow and magenta layers are active leaving behind only red out of white light.

Photographic film is available in several formats. Each format has a characteristic size. Figure 18.9 shows some of the most common film formats available

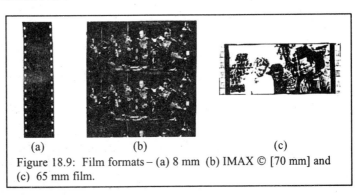

Figure 18.9: Film formats – (a) 8 mm (b) IMAX © [70 mm] and (c) 65 mm film.

18.2.3 Electronic Camera

The invention and proliferation of the photographic film has indeed been a revolutionary event in human history. No longer do we have to rely on people's memories, their ability to describe a scene in words or their ability to depict the scene accurately in a crude sketch or an elaborate painting. We do have photographic evidence of how Abraham Lincoln looked like but only know how George Washington looked like through his portraits. Many significant events in our nations history in past 100 years have been etched in our collective memory as still images or movies (President Kennedy's assassination or Neil Armstrong standing on the surface of the moon, for example). As powerful as this development in imaging has been, it still has one major drawback – one has to *WAIT* for the film to be processed before one sees the result. The so-called instant photography invented by Edwin Land, who launched Polaroid Corporation some 60 years back, simply included the film chemicals along with the film and reduced the time from hours to minutes. But that is still not "instant" photography as we use the term today!

A truly instant photography required the development of electronic light sensors and the rest of the electronic revolution of the past 40 years. The beginning of electronic imaging was the development of television. Instead of relying on the chemical change induced in silver compounds by incident light, the light was converted directly to an electrical signal. This electrical signal was transmitted to another location, where it was converted back into light again by using another device. The early prototypes for television cameras and monitors involved mechanically moving parts that scanned a fine spot of light and very simple electronic parts to convert that light spot into an electrical signal and vice versa. This was done simply because the field of electronics was at a very primitive stage compared to the state of the art of mechanical systems. The mechanical system quickly gave way to evacuated glass tubes in which electrons (tiny subatomic particles carrying electric charge) were generated and scanned. Both television cameras and monitors were constructed using this basic principle. This technology, though superior to film-based technology it was replacing, still had its drawbacks. It was bulky, delicate, expensive and consumed lots of electric power. The early generation video technology, thus, remained confined to professional television studios and did not reach ordinary consumers. In 1960s, a new invention launched the modern age in electronic imaging. This was the invention of the Charge Coupled Device (CCD). A CCD was a solid-state device and used silicon semiconductor elements. Electrons were generated, stored and transferred inside device structures made in silicon instead of inside evacuated glass tubes to perform the image sensing operation. The resulting image sensor is very small, rugged, and relatively inexpensive and consumes less power.

One striking feature of the CCD image sensor is that it consists of a regularly spaced rectangular array of tiny individual image sensor elements. Each element is termed a "Picture Element", or **pixel** for short. Each pixel is responsible for converting the light energy falling over its entire area into electrons and also for storing those electrons in an "electron well". Light could strike anywhere in the pixel and the electrons generated by the light are stored in the same well. So details in an image that are smaller than the size of the pixel are not reproduced. This procedure was discussed in Chapter 4 as the sampling of a continuous image to generate a discrete

---

Solid State Device : It is an electronic circuit made of a semiconductor material. It operates in the *solid state* as there are no vacuum tubes or gas plasmas involved.

representation of the image. After a predetermined length of time (the exposure time) the accumulated charge is moved to the outside circuits via the electrical wires embedded in the CCD array. This moving charge is called an electric current and is our desired signal. By increasing the exposure time, more charges are accumulated in the well for the same amount of available light, the CCD camera will be able to "see" better in low light scenes. The penalty we pay is, of course, in not capturing motion in the scene that is faster than the exposure time. The schematic diagram of a CCD camera with associated electronics is shown in Figure 18.10. In a color CCD camera, the light sensing pixels are organized in groups of three pixels, each with a color filter overlaying. The three adjacent pixels have Red, Green and Blue color filters associated with corresponding to the three primary colors that make up the visible spectrum

The electric current that is the output of a CCD sensor is further processed by the associated electronics. In the CCD video cameras in the camcorders, the image information contained in the time varying electric signal is stored on a magnetic tape as discussed in earlier chapters on data storage. It should be noted that the electric current is allowed to vary in a continuous manner proportional to the amount of light that was incident on the CCD sensor. So the image was sampled in space and time but was not quantized in amplitude. The recent generation digital still and video cameras accomplish the last step of converting continuously varying electric signals into sequence of binary numbers after the analog-to-digital conversion step discussed in earlier chapters. These binary electric signals can now be stored in semiconductor memory chips or on computer magnetic disks. These electronic cameras (still or video) therefore generate a fully digital image. We should emphasize that the difference between the early generation camcorders and the new digital cameras (still and video) is mainly in their electronics and not in their light sensors.

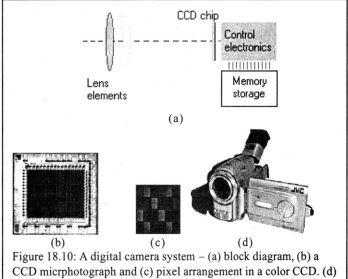

Figure 18.10: A digital camera system – (a) block diagram, (b) a CCD micrphotograph and (c) pixel arrangement in a color CCD. (d) an example of a commercially available digital video / still camera

The performance of a CCD camera is measured by how many pixels it has. The highest quality digital cameras that are currently available have sensors with approximately 3 million pixels divided among the three colors. The size of the sensor and the size of a pixel together decide the total number of pixels in the image that is sensed and stored. Another performance factor is the ability of the camera to record images in low light levels. Ensuring that a large fraction of light incident on the sensor is converted into charge boosts the light sensitivity of a camera. This efficiency is achieved by reducing optical losses due to absorption, reflection and scattering and by special design of the semiconductor device structure to boost the conversion of light into electrons. The image quality is also dependent on the number of bits in the binary representation of each pixel intensity value. Most high quality cameras scientific or industrial employ 8-12 bits per pixel. The number of meaningful bits per pixel depends strongly on the magnitude of electric signal generated by the CCD sensor. In low light level situations, it is not possible to measure the light intensity precisely due to noise and hence number of bits per pixel will be reduced significantly.

Image Resolution of a CCD Camera: One million pixels seem like a very large number and indeed it is. Let us look at the performance of a monochrome CCD digital camera with 1 million pixels. Further assume that the pixels are arranged in an array with equal number of rows and columns. That makes the array 1000x1000. If each detector pixel is approximately 25 microns on a side (a typical size), then the detector chip will be approximately 25 mm on the side (or approximately 1 inch). For a portrait of a person, the typical distance will be 2 meters. If the area that is images on the CCD array is approximately 1 meter square, it means that the one million pixels are covering an object area that is 1 meter on a side. Therefore each detector pixel will cover an area on the object that is 1 mm x 1 mm. So any feature on the object that is 1 mm or smaller will lose all details. For a portrait, this may turn out to be quite adequate.

Now if we use the same camera to shoot a picture at a baseball game, we will be more likely at least 200 meters away from the action (100 times farther). The area that is images on the CCD will also be approximately 100 times larger on a side, or 100 meters x 100 meters. Now the same one million pixels are covering the 100 meter x 100 meter area thereby assigning one CCD pixel for a patch that is 100 mm or 10 cm in size. This patch is large enough that a baseball will be barely larger than a couple of pixels and seeing the details on a player's face are completely out of the question. If we are to maintain the same resolution that we had for portraits, we will need a CCD with 100 million pixels, which is completely unrealistic right now.

How does the human vision solve this problem? It solves it by deciding that we need to achieve high resolution only in the vicinity of the center of the picture and that farther away from the center the details (such as the crowds on the other side) are unimportant. So it samples the center at 5 times finer resolution than the edges. A similar strategy for the CCD would mean that the pixels at the center will be only 10 microns on the side while pixels at the edge could be 50 or even 100 microns on a side. With this design, it will be possible to resolve the baseball in the center of the field well while maintaining the large region of the field simultaneously in the field.

Monochrome: Monochrome means literally *of a single color.* A monochrome camera, can only capture black and white images.

14

18.2.4 Comparison of Different Imaging Systems

The three imaging systems described in this section (film camera, electronic camera and the eye) have the same basic design. They all have imaging lens, a variable aperture to control the amount of light, an effective shutter mechanism to control the exposure time and a light sensitive screen. The biggest difference between them, however, occurs in the nature of the light sensitive screen.

The retina of the human eye is the light sensitive screen, which primarily records light intensity by a complex biochemical reaction induced by incoming light. The human eye is a difficult sensor to analyze since it works very closely with the further processing that takes place in the brain. However, a few unique features should be emphasized. The size of the individual light sensitive cells is much smaller than the finest spot size the lens in the eye can produce. Furthermore, the two different types of light sensitive cells (rods and cones) are distributed across the retina in a highly varying manner. This also affects our visual acuity (ability to see small features), which is greatest directly on-axis and gradually drops off towards peripheral vision. The color perception in humans is an extremely complex phenomenon, which involves behavior of cone cells as well as further processing in the brain. Suffice it to say that cones are the ones responsible for color vision and they are sensitive under bright light conditions. Furthermore, the nature of color perception in humans allows us to break up an arbitrary color as a summation of three primary colors (red, green and blue), which simplifies image sensing as well as displays.

In photographic film the recording and storage is accomplished by light induced chemical changes in the active material (silver salt) in the film. Furthermore, it needs chemical processing to finish the recording and make the image permanent. The photographic film, therefore, acts as a sensor, processor and the storage medium for the image. The finest silver salt crystals are much finer than the finest spot that a camera lens can produce. Furthermore the crystals vary in size and are arranged randomly throughout the film. So in a photographic image, the individual silver salt crystals are seldom visible unless viewed at extreme magnification (500x). To achieve color effects, layers sensitive to different colors are arranged on top of each other achieving the desired color effect by subtracting different colors from the incident white light.

In electronic cameras, the recording and (temporary) storage is accomplished by converting the incoming light to electrons, which are then stored in a semiconductor device structure (electron wells). There is no further processing required to finish recording and temporary storage. However, the light-generated electrons need to be removed from their temporary storage in a pixel to an external circuit for permanent storage. The size of a pixel in an electronic image sensor is usually far bigger than the smallest spot size that good quality lenses can produce. Furthermore, these pixels are arranged in a regular 2-D array. So in an electronic image, the individual pixel structure is readily visible especially when viewed closely. To achieve color effects, adjacent groups of three pixels are provided with red, blue and green filters to detect light signals of those colors. These pixels are usually small and when viewed from a correct distance, the eye perceives them as a single small region of desired color.

Most of the times, the object of the various imaging systems described is to produce images (still or movies) that are for our viewing pleasure, the specific features of our visual system need to be considered when designing these systems. For example, the visual persistence implies that if the images are recorded and displayed at rates greater than 50 to 60 cycles / second, we perceive them as smooth and continuous motion. This has made the standard for electronic and film based movie cameras being operated such that the next image in a sequence appears after $1/60^{th}$ of a second (or $1/50^{th}$ of a second in European countries). Similarly, the human visual color perception allows us to break arbitrary colors into their red, blue and green components to sense and display them accordingly.

18.2.5 Image Scanners

An image generated by a film camera can be digitized none-the-less by following the three-step procedure. The first step is to sample the spatially continuous photograph. This is accomplished in an image scanner shown in Figure 18.11. The scanner consists of a flat plate of glass and a cover. The photograph to be scanned is placed on the glass. The mechanism of the scanner it self involves a mechanically moving light source and a linear (1-D) CCD image sensor array. The photograph is imaged on the CCD sensor one line at a time. By moving the CCD sensor along perpendicular dimension the full 2-D photograph or a line drawing is converted to an electronic signal. Since a CCD sensor has discrete pixels, it performs the function of sampling along its length. Making the light measurements at discrete position in time as the CCD is mechanically scanned accomplishes the sampling along perpendicular dimension.

The scanner performance is most frequently quoted in terms of dots per inch (dpi) for image resolution and number of bits assigned to each pixel. Again, three separate pixels are assigned to measuring red, blue and green components of an image. Since the source of light is built-in to the scanner, light intensity level incident on the CCD sensor is significantly high. So in scanners, it is customary to quantize each pixel intensity value to 12 bits. In specifying the resolution of a scanner, a distinction is often made between "optical resolution" and "scanning resolution". The optical resolution corresponds to the number of CCD sensor pixels per inch and the quality of the optical imaging system. These numbers are typically low (300 to 600 dpi). The effective resolution of the digitized image can be enhanced significantly by employing the image processing techniques described in previous chapters.

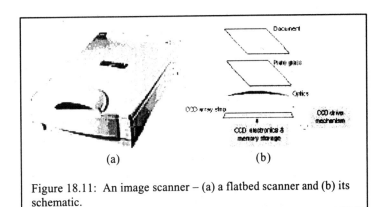

Figure 18.11: An image scanner – (a) a flatbed scanner and (b) its schematic.

Exercise 18.3

- A flatbed scanner can scan a standard 8 ½inch x 11 inch page at 300 dpi spatial resolution and 12 bits of quantization. Calculate the file size in number of bits for a digital image generated by the scanner.

- If the height of the CCD linear array is 100 microns and it takes 1/3 second to acquire one line of image data, how long will it take to scan a full 11 inch high document?

Section 18.3: Displaying Digitized Images

The most common use of images is for people to view them in different activities. These activities span the gamut of entertainment (movies, video games), medicine (X-rays, MRI images), scientific (images of moon and mars) and defense (surveillance, targeting). So the final section of this chapter will deal with ways of displaying image information. In this operation, images sensed by film-based cameras or electronic cameras and stored in analog or digital fashion are converted back into optical waves carrying the image information. The human eye then senses these optical waves to receive the image information.

Image displays also come in analog and digital variety. The simplest and most common image display device is photographic film. As we remarked earlier, a photographic film is not only used to sense the image and store it, but it also serves as an image display device. The film could be in the form of a photograph where light is reflected by the photograph, which selectively absorbs different colors in different amounts to impart image information on the reflected light wave. In case of a slide film, the medium is transparent and light is selectively absorbed in different colors to impart the image information on the transmitted beam. This image carrying light beam can be directly viewed or can be projected on a screen to be viewed. The image is perceived as continuous in space and having smoothly varying light intensities across the image. This is the typical analog display.

18.3.1 Cathode Ray Tube (CRT) displays

Film-based analog displays are inexpensive and high quality and are particularly useful in serving large audiences (movie theaters). In personal use, other forms of displays that have distinctly different characteristics are gradually replacing them. The next most common display device is the ubiquitous Cathode Ray Tube **(CRT)** that we are so familiar with as a television set or a computer monitor. CRTs involve evacuated glass tubes (called picture tubes) with electrons flowing inside them being accelerated by high voltage and striking the special chemical compounds (called phosphors) coating the inside of the glass face. When struck by electrons, these phosphors glow with characteristic color generated the image that we view on the glass face. Inside the CRT are magnetic coils and electrodes, which serve to deflect the electron beam sequentially across the glass face to "paint" the image row-by-row. The phosphors themselves decay slowly once activated by high-energy electrons. This slow decay gives a smooth appearance to the image as it is sequentially scanned across the faceplate.

Interesting fact :
The human eye *sees* light waves in frequencies from 4 $\times 10^{14}$ Hz (red) to 7.5$\times 10^{14}$ Hz (violet). This range is called the visible spectrum.

The number of electrons striking the phosphor (electron current) determines amount of light that is emitted by the phosphor. When displaying analog images that have continuously varying signal amplitudes, the input signal controls the electron current, which in turn will determine the intensity of the light produced by the phosphor. When displaying a digital image, the binary number associated with a sample is first converted into a continuous valued signal (analog signal) by associated electronic circuitry. This analog electric signal then modulates the electron current to produce the desired intensity.

Having fine phosphor dots, which emit red, green and blue light, interlaced within the image, generates full color images. This is essentially the inverse principle of color CCD cameras. On conventional CRTs, the images are updated every 1/60th of second. Due to persistence of vision in our eyes, this leads to smoothly varying motion. Figure 18.12 shows a picture and the schematic diagram of a CRT tube. The quality of a CRT display is indicated by how many lines of image it can produce as well as how sharp the electron beam remains across the faceplate, especially in the corners. A computer monitor is also specified in terms of the size of the smallest spot it produces on the faceplate. A smaller spot size implies a sharper image. A subjective evaluation of the quality often depends on the nature of the color the CRT phosphors produce.

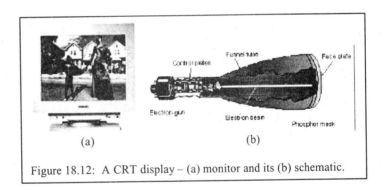

(a)   (b)

Figure 18.12: A CRT display – (a) monitor and its (b) schematic.

CRTs can be classified as light emissive displays since they give out light that is then viewed by the users. Over the years, the largest size picture tubes that are available on the market have increased in size gradually. Currently, TV sets are readily available that have CRT picture tubes that are 35 inches along the diagonal. However, it is not practical to produce picture tubes that are much larger than this.

In sport arenas and other public places (e.g., Time Square in New York city) there is a need for image displays that are much larger (10's of feet across) that are also light emissive. In recent years a new giant display technology has emerged. The display screen contains hundreds of thousands of bright, light emitting diodes (**LEDs**) that are arranged in a regular array. Again, to produce full color picture, red, green and blue LEDs are interlaced within the array. These LEDs are controlled by associated electronics that supply currents to the LEDs that make them glow with the desired brightness. Figure 18.13 shows the photograph of one of these "Jumbotron" displays in a stadium.

Figure 18.13: A very large format display by Lighthouse

Exercise 18.4

- The electron beam moves across the face of a CRT to excite the phosphors and "paint" an image. Suppose the image on a CRT is 29 inches wide and it contains 525 lines. The entire image needs to be written in 1/60th of second. How fast is the electron beam moving across the face of the CRT screen? Assume that the beam returns to the beginning of the next line after finishing scanning the previous line without any time delay (retrace time is zero).

18.3.2 Light Modulating Display

CRT displays have high performance and are inexpensive. Nonetheless, they have some significant drawbacks. They are bulky, fragile and consume significant power. Hence they are not well suited for portable device applications. Over the past 15 years, a new display technology has made impressive progress and is now rapidly replacing the bulky CRT tubes in computer and television displays. The flat panel displays based on Liquid Crystal (LC) are smaller, lighter, more rugged (no evacuated glass tubes) and consume much less power than their CRT counterparts. Due to their compact nature and low power consumption, LC displays are the only choice for portable electronic devices.

The principle difference between LC display and the CRT is that the LC display itself does not generate light but modulates light generated by an external source. LC displays consist of regularly spaced 2-D arrays of discrete pixels. Each pixel contains a liquid crystal (a specially synthesized organic compound) and electrical drive circuitry. The liquid crystals are designed to absorb light according to the (very small) electrical signal that is provided by the electrical circuitry. The level of darkness achieved by each LC pixel depends on the magnitude of the electric drive signal. The electrical circuitry drives each pixel sequentially to write the entire image within the required frame rate ($1/60^{th}$ of second). The LC modulator at each pixel slowly decays in its ability to modulate light till it is refreshed during the next write cycle. Again, color is produced by incorporating red, green and blue color filters with individual pixels and arranging the pixels in groups of three. The peculiar properties of liquid crystal modulator make it viewable only over a limited angle before the contrast of the display (defined as a ratio of the brightest to the darkest value of a pixel) suffers. Liquid crystal molecules are also slow to respond to rapidly changing electrical signals. Hence LC displays tend to smear fast moving pictures. Figure 18.14 shows the schematic diagram of a LC display device. A LC display quality is often specified in terms of total number of pixels it contains, the angle over which good contrast is obtained, and the reduction in smearing of fast moving images.

The LC display can work by modulating the ambient light that is reflected by the display. Most calculators, watches and other portable displays operate on this principle. Since the ambient light is not under our control, the LC displays that work with ambient light are usually monochrome (black-and-white). In many applications, we cannot rely on ambient light. In those cases, the LC display comes with its own source of light in the form of a thin fluorescent light panel. White light generated by this panel is modulated by the liquid crystal pixels in producing the desired image. Since the fluorescent light panel consumes significant amount of electric power, such displays do not have as long a battery life as do LC displays that work with ambient light.

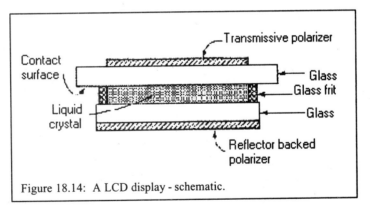

Figure 18.14: A LCD display - schematic.

A recent development in displays that works on the principle of modulating an external source of light instead of generating its own light is a device called Digital Micro-mirror Display device produced by Texas Instruments. It is different from the other displays discussed so far in that it is a ***digital*** device itself. The principle of operation of this device is familiar to anybody who has taken a small mirror and tried to reflect sunlight in somebody's eyes by controlling its angle. This principle is seen in Figure 18.15. A tiny mirror that corresponds to one pixel of a large display reflects the projector light, which is the bright source of light. The mirror has three positions - flat, +10 degrees and -10 degrees tilt. The light reflected by the mirror is then captured by a lens and projected on a screen. The position and the size of the lens are designed such that only when the mirror is tilted by -10 degrees, will the lens capture the light reflected by it. In the other two positions, the light escapes from the imaging system and that pixel appears dark on the projection screen.

By rapidly switching the mirror between the +10 and -10 degree tilt positions, the light on the screen can be made to change between dark and bright values. The tilt of the mirror is controlled by an electric signal that carries the pixel information. It should be noted that the mirror has only two distinct positions that it is switched between and can therefore only display binary signals. This device will therefore directly accept the binary representation of the pixel value. A range of pixel intensities is obtained by controlling the length of time that the mirror is held in the bright position. When the mirror is being driven by binary signals that represent bits of higher significance, the mirror is held in its on position for a longer time. This technique achieves the binary to analog conversion in the device itself. Color images are obtained by employing a color wheel with red, green and blue filters embedded in it. The red, green and blue images are sequentially displayed on the device while the wheel is in the correct position.

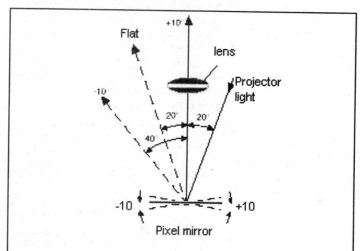

Figure 18.15: DMD operation – each pixel mirror is deflected by electronic signals according to the pixel information to be displayed.

The size of each mirrors (~17 microns) and its thickness are extremely small making these mirrors ultra-light. So these mirrors can be switched in a few microseconds (millionth of a second). It is this high speed that allows the time modulation to be employed in displaying varying intensity information while still updating the whole image in 1/60th of a second for large (>1024x1024) arrays of pixels. The micro-mirrors are fabricated using special processes in single crystals of silicon, the same material that is used to fabricate electronic chips. This allows the control electronics and the micro-mirror arrays to be integrated in the same device making it compact and easier to manufacture. Figure 18.16 shows the picture of one element of such an array along with the full array. The whole system is termed Digital Light Processing system.

The DLP system uses high intensity lamps which allows the images to be projected on large screens suitable for movie projection. A few movie theaters in the country are indeed based on the DLP system. The digital nature of the signals accepted by DLP allows one to achieve a fully digital imaging system. For example, a movie that has been converted to a sequence of binary bits can be directly transmitted to the DLP projector and displayed on the screen. The only analog step in the entire sequence is the one that takes place inside the human visual system where our eye-brain combination integrated light intensity over time to produce the sensation of varying light intensity pixels.

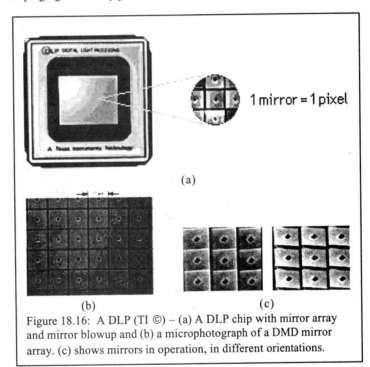

Figure 18.16: A DLP (TI ©) – (a) A DLP chip with mirror array and mirror blowup and (b) a microphotograph of a DMD mirror array. (c) shows mirrors in operation, in different orientations.

18.3.3 Hard Copy Output Devices - Printers

The image display devices discussed thus far are most suitable for displaying moving images. Although they can also display individual images, they are not best suited for that purpose. For sharing still images, different forms of media are employed. Again, photographic film has been one of the most common media for displaying still images. However, in the recent years other means of generating permanent records of still images have been rapidly improving in performance, cost and ease of use. Computer printers have evolved from crude images where individual black dots could be easily seen with naked eye to the current stage where the outputs of high quality color printers are virtually indistinguishable from photographs. The functional description of all printers is the same - they take binary electrical signals output by a computer and produce on a piece of paper, the images that correspond to that particular sequence of binary numbers. Modern computer printers are mainly divided into two major types - inkjet and laser printers. , There are other types of printers with subtle differences in operation. However, in this section we will briefly discuss the operation of these two primary types of printers. All of the printers are essentially binary devices similar to the DLP device. It realizes the gray values for individual pixels in a unique way that is called "half tone encoding". The topic of half tone encoding is much more broadly applicable than computer printers and hence is discussed separately in the end.

Inkjet printers operate by placing on a piece of paper a tiny ink dot according to the control signals sent by the computer. By combining different color inks, full color output can be realized. Figure 18.17 shows the schematic diagram of the print head of an inkjet printer and the photograph of a complete printer. The inkwell contains an element that is heated when electric current passes through it. When heated, the nozzle ejects small amount of ink out of its precisely machined nozzle. This ink drop is then deposited on a piece of paper. An inkjet printer is typically specified in terms of how many binary dots per inch it is capable of printing. A high quality printer is capable of depositing 600 ink dots per inch.

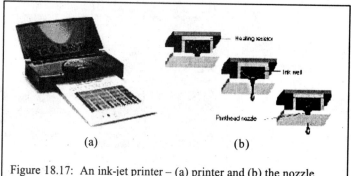

(a)   (b)

Figure 18.17: An ink-jet printer – (a) printer and (b) the nozzle head.

A laser printer operates on a very different principle. Instead of liquid ink, the main active material is a fine powder called "toner". A critical part of the laser printer is a transfer drum. The transfer drum is coated with light sensitive material. The transfer drum operates in such a manner that whenever it is exposed to a laser beam, it causes the toner particles to stick to it by electric forces. These toner particles are then transferred to the paper and subsequently fused on it by thermal fixing stage. The finest dot that a laser printer can produce depends on the smallest spot that a laser beam can be focused to. Since a laser beam can be focused to very small spots, the laser printers have a much superior performance in terms of dots per inch than the inkjet printers. High quality desktop laser printers have 1200 dots per inch. In printing and publishing businesses, this performance can be even higher and laser printers that have as high as 3360 dots per inch performance exist. The schematic diagram of a laser printer is shown in Figure 18.18.

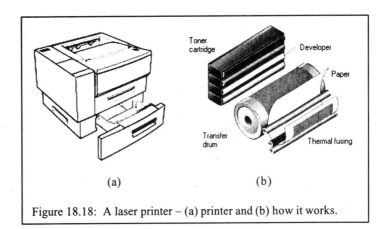

Figure 18.18: A laser printer – (a) printer and (b) how it works.

18.3.4 Half-tone Encoding

In discussing DLP, we saw how the DLP uses its high speed of operation to achieve variable perceived intensity of projected image pixels by modulating the length of time for which the binary pixels are on (bright). For hard copy devices, the time element is not available but we can use a similar idea by exploiting the ability of the printers to produce binary spots far finer than our unaided eye can resolve. This particular technique is termed half tone encoding. Half tone encoding was first developed for new paper printing industry and predates the digitization of imaging system.

Halftone printing achieves varying amounts of gray values by controlling the size of the black dots printed in a region. For example, a 1 mm square area of a newspaper image could contain a 6x6 array of black dots. A completely white area is represented by leaving the black dots out completely (i.e., the area is clear). On the other hand, a completely black area can be depicted by making the black dots so big that they are touching each other thus filling the entire 1 mm square area with black ink. Intermediate levels of blackness can be achieved by changing the size of the black spots and therefore the ratio of the dark space over the clear space in that 1 mm square area. For example, if we want to create a sensation of a pixel with a gray value that is half way between black and white, we can make the black dots with a diameter that is approximately 1/2 of the spacing between the dots. When we read the newspaper from a typical distance of 30 cm or more, our eyes are unable to resolve the individual black dots and the sensation is created of continuously varying gray values. Color printing can be done using three different color dots (red, green and blue) and mixing their sizes to create arbitrary color. Figure 18.19 shows a half-tone encoded image. In part (b) of the figure, one section of the image is expanded such that we are now able to resolve the black dots individually.

In this particular example, the black dots are circular in shape. However, the shape of the dots itself is of no particular consequence as long as the ratio of the black to white areas is maintained. As a matter of fact, each pixel in such half tone image it self can be a small image. In a large poster that is viewed from far away, the pixel size itself can be as large as several cm on the side. In this case one can create an image within that pixel that, when viewed from far, creates the sensation of the desired gray value. This principle is not limited to just black-and-white posters but can also be extended to color posters. Such posters are now commonly found where the poster contains a mosaic of several hundred small images. This mosaic when viewed from a distance leads to the desired color image.

Figure 18.19: A half-tone picture with a small section shown expanded. The black and white dots are now individually resolved.

Exercise 18.5 Half-tone screen frequency

- Half tone encoding is specified by a screen frequency which corresponds to the number of composite pixels the image has. The screen frequency is given to be 150 dpi. If we desire to create 256 distinct gray levels, how many binary dots per inch do we need to produce out of the printer?

VAB Experiment 18.1: Custom half-tone screen pattern design

- The goal of the experiment is to create a half-tone pattern that gives the best gray scale rendition without distracting artifacts. The super pixel contains 8x8 binary pixels. We wish to create a map for 16 gray levels. Design spatial patterns within the 8x8 array to correspond to individual gray levels that range from 0 (complete black) to 15 (complete white).

Appendix A:
Imaging System

The process of forming an image is a multi-step process. So an image sensor has several components to perform different operations involved in an imaging operation. First and foremost, imaging requires an object that is to be imaged. An object can give out light itself (a *self-luminous* object), such as a light bulb, a candle flame or a glow-in-the dark sticker. If it does not give out light, then some source of light is required for an image to be formed. The most common natural source of light is of course the sun. Over the past thousands of years, humans have supplemented sunlight with many other sources of light; fire, lamps and finally electric light. Light from these sources is reflected by the object. This reflected light carries information about the object – whether it is light or dark, how does the light and dark patterns vary across the object, what colors does it absorb and what colors does it reflect. If we are able to capture and record the reflected light, then we can learn something about the object(s) that reflected the light.

But just having a source of light and an object to reflect it is not sufficient to form an image. The light reflected by the object spreads out in every direction quickly, becomes weak and loses all its shape and form. We need an additional element that will keep the light from spreading and will reproduce how the reflected light looked right after it was reflected by the object. That additional part is a simple glass lens that we are all familiar with. Light rays bend as they enter and leave the glass. It is the shape of a lens surface, which controls how the light rays are bent and by what amount. If the shape of a lens is chosen correctly, then light rays generated from a single point on the object come back together some distance behind the lens forming an image of that point. We say that the lens focuses light reflected by the object and forms an image. The simplest shape for the lens surface that will create an image is sphere. The lens surface can be convex (bulging out) or concave (curved inwards). A lens where both surfaces are convex spherical will focus the spreading light from a point on the object to a corresponding point on the image. Figure 18.20 shows the difference in what is seen on the screen without and with a properly positioned lens between the object and the screen. This image formation by a lens occurs only when the object and the viewing screen are placed at specific position.

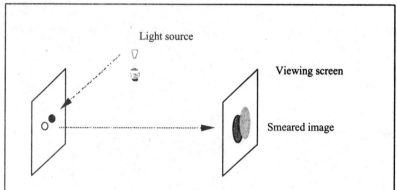

Figure 18.20(a): An object consisting of a bright and a dark point is illuminated by a light source. The reflected light is collected on the viewing screen. The two points are smeared and run into each other.

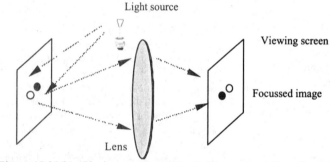

Figure 18.20(b): Placing a convex spherical lens causes the reflected light to come together to form an image on the viewing screen.

18.A.1 Imaging Properties of a Spherical Lens

In order to determine the imaging distance behind a lens, it is important to determine one particular property of a convex spherical lens called **focal length**, usually indicated by letter "f". A focal length is the distance behind a convex spherical lens where parallel rays of light entering a lens come to a focus. Another special rule for a convex spherical lens (especially where the lens thickness is considered to be small) is that a ray of light going through the center of the lens passes through straight without any bending. These two condition together allow us to determine where the light rays diverging from one point on an object will come back to form an image behind a lens. This construction is shown in Figure 18.21. By applying simple rules of geometry (similar triangles), one can show that the distance an object is placed in front of a lens (u), the focal length of that lens and the distance behind the lens where the image is formed (v) satisfy the following relation:

> **Focal length :**
> It is the distance behind a spherical convex lens where parallel rays of light entering the lens come to a focus.

28

$$1/u + 1/v = 1/f \qquad (5.1)$$

The application of similar triangle also tells us the size of the image with respect to the size of the object. If the height of the object is l_o and height of the image is l_i, then their ratio, (l_i/l_o), is called the lateral magnification of the image (indicated by M). "M" is related to "u" and "v" by the following equation:

$$M = -v/u \qquad (5.2)$$

The negative sign indicates that the image is up–down and left–right inverted relative to the object. The Table 18.1 given below describes the magnification and the image distance for some special conditions. The distances are measured in terms of the focal length of the lens.

Object Distance	Image Distance	Magnification
Infinity (far away)	f	Undefined
4f	4/3 f	-1/3
2f	2f	-1
2/3 f	3f	-2
f	Infinity (far away)	Undefined

Table 18.1 Image distance and magnification for specific values of the object distance.

The first and last entries of the table need further explanation. When object is far away, the rays from it are parallel and therefore are focused by the lens one focal distance away. Since parallel rays are focused to a point, the size of the image is infinitesimally small and the magnification is undefined. The last entry discusses an exactly reverse situation. A point object located one focal length away from the lens gives out diverging rays, which are made parallel by the lens. These parallel rays never meet (or in other words, meet at an infinite distance from the lens) and therefore the magnification is undefined. It can be seen from the other entries that as long as the object distance, u, is greater than 2f, the image magnification is less than 1 (image is reduced) whereas when the object distance is between 2f and f, the image is magnified. The focal length of a lens is determined by how curved the lens surface is. A shorter focal length is achieved when the radius of curvature of the spherical surface is small and vice versa. Therefore very short focal length lenses are fat and bulging while long focal length lenses tend to be more flat and thinner. Another way to change the focal length of a lens is to employ different materials. The ability of a transparent material to bend light is termed its "refractive index". Refractive index of vacuum is 1, light does not bend while propagating in vacuum. The refractive index of air is so close to 1 that for all practical purposes it is considered to be 1. Higher refractive index implies more bending power for the material. Refractive index for common glass is 1.5 whereas for diamond it is 2.6. A lens with the same radius of curvature for the surfaces but higher refractive index material will lead to lenses with shorter focal length.

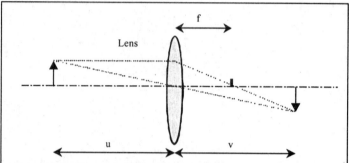

Figure 18.21: The definitions of image and object distance for an optical system with lens of focal length 'f'.

Example 18.1.1: A 35 mm camera has 50 mm focal length. The distance between the lens and the film can be varied from 50 mm to 60 mm. What will be the range over which it can form a clear image?

Example 18.1.2: A slide projector has a 100 mm focal length lens. The slide is mounted 120 mm in front of the lens. How far will be screen have to be to form an image? What will be the magnification of the final image?

18.A.2 The Rest of the Imaging System

An imaging system needs a light source and a lens to form an image that can be viewed by a human observer on a screen. But if we want to record the image in order to share it with somebody else or save it for our collection, we need a light sensitive material at the viewing screen and some sort of a memory material that can remember the image. In addition to the light sensitive screen, we need to be able to control the amount of light that falls on the screen. One way to do it is to control the brightness of the light source. But that is not always feasible if we are using natural light (sunlight) or using ambient light (room lights) that we cannot control readily. In either case, it is necessary to modify the basic imaging system shown in Figure 18.20(b) to include parts that can control the total amount of light that falls on the light sensitive screen. The light sensitive material responds to the total light energy incident on it. That can be controlled by controlling the optical power (energy per unit time) or by controlling the exposure time.

One can see from Figure 18.20(b) that the diameter of the (usually) circular lens will control the total amount of reflected light that is captured and then focused on the light sensitive screen. It is possible change the effective diameter of the lens by placing a circular aperture in front (or behind) the lens and vary its diameter. The amount of light admitted in the imaging system will therefore be proportional to the area of the circle ($= \pi r^2$). An ingenious mechanical design consisting of multiple overlapping metal leafs can create a variable aperture whose diameter can be controlled in a smooth manner to the desired value by rotating the tab on the aperture. The variable aperture, therefore, can be used to control optical power incident on the light sensitive screen.

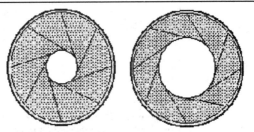

Figure 18.22 : Variable aperture shown at two different opening sizes. The metal leaf structure allows a smooth control of aperture size.

Aperture :
It controls the amount of light energy falling on the screen.
Exposure time :
This controls brightness of image recorded on the light sensitive screen.

Alternatively, a shutter can be installed in the imaging system that controls the amount of time the light sensitive screen is exposed to the reflected light. The timer mechanism can be mechanical or electronic. But in either case, it is possible to vary the exposure time for the imaging system. A combination of the aperture control and the exposure time control will enable us to control the total amount of light energy incident on the light sensitive screen to the optimum value. The light sensitive screen usually has a minimum value for light energy at which it start responding and a maximum value beyond which it stops responding further. The purpose of adjusting the aperture and the exposure time is to keep the total light energy falling on the screen within this range.

Many examples of such an imaging system are found, some natural and some engineered by humans for specific applications. In spite of many differences between these imaging systems, they all share all the functional components shown in Figure 18.23. The specific characteristics of these parts as implemented in different systems give the imaging systems themselves a unique operational behavior. In the following section we will explore some of the most common imaging systems that are relevant to a multimedia system – a traditional film-based camera, an electronic camera (still and video) and the ultimate imaging system, the human eye.

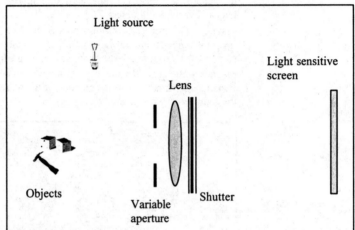

Figure 18.23: Schematic diagram of a complete imaging system with all of the parts (light source, object, lens, aperture, shutter, and a light sensitive screen).

The Infinity Project

Appendix A: Infinity Technology Kit^{SM}

Section 1.1: Infinity Technology Kit^{SM}

The Infinity Technology Kit^{SM} is a multimedia hardware and software system for converting standard PCs into easy to use modern engineering design environments with a wide array of sophisticated capabilities. These capabilities bring to life the engineering concepts taught in the Infinity Project's engineering curriculum. The pre-designed lab experiments that come with the software allow students to see first hand the full range of engineering experiences of envisioning, designing, and testing modern technology.

The technology used in the Infinity Technology Kit^{SM} is based upon Texas Instruments' advanced Digital Signal Processor (DSP) chips and a new and innovative graphical programming environment, called Visual Application Builder (VAB^{TM}), designed and developed by one of the INFINITY Project's partners, Hyperception.

Capabilities: A PC with the Infinity Technology Kit^{SM} is capable of an extremely wide array of "real-time" engineering applications ranging from sophisticated audio, image, and video processing, to a wide array of advanced graphical mathematical operations incorporating real data sets. This system has been designed to acquire data in real time from both microphones and video cameras and simultaneously execute advanced mathematical operations at a rate of 30 million mathematical operations per second. This power allows students to create and test for example new Internet technologies, or develop and evaluate new wireless communication systems.

Components: The Infinity Technology Kit^{SM} contains the components listed below:

- DSP Board: Texas Instruments DSP hardware board with TMS320C31 Digital Signal Processor
- DSP Software: Visual Application Builder graphical component-oriented DSP software
- Accessories include: PC Powered Speakers, PC Microphone with preamplifier, Jack Converter, Audio Cable, AC Power Supply, Anaglyph Glasses (Red/Blue), 9 Volt battery
- Optional USB-based PC Color Video Camera.

How to obtain the Infinity Technology Kit[SM]

The Infinity Technology Kit[SM] can be obtained from Hyperception, Incorporated. The Kit is available in two versions. If your PC already has a video camera then you can order part number HSKT0411A. For those with PCs that do not already have video cameras, the Infinity Technology Kit[SM] may be ordered with a USB-based color camera (part number: HSKT0411B).

HSKT0411A: Infinity Technology Kit[SM]
HSKT0411B: Infinity Technology Kit[SM] with USB-based PC color video camera

Contact Information: To order the Infinity Technology Kit[SM] contact Hyperception for details.

Hyperception, Inc.
9550 Skillman St., Suite 302
Dallas, TX 75243
Phone: 214-343-8525
Fax: 214-343-2457
e-mail: info@hyperception.com
web: www.hyperception.com

What is the VAB?

The Visual Application Builder (VAB[TM]) is an easy to use software program that enables students to run example worksheets that demonstrate the engineering concepts being discussed in this textbook. The VAB software is included with the Infinity Technology Kit[SM]. VAB Lab Exercises are included and allow users to have a "hands-on" experience with powerful Digital Signal Processor hardware and a PC-based video camera. In addition to running the VAB Lab Exercises students can easily create their own signal processing and multimedia applications.

VAB uses a methodology of developing DSP and image processing algorithms and systems graphically by simply connecting functional block components together with a mouse. A user only needs to choose the desired functions, place them onto a worksheet, select their parameters interactively, and describe their data flow using line connections. This method of design is quite similar to drawing a 'block diagram' of the system being designed. A visual design is a more natural design methodology, and is the perfect paradigm of the old saying "A picture is worth a thousand words".

How to create a VAB worksheet

This section is designed to get you up and running with the VAB software as quickly as possible. You may find it helpful to follow the tutorial along and re-create the demonstration presented. This will enable you to learn how to use VAB more easily and quickly. If you want to re-create the demonstration you should first verify that the DSP hardware is properly communicating with VAB before proceeding.

Starting a VAB Session

The first step in creating a block diagram design with VAB is to start a new worksheet. A worksheet is a window which serves as a work space for your applications, and may be thought of being like a piece of paper in which a block diagram may be drawn. Each worksheet can contain different block components that are arranged in a structured block diagram algorithm. Many worksheets can be opened and maintained while in the VAB environment. Choosing the File New menu command will cause an empty worksheet to be loaded. It is in this window where you will place your block components. To re-create the demonstration shown here you should select VAB's File New command to bring-up an empty worksheet. This will produce a worksheet similar to the one shown below:

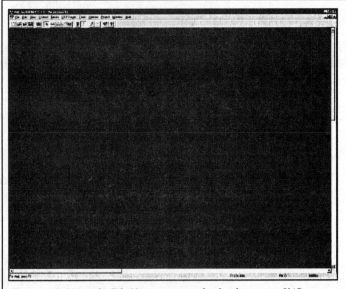

Figure A.1: Using the File New... command to load an empty VAB worksheet

Selecting Block Components

Block components can be placed onto the active worksheet by using the Block Function Selector tool, or by using VAB's block component Toolbars. Choosing the VAB Blocks menu Select Blocks... command will cause the Block Function Selector dialog box to appear. Components are arranged into many different group categories and are shown in the Group

List box. Choosing a function group will cause all functions within that group to be added to the Function list box. De-selecting the group will remove all corresponding block functions from the list.

By selecting the 'Add to Worksheet' button all blocks in the Function List will be added to the active worksheet. Single block functions can be selected into the worksheet by double-clicking the block function name, or by simply dragging the name of the block component into the worksheet. Upon selection of the desired block components their associated icons will appear in the worksheet.

When you are done selecting blocks, you should click on the close button. You may add more blocks later so you do not have to choose all blocks at once. In the figure below we have added several block components into the worksheet by dragging them into position from the Block Function Selector. The blocks used here are the RT Sine Generator, RT DSP-to-PC Upload, DSK D/A, and the 1 Channel X Display. If you are not using a DSK, use the appropriate D/A block that is provided for the hardware you are using. For this tutorial, we will be using a DSK.

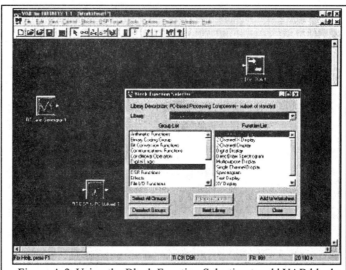

Figure A.2: Using the Block Function Selection to add VAB block components into the worksheet

Establishing a Data Flow

Once you have selected the block components onto the worksheet you will need to establish the block diagram's data flow. This is accomplished by using the mouse to arrange the block icons in the worksheet to form an algorithm or process. The business of establishing the data flow relationship among blocks is typically referred to as "connecting" blocks.

When the blocks have been positioned on the worksheet, you need to connect the blocks to form the algorithm. This is done while operating in the Connect mode. VAB can be placed into Connect mode by selecting the Mode menu connect command. When this mode is selected, the mouse cursor will change to a target cursor labeled either 'SOURCE' when choosing a source block icon, or 'DEST' when choosing a destination block icon. Positioning the target cursor over a block component and clicking the mouse button will cause a connection to be made. Placing the mouse cursor in the right-half of a block will specify a Source connection; a cursor placed into the left-half of a block will define a destination connection.

Once the blocks in the worksheet have been connected, the worksheet needs to be compiled. Compilation determines the order in which individual blocks will execute, and can be performed through use of the Compile menu command. Compilation of the worksheet should be done after all block components have been positioned and connected.

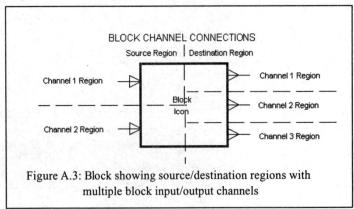

Figure A.3: Block showing source/destination regions with multiple block input/output channels

As the figure above shows, each block icon, or bitmap, is split into the source side (left half) and the destination side (right half). You may connect from source to destination or from destination to source; this is dependent upon which half of the block is clicked first. The cursor will change and be labeled either 'SOURCE' or 'DEST' depending on whether the next connection is for a source block or a destination block. It is possible to connect to multiple destinations by first left-clicking on the source block and then using the right mouse button to click on the destination side of each block intended to be a destination. It is important to follow the correct connection scheme. For example, an input signal generator block connected to an output single-channel display block would use the 'SOURCE' cursor to select the signal generator block, and the 'DEST' cursor to select the single-channel display block.

Changing a Block Component's Parameters

In most cases you will want to select the setup parameters for the block functions in your worksheet. You can modify the setup parameters for a block function by first double-clicking the block function icon with the left mouse button. This will cause the block's setup parameter dialog box to appear (if the box has any parameters which can be changed). The parameters that can be entered are dependent upon the block function chosen. Selecting the dialog box OK button will keep any changes made.

Running the Worksheet

After you have arranged the block icons and established the data flow by connecting the blocks together you can run the worksheet. The VAB run commands can be located in the Run menu. The Run menu command will cause the active worksheet to be executed. VAB will continue to run the active worksheet until you halt its execution through use of the Stop menu command.

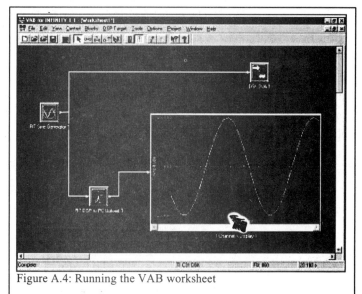

Figure A.4: Running the VAB worksheet

As seen in the figure above, running the worksheet demonstration built in the previous section will result in the graphical display of a sine waveform, and the continuous generation of a tone at the DSK's D/A output. If you have a speaker connected to your DSK then you will hear the output as the worksheet is running.